Warfare in the Middle East since 1945

The International Library of Essays on Military History
Series Editor: Jeremy Black

Titles in the Series:

Modern Counter-Insurgency
Ian Beckett

Warfare in Europe 1650–1792
Jeremy Black

Warfare in the Middle East since 1945
Ahron Bregman

The English Civil War
Stanley D.M. Carpenter

Warfare in Latin America, Volumes I and II
Miguel A. Centeno

United States Military History 1865 to the Present Day
Jeffery Charlston

Medieval Warfare 1300–1450
Kelly DeVries

Medieval Warfare 1000–1300
John France

Warfare in the Dark Ages
John France and Kelly DeVries

Naval History 1500–1680
Jan Glete

Byzantine Warfare
John Haldon

Warfare in Early Modern Europe 1450–1660
Paul E.J. Hammer

Naval History 1680–1850
Richard Harding

Warfare in Europe 1919–1938
Geoffrey Jensen

Warfare in Japan
Harald Kleinschmidt

Naval History 1850–Present, Volumes I and II
Andrew Lambert

African Military History
John Lamphear

Warfare in China to 1600
Peter Lorge

World War I
Michael Neiberg

The Army of Imperial Rome
Michael F. Pavkovic

The Army of the Roman Republic
Michael F. Pavkovic

The American Civil War
Ethan S. Rafuse

The British Army 1815–1914
Harold E. Raugh, Jr

The Russian Imperial Army 1796–1917
Roger Reese

Medieval Ships and Warfare
Susan Rose

Warfare in Europe 1792–1815
Frederick C. Schneid

The Second World War
Nick Smart

Warfare in China Since 1600
Kenneth Swope

Warfare in the USA 1784–1861
Samuel Watson

The Armies of Classical Greece
Everett Wheeler

The Vietnam War
James H. Willbanks

Warfare in Europe 1815–1914
Peter H. Wilson

Warfare in the Middle East since 1945

Edited by

Ahron Bregman

King's College London, UK

ASHGATE

Published by
Ashgate Publishing Limited
Wey Court East
Union Road
Farnham
Surrey, GU9 7PT
England

Ashgate Publishing Company
Suite 420
101 Cherry Street
Burlington, VT 05401-4405
USA

Ashgate website: http://www.ashgate.com

British Library Cataloguing in Publication Data
Warfare in the Middle East since 1945, - (The international
library of essays on military history)
1. Military art and science - Middle East - History - 20th
century 2. Arab-Israeli conflict 3. Middle East - History,
Military - 20th century
I. Bregman, Ahron
956'.04

Library of Congress Control Number: 2007943624

ISBN 978 0 7546 2479 0

Mixed Sources
Product group from well-managed forests and other controlled sources
www.fsc.org Cert no. SGS-COC-2482
© 1996 Forest Stewardship Council
FSC

Printed and bound in Great Britain by
TJ International Ltd, Padstow, Cornwall

Contents

PART II GULF WARS

The Iran–Iraq War 1980–1988

The Gulf War, 1990–1991

The 2003 War in Iraq

PART III INSURGENCIES

Yemen

Intifadas in the Occupied Territories

Hezbollah in Lebanon

Acknowledgements

The editor and publishers wish to thank the following for permission to use copyright material.

Foreign Affairs online for the essay: Gal Luft (2002), 'The Palestinian H-Bomb: Terror's Winning Strategy', *Foreign Affairs*, **81**, pp. 2–7. Copyright © 2002 by the Council on Foreign Relations, Inc. www.ForeignAffairs.org

Journal of Military History for the essay Uri Bar-Joseph (2008), 'Strategic Surprise or Fundamental Flaws?: The Sources of Israel's Military Defeat at the Beginning of the 1973 War', *The Journal of Military History*, **72**, pp. 509–30. Copyright © 2008 by the Society for Military History, all rights reserved.

John Yoo for the essay: John Yoo (2003), 'International Law and the War in Iraq', *The American Journal of International Law*, **97**, pp. 563–76.

MIT Press for the essays: Avner Yaniv and Robert J. Lieber (1983), 'Personal Whim or Strategic Imperative?: The Israeli Invasion of Lebanon', *International Security*, **8**, pp. 117–42. Copyright © 1983 by the President and Fellows of Harvard College and of the Massachusetts Institute of Technology; Thomas L. McNaugher (1990), 'Ballistic Missiles and Chemical Weapons: The Legacy of the Iran–Iraq War', *International Security*, **15**, pp. 5–34. Copyright © 1990 by the President and Fellows of Harvard College and of the Massachusetts Institute of Technology; Daryl G. Press (2001), 'The Myth of Air Power in the Persian Gulf War and the Future of Warfare', *International Security*, **26**, pp. 5–44. Copyright © 2001 by the President and Fellows of Harvard College and of the Massachusetts Institute of Technology.

The Middle East Journal for the essay: F. Gregory Gause, III (2002), 'Iraq's Decisions to Go to War, 1980 and 1990', *Middle East Journal*, **56**, pp. 47–70.

Taylor and Francis for the essays: David Tal (2005), 'The Historiography of the 1948 War in Palestine: The Missing Dimension', *The Journal of Israeli History*, **24**, pp. 183–202. Copyright © 2005 Taylor & Francis; Kenneth M. Pollack (2005), 'Air Power in the Six-Day War', *The Journal of Strategic Studies*, **28**, pp. 471–503. Copyright © 2005 Taylor & Francis Group Ltd; Elie Podeh (2002), 'The "Big Lie": Inventing the Myth of British–US Involvement in the 1967 War', *The Review of International Affairs*, **2**, pp. 1–23; Dima Adamsky and Uri Bar-Joseph (2006), '"The Russians are not Coming", Israel's Intelligence Failure and Soviet Military Intervention in the "War of Attrition"', *Intelligence and National Security*, **21**, pp. 1–25. Copyright © 2006 Taylor & Francis; Hakan Tunç (2005), 'What Was it All About After All? The Causes of the Iraq War', *Contemporary Security Policy*, **26**, pp. 335–55. Copyright © 2005 Taylor & Francis Group Ltd; Avi Kober (2005), 'Does the Iraq War Reflect a Phase Change in Warfare?', *Defence and Security Analysis*, **21**, pp. 121–42. Copyright ©

2005 Taylor & Francis Group Ltd; Sergio Catignani (2005), 'The Strategic Impasse in Low-Intensity Conflicts: The Gap between Israeli Counter-Insurgency Strategy and Tactics during the *Al-Aqsa Intifada*', *The Journal of Strategic Studies*, **28**, pp. 57–75. Copyright © 2005 Taylor & Francis Group Ltd (Taylor & Francis Ltd, http://www.tandf.co.uk/journals).

University of California Press for the essays: Yezid Sayigh (1983), 'Palestinian Military Performance in the 1982 War', *Journal of Palestine Studies*, **12**, pp. 3–24; Augustus Richard Norton (2000), 'Hizballah and the Israeli Withdrawal from Southern Lebanon', *Journal of Palestine Studies*, **30**, pp. 22–35.

Every effort has been made to trace all the copyright holders, but if any have been inadvertently overlooked the publishers will be pleased to make the necessary arrangement at the first opportunity.

Series Preface

War and military matters are key aspects of the modern world and central topics in history study. This series brings together essays selected from key journals that exhibit careful analysis of military history. The volumes, each of which is edited by an expert in the field, cover crucial time periods and geographical areas including Europe, the USA, China, Japan, Latin America, and South Asia. Each volume represents the editor's selection of the most seminal recent essays on military history in their particular area of expertise, while an introduction presents an overview of the issues in that area, together with comments on the background and significance of the essays selected.

This series reflects important shifts in the subject. Military history has increasingly taken a cultural turn, forcing us to consider the question of what wins wars in a new light. Historians used to emphasise the material aspects of war, specifically the quality and quantity of resources. That approach, bringing together technological proficiency and economic strength, appeared to help explain struggles for mastery within the West, as well as conflicts between the West and non-West. Now, the focus is rather on strategic culture – how tasks are set and understood – and on how resources are used. It involves exploring issues such as fighting quality, unit cohesion, morale, leadership, tactics, strategy, as well as the organisational cultural factors that affect assessment and use of resources. Instead of assuming that organisational issues were driven by how best to use, move and supply weapons, this approach considers how they are affected by social patterns and developments.

Former assumptions by historians that societies are driven merely by a search for efficiency and maximisation of force as they adapt their weaponry to optimise performance in war ignored the complex process in which interest in new weapons interacted with the desire for continuity. Responses by warring parties to firearms, for example, varied, with some societies, such as those of Western Europe, proving keener to rely on firearms than others, for example in East and South Asia. This becomes easier to understand by considering the different tasks and possibilities facing armies at the time – when it is far from clear which weaponry, force structure, tactics, or operational method can be adopted most successfully – rather than thinking in terms of clear-cut military progress.

Cultural factors also play a role in responses to the trial of combat. The understanding of loss and suffering, at both the level of ordinary soldiers and of societies as a whole, is far more culturally conditioned than emphasis on the sameness of battle might suggest, and variations in the willingness to suffer losses influences both military success and styles of combat.

Furthermore, war is not really about battle but about attempts to impose will. Success in this involves far more than victory on the battlefield; that is just a pre-condition of a more complex process. The defeated must be willing to accept the verdict of battle. This involves accommodation, if not acculturation – something that has been far from constant in different periods and places. Assimilating local religious cults, co-opting local élites, and, possibly, today, offering the various inducements summarised as globalisation, have been the most important means of achieving it over the years. Thus military history becomes an aspect of total history; and victory in war is best studied in terms of its multiple contexts.

Any selection of what to include is difficult. The editors in this series have done an excellent job and it has been a great pleasure working with them.

JEREMY BLACK
Series Editor
University of Exeter

Introduction

In the aftermath of World War II, as France and Britain gradually withdrew from their overseas empires, two extra-European powers – the US and the USSR – began competing for influence in the strategically-important and oil-rich Middle East.[1] Thus, from the end of the war and throughout the era that came to be known as the Cold War, the Middle East was a battleground for Great Power rivalries. While these same rivalries affected all regions of the world, it was in the Middle East where the effects were, as Shlaim and Sayigh rightly observed, 'particularly pervasive and profound'.[2]

In the Middle East, as elsewhere, the two superpowers competed to 'collect' client states and add them to their camps. Generally, in the post-World War II and Cold-War era, the Arab nations of the Middle East joined the Soviet camp (though there were some exceptions, notably Jordan which often sided with the US), while the non-Arab nations – Turkey, Israel and, until the 1979 Islamic revolution, Iran – were in the US camp. To reward their clients and prevent them from switching camps, the superpowers doled out economic and diplomatic support, and provided privileged access to their advanced weapons systems. So it was that the nations of the Middle East found themselves armed to the teeth, and in a good position to use their arms to sort out their differences over control of oil, water and land.

Post-World War II conflicts in the Middle East took many shapes. They were fought between Israelis and Arabs, Arabs and Iranians, Arabs and Arabs, and also between regional players and outside powers – as was the case in the 1956 Suez Crisis, when France, Britain and Israel collaborated together against Egypt, whose leader Gamal Abdel Nasser had nationalized the Suez Canal. But of all the Middle Eastern confrontations, it is, as Beverley Milton-Edwards and Peter Hinchcliffe have correctly observed, the one between Arabs and Israelis that has proved 'the most dominant and enduring'.[3] And of all the wars fought between the Arabs and the Israelis, it is widely agreed that the Six Day War of June 1967 was the most dramatic and critical in reshaping the geography of the region and Israeli–Arab relations.

After launching a devastating pre-emptive air strike on the morning of 5 June 1967, Israeli ground forces proceeded to take over the Sinai Peninsula and the Gaza Strip from Egypt, the West Bank and Jerusalem from Jordan and the Golan Heights from Syria. In this war, as had previously been the case in the Korean peninsular, Vietnam and the Caribbean, what started as a regional tussle between client states led to growing tensions between the patron superpowers themselves: superpower commitment in the Middle East carried a real danger of a direct superpower clash.

---

1 In the following pages, the term *Middle East* refers to the region from Turkey in the north to Yemen in the south and from Egypt in the west through Iran in the east.

2 Avi Shlaim and Yezid Sayigh, *The Cold War and the Middle East* (Oxford, 1997), p. 2.

3 Beverley Milton-Edwards and Peter Hinchcliffe, *Conflicts in the Middle East since 1945*, (Routledge, 2001), p. 1.

However, the added danger of potential conflict between the superpowers themselves also had a converse, stabilizing effect. For, while it is widely agreed that superpower rivalry exacerbated conflict in the Middle East through arms sales, diplomatic and ideological rivalry, both the US and the USSR also played a major role in *managing* crises when they occurred, working mainly through the UN Security Council to manage and limit conflicts – lest they spill over to create a wider conflict with more devastating consequences.

The Arab–Israeli conflict was not, however, the only flash point in the post-war Middle East. In the Gulf region, tensions between Iran and Iraq had been simmering for decades, focusing mainly on border disputes (notably over control of the Shat el Arab waterway), as well as control of oil and the basic desire of both nations to gain regional hegemony. In 1980 these tensions boiled over into a cruel war that lasted until 1988 and cost an estimated one million lives. However, unlike during the Arab–Israeli wars, Moscow and Washington were this time on the same side, both supporting Saddam Hussein against Ayatollah Khomeini's Iran throughout all eight years of the Iran–Iraq war.[4] It demonstrates something of the limits of superpower influence that Khomeini survived this. The ceasefire that concluded the Iran–Iraq war did not, however, put an end to conflict in the Gulf area as in August 1990, just a short time after the guns fell silent, Iraq invaded her neighbour Kuwait. Tension between the US camp and Iraq then came to a head as a UN-led coalition ejected Iraq's army from Kuwait in 1991. However, this did not settle the matter, and in 2003 the US launched an all out invasion of Iraq to topple Saddam Hussein on the basis of his alleged Weapons of Mass Destruction (though none were ever found).

The post-World War II Middle East was also the scene of several intense civil wars and insurgencies. After the 1967 war with Israel, Palestinian guerrillas clashed with King Hussein's forces in Jordan, as they attempted to topple the King and turn Jordan into a hinterland for attacks on Israel. Supported by the US and Israel, King Hussein eventually confronted the Palestinian guerrillas in 1970, in what came to be known as Black September, and successfully banished them to Lebanon. There, led by Yasser Arafat, the Palestinians established themselves in the capital Beirut and along the international border with Israel. The Palestinians' continued attacks on Israel (which often led to harsh retaliation) and interference in internal Lebanese affairs, caused growing tensions in Lebanon, which in 1975 exploded into a bloody civil war that lasted until 1990. During this war, and after it, Israeli forces also intervened in Lebanon. In June 1982 Israel invaded Lebanon in an attempt to eradicate the Palestinian faction and thus end their cross-border attacks against Israel. Israeli forces successfully ousted Arafat and his colleagues to Tunisia, but ended up occupying southern Lebanon for eighteen years. There they attracted the attention of Iranian-backed Hezbollah militants, who fought a guerrilla campaign that eventually saw Israel's withdrawal from Lebanon in May 2000. Elsewhere, in the Gaza Strip and West Bank territories occupied by Israel in 1967, an all-out Palestinian uprising broke out in 1987. This first intifada lasted until 1993, and after a short respite it rekindled in September 2000, erupting in a much more violent and deadly form, coming to be known as Al Aqsa intifada. At the time of writing, it is still underway.

***

4 It could be argued that the Cold War in the Middle East came to an end not in 1989–90, but a decade earlier, with the Iranian revolution in 1979 and the onset of the Iran–Iraq war in 1980 and the siding of both superpowers with Iraq.

The essays in this collection focus on some of the most important facets of conflict and warfare in the Middle East since the end of World War II. I have organized the essays into three parts. Part I focuses on the Arab–Israeli conflict, with essays covering the period from the 1948 war that followed Israel's creation to the 1982 Israeli invasion of Lebanon. Part II focuses on the Gulf Wars, with essays covering the period from the opening of the Iran–Iraq War in 1980 to the 2003 US invasion of Iraq. Part III concentrates on insurgencies in the Middle East and it includes essays on insurgency in Yemen in the early 1960s, Palestinian insurgency during the 2000 intifada in the Occupied Territories and Hezbollah's insurgency against Israel in the period after the 1982 Israeli invasion of Lebanon.

Arab–Israeli Conflict

The 1948 War

The first section opens with an essay by David Tal (Chapter 1) on the first Arab–Israeli war of 1948. While the vast majority of studies on this war, particularly those in the English language, tend to focus on the diplomatic history of this conflict, Tal concentrates on its military dimension. Looking at the fighting between Israel and Jordan, Tal demonstrates that the two armies clashed only because of mutual misperceptions and a lack of clear channels of communication. He also analyses the Palestinians' success in inflicting severe blows on Jewish forces in late March 1948, and looks into the Israeli attack on the Egyptian forces at Isdud (now Ashdod) in June 1948.

The June 1967 Six Day War

In this section are three essays on the decisive Six Day War of June 1967. Isabella Ginor's essay (Chapter 2) is an original attempt at unearthing the crucial – and surprising – role played by the USSR in instigating the 1967 war. The Soviet warning to Egypt in mid-May 1967 about supposed Israeli troop concentrations on the Syrian border has long been considered a blunder that precipitated a war which the USSR neither desired nor expected. Ginor uses new evidence from Soviet and other Warsaw Pact documents to show that, in fact, the Soviet warning of an Israeli mobilization (which never really happened) was part of a plan approved at the highest level of Soviet leadership to elicit Egyptian action that would provoke an Israeli strike. Soviet military intervention against the 'Israeli aggressor' was intended to follow and was prepared well in advance.

Kenneth M. Pollack (Chapter 3) provides a unique insight into the role played by the Israeli Air Force (IAF) during this war. In the aftermath of the Six Day War the IAF was showered with accolades for the victory, with many believing that it had been the air strike on the morning of 5th June 1967 that had been the decisive element in defeating the Arabs and winning the war. However, as Pollack makes clear, while the IAF did play an extremely important role in pre-empting and destroying the Egyptian air force while it was still on the ground, this was not the decisive element of Israeli victory. Pollack demonstrates that it was the psychological effect of the Israeli air strike, rather than its physical impact, that had the greatest effect on the Egyptians. This conclusion, it is worth noting here, is quite similar to the one advanced by Daryl Press in the second section of this volume in his analysis of the role played by air power during the Persian Gulf War in 1991.

Finally, in Chapter 4, Elie Podeh looks at how the Israeli victory has been depicted in Egyptian and other Arab textbooks, where it has often been claimed that Israel was not fighting on its own but was aided by American troops and pilots. Of course, this is not true – Israel fought and won the war on her own – but the pervasiveness of the myth in Arab thinking demands some explanation, which is where Podeh has stepped in.

The 1969–1970 War of Attrition

The two essays in this section focus on the relatively forgotten War of Attrition between Israel and Egypt along the Suez Canal from 1969 to 1970. One of the most intriguing elements of this war was the Soviet intervention in the fighting: on one occasion, the Israeli air force shot down five Egyptian jets manned by Soviet pilots. Dima Adamsky and Uri Bar-Joseph (Chapter 6) analyse the failure of Israel's intelligence to warn that the Soviets might become so involved in the war. Drawing mostly on primary sources, the writers show how the Israeli belief that the Soviets would avoid direct intervention became so dominant that when information refuting it became available to Israeli intelligence in the summer of 1969, it was ignored.

Robert S. Bolia (Chapter 5) shows how this bloody confrontation along the Suez Canal became a testing ground for Egypt and Israel to gauge the effectiveness of weapons systems provided by their superpower patrons, and was in effect a rehearsal for the war that broke out in 1973. Bolia also shows how Israel believed it had won the War of Attrition and assumed that Egyptians knew they had lost it, and thus further assumed (wrongly as it turned out) that Egypt would therefore refrain from attacking Israel again. This Israeli mistake, as Bolia observes, led to complacency and gave Egypt the chance to launch a surprise attack in October 1973.

Indeed, what makes the Yom Kippur War of October 1973 such a fascinating story is the failure of Israeli intelligence to realize in time that Egypt and Syria were planning an all-out attack to recover lands lost in 1967.

The October 1973 Yom Kippur War

In Chapter 8, Uri Bar-Joseph, a leading expert on the Yom Kippur war, focuses on the Israeli intelligence blindness and other failures. Bar-Joseph assesses whether Israel's military defeats at the early stage of the war were purely due to the strategic surprise, or whether it was down to inherited deficiencies in the Israeli Defence Force's (IDF's) preparedness for war. His conclusion is that the main reason for Israel's stumble in 1973 was indeed the element of strategic surprise, but that there were certain weaknesses in the IDF's war planning: over-reliance on intelligence warning and ineffective air–ground coordination; poor force structure with over-reliance on aircraft and tanks; and, in some cases, a low-level of military professionalism.

The 1982 Lebanon War

The last two essays on the Arab–Israeli conflict focus on Israel's June 1982 invasion of Lebanon. In Chapter 9, Avner Yaniv and Robert J. Lieber look at this invasion, which was first called 'Operation Peace for Galilee', then renamed 'The Lebanon War', but was often dubbed 'Arik's War' after the Defence Minister who was the driving force behind the invasion – Ariel 'Arik' Sharon. The writers analyse the often-made claim that this war demonstrates the tendency of observers to identify Israel with the idiosyncrasies of specific individuals. In the 1950s, Israel's conduct was instinctively associated with the virtues and vices of David Ben-Gurion; in the 1960s, it was Moshe Dayan; in the first part of the 1970s, it became a kind of Golda Meir writ large; in 1977 Menachem Begin and Ariel Sharon. Yaniv and Lieber wonder whether such an approach does justice to the subject or merely oversimplifies what was in fact the culmination of a complex strategic sequence.

Another aspect of the Lebanon war is illuminated by Yezid Sayigh in Chapter 10. Writing in 1983, Sayigh looks at lessons from the Palestine Liberation Organization's (PLO's) military performance during the 1970s in general, and in the 1982 war in particular. There has been very little in Western writing, or even in Palestinian literature, on Palestinian military performance in Lebanon, where in some sectors Palestinian fighters offered stiff resistance to the Israeli invasion. Sayigh fills this gap in the existing literature.

Gulf Wars

The Iran–Iraq War 1980–1988

In the second part of this volume, which focuses on Gulf Wars, F. Gregory Gause III (Chapter 11) looks at Saddam Hussein's decisions to invade Iran in 1980 and Kuwait in 1990. Gause offers an interesting angle by arguing that the driving force behind Saddam Hussein's decisions to invade was his belief that foreign forces were working to destabilize his regime in Iraq, and that by striking across Iraq's borders he would be able to stabilize his position in Iraq proper. In other words, goes the explanation, Iraqi aggressiveness was driven, first and foremost, by Saddam's wish to survive and remain in power.

Focusing on the Iran–Iraq war, Thomas L. McNaugher (Chapter 12) looks at the use of ballistic missiles and chemical weapons, and comes to the interesting conclusion that there has been a tendency to overstate the usefulness of these weapons. While McNaugher agrees that ballistic missiles and chemical weapons produced visible panic in Tehran, he argues that at the same time chemical weapons in particular produced a reaction that was 'out of proportion to their lethality' (p. 278). Another observation McNaugher makes is that 'chemical weapons are generally effective when first introduced into battle against unsuspecting and unprepared adversaries', but they are 'less effective against prepared forces' (p. 278).

The Gulf War, 1990–1991

On Saddam Hussein's invasion of Kuwait in 1990, Ibrahim Al Marashi (Chapter 13) examines how the Iraqi security apparatus, previously used by Saddam as a tool to penetrate every layer of Iraqi society, was now tasked with infiltrating Kuwaiti society to subdue local resistance and make Kuwait accept the Iraqi annexation.

Daryl Press (Chapter 14) analyses the effectiveness of the air power used against Iraqi forces in the 1991 war. He argues that the conventional wisdom among historians, military analysts and foreign policy decision makers is that air power neutralized the Iraqi military before the UN-led ground forces intervened. Such a conclusion – if not based on hard evidence

– is of course dangerous, as it may lead to the incorrect belief that future wars could be won from the air. Press's two main observations are quite surprising: first, although air power played an important role in the coalition's victory over Iraq, its role has been exaggerated and misunderstood and in fact air power was not decisive and did not neutralize the Iraqi forces. Second, the six-week air campaign did not necessarily make it any easier for the UN-led ground forces to achieve their objectives.

The 2003 War in Iraq

In this section on the 2003 invasion of Iraq John Woo (Chapter 15) looks at the legality of this war. He challenges the view of many international legal scholars and foreign governments that the invasion violated international law and he goes on to explain that critics of the invasion have failed to interpret properly existing United Nations Security Council resolutions that authorized the use of force against a state like Iraq. Use of force against Iraq can be regarded as a justified exercise of self-defence as, according to Woo, traditional notions of self-defence, which permit the anticipatory use of force before an offensive attack occurs, have changed to take into account the greater magnitude of destruction of weapons of mass destruction and the behaviour of rogue states.

Hakan Tunç (Chapter 16) analyses the causes of the war, arguing that the Bush administration went to war with Iraq on the basis of several rationales, all of which were directly related to fighting the war on Islamist terror after 9/11. According to Tunç, taken together these rationales constitute a comprehensive, even sophisticated, reasoning for going to war. But, according to Tunç, this does not mean that the assumptions underlying the various rationales are valid. For instance, it is debatable whether Saddam was as irrational as the war hawks claimed, and it is also disputable that a demonstration of US power and resolve in Iraq would deter potential terrorists and the regimes that sponsor them.

Finally, Avi Kober (Chapter 17) sheds light on the patterns of historical change in warfare. He identifies an important shift in warfare in 2003 Iraq. At the heart of this is the unprecedented capability of the military to wage surgical wars and the dramatic strengthening of firepower at the expense of the traditional concentration of forces.

Insurgencies

Yemen

Not all post-war Middle Eastern confrontations have occurred between states; civil wars and insurgencies were also rife. David M. Witty (Chapter 18) opens the third part of this volume with an examination of the way the Egyptian army operated in North Yemen in the early 1960s. This is a particularly unique essay, as there is so little literature on this almost forgotten episode, which saw the Egyptian army intervene in a civil war in North Yemen in October 1962 to protect a newly-established revolutionary government against opponents who were seeking to reinstate the country's traditional, oligarchic government. Egyptian political and military leaders expected to achieve their goal in North Yemen in three months, but they found themselves fighting tribal insurgents in a drawn-out war that they did not fully understand and for which they were not prepared. Five years later, when the Egyptians were forced to withdraw from Yemen following Israel's victory in the 1967 war, Egypt appeared no closer to victory in this murky war than they had been in 1962. Witty shows that Egypt's performance in Yemen illustrates the problems that face a regular army in a low intensity conflict. Among these problems are the adaptations a regular army must make to conduct counter-insurgency operations; the effect that a foreign intervention force has on the credibility of the government it is supporting; the importance of cultural understanding; the role of superior technology in fighting guerrillas; and the impact of a prolonged counterinsurgency campaign on a regular army's ability to fight a conventional war. The Egyptian experience in Yemen, like the US and Soviet experiences in Vietnam and Afghanistan, affirms that a regular army must make fundamental changes to its conventional doctrine if it is to succeed in a war against insurgency.

Intifadas in the Occupied Territories

Sergio Catignani (Chapter 19) focuses on Israeli counter-insurgency strategy and tactics during the Al-Aqsa intifada from 2000. He argues that Palestinian insurgency has posed serious challenges to the Israel Defence Forces, and although the IDF has been able to adapt its tactics successfully to counter Palestinian terror and urban guerilla warfare, they have not been able to achieve a victory over the insurgency. The ultimate cause of this, Catignani argues, is the nature of the Israeli–Palestinian conflict, which ultimately necessitates a political rather than military solution.

Another aspect of Palestinian insurgency is covered by Gal Luft in Chapter 20. Focusing on the Palestinians' use of suicide bombing, Luft argues that increasingly Palestinians are going to see suicide attacks as a strategic weapon – a poor man's 'smart bomb' – that can balance Israel's technological prowess and conventional military dominance. According to Luft, Palestinians appear to have decided that suicide bombings used systematically in the context of a political struggle give them something no other weapon could: the ability to cause Israel devastating and unprecedented pain. The dream of achieving such strategic parity is more powerful than any pressure to cease and desist, and Luft's gloomy prediction is that it is unlikely that Palestinians will abandon this weapon.

Hezbollah in Lebanon

In this final essay, Augustus Richard Norton (Chapter 21), focuses on the role played by Hezbollah in driving the Israeli army out of Lebanon after 18 years of occupation. Using the actual Israeli withdrawal as his starting point, Norton traces Hezbollah's development from a radical resistance group known for suicide bombings and kidnapping, to a highly professional guerrilla force and political party with a broad constituency and pragmatic leadership. Norton examines Hezbollah's entry into politics in the early 1990s, its evolving relations with state actors, especially Syria and Iran, its conduct during and following the withdrawal, and its future prospects.

***

The essays collected in this volume focus primarily on the military dimension of Middle Eastern conflicts and war. As the reader will see, warfare in this strategically important area has been conducted by nations and groups in order to gain tangible resources – oil, land and so on – but also for less tangible assets, such as ideology. As is so often demonstrated in the following essays, the study of warfare and war in the Middle East can only be properly understood in the context of international relations, particularly superpower relations. It is hoped that the essays that follow will provide the reader with a good introduction to the subject, and show how warfare since 1945 has shaped the modern Middle East.

Ahron Bregman,
War Studies, King's College London
December 2007

Part I
Arab–Israeli Conflict

The 1948 War

[1]

The Historiography of the 1948 War in Palestine: The Missing Dimension

David Tal

While the "New Historians" of the 1948 Palestine War have deepened our understanding of those events, they have primarily been interested in its diplomatic history. What is still missing is a thoroughly documented military history of the war, which can help us treat the war as a war, *and not as a political or ideological event, and thus address issues that have so far been neglected. This article shows how such a history can illumine the diplomatic and political aspects of the war through the study of three test cases: the Israeli–Jordanian fighting, which reveals that the two armies came into combat only because of mutual misperceptions and a lack of channels of communication; the Palestinians' success in inflicting severe blows on the Haganah in late March 1948, which points to the role of the Haganah's strategy in its failure; and the story of the unsuccessful Israeli attack on the Egyptian forces at Isdud on the night of 2/3 June 1948, which sheds new light on the meaning and nature of the Jewish victory in the war.*

The historiography of the 1948 Palestine War became a source of controversy following the publication of Benny Morris's article in *Tikkun* in 1988, in which he claimed that a new kind of historian had emerged who, unlike the old historians, was not committed to the old Zionist story and was ready and able to tell the true story of the 1948 war in Palestine.[1] Now that the dust created by the subsequent debate has settled, and Morris himself has since adopted a rather different position, it is time to take another look at the historiography of the 1948 Palestine war, to see what is new, and what is still missing. The emergence of the New History signaled above all the entry of the historiography of the 1948 war into its "archival" phase. After 30 years, historians were given access to the documents pertinent to the 1948 war deposited in the relevant Israeli archives and the British Public Record Office. It is hence hardly

David Tal is a lecturer in the Department of History, Tel Aviv University. Correspondence to: David Tal, Department of History, Tel Aviv University, Tel Aviv 69978; E-mail: tdavid@post.tau.ac.il

surprising that the first studies of the new historians that were based on archival sources, such as those of Ilan Pappé, Benny Morris and Avi Shlaim, were published during the second half of the 1980s, and this trend has been growing ever since. Earlier studies, like those of Nethanel Lorch and Elhanan Oren, were also based on archival sources, but these were exclusively Israeli documents that strongly reflected the authors' affiliation to the Israeli security establishment.[2]

Hence, the major contribution of the new historians lay in bringing the 1948 war historiography to its archival stage, which allowed them to deepen our understanding of the war in a way that had previously been impossible. However, the new historians were primarily interested in the war's diplomatic history. Military historians such as Trevor Dupuy, Chaim Herzog, Martin van Creveld and Ahron Bregman did explore the military dimension of the war, but as they relied mainly on secondary, mostly Israeli, sources and not on archival documents, their work is far from complete.[3] Hence, the full military history of the war still remains to be written. Examining the course of the war from the perspective of military history can also deepen our understanding of events and developments that works of diplomatic history have failed to explore. One example is the Israeli–Jordanian war. As will be seen, a careful study of the Israeli–Jordanian fighting reveals that the two armies had no reason to fight, and it was only because of misperceptions and a lack of channels of communication that the two armies came into combat. It is precisely the military, rather than the diplomatic, history of the war that can shed light on such developments.

Examining this missing dimension in the historiography of the war can also help us treat the war as a *war*, and not as a political or ideological event, and thus to address issues that have so far been neglected. One example is the Palestinians' success in inflicting severe blows on the Haganah in late March 1948, when their forces destroyed five Haganah convoys in various parts of the country and placed Jerusalem under siege. As we will see, the research has concentrated on the Haganah's reaction to the crisis, and not on the way the Palestinians succeeded in turning the tide of war. What enabled the Palestinians to achieve such unprecedented successes? Did the Haganah's strategy play a role in its failure? Similarly, the story of the unsuccessful Israeli attack on the Egyptian forces on the night of June 2/3 is also revealing, mainly because of the gap between the way the battle and its consequences are presented and its real significance. The story of the Isdud battle can thus shed new light on the meaning and nature of the Jewish victory in the war.

The War on the Roads: The Turning Point in the Jewish–Palestinian War

After four months of fighting, during which the Palestinians had tried in vain to win control over Palestine's roads and to cut off Jewish settlements throughout the country, especially Jerusalem, a change occurred at the end of March 1948 when they succeeded in destroying five Haganah convoys within one week. Indeed, the Palestinians now seemed to be winning the "War on the Roads," whose high point was their success in cutting off the Jerusalem–Jaffa road and laying siege to Jerusalem. Nonetheless, all David Ben-Gurion had to say of these successes was that "the Arabs closed the [Jerusalem–Jaffa] road," while both "old" and "new" historians have treated the issue in a similarly cursory manner.[4] Nor has there been any serious attempt to analyze the Jews' response to the crisis and how they overcame it. An understanding of what really happened during this last week of March 1948 necessitates a comprehensive scrutiny of the Jewish strategic position, the decision-making process in the Palestinian camp, the inter-Arab politics of Haj Amin al-Husayni, the ex-Mufti of Jerusalem, and the military developments before, during, and after that week.

Two determining factors underlay the Palestinians' response to the United Nations Partition Resolution. The first was their inability to conduct their struggle against the resolution, which they completely rejected, on their own, and their overwhelming dependence on the assistance of the Arab states. The second was the Arab governments' refusal to let the Palestinians fight their own war. The Arab leaders, particularly in Syria, Jordan, and Iraq, were unwilling to allow al-Husayni to lead the struggle for Palestine. In late 1947 the Arab League's political committee established a military committee, consisting of representatives from all the Arab League's states, under the command of Iraqi General Ismail Safwat. Safwat was put in command of both the Palestinian military groups in Palestine and the Arab Liberation Army, set up by the League in December 1947.[5] The ex-Mufti was fighting two wars: one against the Jews and one against the Arab League, which could be seen as usurping the conduct of the war from the Palestinians. He invested much effort in the attempt to install his nephew, Abed al-Qadir al-Husayni, as commander of the forces fighting in Palestine, instead of Safwat, but failed. This state of affairs created a duality that hampered the Palestinian conduct of war, since the arms contributed by the Arab states to the Palestine war according to the Arab League's decision were given to Safwat, who was responsible for distributing them. Another conflict was created by the ex-Mufti's attempts to establish his control within Palestine by nominating his protégés to commanding positions in Palestine, in defiance of Safwat's claim that he was the only one to authorize such nominations.[6] Thus, while in some places men loyal to the ex-Mufti were in control of military affairs, in other places Safwat's Military Committee gained control by installing foreign commanders, who came with volunteers sponsored by the Arab League.[7] The latter phenomenon became increasingly frequent as time elapsed. Thus the Palestinians entered the crucial stage in the struggle for Palestine internally divided and without influence in the inter-Arab and international arenas.[8]

The Palestinians were unable to overcome their initial weakness and lack of sufficient trained and committed fighters. They also had no central command to run the war, mainly because the Arab League sought to take control of the war. Hence, despite the Palestinians' attacks, Jewish convoys traveled to and from Jerusalem almost on a daily basis until late March 1948.[9] To make the campaign possible and effective, the Palestinians needed to establish at least some form of central command, in which both the Palestinians and the volunteer guerrillas would take part. They also needed to significantly improve their combatants' military capabilities. To that end the ex-Mufti and his two senior commanders traveled to Damascus in early February to meet the Syrian president. At this meeting the argument over the division of power in Palestine was settled: it was agreed to establish four command-areas in Palestine: east, west, north and south, each area commander to be nominally subordinated to Safwat. The northern area commander would be the Syrian Fawz al-Din al-Qawukji, who was at that time an instructor at the Iraqi military academy, the eastern area commander would be Abed al-Qadir al-Husayni, and the commander of the western sector, Hassan Salame—both Haj al-Amin al-Husayni men.[10] One result of the agreement was that while unorganized Palestinian attacks against Jews continued, more and more components of the Palestinian struggle came under some form of central command, even if loose and preliminary. The reorganization and reinforcement of the Palestinian camp with trained Palestinians and volunteers enabled the successes of the Palestinians' March campaign.[11]

It was also agreed in Damascus that the ex-Mufti would receive financial assistance which would allow him to pay the wages of local and foreign "volunteers," while Abed al-Qadir al-Husayni and Hassan Salame reportedly received a monthly payment of 500 Palestine liras each.[12] Haj Amin and his commanders thus returned at the head of groups of "volunteers" who had undergone military training, which provided them with capabilities hitherto unseen among the Palestinians. British police reported that the trainees' "discipline and tactics were far in advance of anything yet encountered."[13] Upon their return to Palestine the ex-Mufti instructed Salame and Abed al-Qadir to carry out a major campaign aimed at stopping Jewish traffic on the roads, and in particular the road to Jerusalem,[14] and during March the tide of the fighting changed. In what seemed to be a coordinated Palestinian campaign on a national scale, over 100 Haganah men were killed in one week while escorting five convoys that were intercepted by the Palestinians: at Atarot (14), Hartuv (11), Nebi Daniel (12), Yehiam (46), and Hulda (24). To these casualties should be added the loss in equipment, arms, and vehicles.[15] Palestinian successes were the result of their ability to act in an orchestrated manner, exploiting to the utmost Palestinian advantages and Jewish shortcomings. The reinforcement of the Palestinian camp by trained guerrillas and the effective use of the *Paza'a* system—by which local villagers were mustered at a certain time to await the arrival of a convoy, and which enabled the Palestinian leaders to gather together hundreds of rifle-holders for several hours—all bore results.[16]

The Haganah was surprised by the success of the Palestinians' concerted campaign, even though there had been many indications that this phase was imminent. Reports about the planned campaign had appeared in the Arab press,[17] and numerous intelligence reports had accumulated in the Haganah files about the Palestinians' intention to intensify their struggle over the roads and to mount a massive campaign to halt Jewish transportation, mainly to isolated locations, first and foremost Jerusalem.[18] Despite the unhindered flow of information from the Haganah intelligence branch to the field units, Haganah deployment did not change, and the convoys did not change their routes.

For the Jews, the most worrying aspect of the Palestinians' campaign was their success in cutting off Jerusalem from the coastal plain. This was indeed a significant achievement. However, the convoy crisis was not only a problem, it was also a symptom, indicating that the strategic plan that dictated the Jews' moves—the May 1946 Plan ("Plan C")—was no longer adequate either in its underlying assumptions or in its resultant mode of operation. This plan was based on the assumption that any military conflict between the Jewish and Arab communities would be similar to hostilities that had erupted while the British were still in the country. While acknowledging that the Arabs would be better equipped and organized, it nonetheless envisaged that they would confine themselves to perpetrating local terror acts against industrial and economic targets. The Haganah's reaction, according to the plan, should be to protect those targets and take measures to quell the terror campaign. The counter-measures would be restricted to retaliation attacks against political leaders ("personal terror"), public sites, vital Arab economic and industrial facilities, and so forth.[19] This was essentially a defensive plan that set no offensive missions or goals for the Haganah. Nothing was said about the possibility of seizing territory held by the Arabs, for example, not to mention a full-scale assault against the Palestinian community or the takeover of the country. Jewish reaction to the outbreak of the hostilities in December 1947 was the implementation of the May 1946 Plan. The leadership of the *Yishuv* (Jewish community in Palestine) had assumed that the Palestinians would eventually accept the UN resolution, that a sharp and decisive yet local response to Arab attacks would prevent escalation, and the riots would subside.[20] Haganah intelligence experts claimed that since the majority of Palestinians had not joined in the fighting, nothing should be done to provoke them. This cautionary note was widely accepted by the *Yishuv* military and political leadership.[21]

Until the March crisis, the Haganah successfully countered the Palestinians' challenge, but the improvements in the Palestinians' organization, structure, training, and arms enabled them to fully exploit their geographical dispersion and to gain advantage over the Jews in the War on the Roads. Nonetheless, the May 1946 plan was another factor that eventually worked to the Palestinians' advantage. Its defensive strategy of reaction led to the dispersion of Haganah forces throughout the country, which was able to prevent the fall of Jewish settlements but no longer sufficed for withstanding the Palestinians' orchestrated campaign. It thus became obvious that the convoy system was no longer adequate.[22]

The conceptual basis of the May 1946 Plan also seemed to have lost its validity. By March 1948 it was obvious that the Palestinians would not accept the Partition Resolution, and the attempts to quell the Palestinian resistance through painful strikes had failed. Given this situation, the approaching end of the British Mandate and subsequent declaration of Jewish independence, which would probably be followed by the invasion of the Arab armies, gave a whole new meaning to the Palestinian challenge. It was only natural to link the hostile activities of the Palestinians, mainly those living inside the designated state of Israel, to the external threat to the Jewish state presented by the invading Arab armies. In order to address this challenge, the

Haganah Command issued Plan Dalet ("D") in March 1948. Although the meaning and implications of this plan are beyond the scope of the current discussion, it is important to note that its working assumption was that the Haganah should make it possible to establish a Jewish state in accordance with the UN resolution, and secure its existence against hostile forces. According to Plan Dalet, the Haganah brigades were to launch offensive campaigns with the goal of seizing territory and assuming Jewish control over the territory allocated to the Jewish state by the United Nations. Hence, when the Haganah assembled its much better equipped and trained troops and turned to a strategy that was based on seizure of territory, the Palestinians proved to be no match. As soon as the Jews took the offensive the Palestinians had no chance, and during April and May they lost garrison after garrison. The Jewish–Palestinian war was decided in those months. The Palestinian defeat was complete, and it was followed by the declaration of the establishment of the State of Israel and the Arab armies' invasion of Palestine.

The (Superflous?) Israeli–Jordanian Fighting

At first sight the Israeli–Jordanian fighting seems to have been superfluous—not because of the collusion theory, according to which Jordan's King Abdullah and the Zionist leaders conspired to divide Palestine among them, at the expense of the Palestinian Arabs,[23] but because the two sides had nothing to fight about. The one point of friction was the Jewish Quarter in Jerusalem's Old City, but even that could not be counted as a reason for fighting. All that Abdullah wanted was to occupy the area to the west of his kingdom that had been allocated by the United Nations to be part of the Arab-Palestinian state, an area that the Jews had no intention of acquiring, while the Jews wanted to make the Jewish part of Jerusalem part of their state, even though the United Nations had decided that that area would be internationalized. It seems that the two sides fought so fiercely, with intervals, from May to July 1948 simply because each side had no idea what the other side's intentions were and consequently acted to frustrate what it assumed was the other side's plan.

Abdullah's attitude towards the Jewish–Arab conflict in general, and the prospect of Jewish and Arab-Palestinian states in Palestine in particular, was influenced by his vision of Greater Syria. Since his forced expulsion from Syria and his assumption in 1922 of the throne of the Emirate of Transjordan (which in May 1946 became the Hashemite Kingdom of Jordan), Abdullah had nurtured hopes of widening his tiny fiefdom to include Syria and Palestine. Of these two, the possibility of swallowing Palestine was more plausible, and with the advent of the political process toward the end of the British Mandate and the introduction of an alternative solution in Palestine, Abdullah acted to make his dream come true. However, fearing that a unilateral act on his part would provoke Arab rage against him, even to the point of Jordan's expulsion from the Arab League, he sought ways to legitimize his Palestine campaign. To that end he tried to gain support amongst the Palestinians, in order to justify his control over the Palestinian territory, and he also sought the participation of other Arab states in the Palestine campaign, although not within the framework of the unified Arab League's command and aims.[24]

That command was created in April 1948, led by Iraqi General Mahmud Nur a-Din. In early May the Arab chiefs of staffs or their representatives met in Damascus to assign each army its missions, according to a unified plan prepared by Nur a-Din. The focus of his plan was to be northeastern Palestine, where the invading forces would try to sever the eastern Galilee, from the Huleh Valley to Lake Kinneret (Sea of Galilee), from the territory of the Jewish state. This was to be achieved through the advance of the Syrian and Lebanese forces from the Bint Jubayl area in Lebanon through Malkiya towards Safed, Tiberias and Nazareth; Iraqi and Jordanian units would push westward towards Afula and Nazareth and await the next stage; and the Egyptian army would move northward, to Yibna (over 30 km. south of Tel Aviv), which was inside the designated Arab state, in order to "pin down and destroy Jewish forces and thereby aid the success of the [northern] operations." The Egyptians were not to advance automatically into the territory of the Jewish state but, at least in the first stage, to create a diversion and lure Jewish forces into their sector, thus removing pressure from the north, where the Arabs would make their main push.[25]

The unified campaign idea did not materialize, as King Abdullah made his own plans for the Palestine campaign, which emasculated the unified command plan and made it meaningless. According to Abdullah's plan, the Arab Legion would carry out operations in the central area of Arab Palestine. It would cross the Jordan River after 15 May and advance towards Beer Sheva, Hebron, Ramallah, Nablus, and Jenin, with forward elements in Tul Karem and the area just south of Lydda—all within the area allocated to the Arab Palestinians.[26] What is missing from this description of Transjordan's war aims is Jerusalem. The available sources indicate that initially there was no intention of sending the Legion to Jerusalem. Moreover, British Minister in Amman Alan Kirkbride described Abdullah's great distress at the reports on the dire situation of the Palestinians in Jerusalem. It was only after receiving the reports, along with phone calls and delegations from Palestinians that Abdullah instructed his army to move into Jerusalem.[27]

Even at this late stage, there seemed to be a slight chance that an Israeli–Jordanian war could be prevented. On 11 May Abdullah met with Golda Meyerson (Meir) of the Jewish Agency for the second time (the first meeting had been in November 1947). This time their meeting was tainted by the events of the preceding month: the massacre of Arab villagers at Deir Yassin, the total collapse of the Palestinians in the face of the Jewish offensive and the ensuing massive flight of Palestinians, mainly to Jordan. All these developments stiffened Abdullah's attitude, and Meyerson indeed found him sounding tougher than before. The discussion, he said, should proceed from the assumption that one state would be established in Palestine, with the Jews to be granted autonomy in the areas where they had a majority, and the country to be annexed to Jordan after one year. Meyerson rejected this out of hand, to which the king retorted that "he had been in favor of peace all along, but the only way to avert a war is to accept his proposal."[28] The conversation was doomed from the outset.

While Abdullah was in no position to make meaningful gestures to the Jews, he was well aware that the Jews would not accept his statement that "the only solution to the Palestine problem was an Arab state which included areas over which the Jews exercised local autonomy." In fact, he had told Kirkbride that "he did not expect the Jews to accept the proposal," and he knew that "the Jews reacted sharply" to the public statement he had recently made to that effect.[29] However, he wanted the meeting with Meyerson as he wished "to keep the door open for negotiations when both sides were in a more reasonable frame of mind."[30] Furthermore, given the Arab Legion's limited goals and its intention to refrain from invading territory allocated to the Jewish state, war between the two armies was not unavoidable. This was the gist of the talks on 2 May between representatives of the Haganah and a Jordanian delegation headed by Colonel Goldie, assistant to John Glubb, the Arab Legion commander, at which Goldie claimed that the Arab Legion "[did] not desire to fight" the Jews.[31]

After hearing Meyerson's account of the 11 May meeting, Ben-Gurion ordered the Haganah high command "to plan a campaign against a general Arab invasion" and "to issue an order to the brigades about an expected invasion by Transjordan." He consulted with his senior advisers on whether to attack the Jordanian forces the moment they crossed the Jordan River or to wait until they entered territory allocated to the Jewish state.[32] As far as Israel was concerned, then, the Arab Legion was poised to join the invasion of the Jewish state. Shertok indeed acknowledged that Abdullah intended to take over only the Arab parts of Palestine, but as Shertok told the American Secretary of State, George Marshall, "we were by no means certain whether all these assurances could be taken on their face value."[33] Israel's deployment in the country's central sectors reflected that assessment. Four Haganah brigades were stationed in May 1948 opposite the Jordanian front, more troops than in any other sector, even though numerically the Arab Legion was smaller than the Egyptian contingent, for example.[34]

The Jewish–Jordanian war "officially" started only after the termination of the British Mandate and Ben-Gurion's declaration of the establishment of a Jewish state on 14 May. On that night the Legion's forces, some of which had just left Palestine, crossed the Jordan River on their way westward.[35] After crossing into Palestine over the Allenby Bridge, the troops turned north, instead of west to Jerusalem, and went to Nablus via a third-grade road, and from there to Ramallah. The reason for the diversion was Glubb's reluctance to engage the Jewish forces in Jerusalem.[36] Since the Legion's aim was the major cities of Samaria and Judea, the Jordanian troops advanced and positioned themselves there.[37] The Jordanians' takeover of the parts west of the River Jordan that had been allocated to the Palestinians was completed on 20 May, with the appointment of Ibrahim Pasha Hashim, a former prime minister, as military governor of all parts of Palestine occupied by the Arab Legion, who appointed deputies in the main cities in the occupied territories.[38]

While drawing up the invasion plans, Glubb's attitude to Jerusalem had been mixed. On the one hand, he wanted to avoid fighting in Jerusalem. He was afraid that urban warfare in Jerusalem against the Jews would lead to many casualties among his troops, who, unlike the Jews, had no experience of such fighting. Defeat in Jerusalem, Glubb feared, could lead to the loss of the positions the Arab Legion already held in Samaria, Judea and the Hebron area. On the other hand, if the Jews succeeded in holding all of Jerusalem, they would gain free passage to Jericho, which might expose the Arab Legion on the West Bank to the danger of being cut off from its bases in Jordan. Glubb believed that the best solution would be a cease-fire in Jerusalem, which was being negotiated at that time and which would maintain the existing status quo. It would give advantage to neither side and would not jeopardize the Arab Legion's position.[39] The Iraqi military governor of Arab Jerusalem, Fadil Bey, was inclined to accept a cease-fire, but he referred the matter to King Abdullah.[40] On 15 May, talks were conducted between Arab representatives on the one hand, and the Security Council Truce Commission, the High Commissioner and the Chief Secretary for Palestine on other hand, resulting in an agreement on a cease-fire in Jerusalem. However, the agreement did not come into effect as immediately upon the termination of the Mandate, the Jews launched Operation "Kilshon" (Trident), whose goal was to impose Jewish control over the western part of the city, where Arab enclaves separated Jewish neighborhoods from one another.[41]

The Jews' offensive in Jerusalem put the Arabs of Jerusalem in a dire situation, and Amman was deluged with desperate pleas for help, by telephone and from delegations to the King's palace. Abdullah was unable to ignore the outcry. Glubb tried to postpone a decision on this matter in an attempt to avoid fighting in Jerusalem. The Jordanian government was so disturbed by his delaying tactics that the Jordanian prime minister checked with Kirkbride to find out whether the British government had ordered Glubb not to go to Jerusalem or whether he was acting on his own initiative. In an attempt to avoid clashes with the Jews, Glubb decided to limit the Legion's operation to the occupation of the Old City and the Arab part of Jerusalem, and to avoid occupying all of Jerusalem, so that King Abdullah could become the savior of the Holy Places without risking an all-out war with the Jews. The Jordanians also assumed that operating in Jerusalem would not be considered a violation of Abdullah's commitment not to invade the Jewish state, as Jerusalem and its approaches were beyond the boundaries of the State of Israel.[42] In any case, with the change in plans, the Arab Legion did not intend to do more than protect the Old City, and its subsequent attacks were also designed to serve this purpose.[43]

The controversy over sending the Legion to Jerusalem ended on 17 May when King Abdullah personally ordered Major Abdullah al-Tal, the 6th Battalion commander, to send forces to the Old City's rescue. Al-Tal immediately sent three of his battalion's companies, and upon his arrival in Jerusalem on the following day, he himself took command of the Arab forces in the city.[44] This reinforcement sealed the Jewish Quarter's fate, and after a few days of heavy fighting, al-Tal delivered an ultimatum through the Red Cross representative, demanding the Quarter's surrender. The ultimatum fell on receptive ears. Despite the urgings of David Shaltiel, commander of the 4th Brigade, to the forces and residents of the Jewish Quarter to continue the struggle, on the next day the Quarter surrendered.[45]

For the Jordanians, the war was now over. They had achieved all their goals, both those planned beforehand—the occupation of the West Bank—and those set after the invasion to Palestine—the occupation of Jerusalem's Old City. However, the war did not end here because the Israelis had no idea of the Arab Legion's limited goals. All they knew was what they saw, and from the Israeli side it seemed as if the Jordanians were conducting a concerted campaign aimed at capturing all of Jerusalem in order to sever it from the Jewish state. The events in the Old City and north Jerusalem certainly seemed to be proof of this, as were the events to the south of Jerusalem, where Ramat Rachel came under heavy attack and was conquered.[46] Ben-Gurion became convinced that it was most urgent to launch a forceful offensive against the Legion. However, to make such a campaign effective it was first necessary to clear the Tel Aviv-Jerusalem road by destroying Ramle and Lydda and capturing Latrun from the Arab Liberation Army forces that had held it until 14 May (he did not yet know that the Arab Legion had already taken their place), after which he intended to send heavy reinforcements to Jerusalem. Until then, a cease-fire was the only way to relieve the pressure from Jerusalem.[47] Yadin and other senior military men agreed with Ben-Gurion that a cease-fire in Jerusalem was desirable, but for different reasons. Unlike Ben-Gurion, IDF Chief of Operations General Yigael Yadin opposed investing great efforts and forces in the struggle over the road to Jerusalem, and he did so only under heavy pressure from Ben-Gurion. He thought that a cease-fire in Jerusalem might assuage Ben-Gurion's fears and bring him to agree to the different strategy proposed by Yadin.[48] However, a cease-fire proved unattainable, as with the Arab Legion's success in gaining control over east and north Jerusalem, the Jordanians lost interest in a cease-fire, seeking to strengthen their hold in the Arab part of the city and in the Old City.[49]

The Latrun garrison on the Jerusalem-Tel Aviv road had been held by Qawukji's forces, but they withdrew in accordance with an earlier agreement with King Abdullah that upon the Arab Legion's invasion of Palestine they would concentrate north of Ramallah "for a week's recuperation," after which the Arab Liberation Army would act in concert with the Arab Legion. Jewish intelligence intercepted a message sent from the Legion's HQ in Amman to Qawukji, in which the latter was instructed to pull his forces out of Palestine to Transjordan, after the League's invasion.[50] The Jordanian forces in Palestine were unaware of this order, so that Qawukji's abandonment of the garrison came as a surprise to the Arab Legion's commanders. It took the Legion three days to man the strategic place, a critical period which, fortunately for the Jordanians, went unnoticed by the Jews.[51] It seems that the Arab Legion had not planned to go to Latrun. After arriving in Ramallah, the Jordanian 3rd Brigade seized positions along the Ramallah-Latrun road. While positioning itself, a battalion's command noticed that units of the Liberation Army were leaving the Latrun garrison. The move surprised the Jordanians, but even though they were afraid that the Jews would capture the vacant garrison, the battalion was not allowed to make any further moves.[52] Instructions to capture Latrun arrived on 17 May. The Arab Legion was preparing its offensive against Jerusalem, and the Jordanian Brigade commander sent forces to the garrison to prevent the Jews from bringing reinforcements to the city.[53]

Once fighting broke out in Jerusalem between the Jewish and the Jordanian forces, Latrun became the crucial point for both sides. The Jews needed Latrun as it dominated the road to Jerusalem, vital for sending reinforcements to the Jewish forces in the city to enable them to thwart what they believed to be the Jordanian attempts to capture the entire city. The Jordanians needed Latrun to prevent those reinforcements from reaching Jerusalem and enabling what they believed to be the Jews' plans to attack the Jordanian stronghold in Jerusalem and its vicinity. But neither side had the intentions attributed to it by the other. For this reason, the war between Jordan and Israel in 1948 was the result of misconception and misunderstanding. This, however, was also why it was the first front that moved from war to a cease-fire and peace negotiations.

The Isdud Battle, 2/3 June 1948

The missing military dimension of the war is especially salient in the existing description and interpretation of the Israeli counter-attack on the Egyptian stronghold in Isdud on the night of 2/3 June 1948. Both conservative and revisionist historians agree that the Israeli attack was a success. Those who admit that the attack itself was a failure nonetheless argue that its strategic consequences were in Israel's favor.[54] One scholar even argues that the Israelis dealt the Egyptians a humiliating defeat in that battle.[55] Neither version is true. The real story, and the significance, of the events that took place that night have been obscured by ideological biases, negligence of existing documents, and a focus on the narrative of one side. The Israeli attack was in reality a failure that demonstrated the capabilities of the Egyptian command and the shortcomings of the Israelis, while the alleged strategic results, according to which the attack forced the Egyptians to withdraw from their plan to advance further north, were irrelevant as at this stage the commander of the Egyptian expeditionary force had no such plan.

The decision to launch the counter-attack was made in expectation of the cease-fire that was being negotiated by the United Nations' mediator Count Folke Bernadotte. Assuming that a truce might come into effect in short order, the Israeli General Staff instructed the 5th (Givati) Brigade to launch an all-out attack on the Egyptian force at Isdud with the aim of destroying it. This decision is usually explained as arising from the concern that the Egyptians would try to "keep pushing northward towards Tel Aviv, which was only 32 kilometers away."[56] However, no such consideration can be discerned in the attack order, codenamed "Pleshet" (Land of the Philistines), which stated that the Egyptians' "aim is not known.... It must be assumed that this is a diversionary force that has the task of pinning down our forces in order to enable another [Egyptian] force to operate in the area."[57] The identity of the "other force" remains a mystery, but it is perfectly clear that no one on the General Staff seriously thought that the Egyptians were about to push north. Probably the underlying reason for the decision to attack and annihilate the Egyptian column was the assessment that the success of the operation would free forces that could then be rushed to assist in the battle for the road to Jerusalem, which Ben-Gurion considered crucial. Both Ben-Gurion and the General Staff believed—on the basis of the low opinion in which they held the Egyptians' combat capability—that it was possible to achieve the ambitious plan. Afterwards, if the truce failed to come into effect, it would be possible "to plan the operation for the liberation of Jerusalem," Ben-Gurion maintained.[58]

The General Staff's assumption was indeed correct. As mentioned above, the Arab war plan stipulated that the Egyptian expeditionary force was to reach Yibna and stay there, in order to confine Israeli forces to that sector, thereby reducing the number of Israeli forces that could be moved to stop the invasion of the other Arab forces in northern Palestine. The Egyptian expeditionary force crossed the Egyptian–Palestinian international border on 15 May and started its journey north. On 24 May it reached Majdal, 30 km. south of Yibna, where its commander, General Ahmad 'Ali al-Muawi, decided to stop. He assumed that a thrust towards Yibna would expose the flank of the Egyptian column to attacks by the Jewish forces and to the danger of being cut in two, as the Egyptians were advancing along one route, stretching their supply lines almost to the limit. General Muawi was well aware of the weakness of his supply system, which lacked reserves of ammunition and arms depots.[59] He was also concerned by the presence of Jewish settlements near his route of advance, and to avert possible disruptions to his supply lines he decided to concentrate on widening his area of control to the east of his line of supply, thereby removing any threat from that direction.[60] Muawi realized that his connection with the rear echelon ran along a route that was lined with tenacious Jewish settlements, and he was undoubtedly acquainted with the assessments that had been presented to the Palestine Committee concerning the strength of the Jewish forces. The report also stated that "it has already been proved that all the Jewish settlements, from the smallest to the largest, are well fortified... and all of them are fully equipped for defense. Reality has shown that it is impossible to overcome and capture these settlements solely with light weapons."[61] The mission was assigned to the Egyptian 9th Battalion, which was instructed to launch the attack on the Jewish settlements to the east of the Egyptian line of supply on 29 May, but meanwhile the encampment of the Egyptian forces incensed the Jordanians, who were being left to face the Israelis alone. At this time the Israelis were attacking Latrun, and Glubb requested Muawi to push ahead further north, to ease the pressure on the Legion, but the Egyptian commander, fearing that he would encounter Israeli forces, rejected the request.[62] It was only after the issue was treated on a governmental level, with Amman asking the Egyptian government for assistance, that Muawi was ordered to move north and relieve the pressure from the Jordanian forces at Latrun. Against his will, then, Muawi resumed his march north on 28 May, towards the Yibna-Isdud area. The Egyptian force, which consisted (apparently) of three infantry battalions and an artillery battalion, deployed at the ruins of the bridge on the main road leading into the heart of Israel.[63]

When the Egyptians encamped at the bridge next to Isdud that had been destroyed by Israeli sappers a few days before, there were no Israeli forces to block their way and prevent them from continuing their advance. The Givati Brigade sent reports to the

General Staff about the Egyptians' movements and the fact that they had stopped, but did not send troops to confront them. The Givati command was well aware that the destroyed bridge could not prevent the Egyptian commander from advancing if he wished to.[64] However, as noted above, the Egyptians had no intention of moving further north both because their plan stipulated their encampment in Yibna-Isdud and because any further advance northward would have made the column's flank intolerably vulnerable to attack by the Israeli forces and overextended the force's supply lines to the breaking point. The Egyptian advance thus came to a final halt near Yibna, in an area that was still outside the boundaries of the Jewish state according to the November 1947 partition resolution.[65]

The Israeli decision to launch the counter-attack seems to have arisen from an almost instinctive feeling that after several weeks in which the Arabs had held the initiative, it was now time to reverse the trend and to put the Arab side on the defense. The aim of Operation Pleshet was to destroy the "enemy force" by means of frontal attacks on the Egyptian concentration in the Isdud area and to occupy the territory it was holding.[66] The attacking forces did not know exactly what they were up against or where the Egyptian forces were located. In the order all the references to the Egyptian formation were couched in terms of "unknown," "as conjectured," or "we have to assume."[67] The Egyptian 2nd Battalion was deployed in the area between the Isdud train station and the village itself; while the 9th Battalion, considered the best of the Egyptians' units, was based in the area between the Isdud bridge and the train station. A company of medium machine guns took up positions between the 2nd and 9th Battalions, strengthening the Egyptians' lines. The Egyptians' defenses were further beefed up by the deployment of two armored platoons, a light tank platoon, and a battery of 25-pounder and 3.7″ guns. The 9th Battalion had been reinforced with additional medium machine guns. All told, the Egyptian fighting force was about 2,300-strong.[68]

The Israelis had half that number of combat personnel. The Givati Brigade was reinforced by the Negev Brigade's 7th Battalion and by companies from its commando battalion. The soldiers of the 7th battalion were new immigrants who had acquired their military training in battle. The attack was set for the night of 1/2 June, with two forces slated to launch the assault at two separate points. As the Israeli forces were making their way towards the Egyptian lines, they received an order to call off the operation in the wake of reports that the Egyptians had accepted the Security Council's call for a truce.[69] However, it soon turned out that the actual time for the truce to come into effect had not yet been set.[70] It did not take long for the Israelis to realize that the information they had received about the truce was wrong, but in any case Ben-Gurion and the General Staff decided to continue fighting and to implement the assault at Isdud.[71]

Operation Pleshet, which was launched on the night of 2 June 1948, was a failure. The Egyptians repulsed the Israeli forces and in some cases pursued them as they retreated. Over 50 Israeli soldiers were killed or missing and about the same number wounded.[72] A decade later, Shimon Avidan, the commander of the 5th Brigade, regretted having carried out the attack, but there is also evidence that he objected to the plan at the time it was discussed.[73] Indeed, the very plan of the operation doomed it to almost certain failure. The main problem, as the semi-official history of the Givati Brigade shows, was that the planners took into account neither the advantage a defensive force enjoys over an attacking force nor the serious numerical inferiority of the Israeli force. The battle plan failed to address, and hence to exploit, the fact that the Egyptian lines, though dense, were spread over a fairly wide area. Instead of making a concentrated effort to breach the lines at one point or to sever the Egyptians from their rear echelon by attacking the more southern 2nd Battalion, the Israelis split their forces into small units, each of which was assigned a specific sector to breach. In every sector the attacking forces were numerically inferior and possessed less effective firepower. The artillery at their disposal was scant and, according to Avidan, was not utilized intelligently. It did not take a great effort for the dense Egyptian defensive alignment, based on concentrated machine-gun and artillery fire, to break the attack. Disparagement of the Egyptians' combat ability, defective intelligence, and a faulty battle plan were the major factors that led to the failure of the attempt to destroy and rout the Egyptian forces around Isdud.[74]

Nevertheless, Israeli historiography takes pride in the accomplishment of Operation Pleshet, viewing it, despite the failure, as "a strategic victory of supreme and decisive importance."[75] It seems, however, that the battle had little impact on the Egyptians' war plans. Operatively, the Egyptian move had had limited objectives—at most to capture the Arab village of Yibna—and this was also the assessment of Haganah intelligence.[76] As we saw, the reason why the Egyptian column came to a halt at the ruins of the Isdud bridge was not because it was stopped by Israeli forces. Indeed, it should be recalled that Muawi reached Isdud only because he was ordered to, with the aim of pinning down Israeli forces in the southern sector and thus taking some of the pressure off the Jordanians at Latrun; and Ben-Gurion, too, wanted to attack the Egyptians so that he could free more troops to help at Latrun. From their point of view, the Egyptians accomplished their goal, as the invading force was able to repulse an attack and thus compel the Israelis to continue committing troops to the south.

Conclusion

There is some irony in the statement that the missing dimension in the historiography of the 1948 war is precisely the study of its military history. After all, this was where the study on the war started: Nethanel Lorch's history of the war was already published in 1958; many books told the story of the brigades and other units that took part in the war; and Elhanan Oren, one of the most prolific scholars of the war, has enriched us with his studies since the 1970s.[77] However, the shortcomings of these studies, which I mentioned at the beginning of this article, were not addressed in the work of the new historians. In some cases, it is even argued outright that "the historian of the war should pay less attention to its military development, and instead address the political aspects."[78] This article has sought to show the contrary, that it is precisely a detailed military history of the war, based on systematic study of the pertinent documents, that can shed light on political and diplomatic aspects that would otherwise remain unknown or obscure. Studying the military history of the war without bias and political orientation can thus shed light on what really happened in 1948, during the first Arab–Israeli war.

Notes

[1] Morris, "The New Historiography."

[2] Lorch, *Korot milhemet ha-atzma'ut* (1989); Oren, *Ha-derekh el ha-ir* Pappé, *The Making of the Arab-Israeli Conflict*; Shlaim, *Collusion across the Jordan.*

[3] Dupuy, *Elusive Victory*, 3–128; Herzog, *The Arab-Israeli Wars*; Van Creveld, *The Sword and the Olive*; Bregman, *Israel's Wars, 1947–93.*

[4] Ben-Gurion, *Medinat Yisrael ha-mehudeshet*, 76; Gelber, *Garin le-tzava ivri*, 103–6; Lorch, *Korot milhemet ha-atzma'ut* (1958), 121–22; Morris, *Righteous Victims*, 202–4; Oren, "Ha-ma'arakhah," 36; Pa'il, *Mi-"Haganah" li-tzva haganah*, 295, 305–6; Pappé, *The Making of the Arab-Israeli Conflict*, 85–86; Shlaim, *Collusion across the Jordan*, 156; Sela, "Aravyei Eretz Yisrael," 158–59.

[5] British Legation, Damascus: "Political Summary for February and March, 1948," Public Record Office, London (hereafter PRO), FO 371/68808; *Me'ahorei ha-pargod*, 51–52.

[6] *Me'ahorei ha-pargod*, 100–101.

[7] Ibid., 76, 82-85.

[8] Record of Conversation [Between a British Officer of the Arab Legion and a Visitor from the Hebron District], 28 January 1948, PRO, FO 816/116; Sela, "Aravyei Eretz Yisrael," 130–35.

[9] Records of the convoys plying the road from the Coastal Plain to Jerusalem, and reports of attacks on convoys all over the country reached the Haganah command on a daily basis from December 1947. These reports are deposited in the Military (IDF) and Defense Establishment Archives, Givatayim (hereafter IDFA), 464/54/2; report by "Benjamin," 26 February 1948, Haganah Archives, Tel Aviv (hereafter HA), 80/50/21.

[10] Gen. Sir A. Cunningham to the Secretary of State for the Colonies: Weekly Intelligence Appreciation (hereafter WIA), 1 and 14 February 1948, PRO, CO 537/3869; British Legation, Damascus: "Political Summary for February and March, 1948," PRO, FO 371/68808; intelligence reports, 24 March 1948, HA 105/98; Sela, "Aravyei Eretz Yisrael," 154, 156, 158.

[11] Intelligence reports, 3 March 1948, HA 105/98; Gen. Sir A. Cunningham to the Secretary of State for the Colonies: WIA, 8 March 1948, PRO, FO 816/117.

[12] Intelligence reports, 4 March 1948, HA 105/98.

[13] Gen. Sir A. Cunningham to the Secretary of State for the Colonies: WIA, 17 January 1948, PRO, CO 537/3869.

[14] Cable to "Sasha" (Y. Alon) from Y. Yadin, January 1948, IDFA, 661/69/36; cable from E. Denin to E. Shiloah, 13 January 1948, Central Zionist Archive, Jerusalem (hereafter CZA), S/25/9007; Gen. Sir A. Cunningham to Secretary of State Colonies: WIA, 16 and 24 January 1948, PRO, CO 537/3869; cable from "Adina" to "Hilel" (I. Galili), 3 February 1948, IDFA, 464/54/2; cable from "Max," 4 February 1948, CZA, S/25/9007; Gen. Sir A. Cunningham to the Secretary of State for the Colonies: WIA, 7 February 1948, PRO, FO 816/116; from Amman to Foreign Office, 9 February 1948, PRO, FO 371/68366; "The Army for the Salvation of Palestine" [unsigned paper], 13 February 1948, , S25/3999; from the High Commissioner for Palestine to the Secretary of State for the Colonies: WIA, 14 February 1948, CO 537/3869; "Tene," intelligence report, 20 February 1948, HA, 80/50/21. On the training program the "volunteers" had undergone, see J. Palmon's report, 26 February 1948, HA, 80/50/21; "Tene/Yavne,"

"The volunteers in Jerusalem," 3 March 1948, HA, 105/216/1; "Tene/Yavne," intelligence report, 15 March 1948, IDFA, 661/69/36; "Tene/Yavne," "The Arab force in Jerusalem," 16 March 1948, IDFA, 661/69/36; *Me'ahorei ha-pargod*, 82–84; Levy, *Tishah kabin*, 354–55.

[15] Daily report, 1 April 1948, IDFA, 464/54/2; *Sefer toldot ha-Haganah* 3.2: 1443–56.

[16] See also a report on the Palestinians' destruction of the Yehiam convoy: Intelligence reports, 1 April 1948, HA 105/98.

[17] Gen. Sir A. Cunningham to the Secretary of State for the Colonies: WIA, 20 March 1948, CO 537/3869.

[18] Intelligence reports: "Tene/Naim," 3 March 1948; "Tene/Matspa," 5 March 1948; "Tene/Sasha," 7 March 1948; "Tene/Yavne," 10 March 1948; "Tene/Yavne," 15 March 1948; "Tene/Avner," 25 March 1948; "Tene/Naim," 6 April 1948; "Tene/Mitspa," 7 April 1948. All in IDFA, 661/69/36.

[19] Plan B had explicitly assumed that "following the policy conducted by the [Mandate] government, a [Jewish–Arab] war will not exceed the level of 'small war'." "Plan B," September 1945. HA, 73/100. "The May 1946 Plan," HA, 73/140.

[20] Galili in meeting of the *Yishuv* Security Committee, 13 November 1947. CZA, S25/9343; Ben-Gurion in meeting of Mapai Political Committee, 9 December 1947 and 8 January 1948, Labor Party Archives, Beit Berl, Israel, 25/47, and 25/48; Y. Ben-Zvi in meeting of Mapai Political Committee 11 October 1947, in Avizohar, ed., *Akhshav o le'olam lo*, 183. Ben-Gurion on equal rights for the Arab citizens of the Jewish state, speech to the Keren ha-Yesod convention, 29 October 1947, in Ben-Gurion *Pe'amei medinah*, 438.

[21] Minutes of a meeting on Arab Affairs, 1–2 January 1948, HA, 80/50/21; Israel Galili in Gilad, ed., Sefer ha-Palmah, January 1948, xx–xxii; February 1948, xxii–xxiv; Ben-Gurion, letter to M. Shertok and G. Meyerson, 14 March 1948, Memoranda files, the Ben-Gurion Research Center, Sde Boker Campus, Israel (hereafter BGA).

[22] Summary of the Military Situation, 1 April 1948, IDFA, 922/75/595; I. Galili report, 4 April 1948, HA, 80/50/18.

[23] Ber, *Bitahon Yisrael*, 125–26; Rogan, "Jordan and 1948," 109–10; Pappé, *The Making of the Arab-Israeli Conflict*, 119–21; Shlaim, *Collusion across the Jordan*, 1–2.

[24] Sir A. Kirkbride, Amman, to the Foreign Office, 6 October 1947, PRO, FO 371/61497; Mr. Evans, Beirut to Foreign Office, 12 October 1947, PRO, FO 371/61530; Nevo, *Abdallah*, 50–51.

[25] *Me'ahorei ha-pargod*, 23, 148–50; *Be-einei oyev*, 82, 89–90; and see the testimony of Gamal Abd el-Nasser quoted in Ayalon, Hativat "Givati", 97, 105.

[26] Record of Conversation between Brig. Glubb Pesha and D.M.I. War Office, Jan. 30, 1948, PRO, FO 371/68369; Sir A. Kirkbride, Amman, to the Foreign Office, 13 April 1948, PRO, FO 371/68852; *Me'ahorei ha-pargod*, 145–46, 149–50.

[27] For a fuller discussion, see Tal, *War in Palestine*, 205–14.

[28] Golda Meyerson in a meeting of the People's Administration (Minhelet ha-Am), *Protokolim yeshivot ha-memshalah, 18 April–13 Mai 1948*, 40–43.

[29] A. Kirkbride, Amman, to the Foreign Office, 24 April 1948, PRO, FO 371/68852.

[30] A. Kirkbride, Amman, to the Foreign Office, 23 April 1948, PRO, FO 371/68852.

[31] R. Zaslani to M. Shertok, 3 May 1948, Yogev, ed., *Te'udot mediniyot: Detzember 1947–april 1948*, 721–22.

[32] Ben-Gurion, *Yoman ha-milhamah* 1:409, entry for 11 May 1948; 411, entry for 12 May 1948.

[33] Meeting: M. Shertok, E. Epstein-G. Marshall, R. Lovett, D. Rusk, 8 May 1948, in Yogev, ed., *Te'udot mediniyot: Detzember 1947-april 1948*, 759.

[34] Pa'il, *Mi-"Haganah" li-tzva haganah*, 325–35.

[35] Sir A. Kirkbride, Amman, to the Foreign Office, 15 May 1948, PRO, FO 371/68372; telegram no. 338 from Sir A. Kirkbride, Amman, to the Foreign Office, 16 May 1948, PRO, CO 537/5315.

[36] Telegram from the Sixth Brigade to Yadin, 14 May 1948, IDFA, 922/75/1175; Glubb, *A Soldier with the Arabs*, 82–83; al-Rossan, "Ma'arakhot Bab al Wad," 147.

[37] Sir A. Kirkbride, Amman, to the Foreign Office, 19 May 1948, PRO, FO 371/68506; telegram no. 358 from Sir A. Kirkbride, Amman to Foreign Office, 21 May 1948, CO 537/5315; Glubb, *Soldier with the Arabs*, 105; al-Tal, *Zikhronot*, 65; Levy, *Jerusalem in the War of Independence*, 240; al-Rossan, "Ma'arakhot Bab al Wad," 147.

[38] Telegram no. 357 from Sir A. Kirkbride, Amman, to the Foreign Office, 20 May 1948, PRO, FO 816/120.

[39] Glubb, *A Soldier with the Arabs*, 106–10, 113.

[40] Report by C.V. Herzog, 21 May 1948, in Yogev, ed., *Te'udot mediniyot: Detzember 1947-April 1948*, 46–47.

[41] Mr. Beaumont, Jerusalem to Foreign Office, 17 May 1948, PRO, FO 371/68505; British Legation Amman: Monthly Situation Report on Transjordan for the Month of May, 2 June 1948, PRO, FO 371/68845.

[42] Sir A. Kirkbride, Amman, to the Foreign Office, 17 May 1948, PRO, FO 371/68829; telegram no. 348 from Sir A. Kirkbride, Amman, to the Foreign Office, 19 May 1948, PRO, FO 816/120; conversation with Major Abdullah al-Tal, 24 May 1948, PRO, FO 371/68641; Ibrahim Kandalfat, "After the Mandate" (pt. 7), *Al Bilad* (Jordan), 4 January 1954, IDFA, 922/75/693; al-Tal, *Zikhronot*, 75, 79; Glubb, *A Soldier with the Arabs*, 109–11.

[43] Telegram no. 358 from Sir A. A. Kirkbride, Amman to Foreign Office, 21 May 1948, CO 537/5315.

[44] Telegram no. 343 from Sir A. A. Kirkbride, Amman to Foreign Office, 18 May 1948, CO 537/5315; Sir A. Kirkbride, Amman, to the Foreign Office, 19 May 1948, PRO, FO 371/68506; conversation with Major Abdullah al-Tal, 24 May 1948, PRO, FO 371/68641; al-Tal, *Zikhronot*, 79–81; Levy, *Tishah kabin*, 57.

[45] Mr. Beaumont, Jerusalem to Foreign Office, 18 May 1948, PRO, FO 371/68505; telegram from Etzioni to Yadin, 26 May 1948, IDFA, 922/75/1175; telegram from Etzioni to Yadin, 27 and 29 May 1948, IDFA, 922/75/1175; Sir A. A. Kirkbride, Amman, to the Foreign Office, 29 May 1948, PRO, FO 371/68374; al-Tal, *Zikhronot*, 83–84, 86–87, 88–90, 100–101, 103–7; Shaltiel, *Yerushalayim*, 176–78. A detailed report on the situation in the Old City and the circumstances that led to its surrender is in Mr. Beaumont, Jerusalem to Foreign Office, 30 May 1948, PRO, FO 371/68510.

[46] Lorch, *Korot milhemet ha-atzma'ut* (1989), 287–91.

[47] Ben-Gurion, *Yoman ha-milhamah*, 2:453, entry for 24 May 1948; 457 and 459–60, entry for 26 May 1948.

[48] Ben-Gurion, *Yoman ha-milhamah*, 2:453, entry for 24 May 1948; 457 and 459–60, 453, entry for 24 May 1948.

[49] From C in C, MELF to War Office, 25 May 1948, PRO, FO 371/68621.

[50] Sir A. Kirkbride, Amman, to the Foreign Office, 15 May 1948, PRO, FO 371/68372; telegram from the Arab Legion HQ, Amman to Qawukji, 18/5/1948, HA 105/129; al-Qawukji, "Memoir, 1948," 26.

[51] Al-Rossan, "Ma'arakhot Bab al Wad," 148–50.

[52] Ibid., 148–49.

[53] Glubb, *A Soldier with the Arabs*, 110; al-Rossan, "Ma'arakhot Bab al Wad," 151.

[54] Ayalon, *Hativat "Givati"*, 142; Bar-Zohar, *Ben-Gurion*, 3:771; Lorch, *Korot milhemet ha-atzma'ut* (1989), 327–28, 331.

[55] Pappé, *The Making of the Arab-Israeli Conflict*, 140.

[56] Lorch, *Korot milhemet ha-atzma'ut* (1989), 327; Editor's Preface, in Ben-Gurion, *Yoman ha-milhamah*, 2:423; Bar-Zohar, *Ben-Gurion*, 2:756, 770–71.

[57] Operations Branch Cable to Givati (and others), 1 June 1948 in Ayalon, *Hativat "Givati"*, 113; Ben-Gurion, *Yoman ha-milhamah*, 2:473, entry for 31 May 1948.

[58] Ben-Gurion *Yoman ha-milhamah*, 2:472, entry for 31 May 1948; for a report on the Egyptians' inferior warfare capabilities, see Palmah Memorandum, 16 May 1948, IDFA 661/69/36.

[59] Sir A. A. Kirkbride, Amman, to the Foreign Office, 24 May 1948, PRO, FO 371/68373.

[60] *Hativat ha-negev be-milhamah*, 66, 70–71; Ben-Gurion, *Yoman ha-milhamah*, 2:427, entry for 17 May 1948; 432, entry for 17 May 1948; 438, entry for 18 May 1948. For the fears of the Egyptian commander and his cautious planning of the campaign, see Kimche and Kimche, *Both Sides of the Hill*, 149–50.

[61] *Be–einei oyev*, 80, 89, 96, 99–100.

[62] H.M. Minister, Amman, to Secretary of State, London, 23 May 1948, PRO, FO 816/121; see the testimony of a captured Egyptian artillery officer in Ayalon, *Hativat "Givati"*, 105.

[63] Operation/Intelligence, "The Egyptian Army in Palestine," 26 July 1948, HA 105/138; Ayalon, *Hativat "Givati"*, 97–98, 105; *Be-einei oyev*, 89–90; *Me'ahorei ha-pargod*, 157; Nasser, "Memoirs of the First Palestine War," 14; an Egyptian artillery officer reported that the force heading northward was ordered to take positions north of Isdud. Report by an Egyptian Lt. Col. to the Artillery HQ: "Summary of the First Artillery Brigade in the Palestine Campaigns, 15 May–11 June," HA 105/138; Ben-Gurion suspected that the Egyptian advance was aimed to divert Israeli forces from the Jerusalem front, and to make it easier for the Jordanian Legion to capture the city. Minutes of the Provisional Government meeting, 30 May 1948, ISA.

[64] Givati Cable to Yadin, 29 and 31 May 1948, IDFA 922/75/1175; Ayalon, *Hativat "Givati"*, 548.

[65] *Be-einei oyev*, 80, 89, 96, 99–100; the Egyptian artillery officer made no reference to the destroyed bridge in his report (see n. 63 above). Report by an Egyptian Lt. Col. to the Artillery HQ: "Summary of the First Artillery Brigade in the Palestine Campaigns, May 15–June 11," HA 105/138.

[66] Operations/Southern Section: Operation Pleshet Order, 1 June 1948, in Ayalon, *Hativat "Givati"*, 112–13.

[67] Ibid.

[68] Ibid., 118–21.

[69] Ben-Gurion, *Yoman ha-milhamah*, 2:477, entry for 2 June 1948; Ayalon, *Hativat "Givati"*, 116; Gilad, ed., *Sefer ha-Palmah*, 2:862.

[70] A. Eban and G. Ruffer to M. Shertok, 31 May 1948, in Yogev, ed., *Te'udot mediniyot: Mai 1948–september 1948*, 103; Ben-Gurion, *Yoman ha-milhamah*, 2:481, entry for 2 June 1948; A. Eban to M. Shertok, 2 June 1948, in Yogev, ed., *Te'udot mediniyot: Mai 1948–September 1948*, 113; Ilan, *Bernadotte in Palestine*, 88.

[71] M. Shertok to A. Eban, 2 June 1948, in Yogev, ed., *Te'udot mediniyot: Mai 1948–September 1948*, 112; Ben-Gurion, *Yoman ha-milhamah*, 2:478, entry for 2 June 1948.

[72] "[IDF] Fatal Casualties," IDFA 922/1975/1189; Ayalon, *Hativat "Givati"*, 119–42; Gilad, ed., *Sefer ha-Palmah*, 2:862–63.

[73] Stated by Avidan in 1958. Quoted in Dagan and Yakir, *Shimon Avidan Givati*, 240.

[74] Gilad, ed., *Sefer ha-Palmah*, 2:863; Ayalon, *Hativat "Givati"*, 121 (for a full description of the battle, see 123–40).

[75] Ayalon, *Hativat "Givati"*, 142; Lorch, *Korot milhemet ha-atzma'ut* (1989), 331; Bar-Zohar, *Ben-Gurion*, 2:771.

[76] Intelligence report, "Survey of the Situation in the Negev Ahead of the Ceasefire," 17 June 1948, HA 105/124/1.

[77] A brilliant survey of the literature written during those early days is Bar-On, *Zikaron ba-sefer.*

[78] Pappé, *The Making of the Arab-Israeli Conflict*, viii.

References

Avizohar, Meir, ed. *Akhshav o le'olam lo* (Now or never). Ayanot: Beit Berl, 1989.

Ayalon, Avraham. *Hativat "Givati" mul ha-polesh ha-mitzri* (The Givati Brigade facing the Egyptian invasion). Tel Aviv: Ma'arakhot, 1959.

Bar-On, Mordechai. *Zikaron ba-sefer: Reshitah shel ha-historiografiyah ha-yisraelit shel milhemet ha-atzma'ut, 1948–1958* (Memory in the book: The beginning of the Israeli historiography of the 1948 war, 1948–1958). Tel Aviv: Ma'arakhot, 2001.

Bar-Zohar, Michael. *David Ben Gurion* (Hebrew). Tel Aviv: Am Oved, 1977.

Be-einei oyev (In enemy eyes) Translated by Shmuel Segev. Tel Aviv: Ma'arakhot, 1954.

Ben-Gurion, David. *Medinat Yisrael ha-mehudeshet* (The restored State of Israel). Tel Aviv: Am Oved, 1969.

———. *Pe'amei medinah* (Chimes of Independence). Edited by Avizohar Meir. Tel Aviv: Am Oved, 1993.

———. *Yoman ha-milhamah: Milhemet ha-atzma'ut, 1947–1948* (War diary: The War of Independence, 1947–1948). 3 vols. Tel Aviv: Ma'arakhot, 1982.

Ber, Israel. *Bitahon Yisrael: Etmol, ha-yom, mahar* (Israel's security: Yesterday, today, tomorrow). Tel Aviv: Amikam, 1966.

Bregman, Ahron. *Israel's Wars, 1947–93*. London and New York: Routledge, 2000.

Dagan, Shaul and Yakir Eliahu. *Shimon Avidan Givati* (Hebrew). Givat Haviva: Yad Ya'ari, 1995.

Dupuy, Trevor N. *Elusive Victory*. New York: Harper & Row, 1978.

Gelber, Yoav. *Garin le-tzava ivri* (The nucleus of a Hebrew army). Jerusalem: Yad Yitzhak Ben Zvi, 1986.

Gilad, Zrubabel, ed. *Sefer ha-Palmah* (Book of the Palmah). Tel Aviv: Ha-kibbutz ha-Me'uhad, 1956.

Glubb. *John B.* A Soldier with the Arabs. London: Hodder & Stoughton, 1957.

Hativat ha-negev be-milhamah (The Negev Brigade in war). Tel Aviv: Ma'arakhot, n.d.

Herzog, Chaim. *The Arab-Israeli Wars*. New York: Random House, 1982.

Ilan, Amitzur. *Bernadotte in Palestine,1948*. Oxford: Oxford University Press, 1989.

Kimche, John and David. *Both Sides of the Hill*. London: Secker & Warburg, 1960.

Levy, Itzhak. *Tishah kabin: Yerushalayim bi-kravot milhemet ha-atzma'ut* (Nine measures: Jerusalem in the War of Independence). Tel Aviv: Ma'arakhot, 1986.

Lorch, Nethanel. *Korot milhemet ha'atzma'ut* (Israel's War of Independence). Ramat Gan: Massadah, 1958.

Lorch, Nethanel. *Korot milhemet ha'atzma'ut* (Israel's War of Independence). Tel Aviv: Yedi'ot Aharonot, 1989.

Me'ahorei ha-pargod: Va'adah parlamentarit irakit al ha-milhamah be-Yisrael (Behind the curtain: Iraqi Parliamentary Inquiry Committee on the Palestine War). Translated by S. Segev. Tel Aviv: Ma'arakhot, 1954.

Morris, Benny. *Righteous Victims: A History of the Zionist–Arab Conflict, 1881–1999*. London: J. Murray, 1999.

Morris, Benny. "The New Historiography: Israel Confronts Its Past." *Tikkun* 3, no. 6 (Nov.–Dec. 1988): 19–23, 99–102.

Nasser, Gamal Abdul. "Memoirs of the First Palestine War." Translated and annotated by Walid Khalidi *Journal of Palestine Studies* 2, no. 2 (winter 1973): 3–32.

Nevo, Yosef. *Abdallah ve-aravyei Eretz Yisrael* (Abdullah and the Arab Palestinians). Tel Aviv: Papirus, 1975.

Oren, Elhanan. *Ha-derekh el ha-ir* (The Road to the City). Tel Aviv: Ma'arakhot, 1976.

Oren, Elhanan. "Ha-ma'arakhah al mevo'ot Yerushalayim, April 1948." (The campaign on Jerusalem's outskirts, April 1948). *Ma'arakhot*, no. 243–244 (April–May 1975).

Pa'il, Meir. *Mi-"Haganah" li-tzva haganah* (From the Haganah to the IDF). Tel Aviv: Zmora, Modan, Bitan, 1979.

Pappé, Ilan. *The Making of the Arab-Israeli Conflict, 1947–1951*. London: I.B. Tauris, 1992.

Protokolim yeshivot ha-memshalah, 18 April–13 Mai 1948 (Minutes of government meetings, 18 April–13 May 1948). Jerusalem: Israel State Archives, 1978.

al-Qawukji, Fauzi. "Memoirs, 1948, Part I." *Journal of Palestine Studies* 1, no. 4 (Summer 1972): 27–58.

Rogan, Eugene L. "Jordan and: the Persistence of an Official History." In *The War for Palestine: Rewriting the History of 1948*, edited by Eugene L. Rogan and Avi Shlaim. Cambridge: Cambridge University Press, 1948.

al-Rossan, Rais Mahmud. "Ma'arakhot Bab al Wad" (The Bab al Wad campaigns)(In enemy eyes). Translated by Shmuel Segev. Tel Aviv: Ma'arakhot, 1954.

Ben-Zion, Dinur, ed. *Sefer toldot ha-Haganah* (The history of the Haganah). Jerusalem: Ma'arakhot and Ha-Sifriyah ha-Tziyonit, 1954–72.

Sela, Avraham. "The Arab Palestinians in the 1948 War." In *The Palestinian National Movement: From Confrontation to Reconciliation?* edited by Maoz Moshe and B. -Z. Kedar. Tel Aviv: Ma'arakhot, 1996.

Shlaim, Avi. *Collusion across the Jordan: King Abdullah, the Zionist Movement, and the Partition of Palestine.* Oxford: Oxford University Press, 1988.

Shaltiel, David. *Yerushalayim tasha"h* (Jerusalem 1948). Tel Aviv: [Private Publication], 1981.

al-Tal, Abdullah. *Zikhronot Abdullah al-Tal* (Abdullah al-Tal's Memoirs). Tel Aviv: Ma'arakhot.

Tal, David. *War in Palestine, 1948: Strategy and Diplomacy.* London and New York: Routledge, 2004.

Van Creveld, Martin. *The Sword and the Olive: A Critical History of the Israeli Defense Force.* New York: Public Affairs, 1998.

Yogev, Gedalya, ed. *Te'udot mediniyot ve-diplomatiyot: Detzember 1947–April 1948* (Political and diplomatic documents, December 1947–April 1948). Jerusalem: Israel State Archives, 1979.

———. ed. *Te'udot mediniyot ve-diplomatiyot: Mai 1948–September 1948* (Political and diplomatic documents, May 1948–September 1948). Jerusalem: Israel State Archives, 1981.

The June 1967 Six Day War

[2]

THE COLD WAR'S LONGEST COVER-UP: HOW AND WHY THE USSR INSTIGATED THE 1967 WAR

By Isabella Ginor*

The Soviet warning to Egypt about supposed Israeli troop concentrations on the Syrian border in May 1967 has long been considered a blunder that precipitated a war which the USSR neither desired nor expected. New evidence from Soviet and other Warsaw Pact documents, as well as memoirs of contemporary actors, contradicts this accepted theory. The author demonstrates that this warning was deliberate disinformation, part of a plan approved at the highest level of Soviet leadership to elicit Egyptian action that would provoke an Israeli strike. Soviet military intervention against the "aggressor" was intended to follow and was prepared well in advance.

> "The truth of anything at all doesn't lie in someone's account of it. It lies in all the small facts of the time."
> --Josephine Tey, The Daughter of Time(1)

It is well-accepted in Middle Eastern historiography that the 1967 war's immediate trigger was disinformation fed by the Soviet Union to Egypt in May 1967 about massive reinforcements Israel was supposedly concentrating on its border with Syria. However, new and compelling evidence, including a hitherto secret speech delivered by Leonid Brezhnev shortly after the war, challenges the equally accepted notion that the USSR simply blundered into a war which Moscow neither desired nor expected. This evidence also casts doubt on the conclusion reached in the recent and definitive study of the conflict: "why, exactly, the Soviets acted as they did proved less important than the way the Egyptians reacted."(2)

Even the traditional version of these events could never explain why the USSR blundered or miscalculated. Both Middle East and Cold War historiography have been unable to explain the Soviet "hallucinations," as the Egyptian chief of staff termed the Soviet intelligence on the supposed Israeli reinforcements after inspecting the Syrian front.(3)

In order to reconcile this Soviet provocation with the accepted view that Moscow had no intention to precipitate a war, various theories have been proposed.(4) An especially noteworthy version was offered recently by Karen Brutents, a former CPSU Central Committee counsellor,(5) who claimed that Deputy Foreign Minister Vladimir Semyonov "couldn't control himself" and prematurely revealed yet-unconfirmed intelligence to visiting Egyptian Parliament Speaker Anwar Sadat.(6) According to previous accounts, the information was supposedly transmitted to Sadat at a much higher level by Supreme Soviet Chairman Nikolai V. Podgorny.(7)

Brutents's assertion of Semyonov's "slip" is but the latest example of continuing efforts even in post-Soviet Moscow to obscure the facts of, and responsibility for, the USSR's fateful move. There is no mention at all of the matter in the first Soviet official account of the 1967 crisis.(8) In other instances, it is claimed that Soviet intelligence actually did possess exact knowledge of Israeli intentions to wage war on Arab states(9) and even the supposed dates for this attack.(10) At the other extreme, one recent version quotes the head of the KGB's First Main Directorate (foreign intelligence), Lieutenant-General Aleksandr Sakharovski, as suggesting that Soviet intelligence doubted the accuracy of the information but decided it was their duty to share it with the Egyptians.(11)

The USSR and its satellites, however, were in fact much less conscientious about sharing accurate intelligence with their closest allies in the Arab world, including both Egypt and Syria. Some of their actual activity is revealed by a document discovered in the East German intelligence (Stasi) archives by the German historian Stefan Meining. This protocol of a KGB-Stasi meeting at the Soviet Council of Ministers between April 10 and 15, 1967 (with Sakharovski leading the Soviet delegation) delineates a complex program of "active measures" for 1967. It mandates the continuation of "Operation *Marabu*" aimed at "aggravation of tensions between West Germany and the Arab countries, in particular by [drawing attention to] the politico-military and economic cooperation of West Germany and Israel" by means of "manufactured documents" and rumors.(12) Some of these, as will be detailed below, may have had a direct connection to the May crisis. But overall, *Marabu* demonstrates conclusively that the Soviets systematically disinformed their Arab clients--which provides a plausible motivation for their persistent attempts to cover up the case that exposed this practice most dramatically.

In another extreme example of these cover-up efforts, American diplomat-historian Richard B. Parker was told during a visit to Moscow in September 1990 that the Soviet ambassador in Israel in 1967, Dmitri Chuvakhin, whom he wished to interview on his role in the crisis, had died.(13) Chuvakhin, however, was in fact alive enough to be interviewed subsequently by at least two journalists, including the present writer--although he contributed little to clarifying this affair.(14)

Brutents's attempt to blame the relaying of disinformation to Egypt on a slip of the tongue by a single official is blatantly inconsistent with the standard procedures of Soviet officialdom. If Semyonov had indeed been "notorious for scanning intelligence reports before such meetings [as with Sadat] and then retailing them as the latest gossip,"(15) and in this momentous case acted on his own initiative rather than on instructions from above, he would hardly have been included in the crucial talks with Egyptian Minister of War Shams Badran two weeks later.(16) He certainly would never have been trusted in 1968 with the most sensitive mission as head of the Soviet team to the SALT talks in Helsinki, which has been described by a Foreign Ministry veteran in the following terms: "never did such a high and representative Soviet delegation go abroad."(17)

These various attempts to belittle the warning given to Sadat also disregard the established fact that simultaneously with its transmission to him, the warning was hammered home to the Egyptians through two additional channels: by the Soviet intelligence "resident" in Cairo to Egyptian intelligence(18) and by the Soviet ambassador himself. In the early evening of May 12, a coded message from the Soviet Embassy to the Ministry of Foreign Affairs in Moscow was intercepted "somewhere in Western Europe," stating that "today we passed on to the Egyptian authorities information concerning the massing of Israeli troops on the northern frontier for a surprise attack on Syria. We have advised the

UAR government to take the necessary steps."(19)

As an Egyptian official who took notes during this meeting pointed out at a conference 25 years later, "it was not customary for the Soviet ambassador to deliver such reports... it never happened before, so why did the Soviets choose to act this way? Was it to confirm the seriousness of the situation and to leave no doubt that an Egyptian action is required?" This official, Salah Bassiouny, went on to relate how after meeting Ambassador Dmitri Pozhidayev, the Egyptian undersecretary of foreign affairs sent "an alarming report... based on the Soviet ambassador's repeating... that it should be seriously considered in the light of Syrian fears and the reported Israeli massing of ten to twenty brigades on the Syrian border." But, in line with the Soviet/Russian cover-up attempts described above, a Russian participant at the same conference reported that no evidence of such instructions to Pozhidayev had been found in the Foreign Ministry's archive. He added, correctly, that "without the decision of the Politburo, none of the ambassadors would have been allowed to pass it."(20)

It is therefore of the utmost significance that a document has now emerged showing the Politburo did adopt precisely such a resolution. In his memoirs, Brutents relates being summoned on June 5 to Communist Party headquarters. He was instructed to prepare a speech on the Middle Eastern crisis and newly erupted war for delivery by General Secretary Leonid Brezhnev to the CPSU Central Committee.(21) Brutents added to the present writer: "I and one of my colleagues prepared the urgent plenum. We were called in the evening to Brezhnev and were told that in the morning we had to submit Brezhnev's speech to the plenum of the Central Committee."(22) The meeting was ultimately convened, however, only on June 20.

Brezhnev's speech was never published, but a typescript translation of its text was also discovered by Meining at the Stasi archives in Berlin. While this speech must be viewed cautiously as an apology for Brezhnev's own role and that of his colleagues, it does give away several crucial factual disclosures. One of these startling statements relates to the warning to Egypt:

> In mid-May--and to this I want to draw your attention--reports reached us that Israel was intending to land a military blow on Syria and other Arab states. *The Politburo resolved* to bring this information to the attention of the governments of the UAR and Syria.(23)

This attribution to a Politburo resolution appears finally to negate the claim that the USSR simply stumbled into what was termed recently a "catalytic role" in precipitating the war.(24) The accepted practice in preparing "Politburo top-secret documents... never disclose[d] the authors of the proposals or how the decision was made."(25) The almost offhand mention of a Politburo resolution by Brezhnev seems aimed primarily to deflect responsibility for the resulting fiasco from himself personally and spread it over the collective body. But with most members of the Politburo present as he spoke, Brezhnev could hardly have invented a resolution that was never adopted.(26) Moreover, Brutents states that the draft he prepared for the speech was based on party documents supplied for the purpose.(27)

Brezhnev's assertion that it was the Politburo which resolved to inform Egypt and Syria of Israel's aggressive intent (he made no explicit reference to the supposed "troop concentrations") does not in itself rule out the possibility that the Soviet political leadership-- as distinct from the intelligence apparatus-- believed the allegation. This version is still proposed by some Western analysts, resting on Prime Minister Alexei Kosygin's confirmation of the "troop concentrations" to Minister Badran during the latter's visit to Moscow on May 25-28 and Kosygin's repeating this allegation on June 19 at the UN General Assembly.(28)

However, even before Brezhnev's speech, there were indications from diplomatic spokesmen of the USSR and its allies that they knew the troop concentration story was untrue. As early as June 13, 1967, Bulgarian UN delegate Milko Tarabanov, "the loyal echo of [Soviet Ambassador Nikolai] Fedorenko," asserted at the UN Security Council: "The question of concentration of forces being true or not does not matter at all, as it is known that Israel is able to mobilize within 24 hours."(29) This extraordinary equation of potentiality with fact in order to justify a lie was still maintained recently by a former Soviet official: "Israel...is well organised... so for them to mobilise the reserve is a matter of 24 hours, not more...their army had always been... prepared both for defense and for attack."(30)

But did the Soviet leadership believe the veracity of its warning when it was given to the Egyptians in mid-May? Brutents hinted to the present writer that Soviet intelligence may have foisted fabrications on the political leadership: "Neither you nor I can guarantee that intelligence does not do some things. In my opinion they are capable of anything.... And they have closely guarded secrets which they do not disclose to anyone, even...to their governments."(31) The sheer numbers of Israeli troops supposedly concentrated on the Syrian border--11 to 13 brigades, above the strength of Israel's entire standing army--do seem designed more to impress party leaders, who held Soviet concepts of military scale, than to conform with Middle Eastern realities.(32)

However, it seems improbable that either the KGB or its bitter rival the GRU (military intelligence) could have submitted such false information separately to the Soviet leadership, as each would only have been glad to disprove the other. In view of the disastrous results, collusion of both agencies in this venture without the blessing of a ranking patron would certainly have been punished. In reality, both KGB Chairman Yuri Andropov and Defence Minister Andrei Grechko were promoted to Politburo candidate-membership status near the end of June 1967-- that is, not only following this incident but also after its outcome had become apparent. It might be argued that both of them were fresh appointees to their positions (Grechko in mid-April and Andropov on May 19), and so could not be blamed for the misdeeds of their predecessors. Indeed, Grechko tried to lay the blame for overevaluation of the Egyptian army's preparedness on the previous Defence Minister, Rodion Malinovski, who had died on March 31.(33)

But Grechko (whose official title was then First Deputy Minister of Defense and Commander-in-Chief of Warsaw Pact Forces) is recorded to have assumed, informally but effectively, the duties of the ailing Malinovsky by November 1966.(34) Andropov, as a Secretary of the

Central Committee, functioned as "curator" of the KGB for at least a year before Brezhnev exploited the defection to the West of Stalin's daughter to remove the mistrusted Vladimir Semichastny from formal leadership of the agency.(35) Both new nominees--Grechko and Andropov--as well as Foreign Minister Andrei Gromyko, were or became personal friends of Brezhnev and all three were elevated later to full membership of the Politburo. Given this background, it appears highly unlikely that Brezhnev was misled by his own proteges, and it is much more probable that he was a participant rather than a victim of this scheme.

In the search for some substantive base for the Soviet warning, some have depicted it as an extrapolation from a supposedly aggressive statement toward Syria--the threat of a "lightning strike" to "occupy Damascus [and] overthrow the regime there"--attributed to Israeli Chief of Staff Yitzhak Rabin. Nasser, speaking on May 22 to the troops he had moved into Sinai, referred to "a very impertinent" threat made in Israel on May 12, after which "one simply cannot remain silent."(36) No published text exists for any such Israeli statement, which would have been made around Independence Day when every word spoken by a politician is carefully covered, and Rabin vehemently denied ever making it. Its dissemination has been attributed to a misquote by United Press International from an Israeli Defense Force press briefing on May 11.(37)

However, a different provenance for Rabin's "threat" is now suggested by one of the "active measures" agreed upon by the KGB and Stasi, and confirmed a month before the May crisis as one of the elements comprising Operation *Marabu*:

> ...the KGB will undertake measures in Arab countries (UAR, Algeria, Yemen, Iraq, Syria), around documents manufactured by the [Stasi] and verbal information, exposing the policy of West Germany and the USA against these countries and the common actions of West Germany and Israel against Arab interests, aimed at launching a government crisis [*Regierungskriese*] of Arab countries.(38)

In any event, the purported causal connection between this alleged Israeli verbal threat and the Soviet warning has been definitively contradicted by the posthumously published memoirs of Yevgeni Pyrlin, who at the time was a senior member of the Egyptian desk in the Soviet Foreign Ministry. Pyrlin confirms, for the first time in print, that the information on Israeli troop concentration on the border with Syria was passed to Sadat by Semyonov already on May 12--not May 13 as conventionally accepted.(39) The newly disclosed Politburo decision--for which Brezhnev gave no exact date--must have preceded this, and therefore could not have resulted from Israel's purported threat. Moreover, there is an account of preparations already being made on May 11 by the Soviet Embassy in Cairo (undoubtedly on directions from Moscow) for an "unavoidable" war between Egypt and Israel.(40)

The very fact that the decision to alert Egypt was brought before the Politburo for final approval means that it was considered a matter of major strategic planning. "By the Politburo's mandate and its established practice," according to an insider's description, "it made decisions only on important, large scale military actions."(41) But was this Politburo resolution taken with full understanding that it would cause a war, indeed with the intent to achieve such a result? Considering the nature and extent of the military preparations undertaken by the USSR well before it triggered the crisis, and that are now coming to light from a wide variety of contemporary witnesses, there can be little question that this was indeed the case.(42)

In his memoir, Pyrlin still tried to downplay the significance of this Soviet move: "The military tension in the Middle East was growing every day in geometric progression and our information, which was passed among other confidential messages to the Egyptian President, was an ordinary message, one of many dozens of communications that were passed then."(43) But this description is contradicted by Pyrlin's own statement to BBC researchers who in 1997 were preparing a documentary on the Arab-Israeli conflict: "In the presence of the translator, he [Semyonov] told it not for the personal information of Sadat, it was rather to be transformed [transmitted] to Nasser. Relative steps were supposed to be taken by the Egyptian side."(44)

In Sadat's judgment, Semyonov's information about "ten Israeli brigades… concentrating on the Syrian border" was urgent enough for him to rush, upon return to Cairo shortly after midnight on May 13, straight from the airport to Nasser's residence in order to report. But he found the president already closeted with Abd al-Hakim Amer, vice-president and deputy supreme commander of the armed forces. Sadat then "realized that the Soviet Union had informed Nasser of this" through other channels as well.(45)

The Soviet warning was thus calculated to produce a specific result, and every precaution was taken to ensure that Egypt followed up with the "relative" or "necessary" steps that the Soviet leadership "advised" and expected. Egyptian diplomat Bassiouny appeared to be glossing over this when he stated, "The fact is that the report was presented to us, whether with instructions or not."(46) As Brezhnev's speech confirms: "They [the governments of UAR and Syria] informed us immediately that they were taking the necessary steps in the military sphere, and their forces were being put on full combat alert."(47) The recent official Russian history states unequivocally that "the Soviet leadership also knew about the war being prepared."(48)

Pyrlin states that three weeks before the war "…several people of the [Foreign] Ministry… were addressed… not with a request but with the order, instruction to prepare…a document which would evaluate the war [between Israel and Egypt] as if it had happened, as if the war were over." Such a report could be ordered only by the party leadership. Its "unanimous" opinion, which according to Pyrlin was shared by the KGB and the military headquarters, was "that the war would end up without anybody winning"(49)--indicating that even a limited Soviet intervention could tip the balance in favor of the Arab side.

Brezhnev did not elaborate in his speech what measures the Soviet leadership "advised" or expected from Egypt and Syria. As it happened, Egypt took three steps: on May 14, its forces started to pour into Sinai; on May 16, the UN force was asked to leave its positions along the border; and on May 22, Nasser declared a blockade on Israeli shipping through the Straits of Tiran.

In his speech, Brezhnev ignored the first move; in relation to the latter two, he took care to disclaim any Soviet collusion:

> I have to say that the government of UAR took a series of steps which were not thought out till the end. As a complete surprise for us, UAR government demanded on May 19 the withdrawal of UN forces from the armistice line. The UAR government did not consult with us on this momentous step, which in the developing situation could have been understood as a step toward its escalation....[On May 23] President Nasser explained to us through the Soviet ambassador that the situation in the Middle East had improved as a result of the determined steps that were taken by the UAR leadership. At this time he also informed the ambassador, as a *fait accompli*, that the UAR government has closed the Aqaba Gulf to Israeli ships and to ships of other countries carrying strategic materiel to Israel. Again, no prior information was given to the Soviet government for this important action, which caused serious results.(50)

But were Egypt's initial responses to the Soviet warning [moving forces into Sinai, removing the UN force and closing the Straits] actually planned by or coordinated with Moscow? The leading contemporary Western analyses found "the evidence is conflicting."(51) The Soviet reaction to those steps caused a U.S. diplomat to remark on May 26: "It almost seemed as though the Soviet Union had been aware in advance of the coming Near Eastern crisis, since Brezhnev had first called for the withdrawal of the Sixth Fleet [from the Mediterranean] on April 24."(52) The American ambassador in Moscow reported asking a "well-informed" Soviet source "point blank whether Soviets knew in advance of Egyptian action in closing [the] Gulf of Aqaba. He was obviously embarrassed...and after a long pause said he thought Nasser had acted on his own."(53)

An anonymous Soviet diplomat, speaking with *Le Nouvel Observateur* in "early July" [1967], admitted *some* collusion: "President Nasser stationed the Egyptian Army on the Sinai-Israeli frontier in agreement with the Soviet Union, in order to prevent an Israeli attack on Syria. The other two serious decisions however--to demand the evacuation of the UN Forces and to close the Sraits--Nasser took on his own and only told us about them afterwards."(54)

Russian sources remain extremely reticent on the question to this day. "The Soviet leadership did not react in any way to Egypt's steps to close Aqaba and Tiran, apparently, not wishing to 'annoy Nasser,'" wrote Pyrlin in his memoir. Interestingly, he cited legalistic arguments--including Israeli trade statistics--which the Soviets apparently had at the ready immediately upon Nasser's announcement, in order to prove that closure of the Straits could not constitute a legitimate *casus belli* for Israel. Nonetheless, Pyrlin contends the USSR might have dissuaded Nasser had it been consulted.(55)

Earlier, former Deputy Foreign Minister Georgi Korniyenko, who at the time was a senior member of the U.S. desk, gave rather conflicting explanations. On the one hand, he claimed, "no consultations took place between Cairo and Moscow about the withdrawal of UN forces, about the Strait of Tiran, and so on" and attributed the lack of Soviet objection to the UN force's removal to "the situation...developing too fast for our [Soviet] bureaucratic machine to react properly." On the other hand, he cited "ideological, political considerations" of "solidarity... in general and not on this particular question."(56) Soviet collusion with Egypt's initial steps is indicated, however, by the fact that Nasser did deem it nessessary to obtain Soviet approval (through Minister Badran's mission to the Kremlin on May 25-28) only for the further escalation of a pre-emptive strike against Israel.

A plan including two of the initial Egyptian measures (removal of UNEF and closure of the Straits) was wired to Nasser by Marshal Amer from a visit in Pakistan in early December 1966.(57) The three leaders of this delegation to Pakistan--Amer, Badran and intelligence chief Salah Nasr--arrived there one week after talks in Moscow with the Soviet military command headed by Grechko. Indeed as far back as 1975, the Israeli regional expert Avraham Ben-Tzur, in a brilliant analysis of Arab and Soviet documents and press reports, concluded that even if this plan was not directly suggested by Grechko to his Egyptian guests, he "provided the inspiration" for its inception.(58) By April 18, according to a CIA report, Nasser himself was telling a senior Egyptian diplomat of his own desire to get UNEF out of Sinai and close the Straits.(59)

While Ben-Tzur makes a persuasive case for Grechko's original authorship of the "Amer Plan," its promotion appears to have been adopted as Politburo policy well before the mid-May decision to trigger its implementation. When the Soviet foreign minister paid an unexpected visit to Cairo during the last week of March, "The only concrete detail leaked out in the Cairo press was that Gromyko would also discuss the problems of the UN peace-keeping force in Gaza."(60) However, at the height of the crisis, the third-ranking official of the Egyptian Embassy in Moscow confided to an American counterpart that "the purpose [of] Gromyko's visit to Cairo" was to give Nasser a "larger commitment than anyone...had realized...[the] absence of Soviet public endorsement of UAR position on Aqaba [is] not important because Soviets [are] supporting UAR 'in other ways.'"(61)

According to one of his subordinates at the time, "Gromyko did not resolve a single tiny question: he would not fulfill a decision or make up his mind without getting the approval of the Politburo first."(62) Gromyko indeed told the Egyptians "that he came, not in his capacity as minister of foreign affairs, but as candidate member of the Politburo....Soviet relations with Egypt were regarded as so important that they remained the concern of the Politburo."(63)

The next ranking CPSU official to visit Egypt (April 11-26) was Moscow city party boss Nikolai G. Yegorychev, who came as the guest of the Arab Socialist Union (ASU), Nasser's political party. The formal host was Yegorychev's counterpart as head of the ASU in Cairo, who was also Nasser's chief of staff. In a recent interview with the present writer, Yegorychev declined to discuss the content of a confidential document which he subsequently presented to the Central Committee. He denied later reports that he endorsed direct military support for Egypt or Syria during the war.(64) But it is noteworthy that Nasser pointed out on May 26: "I was authorized by the Arab Socialist Union's Higher Executive to implement this plan [moving forces into Sinai, removing UNEF and closing the Straits] at the right time. The right time came when Syria was threatened with aggression."(65) Yegorychev was sacked following the June session of the Central Committee, where he "infringed on the General Secretary's personal bailiwick" by asking whether the defeat of the USSR's Arab allies did not cast doubt on its capability to defend its own territory.(66)

Brezhnev's speech confirms that following the Egyptian moves, "in the United Nations we did everything that depended on us, comrades, to lessen the pressure of the Western superpowers on the UAR in connection with the question of the free passage in the Aqaba Gulf and in order to frustrate the plans for military provocation against UAR."(67)

Indeed, at the UN Fedorenko so diligently stalled proposals to lift the blockade (even to convene the Security Council) that his Canadian and Danish colleagues told him they had "A nasty feeling [that the] USSR [was] playing [a] game of allowing crisis to build to force Israel to act."(68)

This appears, however, to have been the limit of prior coordination with Egypt, and it was almost disrupted on May 25 when Badran arrived in Moscow to seek consent for an Egyptian first strike against Israel. In his Central Committee speech a month later, Brezhnev described Badran's mission as limited to a "request of military assistance that was sent by the UAR leadership to the Central Committee;" this request to bring forward military supplies scheduled for 1968-69 was accepted, with delivery dates set to begin in June.(69) But a participant at the actual talks, Pavel Akopov, an Egyptian desk officer at the Soviet Foreign Ministry, denied this: "As far as the Egyptians requests to increase the armament supplies for the Egyptian army, at the talks where I was present this matter was not raised." According to his account, Badran addressed "the Soviet leadership with their [Nasser's] request to approve of pre-emptive measures... early strikes to prevent Israel from being the first [to attack]." The exchange continued for another two days of "sharp" conversation until "apparently he [Nasser] accepted the argument of the Soviet leadership."(70)

Brezhnev stated in his speech that on May 26 the Soviet government sent Nasser a message "which spoke of the necessity 'to do everything possible in order to prevent military conflict.'"(71) This must refer to the talks with Badran, which were subsequently portrayed by Soviet spokesmen as an attempt to dissuade the Egyptians from any military action. Brutents, for example, told the present writer: "Kosygin was saying 'Nyet' to him...these talks ended with a 'sour reaction' of Badran. He received an instruction from Nasser...:If our Soviet friends do not support us, if they oppose, we do not have a choice."(72)

But Brezhnev did note that upon receiving the message Nasser expressed full aggreement with the Soviet considerations it outlined, and "stated to our ambassador... that the UAR never will start *first* the armed conflict."(73) This position was confirmed by Nasser on May 31 or June 1 to President Johnson's special envoy to Cairo, Robert B. Anderson, former secretary of the treasury, by stating that "he would wait until the Israelis had moved."(74)

Describing the talks with Badran, Akopov adds: "the first thing he [Kosygin] marked and sounded from the very beginning without any diplomatic mannerism, that we, the Soviet Union, cannot give you our consent for your pre-emptive strikes against Israel. This would contradict our policy and our position. Should you be first to attack, you will be aggressors, and once you are aggressors...we cannot support you."(75)

The recent official Russian military history confirms that the Arab attack was, at Soviet behest, not cancelled but only restored to its original design as a counterstrike:

> The Arabs planned to open the offensive *first* but because of some difference of opinion within their leadership (and perhaps the Soviet warning had its effect), the date for the beginning of the 'decisive actions' was postponed.... Soviet experts were of opinion that in the impending war...an important role will be played also by who will appear as direct initiator of the offensive.(76)

Akopov notes: "Kosygin probably has reported it to the Politburo...because it was a question of war and peace and our involvement, naturally it was discussed at the Politburo."(77) While the talks with Badran were in progress, Kosygin sent messages to the leaders of the United States and Britain, which must have been approved at Politburo level, assuring them that the measures taken by Arab states were "of a defensive nature" but warning that "if Israel commits aggression and military action begins then we would render assistance to...the victims."(78) In his speech, Brezhnev mentioned only "the Soviet message from May 26 to [Israeli] Prime Minister Levi Eshkol that included a warning to the Israeli government not to increase the tension and not to escalate the situation to the point of letting the arms speak"--an expression apparently used in the Politburo discussion as it is virtually identical with that used in Kosygin's letters to the Western leaders.(79)

At 2:10am on May 27, Chuvakhin woke up Eshkol and handed him Kosygin's message--reiterating also the original Soviet allegation: "I urged Eshkol to stop the escalation, stop the concentration [of forces] on the Syrian border and start negotiations with Arab states."(80) In reply, Eshkol offered to meet the Soviet leadership, but this--according to Pyrlin--"was left unnoticed." Pyrlin explains that the Soviet leadership needed approval from Nasser for meeting the Israeli prime minister,(81) which in itself suggests very close coordination between Soviet and Egyptian leaders, who "during past year... maintained direct dialogue... and handled some business directly."(82) But more significantly, Pyrlin says that Foreign Ministry officialdom did not dare "to push the leadership towards some constructive solution or response. We could have received a reproach for that but...the absence of a reply is a reply itself."(83)

The Soviet bureaucracy was thus under the impression that the leadership desired to precipitate a crisis, not to prevent it. This was also the reading of the U.S. ambassador in Moscow, who cabled on the final day of Badran's mission, based on a warning from the Egyptian Embassy's political counselor: "[The] Soviet objective is to transform Arab-Israeli struggle into [a] showdown between Communists and anti-Communists for control of Middle East, and [the] Soviets are succeeding. If Nasser wins this one, monarchies and Western oil interests will go."(84)

Years later, Pyrlin confirmed that this was exactly the USSR's strategic aspiration: "it was possible to hope and to count on the fact that the distribution of political forces on the Middle East would be considerably changed due to this war, and that events will take place similar to post-1956 events when the whole chain of revolutions took place in the Arab world, and a number of regimes which were pro-Western were replaced by nationalistic regimes...."(85)

The determining test of the Soviet leadership's intentions must remain in the facts of its preparation for military action. Pending the unlikely declassification of the directly relevant Soviet documents, the full scope and content of the Soviet operational plan can only be approximated by piecing together a growing number of partial but revealing and complementary accounts from participants. These outline deliberate and complex military moves which predated and parallelled the political-diplomatic efforts designed to manipulate Egypt into provoking Israel to launch a first strike, following which the USSR would intervene to support the Arab side against the "aggressor."

Michael Oren cites numerous sources to establish that already in 1966, with Egypt ostensibly barred from deploying substantial forces in Sinai, the Soviets devised a master plan for such deployment codenamed "Conqueror." More revealing, perhaps, is the description of this plan's strategy as "shield and sword"--the motto and emblem of the KGB. One of this plan's basic features (a lightly defended front line) was specifically designed "to serve as bait for luring the Israelis into a frontal assault."(86)

Perhaps even more significant is that "Conqueror" was originated at the same time that Egypt signed its defense treaty with Syria, which was invoked by the warning of May 1967. Syria's role in the Soviet instigation of the 1967 crisis has not been adequately explored, partly because Russian sources--not to mention Syrian ones--are absolutely silent on it. The Egyptian-Syrian pact was signed in November 1966, just before Amer's visit to Moscow; but the USSR began pressing for its conclusion shortly after the coup on February 23 which put Damascus firmly in the Soviet camp and provided a test case for the activist foreign policy formulated at the Soviet Communist Party's congress a few weeks later. This official CPSU doctrine asserted "unity of the three revolutionary trends in modern times--global Socialism, national-liberation struggle of enslaved peoples and the international workers' movement."(87)

Syrian leaders, including the new prime minister and defense minister, were flown to Moscow in a Soviet military plane on April 18. On May 2, a treaty was signed between Syria and the USSR.(88) Then, as documented by Walter Laqueur, "During his May [10-18] 1966 visit to Cairo Kosygin persuaded the *Rais* [Nasser] that a mutual defense pact between Cairo and Damascus (to be guaranteed by Moscow) would be in the best interests of all those concerned."(89) Kosygin's mission and his speech to the Egyptian National Assembly on May 17, stressing "the important role of your country also in the Arab peoples' struggle for the solution of the Palestinian question"(90) must have been cleared by the Politburo with a clear view of what was to follow. By that time the Soviet plan, at least in its political aspect, might have begun to take shape. It conformed with the overall change toward an activist strategy against the United States.

As recalled by Aleksandr Bovin, a member of Andropov's think-tank from the latter's pre-KGB days (and later Brezhnev's speechwriter), "[By] about the middle of 1966 there began to ripen within the Soviet leadership an intent to stamp its foot, to scare the Americans, to put them in their proper place."(91) The Soviet ambassador in Washington at the time, Anatoli Dobrynin, writes that such a trend "was reflected during the 23rd

Party congress...which devoted much attention to Soviet-American relations and to criticism of the USA's policy in Vietnam."(92) Ben-Tzur points to the meetings held by the Soviet military with Arab and other delegations to the 23rd congress, in all of whose countries violence erupted in short order.(93)

Syria's role in the CPSU's new global concept can be extrapolated from contemporary Soviet propaganda, according to its standard methods: "In Soviet diplomatic practice, every important foreign policy step...was accompanied by a number of propaganda actions--including publication of articles, statements by 'independent' organizations and public figures supporting the Soviet position, invitation...to participate in a Soviet V.I.P.'s visit, praising the latter and his position, and so on.... these expensive operations were called 'propaganda insurance' or 'propaganda backing.'"(94)

Such a campaign was begun already on May 8, 1966 in Izvestia, which for the first time claimed that Syria "became a central object of military blackmail and provocation by Israel."(95) On the same day, a TASS cable from Damascus made the first mention of "a suspicious concentration and movement of Israeli troops sighted lately on the border with Syria."(96) This report, predating Kosygin's trip to Cairo, appeared only in the provincial *Sovietskaya Kirgizia*. By May 21, upon his return, the national *Sovietskaya Rossiya* was charging that "about a third of the Israeli army, after marching to music through the streets of Haifa, was immediately following the parade transferred to the Syrian border."(97)

On this background the first official Soviet protest about these troop concentrations was delivered on May 25, 1966 by the same Semyonov to Israeli Ambassador Katriel Katz. "The Soviets appear to have had an obsession about such troop concentrations," writes Parker; Israel counted at least eight such warnings before the last one actually touched off the war.(98) An apter description is probably that given by Solomon M. Schwartz, one of the first researchers on the subject: "The legend about energetic preparation by Israel for attacking Syria became from the summer of 1966 an integral part of the Soviet propaganda in the Middle East."(99) In the summer of 1966, this was indeed a legend. Despite recurring firefights on the frontier, Israel's entire defense line from Lake Tiberias northward was held by one company of paratroops with minimal auxiliary units.(100)

The recurrence of these Soviet warnings is customarily invoked to support the thesis that the May 1967 disinformation was merely a routine exercise that happened to get out of hand. But this is purely speculative, and certainly no better founded than an alternative interpretation: that the repetition of these charges, together with increasingly acrimonious Soviet statements and the encouragement of Syria to undertake actions (which indeed provoked a forceful Israeli response climaxing on April 7), were part of a deliberate escalation designed to prepare the ground for harnessing Egypt to the military confrontation being prepared and to draw an Israeli strike against Egypt as well.

On April 22, 1967, in Berlin on his way to a gathering of Communist bloc leaders at Karlovy Vary, Brezhnev signalled that this regional build-up was approaching its global objective: a strike at the United States via its Israeli client. He notified his counterparts, East Germany's Walter Ulbricht and Poland's Wladislaw Gomulka, of a "decisive blow" that was about to be dealt to American interests in the Middle East--even at the cost of sacrificing Nasser.(101) After the mid-May warning was transmitted to Egypt, Soviet "propaganda insurance" concentrated on "pushing the United States into the forefront of the Middle East crisis by making Washington responsible for Israel's actions"(102) in its forthcoming assault on Syria--precisely according to the guidelines now revealed for Operation *Marabu*.

Preliminary details of the Soviet naval landing on Israeli shores with air support, which was aborted after being put in motion on June 10, were first published by the present writer in *MERIA Journal* three years ago, based largely on reminiscences of participants.(103) In response to the recent publication of these findings in Russian, Academician Aleksandr K. Kislov has added the hitherto unconfirmed fact that the landing force included (in addition to improvised platoons from warship crews) "*desant* [landing] ships with well-prepared marines."(104) Since the original publication, other substantial corroborating evidence has emerged, which, among other aspects, indicates that planning of this operation began well before mid-May 1967.

This was already described in 1996-1997, in mostly unused portions of several interviews conducted by a BBC team for a documentary series. In denying any Soviet intent to intervene in the 1973 Yom Kippur War, former KGB "resident" in Cairo Vadim Kirpichenko said: "this was not...1967...when we were insisting and we were prepared for some decisive actions."(105) And in a portion of his interview that was not included in the broadcast series, Foreign Ministry officer Pyrlin stated specifically: "*As far as the invasion readiness goes, yes, the order was given* but there was no order to bomb or to attack Israel--Grechko would not be able to issue such an order without a Politburo decision."(106)

Preparations for the prospective landing were one feature of the unprecedented reinforcement of naval units in the Mediterranean which had been in full swing from January 1967, when Navy commander Gorshkov visited Egypt.(107) According to accepted Soviet procedure, military moves on this kind of scale required Politburo approval:

> ...such as mobilizations general or partial; substantial movements of troops, particularly from one military district to another; large maneuvers, especially unplanned ones; deployment and use of any type of weapons of mass destruction; putting on alert all Soviet armed forces, or forces in one or several military districts; and some other matters.(108)

The Politburo thus had to approve the "first large-scale movement of Soviet naval units into the Mediterranean at the end of February" which began shortly after Amer submitted his plan to Nasser and Gorshkov's visit to Egypt.(109) The ships were drawn from the Black Sea and Northern Fleets. Two of the participants in the projected landing operation give their respective dates for starting the voyage to the Mediterranean as May 3, from the Baltic and May 20, from an Arctic base.(110) At least one Soviet nuclear submarine (K-131) was sent from the Barents Sea into the Mediterranean "on the eve of crisis...by decision of the leadership."(111) Another Soviet nuclear submarine, based in Alexandria, received orders to fire nuclear missiles at Israel if the latter should use nuclear arms against the Arabs.(112)

Soviet warships carrying nuclear weapons were also dispatched to the Red Sea before the hostilities started, ostensibly because "...there existed in Moscow a concern that in a turn of events unfavorable for it, Israel could use certain kinds of WMD [weapons of mass destruction], the existence of which never was denied by official Tel Aviv." On June 8, this squadron was at close enough range to arrive "partially for deterrence, to the Red Sea shores of Egypt"--a move triggered, according to the official Russian history, by the Israeli attack on the U.S.S. *Liberty*.(113)

After another 10 ships passed from the Black Sea into the Mediterranean on May 31,(114) the Soviet *eskadra* in the Mediterranean consisting of 40 battle units, including 10 submarines, was put on battle alert on June 1.(115) On June 4, it was given 12 hours to reach full battle alert.(116) The head of Israeli SIGINT (signal intelligence) in 1967 told the present writer that his unit tracked radio messages from 43 Soviet vessels in the eastern Mediterranean but was unable to crack their code.(117) In any event, these signals would not have given away the Soviet *desant* operation, as orders to raise a landing party on each ship were given orally to the captains on board the flagship.(118) The Soviet military interpreters into Arabic, who were confined to the Soviet Embassy in Cairo since May 11, were transferred at some point to Alexandria. There they were supposed to be embedded with the landing forces "for liason with Israeli Arabs."(119)

As these preparations could not have been detected by U.S. or Israeli intelligence unless revealed in signals, this belies subsequent Soviet and Russian attempts to present this operation as purely deterrent. The same applies to the combat alert ordered, according to Pyrlin, for Soviet land and air forces "in the Transcaucasus,...in the Transcaspian, all the districts oriented at the Middle East. It was publicly known that these military districts are responsible for the situation... in the Middle East region." Pyrlin claims "the fact that they were raised to alarm--it was well known, and from that various conclusions could be drawn: either we are about to initiate some military steps or whether it was going to be the demonstration of force....It was meant as a demonstration of power."(120) There is, however, no evidence that U.S. or any other Western intelligence was aware of this.

On Sunday [June 4], in Ukraine, oral orders were also given to deploy a "regiment" of strategic bombers to vantage points on the USSR's southern fringe, "from where they could reach Sinai." With the possible exception of the first order, all the instructions were delivered from Moscow over the phone to save time over the decoding, according to the Air Force Corps commander, Col. General (ret.) Vasili Reshetnikov. A day before, the pilots were assigned pre-determined targets in Israel: "The objects...were named to us--that strikes had to be delivered against: they were marked by the geographical terms on the map; and we were particularly interested about the anti-aircraft defense systems, the Hawk complexes."(121)

This reference to Israel's U.S.-supplied anti-aircraft missiles appears to connect the Soviet operational plans to one of the USSR's central concerns regarding Israel--its nuclear capability. The first Hawk batteries were deployed in 1965 around the nuclear plant at Dimona.(122) The KGB "resident" in pre-1967 war Israel mentions being ordered to check "the reliability of existing information about the progress of works conducted in Israel for producing nuclear weapons."(123) Dimona is known to have been, at least on two occasions (May 17 and 26, 1967), the target of Egyptian high-altitude aerial photography sorties which the Hawks either did not attempt or failed to intercept. A Soviet military historian who has specialized in the Middle East, Col. Valeri Yaremenko, wrote that: "this lack of action [response] pushed Nasser and Commander-in-Chief Marshal Amer to reach, in absolute secrecy, a decision to destroy the Israeli reactor before it would be able to produce nuclear weaponry. Intensive training flights were started, with live bombing of a 'full-scale Dimona model' in the Egyptian desert....In the beginning of June [1967] Amer decided to bomb Dimona in the period of June 7-10."(124) In defense of Nasser's conduct, Egyptian diplomat Tahsin Basheer said he "miscalculated badly, but he defended the area against atomization."(125)

Yaremenko states that "Moscow remained a passive observer" of this activity, but adds, "according to the opinion of the then foreign minister, nuclear war in the Middle East could have been beneficial for the USSR."(126) Gromyko himself is recorded as telling his subordinates 14 years later that "Amer--a decisive and even aggressive person--gave an order to bomb Dimona and other important objects on Israeli territory. But at our behest Nasser cancelled this order." According to the source of this account, former Soviet diplomat Oleg Grinevski, Gromyko claimed that:

> the Soviet leadership did not know then about the Egyptians' plan to liquidate Israel's nuclear potential. We knew only about the intent to strike a sudden blow upon important objects on Israeli territory in general without any concretization. This is why we sent a note to Nasser, in which we very insistently advised not to start this war....I think that if we had clearly envisaged then that the main goal of this strike [was] to destroy the nuclear potential of Israel, we would not have chosen to convince Nasser to avoid it.(127)

However, Akopov relates that Badran, when questioned by Kosygin, did disclose the details of Amer's proposed targets.(128) Given that in following years all such deep-penetration reconnaissance of Israel was performed on behalf of Egypt by Soviet aircraft and personnel,(129) it seems unlikely that the Soviets were totally unaware of the Dimona missions. In this context it is noteworthy to mention that in April 1967, 15 Soviet SU-7 bombers were delivered to Egypt, but "the Egyptian pilots did not have time to master them."(130) Gromyko's statement may actually indicate there was an active Soviet interest in taking out Dimona--as Reshetnikov's account appears to suggest.

While Moscow itself treated the forthcoming war in the Cold War context, it took pains to present it as a local conflict, and to camouflage its participation. "There was serious warning against any losses and casualties, because every loss of any plane could unfold the essence and the meaning of our race [raids], our Soviet aviation." To prevent formal identification of the Soviet air intervention, "all the documents were taken from the pilots and the crew in case some plane is burning in the desert."(131) The idea was to "let others guess who fell down and why they were there, what happened."(132) Reshetnikov notes that "we had to work under the colors of the Egyptian flag."(133) There was a logistic problem with repainting the planes in Egyptian markings, because "no one knew what these signs should look like;" being Sunday, it was very difficult to obtain from closed factories the needed paint, and "...we needed time to let it dry,...but in fact we were putting the colours on and flying straight away, flew immediately, and new planes were ready to take off as well."(134) When "the first group was [at] the launching airfield," Reshetnikov was instructed to await further orders.(135) "In 1967 and in 1973 he [Grechko] was not able to issue the order without Brezhnev's decision," stated Pyrlin, equating the Politburo with the secretary-general.(136)

The Politburo, then, had to approve the preparations and preliminary stages of the operation up to the actual implementation, and its mid-May resolution approving the transmission of a warning to Egypt was but the continuation of a series. It certainly had to approve Brezhnev's major overt move in Mediterranean naval matters: his demand, at Karlovy Vary on April 24, for removal of the U.S. Sixth Fleet. As it happened, this was almost achieved at the the outset of the war, when the Sixth Fleet was withdrawn westward in order to disprove Arab charges that its aircraft were assisting Israel's air offensive. From then until June 10, the Soviet navy enjoyed a virtual monopoly on the eastern Mediterranean--the best conditions Brezhnev might have desired for the projected intervention.

The scope of this paper cannot include the sequence of events that prevented the activation of the Soviet operation on behalf of Egypt in the opening days of the war, and then brought about its restart as a declared, deterrent move to stop Israel's subsequent onslaught on Syria. But it merits mention that Brezhnev's speech confirms, for the first time in an official Soviet document, that the latter action was undertaken--and hence that it had been prepared: "On June 10...all Soviet warships in the Mediterranean, including missle launchers, were given an order to turn and under escort of submarines [steam] to the Syrian coast."(137)

While there is as yet no direct evidence that Brezhnev and Grechko were personally involved in the planning of a naval landing, their joint authorship of such a scheme appears very much in character. Consider the following account of such an operation:

> Landing from the sea... would be an absolute surprise...a plan that is deciphered by the opponent, as is known, is half-destined to fail. Therefore, the first task was to ensure absolute secrecy. We forbade any correspondence in connection with the operation being prepared. For its development only a severely limited circle of people was drafted....In order not to disclose our intentions its [intelligence gathering] was conducted across a broad front. Work upon the disinformation of the opponent was conducted, suggesting the '*desant*' would take place [elsewhere].(138)

This description corresponds exactly to the preparation of the Soviet Mediterranean operation in 1967: all orders for the landing were delivered orally, only ship captains were informed until the actual implementation, and total radio silence was observed--not to mention the use of disinformation. But the quotation actually refers to a landing at Novorossisk on the Black Sea during World War II. It is taken from Brezhnev's memoirs, which describe his service as the political officer of the 18th Army as a lifelong defining experience, and take credit for this successful operation together with the Army's commanding officer--Grechko.(139)

These two old comrades-in-arms appear to have reverted in 1967 to the victorious tactic of their joint heroic memories. Brezhnev's tendency to see the two conflicts in the same context is further indicated by his use, in the June 1967 speech, of the term "treacherous" to describe Israel's pre-emptive attack--an epithet usually reserved in Soviet parlance for the German attack on the USSR in 1941.(140) A landing operation would in any case not seem far-fetched to Soviet brass: as late as 1969, such an assualt (in this case a paratroop drop) was proposed in order to take Beijing following the Soviet-Chinese border clashes.(141)

As for Grechko's input, his characterization by Yegorychev as a *soldafon* (rough soldier) is borne out by accounts that his "self-will, capriciousnes, roughness and rudeness" went as far as suggesting a conquest of Western Europe as revenge for the Cuban debacle of 1962. He was reportedly moderated somewhat by promotion to ministerial rank, but still "would not hesitate to demonstrate the superiority and might of the Soviet armed forces."(142) Considering another of Grechko's defining youthful experiences--his service, during the Civil War and after, in the notoriously and murderously anti-Semitic army of Semyon Budyonny--it is hardly surprising that even as minister "sometimes he would wave his fists, threatening to liquidate imperialism and Zionism."(143) According to a former Soviet officer, who in 1967 was in the graduating class of cadets:

> In the second half of May 1967...the Middle Eastern situation was deteriorating, war between the Arab states and Israel was considered inevitable, indeed imminent. The war's result was predetermined, as everyone in the USSR believed...In order to prevent the West coming to Israel's defense, combat readiness was raised...for this to be better understood by officers and upper-class cadets, they were read a statement by Minister of Defense Marshal Grechko: "The fiftieth year of the Great October Socialist Revolution will be the last year of the existence of the State of Israel."(144)

While the destruction of Israel was not an officially stated goal of Soviet policy, there are numerous other instances indicating that the idea pervaded Soviet thought and parlance, particularly among the military. One example of many for this indoctrination is provided in the memoir of an officer who was dispatched to Egypt shortly after the war: "The Arabs had decided to reestablish Palestine on the area that had already been captured by Israel. With this purpose, led by the UAR under the leadership of...Nasser, [they] deployed armed forces, leaning on the assistance of the Soviet Union."(145) Ambassador Chuvakhin, while proclaiming to Eshkol the USSR's peaceful intent if Israel did not attack, was evidently more candid with the leader of Israel's Communist party (MaKI), Dr. Moshe Sneh: "The war will last 24 hours only and no trace of the State of Israel will be left."(146)

Akopov, however, recalls his diplomatic colleagues demurring at least from the feasibility of this aim: "If we put the task of an offensive...liberation of earlier occupied territories--then we estimated it differently: we [at the Foreign Ministry] thought that the Egyptian army is not capable of such operations....Our military believed, thought that the Egyptian army could fulfil these tasks."(147) "Earlier occupied territories" might apply to all of pre-1967 Israel, or at least to those parts not included in the Jewish state by the 1947 Partition Resolution. Official Soviet foreign policy did not accept the 1949 armistice lines as final borders between Israel and its neighbors.(148)

Restoring the Partition borders might well be "the unavoidable weakening of Israel's positions" which Pyrlin mentions as the expected outcome of the war, while qualifying that "of course there could not be any consideration of its absolute liquidation, as called for by some hot Arab heads." Such a result, "could have constituted according to this [Soviet] way of thinking a serious blow to the prestige of the USA, Israel's main ally which was at that period getting bogged deeper and deeper in the Vietnam war."(149)

It also might have been a fitting gift for November 7, 1967. The approaching anniversary of the revolution provides an element of timing and motivation for the Soviet initiative that merits further investigation. Several references from other sources indicate that Grechko was not alone in seeking a dramatic deed to mark the event, such as a blow on the "imperialist forces" which would crown Soviet leaders with an historic Leninist achievement. "Brezhnev," according to his speechwriter at the time, "began by May [1967] to show his interest in the 50th anniversary...at the beginning of June [before the war] we met at Gor'ki's dacha and were improvising the approximate plan for celebrations."(150)

The KGB-Stasi meeting in Moscow in mid-April specifically stressed the importance of "active measures" for commemorating the jubilee.(151) At the height of the Middle Eastern crisis, an unnamed Soviet diplomat at the UN appeared to betray this preoccupation in an inverted form, by "saying they would...not get involved in a war on their 50th anniversary."(152)

Ironically, when the timing, character and success of Israel's pre-emptive strike surprised the Soviets and obviated their planned intervention, it also put a damper on the festive occasion: "This interest [in celebrations] waned with the Six-Day War,"(153) which instead necessitated a meticulous cover-up that continues to this day.

*The author, a Fellow of the Harry S. Truman Research Institute for the Advancement of Peace at the Hebrew University of Jerusalem, thanks the Institute for a research grant on the Soviet military involvement in the Arab-Israeli conflict, which facilitated the research for and writing of this paper. She thanks Dr. Stefan Meining of Munich for granting access to Stasi documents he uncovered; Brook Lapping Productions for permission to quote from material relating to "The 50 Years War: Israel and the Arabs," a six part television documentary made by Brian Lapping Associates, 1998; and the Trustees of the Liddell Hart Centre for Military Archives, King's College, London for granting of access to, and permission to quote, interview transcripts from this material (henceforth referred to as "transcripts;" spelling and grammar reflect the text of the original English translation). She is also the author of "The Russians Were Coming: The Soviet

Military Threat in the 1967 Six-Day War" which appeared in the December 2000 issue of MERIA.

NOTES

1. Josephine Tey, The Daughter of Time (London: Arrow Books, 1997), p. 111.
2. Michael B. Oren, Six Days of War (New York: Oxford University Press, 2002), p. 55.
3. Nadav Safran, From War To War (New York: Pegasus), p. 274*n*, quoting deposed Egyptian Minister of War Shams al-Din Badran at his trial, according to al-Ahram, February 25, 1968. Also Jerusalem Post, February 28, 1968. The United Nations Truce Supervision Organization also "reported to the Secretary-General that there was no evidence of any Israeli build-up" [F.T. Liu, at the time Senior Advisor to UNTSO, in Richard B. Parker, ed., The Six-Day War: A Retrospective (Gainsville: University Press of Florida, 1996), p. 99.]
4. For a comprehensive overview of all these theories see Richard B. Parker, The Politics of Miscalculation in the Middle East (Bloomington: Indiana University Press, 1993), Chapter 1: "The Soviet Warning," pp. 3-20.
5. Brutents is one of several Soviet officials active in 1967 who still vehemently defend the USSR against "American and Israeli writers [who] tried to prove that the Soviet Union blessed, so to speak, the events that led to the Six-Day War" and accuse them of "attempts… to justify the provocative Israeli behavior by the actions of the Soviet side that on May 12 informed Nasser about dangerous concentration of Israeli troops on the border with Syria." See Karen Brutents, interview on CNN, August 1997; and Yevgeni Pyrlin, Trudny I dolgiy put' k miru (Russian: The Difficult and Long Road to Peace) (Moscow: ROSSPEN (Rossiiskaya Politicheskaya Entsiklopediya), 2002), p. 56.
6. Oren, op. cit., p. 54.
7. Galia Golan, Soviet Policies in the Middle East from World War Two to Gorbachev (Cambridge: Cambridge University Press, 1990), p. 58; Mohamed Heikal, 1967--Al-Intifijar (Arabic), Cairo, 1990, p. 447, cited in Parker, Politics, p. 6-7, p. 247 *n*.9.
8. Maj. Gen. Vladimir A. Zolotaryov et al, Rossiya (SSSR) v lokal'nykh voynakh I vooruzhennykh konfliktakh vtoroy poloviny XX veka (Russian: Russia (USSR) in Local Wars and Military Conflicts in the Second Half of the 20th Century) (Moscow: Institute of Military History, Ministry of Defense of the Russian Federation, 2000), p.181. This first official history of the Soviet involvement in local wars does not mention passing the information to Egypt at all, starting its chronology of the Middle Eastern crisis on May 18, 1967, with the removal of UNEF by Egypt.
9. A vague and sparse chapter in the official history of the Foreign Intelligence Agency of the Russian Federation on its website states: "In the [1960s], foreign intelligence received information about Israel's preparations for new aggression against Arab countries, including the date for it to attack Egypt and Syria in 1967. This intelligence was passed on to the leaderships of Arab countries, who, however, undervalued it and overvalued the military potential of their countries." <http://svr.gov.ru/history/stage09.htm> (Russian), n.d. Gromyko recalled years later that during the "worrying days of May 1967… our military was apprehensive that Israel any moment would attack Syria"; Oleg Grinevski, "Atomnaja bomba I Blizhnij Vostok" (Russian: "The A-bomb and the Middle East"), Dipkur'er (supplement to

Nezavisimaya Gazeta, Moscow), March 1 2001; Grinevski: Stsenarii dlya tret'ey mirovoy voyny (Russian: The Script for World War III) (Moscow: Olma-Press, 2002), p. 112.

10. The KGB "resident" in Israel in 1967 stated as recently as five years ago: "From the regularly incoming information it transpired that a war in the Middle East was approaching and could break out at the end of the 1966 or in the first half of 1967.... Our efforts and means were concentrated in gathering reliable secret information." Ivan Dedyulya: "Na Zemle Obetovannoy: dejatel'nost' sovetskoj rezidentury v Izraile v 60-x godax" (Russian: "In the Promised Land: the Activity of the Soviet *Residentura* in Israel in the '60s"), NVO (military supplement of Nezavisimaya Gazeta) Vol. 20, Moscow 1998.

11. Leonid Mlechin: Mossad: sekretnaya voyna (Russian: Mossad: The Secret War) (Moscow: Centrpoligraf, 2000), pp. 246-247. According to this source, the KGB resident in Cairo warned about up to 12 brigades concentrated on the Syrian border.

12. MfS-S.d.M-1465, Protokoll ueber Verhandlung zwischen Vertretern des MfS der DDR und des KfS beim Ministerrat der UdSSR ueber gemeinsame aktive Massnahmen fuer das Jahr 1967 (German: Protocol of Negotiation between Representatives of the MfS of the GDR and the KGB at the Council of Ministers of the USSR over Common Active Measures for the Year 1967), pp. 8-9.

13. Parker, Politics, p. 21.

14. Isabella Ginor, "Adayin oneh bizehirut" (Hebrew: "Still Answering Cautiously"), Ha'aretz, July 5, 1991.

15. Parker, Politics. p. 248 *n*12. This characterization of Semynov was presented in 1992 during a conference on the 25th anniversary of the June War convened at the Center for the Study of Foreign Affairs at the Department of State's Foreign Service Institute by former Deputy Foreign Minister Georgy Korniyenko and Vitaly Naumkin of the Oriental Studies Institute. It also appears to conform with the attempts to downgrade the echelon involved in the disinformation incident and the motivation behind it.

16. Pavel Akopov, transcript, p. 4.

17. Oleg Grinevski, Sekrety sovetskoy diplomatii (Russian: Secrets of Soviet Diplomacy), (Moscow: Vagrius, 2000), p. 6; Korniyenko was also a member of this delegation and so must have known the position of trust that was conferred on Semyonov.

18. Mlechin, op.cit., pp. 246-247.

19. Michael Bar-Zohar: Embassies in Crisis, (Englewood Cliffs, N.J.: Prentice-Hall, 1970), p. 1. In a telephone interview from his home in Tel-Aviv on November 1, 2002, Bar-Zohar confirmed this quotation and gave his source as "probably a British intercept."

20. Richard B. Parker, ed., The Six-Day War: A Retrospective (henceforth SDW), (Florida: University Press of Florida, 1996), pp. 37, 40, 42. This book presents the proceedings of the 25th anniversary conference (see note 16). In June 1967, Bassiouny was a special assistant in the office of Undersecretary Feki. The Russian participant quoted is Naumkin; a similar view was presented by Korniyenko.

21. Karen N. Brutents: Tridtsat' let na Staroy ploschadi (Russian: Thirty Years on the Old Square), (Moscow: Mezhdunarodnye Otnoshenija, 1996), p. 230.

22. Telephone interview with Brutents (Moscow), October 17, 2000.

23. Emphasis added. Der Rede Von L.I. Breschnev auf dem Juniplenum des ZK

der KpdSU, Uber die Politik der Sowjetunion im Zusammenhang mit der Agression Israels im Nahen Osten, 20.06.1967 (German: The Speech of L.I. Brezhnev at the June Plenum of the CP of the USSR in Connection with the Israeli Aggression in the Middle East). SAPMO=NA ZPA IV 2/1/362. The authenticity of Brezhnev's reference to a Politburo resolution is confirmed by a Polish report of the same speech (AAN KC PZPR 2632), which was found and partly published by Uri Bar-Noi, Notes from the Chaim Herzog Center for Middle East Studies and Diplomacy (Beersheba: Ben Gurion University), Vol. 6, May 2001. Bar-Noi, however, considers that Brezhnev's speech "does not shed light on the controversial information about concentration of Israeli troops,"and appears to attach no significance to the mention of the Politburo -- a conclusion disputed by the present writer.

24. Uri Ra'anan, "Not Just Six Days, Not Just a War," Bostonia (Boston University), Fall 2002. <http://www.bu.edu/alumni/bostonia/2002/fall/essays/raanan/raanan-02.html> Ra'anan and Bar-Zohar (in Embassies in Crisis, p. 2) were among the first who pioneered the hypothesis of a deliberate disinformation maneuver on the part of the USSR. However, Ra'anan suggested that the Soviets believed the Egyptian response would not cause an actual war, and Moscow could then take credit for preventing an Israeli attack on Syria that was never going to occur anyway. "Soviet Global Policy in the Middle East," Naval War College Review, September 1971, pp. 25-26.

25. Victor Israelyan, Inside the Kremlin During the Yom Kippur War (University Park, Pennsylvania: Pennsylvania State University Press, 1995), pp.29-30.

26. Kosygin was away for the UN General Assembly's extraordinary session and his meeting with President Johnson in Glassboro--where he was evidently constrained by collective instructions. Johnson was very frustrated when "each time I mentioned missiles, Kosygin talked about Arabs and Israelis." Lyndon B. Johnson, The Vantage Point: Perspectives of the Presidency, 1963-1969 (New York: Henry Holt, 1971), p. 483.

27. Brutents, Thirty Years, p. 374.

28. Parker, Politics, p. 130.

29. On June 13, 1967. Solomon M. Schwarz, Sovetskii Soyuz i arabo-izrail'skaya voyna 1967 goda (Russian: The Soviet Union and the Arab-Israeli War 1967) (New York: American Jewish Workers' Committee, 1969), p. 72, quoting Security Council minutes S/PV, 1358 p. 147-150. On Tarabanov as a mouthpiece for Fedorenko, see Arthur Lall, The UN and the Middle East Crisis, 1967 (New York: Columbia University Press, 1968), p. 88.

30. Pyrlin, transcript, p. 5.

31. Telephone interview with Brutents.

32. In a speech on May 22, 1967, Nasser said: "On 13 May we received accurate information that Israel was concentrating on the Syrian border huge armed forces of about 11 to 13 brigades. These forces were divided into two fronts, one south of Lake Tiberias and the other north of the Lake." Radio Cairo, May 22, 1967, quoted by BBC, May 24. Middle East Record (henceforth MER) 1967, p. 190. More detail on the "intelligence" provided by the Soviets was apparently disclosed to a U.S. Embassy official in Paris by a "well-connected Arab diplomat": "top secret Israeli plans for [an] 8 brigade 'retaliation' attack on Syrian frontier position" on May 15, which Nasser "had foiled" by moving troops into Sinai. Here too the purported

scope of Israeli action is wildly disproportionate. Department of State incoming telegram 023378, Embassy Paris to Secretary of State, secret, May 23, 1967.
33. Arab sources quoted in MER, p. 17.
34. Al-Gumhuria and Al-Akhbar dailies in Arabic, both from November 26 1966, cited in Avraham Ben-Tzur, Gormim Sovietiim ve Milhemet Sheshet-Ha'yamim (Hebrew: Soviet Factors and the Six-Day War) (Tel-Aviv: Sifriyat Poalim, 1975), pp. 161-162. According to a former GRU Major-General, Grechko was considered "acting" Minister of Defense as early as November 1960 when he hosted an Egyptian delegation headed by Marshal Amer. Sergei Krakhmalov Zapiski voyennogo attashe (Russian: *Notes of a* Military Attache), (Moscow: Rosskaya Razvedka, 2000), p. 76.
35. Vladimir Solovyov and Elena Klepikova: Zagovorschiki v Kremle (Russian: The Kremlin Plotters), (Moscow: Aktsionernoe Obschestvo 'Moskovskii Tsentr Iskusstv', 1991), pp. 20-21. Andropov's promotion to candidate-member of the Politburo was the first for a KGB chief since Stalin's appointee, Lavrenty Beria.
36. Nasser's speech at UAR Advanced Air HQ, May 25, 1967, cited in Walter Laqueur, The Road to War (London: Penguin Books, 1969), Appendix Three, pp. 371-376. This contention was repeated by Mahmoud Riad to Parker as late as 1989: "The proof of Israel's intentions, if any was needed, was a statement by Yitzhak Rabin...on May 12 threatening to occupy Damascus and overthrow the Syrian regime." Parker, Politics, p. 14, p.249 *n*35.
37. MER, p. 187.
38. Protokoll, loc. cit.
39. Pyrlin, Road, p. 56.
40. Aleksandr Khaldeev, "Nesostoyavshiisya Desant" (Russian: "The Landing That Did Not Occur"), Okna (Tel-Aviv), Sept.14, 2000.
41. Israelyan, op.cit., p. 192.
42. Isabella Ginor, "The Russians Were Coming: The Soviet Military Threat in the 1967 Six-Day War," Middle East Review of International Affairs (MERIA) Journal, Vol. 4., No. 4 (December 2000).
43. Pyrlin, Road, p. 56.
44. Pyrlin, transcript, pp. 1, 4.
45. Anwar el-Sadat: In Search of Identity: an Autobiography (New York: Harper & Row, 1977), pp. 171-172.
46. Parker, SDW, p. 65.
47. Brezhnev, Rede.
48. Zolotarev, p.182, citing I.I. Latynin, Opyt primeneniya boevykh takticheskikh grupp v lokal'nykh voynakh na Blizhnem Vostoke (1967-1991) (Russian: The Experience of Using Tactical Battle Groups in Local Wars in the Middle East (1967-1991), PhD dissertation, Moscow, 1997, p. 49.
49. Pyrlin, transcript, pp. 5-6.
50. Brezhnev, Rede.
51. Laqueur, op. cit., p. 97.
52. Department of State Memorandum G/PM:RLGarthoff:pep:5-29-67, confidential. Brezhnev had spoken at a conference of Communist parties at Karlovy Vary, Czechoslovakia.
53. Department of State incoming telegram 029229, Embassy Moscow to Secretary of State, secret, May 27, 1967; the source is identified as "Voslensky," probably Mikhail Voslenski, a historian and interpreter who later defected to the West.
54. Cited in Joseph Govrin, The Six-Day War in the Mirror of Soviet-Israeli Relations--April-June 1967 (Jerusalem: Hebrew University, Soviet and East European Research Centre Research Paper No. 61, December 1985), p. 10.
55. Pyrlin, Road, p. 59
56. Parker, SDW, p. 114. Georgy Korniyenko, The Cold War: Testimony

Isabella Ginor

of a Participant (Moscow: Institute of International Relations, 1994), pp. 129-33, cited in translation by James. F. Leonard; Parker, SDW, p. 72.
57. Testimony of Shams Badran at his trial. Al-Ahram, February 25, 1968, cited in Ben-Tzur, op.cit., p. 165.
58. Ben-Tzur, op.cit., p. 169. While Ben-Tzur's study was published only in Hebrew and drew little direct attention outside Israel, its thesis has been endorsed by leading Israeli actors at the time such as Maj. Gen. Meir Amit, then Head of the Mossad, who cited it both at the 25th anniversary conference and in an interview with the present writer, August 9, 2002.
59. Oren, op.cit., p. 40.
60. Tanjug (Yugoslav Press Agency), cited in Laqueur, Road, p. 53; Moshe A. Gilboa, Shesh Shanim, Shisha Yamim (Hebrew: Six Years, Six Days) (Tel Aviv: Am Oved, 1968), p. 86.
61. Department of State Incoming Telegram 029479, Ambassador Moscow to Secretary of State, confidential, May 28, 1967. Ambassador Llewellyn Thompson qualified the credibility of this information, pointing out the source's "dislike of both Nasser and the Soviets."
62. Akopov, transcript, p. 16.
63. Mohammed Heikal, The Sphinx and the Commissar (New York: Harper & Row, 1978), p. 169.
64. Telephone interview with Nikolai G. Yegorychev (Moscow), November 11, 2000; Christopher Andrew and Oleg Gordievsky, KGB, No location, Nota Bene, 1992 (Russian translation of same title in English, London: Hodder and Stoughton, 1990), p. 503; Yefim Segal and Zinovi Dubrovski, "Ne dolzhny Molchat'" (Russian: "Must Not Keep Silent"), Novosti Nedeli, Tel Aviv, March 2, 2000.
65. Laqueur, *op.cit.*, pp. 98-99.
66. Aleksandr Bovin: XX vek kak zhizn' (Russian: *The* XX Century as a Life) (Moscow: Zakharov, 2003), p. 160.
67. Brezhnev, Rede.
68. Department of State incoming telegram 027005, US Mission UN to Secretary of State, confidential, May 25, 1967.
69. Brezhnev, Rede.
70. Akopov, transcript, pp. 4-8, 10-11.
71. Brezhnev, Rede.
72. Telephone interview with Brutents.
73. Emphasis added. Brezhnev, Rede.
74. Embtel 1517, (Lisbon), Robert Anderson to President Johnson, June 2, 1967, NSF, NSC History, Box 18, LBJL, cited in Avner Cohen: Israel and the Bomb (New York: Columbia University Press, 1998), p. 26; p. 412S *n*27.
75. Akopov, transcript, p.6; Brutents mentions receiving excerpts from those talks for use in preparation of Brezhnev's speech. Thirty Years, p.374.
76. Emphasis in the original. Zolotarev, *op.cit.*, pp. 182-3, citing Latynin, Experience.
77. Akopov, transcript, loc.cit.
78. Kosygin message to Wilson, May 27, 1967, reproduced in Department of State Outgoing Telegram 204008, to American Embassy Moscow from Secretary of State Rusk, Secret, May 28, 1967.
79. Brezhnev, Rede.
80. Ahron Bregman and Jihan El-Tahri: The Fifty Years War (London: Pengiun & BBC Books, 1998), p. 76.
81. Pyrlin, transcript, pp. 6-7, 28.
82. Department of State incoming telegram 029479, American Embassy Moscow to Secretary of State, confidential, May 28, 1967.
83. Pyrlin, transcript, pp. 7, 9.
84. Department of State incoming telegram 029479, American Embassy Moscow to Secretary of State, confidential, May 28, 1967.
85. Pyrlin, transcript, p. 27.

86. Oren, op.cit., p. 65; sources listed in p. 346 *n*10.
87. Mir Pasha Zeynalov: *Nezabyvaemye* vstrechi s predsedatelem Arafatom (Russian: Unforgettable Meetings with Chairman Arafat) (Moscow: Realii, 2002), p. 27.
88. Ben-Tzur, op.cit., p.86; Gilboa, op.cit., p. 85.
89. Laqueur, op.cit., p. 62.
90. Pravda, May 18, 1966.
91. Bovin, op.cit., p. 134.
92. Anatoli F. Dobrynin: Sugubo Doveritel'no (Russian: In Strict Confidence) (Moscow: Avtor, 1997), p. 134. Dobrynin mistakenly dates the congress in May.
93. Ben-Tzur, op.cit., p. 72, *n*13.
94. Israelyan, op.cit., pp. 99-100.
95. V. Kondrashov, interntional commentary, Izvestia, May 8, 1966, cited in Schwartz, op. cit., p. 24.
96. Schwartz, op.cit., p. 25.
97. Schwartz, op.cit., p. 25. Sovietskaya Rossiya, as noted by Ben-Tzur, was considered to be Brezhnev's personal organ. op.cit., p. 73, *n*14.
98. Parker, Politics, p. 11; Gilboa, op.cit., p. 85.
99. Schwartz, op.cit., p. 63.
100. Personal communication from Gideon Remez, a paratrooper then stationed at sector headquarters.
101. Erwin Wiet: Ostblock intern: 13 Jahre Dolmetscher fuer die polnische Partei- und Staatsfuehrung (German) (Hamburg: Hoffman und Campe), 1970, p. 165. Weit, Gomulka's interpreter, understood that Brezhnev was alluding to the war against Israel, into which "Nasser was tempted by the Soviets."
102. Theodore Draper: Israel and World Politics: Roots of the Third Arab-Israeli War (New York: Viking, 1968), pp. 54-55, quoting several such blasts in the flagship Soviet organs Pravda and Izvestiya between May 16 and 22.
103. Ginor, MERIA.
104. Professor Kislov, Head of the Center for Research of Peace Problems in Moscow, was in 1967 "stationed in the Middle East." He makes this statement "based on personal observation," but disputes the present writer's contention that Moscow pre-planned an operation against Israel, and holds that it was to be implemented only "in dire necessity to stop Israeli aggression." A.K. Kislov, "Ne v ladakh s faktami" (Russian: "The Facts Don't Add Up"), USA & Canada Journal, (Moscow: Russian Academyof Sciences, USA & Canada Institute), Vol. 12, No. 396, December 2002, p. 94. This is an afterword to Isabella Ginor, "'Shestidnevnaya voyna' 1967 g. i pozitsiya SSSR" (Russian: "The 'Six Day War' and the Position of the USSR"), loc.cit., pp. 76-91.
105. Vadim Kirpichenko, transcript, p. 28.
106. Emphasis added. Pyrlin, transcript, p. 10.
107. MER, pp. 10, 22. A visit by the Soviet Minister of Fisheries in March also included meetings with top Egyptian commanders.
108. Israelyan, op.cit., p. 192.
109. MER, p. 10.
110. Captain (ret.) Yuri Khripunkov, personal communication to author, July 2001; Fleet Admiral (ret.) Ivan M. Kapitanets: Na sluzhbe okeanskomu flotu 1946-1992: zapiski komandujuschego dvumja flotami (Russian: In the Service of the Ocean Fleet 1946-1992: the Notes of the Commander of Two Fleets) (Moscow: Andreevski Flag, 2000), p. 174.
111. Valeri Mustyats: "Oshibka ekipazha" (Russian: "The Crew's Mistake"), Duel (Moscow) #41(132), October 12, 1999.
112. Yehudit Yeheskeli, "Hikinu lifkuda lishlo'ah til atomi al Yisrael" (Hebrew:

"We Awaited the Order to Launch an Atomic Missile at Israel"), Yediot Aharonot (Tel Aviv), May 8, 1992. When contacted by the present writer a decade later, the source (a former crewman on this submarine, K-125 of the Pacific fleet) retracted his version, claiming that the reporter had misunderstood him (she, however, recorded the interview and stands by its accuracy). The source's name is withheld here as he also professed fear for the safety of his family, despite their present domicile in Israel--which, if well-founded, indicates the even farther-reaching extent of the Russian cover-up.

113. Zolotarev, op.cit., pp. 186-187. This move was confirmed to the present writer by Kapitanets. Telephone interview (Moscow), January 11, 2003.

114. List in Turkish attached to secret Israel Foreign Ministry memo, Minister in Ankara D. Laor to Deputy Director General Y. Tekoa, June 1, 1967.

115. Zolotarev, op.cit., p. 185.

116. Kapitanets, Service, p. 175.

117. Interview with Brig.-Gen. (ret.) Yoel Ben-Porat (Ramat Hasharon), March 8, 2002.

118. Kapitanets, Service, p. 176; Khripunkov, personal communication to present writer.

119. Khaldeev, loc.cit.

120. Pyrlin, transcript, pp. 10-11.

121. Col. General Vasily V. Reshetnikov, Commander of the Second Corps of the Strategic Air-Force in 1967, transcript, pp. 5-8; biography at <http://wwii-soldat.narod.ru/reshetnikov.htm>.

122. Avner Cohen, op.cit., p. 269.

123. Dedyulya, loc.cit.

124. Valeri Yaremenko, "Yadernaya voyna na Blizhnem Vostoke byla by na pol'zu SSSR," (Russian: "Nuclear war in the Middle East Could Have Been Beneficial for the USSR"), Vremya Novostei (Moscow), June 5, 2002.

125. Parker, SDW, p. 250.

126. Yaremenko, loc.cit.

127. Grinevski, "A-bomb"; Script, pp. 112-113.

128. Akopov, transcript, p. 5.

129. Isabella Ginor: "'Under the Yellow Arab Helmet Gleamed Blue Russian Eyes': Operation Kavkaz and the War of Attrition, 1969-70," Cold War History (London: Frank Cass), Vol. 3, No. 1 (October 2002).

130. Anonymous: "Poslevoennaya voyna Sukhogo" (Russian: The Sukhoy's Postwar War), Kommersant-Daily (Moscow), July 31, 1999.

131. Reshetnikov, transcript, p. 4. "Race" is clearly a translation error which appears in the transcript but was corrected to "raids" in the BBC documentary film, which reflects Reshetnikov's audible words in Russian.

132. Ibid., p. 10.

133. Ibid., p. 4.

134. Ibid., p. 6.

135. Ibid., p. 8.

136. Pyrlin, transcript, p. 10.

137. Brezhnev, Rede.

138. L.I. Brezhnev, Vospominaniya, (Russian: Memoirs) (Moscow: Izdatel'stvo Politicheskoy Literatury), 1982, pp. 69-70.

139. Grechko was Brezhnev's immediate superior, according to Grechko's biography in Nikolai Zen'kovich: *Samye zakritye ludi: Entsiklopediya biografiy* (Russian: The Most Classified People: Biographical Encyclopedia) (Moscow: Olma-Press, 2002), pp. 117-120.

140. Brezhnev, Rede.

141. Brutents, Thirty Years, p. 263.

142. Yegorychev, interview; Israelyan, op.cit., pp. 27-28.

143. Biography of Grechko at <http://www.hrono.ru/biograf/grachko.html>. According to Zen'kovich, loc. cit., Grechko was Budyonny's personal protégé.

144. Vladimir B. Rezun, affidavit to present writer, Bristol (UK), January 13, 2001.
145. Vladinir A. Ryabukhin, "V Egypte" (Russian: "In Egypt"), in I.V. Shishchenko, ed., Smolyane-Internatsionalisty (Russian: Internationalists of Smolensk Region) (Smolensk: Smyadyn', 2000), p. 177.
146. Kol Ha'Am (Hebrew, Tel Aviv; MaKI organ), #14, p. 9, cited in Ben-Tzur, p. 237.
147. Akopov, transcript, p. 38.
148. Mikhail P. Popov: Tridtsat' sem' let na Blizhnem Vostoke (Russian: Thirty-Seven Years in the Middle East), Moscow, MGIMO, 2002, p. 20.
149. Pyrlin, Road, p. 54.
150. Bovin, op.cit., p. 160.
151. Protokoll.
152. "Roundup of gossip and attitudes in the UN," Walt Rostow to President, Top Secret, June 8, 1967, National Security File, NSC Histories "Middle East Crisis," Vol. 4, Tabs 128-130. The name and position of the Soviet source were sanitized in the declassified version of this document.
153. Bovin, op.cit., p. 160.

[3]

Air Power in the Six-Day War

KENNETH M. POLLACK

The Saban Center for Middle East Policy at the Brookings Institution, Washington, DC, USA

ABSTRACT In the aftermath of Israel's stunning victory over Egyptian, Jordanian and Syrian forces during the 1967 Six-Day War, accolades for the victory were often showered upon the Israeli Air Force (IAF). Indeed, many believed that it had been Israeli air power that had been the decisive element in the war by first eliminating the Arab air forces and then obliterating the Arab armies in turn. While the IAF did play an extremely important role in the fighting, it was not the decisive element of Israeli victory, and its impact was felt in very different ways from the common perception. Indeed, an appreciation of the true role of Israeli air power in the war reinforces both the importance of psychological factors in combat, and the ability of air forces to have a psychological impact disproportionate to their physical impact.

According to the legends that have grown up around the war, on 5 June 1967 the Israeli Air Force (IAF) obliterated the Arab air forces and thereby won the Six-Day War before it really began. That morning, the first of the war, Israeli Mirages and Mystères swept down on the unsuspecting Egyptian Air Force at its bases and crippled it beyond repair. The IAF then visited the same destruction on the Jordanian, Syrian and finally the Iraqi air forces, gaining complete control of the skies in the first 12 hours of the conflict. Israeli fighter bombers were then free to smash the Arab armies, flying close air support (CAS) missions for Israeli ground units locked in combat with Arab forces, and battlefield air interdiction (BAI) missions to prevent Arab reinforcements from reaching the front. The pervasive Israeli air presence made it impossible for the Arab armies to move or fight in Sinai, on the West Bank or on the Golan. In short, the IAF was the decisive instrument of the stunning Israeli victory.

Correspondence Address: Dr Kenneth M. Pollack, Brookings Institution, 1775 Massachusetts Avenue, Washington, DC 20036, USA. E-mail: kpollack@brookings.edu

This is the legend of Israeli air power during the Six-Day War. It gained wide currency after the war, and is frequently echoed by Arabs, Israelis and Westerners alike. Indeed, in Arab accounts of the Six-Day War, the IAF became a kind of deus ex machina – a supernatural force which the Arab armies could not possibly have been expected to overcome. In the words of King Hussein of Jordan, 'The battle was waged against us almost exclusively from the air with overwhelming strength and continual, sustained air attacks on every single unit of our armed forces, day and night'.[1] To a certain extent, this rationalization allowed the Arab militaries to ignore flaws in their own ground forces by blaming their defeat on Israel's omnipotent air force.

The impression of the might of Israeli air power had an important impact on military developments around the world at the time. The Israelis became so confident in the ability of their air force to act as 'flying artillery' that, after the war, they bought few artillery pieces in favor of more fighter-bombers – much to their regret in the October War of 1973.[2] In the wake of the Six-Day War, armed forces all over the world concluded that a powerful air force could not only supplement, but in some ways even supplant, a powerful army. The poor showing of Argentine ground forces in the 1982 Falklands War (especially when compared to the superb performance of their Israeli-trained air force) was, in part, a product of this mis-learned lesson. Even the United States (US) saw in the performance of Israeli Mirages specially-equipped with 30-mm cannon for tank-killing the need for a dedicated ground-attack aircraft and so developed the A-10 Thunderbolt, armed with its own 30-mm cannon.

The myth of Israeli air power and the Six-Day War persists to this day as an element in the larger debate over air power in US military strategy. Air power enthusiasts regularly cite the Six-Day War as the first time that air power was decisive in a ground war, and one of a quartet of wars in which air power was supposedly decisive – along with the Gulf War, Kosovo and Afghanistan.[3]

The impact of Israeli air power on the outcome of the Six-Day War has been exaggerated, however, and its contributions to the war have been misunderstood. Israeli air power was an important factor in the Israeli victory, but it was hardly all-powerful and it was not the decisive element of the campaign. The *physical* impact of Israeli air power was largely confined to preventing the Arab air forces from participating in the war, and destroying 'soft' military vehicles (trucks, jeeps, etc.) during the various Arab retreats. The IAF played a marginal role in the decisive ground battles, and here its principal impact was not physical, but psychological. As General Uzi Narkiss, one of the principal Israeli front commanders during the war put it, 'The IAF was very important...but not so much for the ground battles. Mostly for morale and for the absence of Arab air strikes that they gave us'.[4]

This is not to dismiss the influence of the IAF on the conflict. Although Israeli air power was not decisive, it still played a very significant role in the course of the war. It hindered the functioning of Arab command and control at all levels, prevented some reserves from reaching the battlefield in a timely fashion, helped to turn the retreat of the Arab armies into a rout and, most important of all, provided Israel's ground forces with the air supremacy that made every other operation considerably easier.

A reassessment of the role of Israeli air power in the Six-Day War not only improves our knowledge of that particular conflict, but also enriches our understanding of how air forces affect ground operations generally. The use of air power in the Six-Day War provides a fuller appreciation for both the extent to which air operations can shape ground operations, and the various environmental factors that can limit this impact. It also has important lessons to teach regarding how air power works its influence on ground forces. In particular, the Israeli experience in the Six-Day War provides a useful counterweight to the experience of US air forces in recent conflicts such as the Gulf War, Kosovo and Afghanistan.[5] Indeed, the success of these US air campaigns were so great that they have become archetypes, from which all planning regarding the future uses of air power are derived. However, not all nations can count on the luxury of time and resources the US enjoyed in the Gulf War, Kosovo and Afghanistan, and the IAF air campaign against the Arab states illustrates what one of the finest air forces in the world can accomplish when forced to operate under stringent limitations (a lesson illustrated to a certain extent more recently by American air power in the invasion of Iraq). Thus the Israeli experience in the Six-Day War serves as a useful alternative model for the employment of air power in land warfare.

Over the past 60 years, most of the academic literature regarding air power has focused on the strategic air power debate – whether strategic bombing alone can coerce an adversary to make concessions.[6] This article addresses another increasingly important debate within the military and policy-making communities, the extent to which air power can be the decisive factor in land warfare.[7] Historically, air forces have supported armies in land warfare because, in the end, the war was always decided by which *army* prevailed. However, the Six-Day War was the first time that air power advocates claimed that it was an air force that had struck the decisive blows, while the Israeli army simply mopped up. Today, in the budget debates in Congress and the battles over strategy in the Pentagon, there is still bickering over whether the US Air Force should take precedence over the US Army as the decisive tactical instrument in waging land warfare. Indeed, the military plan for the invasion of Iraq featured a fairly light ground force (given the size of the undertaking) under the assumption that US air power could carry much of the burden traditionally assigned to ground forces. The Six-Day War has some important lessons concerning this debate.

This article analyzes the role of Israeli air power in the Six-Day War and its relevance to current debates regarding air power and land warfare. It briefly recounts the course of operations during the war. It then disproves the myth that Israeli air power was the decisive factor in the war, and that its decisiveness was derived from its ability to destroy Arab armored fighting vehicles. Next, it draws out generalizations regarding the true impact of Israeli air power on the conflict as a whole. Finally, it culls lessons from the Israeli experience during the war regarding the more general question of the utility of air power. Drawing on the lessons of the Six-Day War, it compares the Israeli experience with US experiences over the past two decades and analyzes how air power affects the conduct of war and suggests how it can most effectively be employed in future conflicts.

The Course of the War: A Synopsis

Israel launched the Six-Day War with a devastating offensive counter-air campaign against Egypt, followed by similar efforts against Jordan, Syria and Iraq.[8] In the early hours of 5 June 1967, the IAF launched an extremely well-planned and executed series of airstrikes against 18 Egyptian air bases. In three hours of constant hammering, the Israelis destroyed over 300 of Cairo's 450 combat aircraft, and killed 100 of 350 Egyptian combat pilots.[9] While the IAF pummeled the Egyptian Air Force, the Syrians, Jordanians and Iraqis launched desultory air raids against Israel – all of which failed to find their targets and none of which did any significant damage – but which provoked Israel to visit the same destruction on them later in the day. The Israelis first turned on Syria, striking the airbases at Damascus, Marj Ruhayyil, Dumayr, Sayqal and T-4, and destroying roughly half the Syrian Air Force including all but four of its cutting-edge MiG-21s. The Syrians also lost four MiG-17s in air-to-air combat with the Israelis, while the IAF lost only one Mystère in the air raids.[10] Although Tel Aviv had hoped to avoid war with Amman, the Jordanian air raid and Jordanian attacks on Israeli positions in Jerusalem, convinced Israel to dispatch eight IAF aircraft against Amman International and Mafraq airfields. The Israelis destroyed 18 of 24 Hunters and badly damaged another four.[11] Finally, at the end of the day (and on into the morning of 6 June), Israeli planes flew

500 miles across the Syrian desert and attacked Iraq's H-3 (al-Walid) airbase near ar-Rutbah, the westernmost Iraqi airfield, destroying 31 Iraqi planes on the ground and in the air.[12]

While the IAF eliminated the Egyptian Air Force, the Israel Defense Force (IDF) launched a major offensive against the Egyptian Army in the Sinai.[13] The Egyptians deployed seven divisions and eight independent brigades with 100,000 troops, 950 tanks, 1,100 APCs and over 1,000 artillery pieces to Sinai in a defense-in-depth along the few main east-west corridors in Sinai.[14] The Israeli army attacked with 70,000 troops and 700 tanks in six armored brigades, an infantry brigade, a mechanized infantry brigade and three paratroop brigades – most grouped into three divisional task forces, or *ugdot*.[15] Beginning at 0800 hours, the three Israeli *ugdot* smashed through the Egyptian infantry divisions in their fortified positions. By the end of the first day, all three Israeli *ugdot* had broken through the fortified defensive lines of the Egyptian infantry divisions, and defeated Egypt's tactical reserves (armored brigades backing up each of the infantry divisions) with only minor losses. Two of the Israeli *ugdot*, those of Generals Tal and Sharon, were already racing for the Suez Canal and the passes through the line of mountains that run north-south in western Sinai. The Egyptians resisted fiercely but not effectively, and the Israelis won quickly and with relatively minor losses. The Israelis found that Egyptian units fought hard from their fortified defensive positions, but would not maneuver against the Israelis, nor would they shift forces once the IDF had broken through their lines or turned their flank. The Egyptian formations rarely counterattacked, and when they did, it was usually a clumsy frontal assault that the Israelis defeated quickly. By the end of the first day, Tal's reconnaissance elements were rolling westward toward the canal. On the second day, elements of General Yoffe's *ugdah* smashed a counterattack by Egypt's elite 4th Armored Division, and then ploughed through Egypt's second line of defense, manned by Cairo's 3rd Infantry Division, before joining Sharon and Tal's forces racing to the passes and the canal beyond. Late on the second day, small Israeli armored forces seized both the Mitla and Giddi passes, effectively sealing two-thirds of the Egyptian Army in central Sinai. Although the Egyptians badly outnumbered the Israelis at the passes, they could not break through, and were decimated in the fighting.

Israel launched its invasion of the West Bank only in response to Jordanian attacks, thus the IDF offensive there did not get started until mid-morning.[16] Despite the excellent defensive terrain of the West Bank, the strong defensive positions the Jordanians had spent 19 years building and the fact that the Jordanians had correctly anticipated Israel's plan of attack and so had their forces deployed almost precisely where Israel intended to attack, the Jordanians had little more luck than the Egyptians. The Jordanians marshaled roughly 45,000 troops, 270 tanks and 200 artillery pieces in the West Bank in nine brigades (seven infantry and two armored) and several independent battalions.[17] Israel attacked with eight brigades (one armored, two mechanized and five infantry) with about 40,000 men and roughly 200 tanks.[18] As in Sinai, on 5 and 6 June, Israeli ground forces cracked the Jordanian frontline infantry positions and then defeated Amman's armored reserves. During the daytime of 5 June, and later that night, Israeli forces broke through the Jordanian lines at Janin, Tulkarm, the hills along the northern face of the Jerusalem corridor, and in Jerusalem both north and south of the Old City. On 6 and 7 June, Israeli armor defeated Jordan's armored reserves at Tel al-Ful north of Jerusalem and Qabatiyah crossroads, south of Janin, allowing IDF units to drive quickly east and seize the crossings over the Jordan, sealing the West Bank and the fate of the Jordanian Army. As with the Egyptians, the Jordanians fought hard from their fortified lines and inflicted heavy casualties whenever the IDF was forced to make a frontal assault against Jordanian positions. However, Jordanian units rarely tried to outflank or otherwise outmaneuver the Israelis, nor would they reorient themselves to meet Israeli maneuvers.

After the destruction of much of the Syrian Air Force on 5 June, Israel mostly ignored Syria to concentrate on Jordan and Egypt. The IAF did fly some airstrikes against Syrian defenses on the Golan in preparation for an eventual ground assault (flying roughly 100 ground-attack sorties against Syria on 6 and 7 June, and 225 sorties against Syria on 8 June).[19] However, not until 9 June, when Egypt and Jordan were soundly defeated, did Israel move against Syria.[20] Damascus deployed about 50,000 men with 500 tanks and assault guns, 500 APCs and nearly 300 artillery pieces in heavily-fortified lines in the forbidding defensive terrain of the Golan.[21] On 9 June, Israel launched about 20,000 troops and 250 tanks in seven tired but confident brigades. One advantage Tel Aviv possessed was that by the time of their attack on the Golan, the IAF was free to participate fully against the Syrians and, despite its busy week, the IAF could still muster over 150 serviceable combat aircraft.[22] In virtually identical fashion to the performance of the Egyptian and Jordanian armies, the Syrians fought back hard when the Israelis attacked, but refused to maneuver against them, to concentrate their forces against Israeli penetrations, or to launch anything but small, sporadic frontal counterattacks which the IDF defeated almost effortlessly. As a result, by the end of 9 June, the Israelis had turned the northern flank of the Syrian lines and the next morning they simply began rolling up the Syrian lines from north to south. By mid-afternoon, the battle was over as Israeli forces took the key town of al-Qunaytarah on the eastern edge of the Golan and the Syrian army fled in disarray.

Was Israeli Air Power Decisive?

Israel's victory during the Six-Day War was defined by Israel's control over the Sinai, West Bank and Golan Heights at war's end. For the IAF to be considered the decisive element of the victory, it would be necessary to prove that it was the IAF that crushed the Arab armies and drove them out of those territories, or else that the IAF effectively enabled the Israeli army to do so – for instance by breaking the morale of the Arab armies such that no real fighting was necessary to defeat them. This was not the case. Important as the contributions of the IAF were, the Six-Day War ultimately was decided on the ground, not in the air. Israel conquered the Sinai, the West Bank and the Golan by smashing the Arab armies entrenched there and, for the most part, it was Israel's army that did so. Israeli air power played an important supporting role, but only a supporting role, in the defeat of the Arab armies.

The most important evidence supporting this argument is that the ground campaigns were effectively decided on 5 and 6 June, and the IAF contributed very little to these battles.[23] For the most part, the IAF did not participate in significant numbers in the Israeli ground campaigns against Egypt and Jordan until 6 June, and did not really bring its full weight to bear in Sinai until 7 June. The campaign against Egypt was decided on 5 and 6 June in the breakthrough battles at al-Ageilah, Umm-Qatef, Qusaymah, Khan Yunis, Rafah and the Jiradi pass, and in the battles against Egypt's operational reserve at B'ir Lafhan and their secondary defense line at Jebel Libni. After that, with the exception of a rearguard action by the remnants of the Egyptian 4th Armored Division at B'ir Gifgafah on 7 June, there were no combat operations by organized, cohesive Egyptian units – just skirmishes between disorganized Egyptian groups trying desperately to escape back to the canal and Israeli forces either blocking their way or pursuing them across the desert. Likewise, the fighting on the West Bank was decided during the breakthrough battles on 5 June, north and south of Jerusalem, at Radar Hill and Latrun on the northern shoulder of the Jerusalem corridor, at Tulkarm and Janin in northern Samaria, and in the defeat of Jordan's 60th Armored Brigade at Tel al-Ful on 6 June. The last element of Jordan's operational reserve – the elite 40th Armored Brigade – was not defeated until 7 June at Qabatiyah crossroads; however, by then, this battle had become something of a sideshow. Thus by the end of the day on 6 June, the contest for the West Bank was also effectively decided (see table 1).

TABLE 1 ISRAELI AIR-TO-GROUND SORTIES BY DAY, BY FRONT, 5–10 JUNE 1967

Front	5 June	6 June	7 June	8 June	9 June	10 June	Totals
Egypt	170	286	321	160	20	8	**965**
Jordan	95	221	233	0	0	0	**549**
Syria	3	107	98	225	299	345	**1,077**
Totals	**268**	**614**	**652**	**385**	**319**	**353**	**2,591**

Source: History Branch, Israeli Air Force, correspondence with the author, 10 September 1997.

Although the Egyptian and Jordanian armies were effectively defeated on 5 and 6 June, the IAF was mostly focused elsewhere. On 5 June, most of the IAF's day was taken up in its counter-air operations against Egyptian, Syrian, Jordanian and Iraqi airbases. The IAF contributed a relatively small number of airstrikes against Arab ground forces on 5 June, flying only 268 ground-attack sorties on all three fronts (compared to 614 on 6 June and 652 on 7 June). Against Sinai, the Israelis flew only 170 ground-attack sorties on 5 June (compared to 286 on 6 June and 321 on 7 June) and probably destroyed no more than 12–15 Egyptian tanks.[24] On 6 June, the IAF was committed more fully to ground support, but Israeli airstrikes on all three fronts started with the Arab forces and infrastructure *farthest* from the frontlines and then slowly worked their way back toward their own advancing ground forces.[25] As a result, most of the IAF airstrikes flown against Arab ground forces were actually interdiction sorties directed against logistics and rear-area formations, rather than the frontline units and operational reserves whose defeat were the keys to Israeli victory. For example, the IAF only became a significant factor against the Egyptian combat forces on the Sinai front on 7 June, when Israeli aircraft hammered the masses of Egyptians surging back to the canal. The official IDF report on the war found that nearly all of the Egyptian armored vehicles destroyed by airstrikes were destroyed during the Egyptian retreat – not during the battles in eastern Sinai where the campaign was decided.[26] Similarly, on 5 June, of the 95 air-to-ground sorties the Israelis conducted against the West Bank, 42 hit targets in the Jordan valley and only 32 struck targets in the key combat zones around Jerusalem and Janin.[27]

The Israeli army broke through the entrenched infantry lines of the Arab armies on all three fronts essentially without air support. In the Sinai, the IDF attacked into the most heavily-fortified and well-defended sectors of the front at Khan Yunis/Rafah/the Jiradi pass in the north, and at Abu Ageilah/Umm Qatef/Qusaymah in the center.

Similarly, on the West Bank, the IDF attacked into the most heavily-fortified and well-defended sectors of the Jordanian front at Latrun, Radar Hill, Ammunition Hill and Abu Tor. In every case, they received little or no support from the IAF and yet in every case they punched through the Arab lines quickly (in a matter of hours) and with very few casualties. Even the worst losses the Israelis took in these battles, such as at Ammunition Hill, barely affected the combat power of the Israeli assault forces or their ability to continue to perform their missions. Only on the Golan did the IAF contribute in a meaningful way to the breaching operation by laying on (along with Israeli artillery) a heavy preliminary bombardment of the forward Syrian positions. In this case, the airstrikes had little physical impact on the Syrian forces as the Israeli ordnance generally could not penetrate the stout Syrian bunkers. However, it did serve an important purpose by suppressing Syrian fire while Israeli combat engineers laboriously cleared the extensive Syrian minefields.

The battles against the Arab infantry formations manning the fortified lines on all three fronts were the most important of the war. The Arab armies had very little ability to conduct fluid battles of maneuver. Consequently, Arab armored and mechanized formations were at a severe disadvantage against their Israeli counterparts, resulting in the lopsided Israeli victories against (usually larger and better-equipped) Arab armored forces at Rafah, Abu Ageilah, B'ir Lafhan, B'ir Gifgafah, the Dotan Valley and Tel al-Ful. However, the Arab infantry formations were tough opponents when conducting static defensive operations from fixed positions. Although Israeli units were competent in set-piece offensives, the greatest strength of the IDF – when it displayed its true tactical genius – was in unstructured maneuver battles. Thus, the initial breaching operations pitted the Arab armies at their best against the Israelis at their worst: the IDF was forced to conduct set-piece, frontal assaults against entrenched Arab defenders. If the Arabs were going to beat the Israelis, it was in these battles.

The fact that the Israeli Army prevailed so handily in these attacks, and that the IAF contributed so little to their outcome, makes two points apparent. First, that the Israeli ground forces were so far superior to their Arab counterparts in every aspect of military operations that the war's outcome was probably a foregone conclusion. And second, that this superiority was so great that the contribution of the IAF was not critical to securing an Israeli victory on the ground.

Numerous Arab sources frequently assert that Israeli airstrikes decided particular ground battles. These accounts are almost entirely inaccurate and reflect a tendency among some Arab writers to ascribe the defeat of their armies to the supernatural powers of the IAF. In nearly every case, the IAF played only a secondary role in these battles, and in a number of cases did not participate at all. For example, even Samir Mutawi, Jordan's most insightful and objective commentator on the war, claims that one reason the Israelis prevailed at Ammunition Hill was because of airstrikes. However, the Israeli accounts, including those of General Uzi Narkiss (the front commander) and Colonel Mordechai 'Motta' Gur (the commander of the 55th Paratroop Brigade, which conducted the attack), point out that there was a major debate among the Israeli commanders over whether to launch the attack at night without air cover or wait until morning when the IAF could participate. Ultimately, Narkiss and Gur decided to launch the attack at night – against the judgment of the Israeli General Staff – and so no IAF aircraft participated in the battle at all.[28] There are many other instances of this phenomenon, particularly for battles on 5 June, when the IAF was so busy destroying the various Arab air forces that it conducted only 95 ground-attack sorties against the West Bank, yet Jordanian sources claim their troops, 'were subjected to [air] attack almost every time they moved'.[29]

The limited role of the IAF in the crucial breakthrough battles against the Arab infantry is reinforced by the experience of the subsequent armor battles in which the IDF defeated the Arab operational reserves. The IAF did not participate in the armor battles against Egypt at Jebel Libni and B'ir Gifgafah, or against Jordan in the Dotan valley. Likewise, Israeli sources make clear that at Tel al-Ful on 6 June, the IDF did not receive any air support despite Jordanian claims to the contrary.[30] The Jordanians had every advantage in this battle – equipment, numbers, terrain – but were still defeated. The Jordanians had a battalion of modern M-48 Pattons and a battalion of mechanized infantry in M-113s pitted against a reinforced company of Israeli Super Sherman tanks (old World War II tanks which the Israelis had refurbished and upgraded) and a battalion of mechanized infantry in World War II-vintage M-3 halftracks. The 76-mm guns on the Super Shermans could not even penetrate the frontal armor of the Jordanian Pattons.[31] Nevertheless, the Israelis still defeated the Jordanian armor handily and drove them off Tel al-Ful and its nearby village while suffering only light casualties.[32] The only contribution of Israeli air power to this battle was that the Jordanian armored brigade had been hit by Israeli airstrikes twice the night before. However, these attacks destroyed only two to three tanks and a few other vehicles and only delayed the brigade's advance by a few hours.[33] In addition, any psychological impact from the attacks should have worn off by the time of the battle, as historical experience of ground forces under air attack from World War II to the Persian Gulf War makes clear that the psychological impact of airstrikes wears off quickly unless applied repeatedly over the course of many days or weeks.[34]

At B'ir Lafhan on 6 June, where two battalions from General Yoffe's *ugdah* defeated the main Egyptian counterattack by two brigades of the 4th Armored Division, the Israelis did conduct airstrikes during the battle itself, and these no doubt contributed to the defeat of the Egyptian armor, but here as well, the evidence confirms that Israeli ground operations were decisive.[35] Of particular relevance is the account of the senior Israeli military officer present at the battle, (then) Colonel Avraham 'Bren' Adan who was Yoffe's chief-of-staff and led the Israeli forces at B'ir Lafhan. Adan has an unrivalled reputation for objectivity and ruthless honesty – even to the extent of criticizing his own conduct. According to Adan, the IDF received only a small number of airstrikes (perhaps a dozen sorties) during the battle, and these destroyed very few Egyptian tanks or APCs, although they did destroy a number of Egyptian trucks. This contention is supported by the findings of the postwar Israeli and US assessments of the war, which found that very few Egyptian tanks had been killed by airstrikes on any front.[36] Likewise, other sources confirm Adan's assessment that the battle was won by the two Israeli tank battalions, one of which pinned the two Egyptian brigades while the second caught them in the flank. The two battalions inflicted roughly 30 per cent casualties on the Egyptian armor, causing the rest of the force to flee.[37]

The most important role of the IAF in the ground battles was probably at Qabatiyah crossroads on 7 June, where an Israeli mechanized infantry brigade and an infantry brigade eventually defeated Jordan's elite 40th Armored Brigade, led by its best field commander, Brigadier Rakan al-Jazi. Even in this battle, Israeli air power played an important role in the outcome of the battle, but its contribution was not decisive. In this battle, the Israelis made two unsuccessful attempts on 6 June to knock the Pattons and M-113s of the Jordanian 40th Armored Brigade off a ridge overlooking the critical Qabatiyah road junction controlling movement south into the West Bank. Both efforts were frontal assaults by a tired and understrength battalion of Israeli Super Shermans without infantry, artillery or air support that were handily beaten back by the Jordanians. The next morning, the Israelis bombarded the ridge for 15–30 minutes with artillery and airstrikes, and then used a combined-arms team of Super-Shermans and mechanized infantry, coupled with leg infantry, to pin the Jordanian forces on the ridge while a second combined-arms team outflanked them. This maneuver forced the Jordanians to try to fall back from the ridge, at which point they were caught and demolished by Israeli armor and aircraft. In this case, IAF airstrikes were important to the Israeli victory, but not as important as the Israeli ground attack.

Vicious though the preparatory aerial bombardment may have been, there is no reason to believe that this would have forced the Jordanians off the ridge alone. It was the fear that the Israeli flanking force was enveloping his brigade that convinced al-Jazi to try to fall back. Absent the airstrikes, al-Jazi might have been bolder and attempted to block the Israeli envelopment – something he had successfully done in the fighting on 6 June. However, by then, the IDF outnumbered him on the ground (two brigades to one), a third Israeli brigade was already in his rear on the Janin-Tubas highway and could easily have been moved north to take his force from behind, and the rest of the Jordanian army was collapsing all around him. Thus the Israeli airstrikes were hardly the decisive factor in forcing al-Jazi to retreat. Instead, he fell back because of the combination of the immediate tactical problem of the Israeli flanking attack and the larger operational problem that his position was untenable even if he solved his tactical problem.[38]

Israel's command of the air had a heavy impact on the morale of the Arab ground forces, but not until after the decisive ground battles in Sinai and the West Bank had been fought and won by the Israeli army. As noted above, the IAF was generally absent from most of these battles. Nor could the shock of having all of the Arab air forces destroyed in one fell swoop – leaving Israel with air supremacy – have affected the Arab defenders prior to these battles, since nearly all of them occurred *before* the Arab ground commanders were even aware of the impact of the Israeli airstrikes. Specifically, the Egyptian Air Force high command did not even tell President Nasser of the outcome of the strikes until about midday on 5 June, while the army in Sinai, including its commander, General Murtagi, were not told of the results until 1200 hours on 6 June – after all of the breakthrough battles were over and the IDF had defeated the counterattack by the 4th Armored Division at B'ir Lafhan.[39] Indeed, several Egyptian generals I interviewed indicated that the persistent query that they and their men had during the war was, 'where is our air force to help us?'. All indicated that it was not until after the war that they learned of the Israeli air raids and the destruction of their air force.[40] Indeed, in the battles on 5 and 6 June, the Arab units generally fought very hard and showed no indications of demoralization. There were some exceptions to this rule – an Egyptian brigade at Qusaymah that fell apart on first contact and Jordanian forces on Radar Hill who disintegrated even though they were defending terrain so good that the Israelis thought they would be making a suicide charge. However, in the vast majority of cases the Egyptian and Jordanian ground forces showed no signs of demoralization: they fought very hard, just not very well.

The greatest contribution of Israeli air power to the various ground campaigns came during the pursuit phases. Because the IAF concentrated first on destroying the Arab air forces and then on disrupting Arab rear areas, the Arab combat forces defending their forward lines suffered least from airstrikes, the operational reserves only slightly worse, while the logistical support units and second-echelon forces suffered the most from air attack. To some extent, only when the Arab combat forces began retreating did they come into the kill zones being worked over by Israeli fighter-bombers. At this point, the IAF began to cause severe damage to Arab forces. Between the demoralization of their defeats at the hands of the Israeli army, and the channeling effect of the terrain in all three theaters, the Arab ground forces were physically and emotionally ripe for destruction by air when they began their retreats. Consequently, Israeli airstrikes against Egyptian and Jordanian forces beginning late on 6 June and Syrian forces early on 10 June had an ever greater impact, destroying vehicles, delaying and dispersing Arab formations to ease their destruction by Israeli ground forces, and causing huge numbers of Arab personnel to simply abandon their equipment and run. Nevertheless, by this point, the war had been decided, and all that remained was to determine the final score.

How Many Tanks did the IAF Destroy?

The IAF's primary impact on the decisive ground campaigns of the Six-Day War was psychological, not physical. All available evidence indicates that Israeli airstrikes caused little physical damage to Arab combat forces during the Six-Day War. Indeed, the official US survey team sent to Israel after the war to collect data for US planners concluded that the role of the IAF in the Israeli victory had been greatly exaggerated. After examining in minute detail a sample of 30–40 per cent of all tanks (Israeli and all Arab armies) destroyed in the fighting, they concluded that airstrikes had been only a minor cause of damage to armor.[41] They found that the 20- and 30-mm cannons which were the primary ground attack weapon on Israeli aircraft did 'uniformly slight damage' to Arab tanks. Indeed, only eight per cent of the Arab tanks were even hit by aerial munitions of any kind, and only 2–3 per cent of the Arab tanks were actually destroyed by these hits. The survey team concluded that, 'This type of data tends to refute the contention that the Israeli Air Force was directly responsible for the damage to the majority of the Arab tanks and shows conclusively that ground weapons were, in fact, responsible for practically all damage to tanks under the strategical [sic], tactical, terrain, and weather conditions of the June 1967 war'.[42]

Even the official Israeli survey – which was far more favorable to the IAF than the later US report – concluded that only 15 per cent of Egyptian tank losses could be attributed to IAF airstrikes. The Israeli damage assessment teams that canvassed the battlefields in Sinai assessed that, at most, the IAF knocked out 75–90 Egyptian tanks and self-propelled guns during the entire course of the war. The survey also revealed that Israeli pilot claims regarding destroyed Egyptian armored vehicles were three or four times higher than the actual number of kills.[43] Moreover, anecdotal accounts of Israeli airstrikes, the assessments of Western personnel who toured Sinai after the war, and the recollection of Israeli ground officers, all fully concur with the conclusions drawn by the American and Israeli studies that the IAF had had comparatively little impact on Arab combat formations.[44]

Although the IAF did little damage to the 'teeth' of the Arab armies, it did considerably more damage to their 'tails'. Israeli fighter-bombers took a far greater toll of Arab trucks and other thin-skinned vehicles such as jeeps and cars (and even a few military trains) than armored fighting vehicles. Because Israeli interdiction operations began at the very rear of the Arab defensive zones and then worked their way forward, most of the IAF airstrikes flown against ground forces were actually directed against logistics and rear-area formations. Israeli pilot reports from Sinai claimed to have destroyed more than twice as many unarmored vehicles than tanks and self-propelled guns.[45] Again, anecdotal reporting of IAF strikes, as well as the accounts of IDF participants and Western observers, corroborate the statistical conclusions that Israeli aircraft did more damage to Arab trucks, jeeps and cars than to tanks and APCs. When Tel Aviv's fighter-bombers did attack Arab maneuver units, most of the damage they did was to the organic logistical elements of the unit, rather than to the combat formations themselves. However, because the IAF initially concentrated on ground targets in the Arab rear areas, most of the thin-skinned vehicles that the Arabs lost to airstrikes in the early days of the war were part of supply and transport companies attached to higher level formations and not to the frontline units. As a final note, the IAF claimed to have destroyed 783 trucks and other unarmored vehicles in Sinai. This is not very many given that the Egyptian Army in Sinai possessed well over 10,000 such vehicles.[46] Again, this strongly indicates that the physical damage from Israeli airstrikes was limited, and much less than suggested by many of the Arab, or even some of the more bombastic Israeli, accounts.

Overall, this pattern of damage wrought by the IAF supports the claim that the IAF was an important secondary factor, but not the decisive element in the defeat of the Arab armies. First, the IAF flew relatively few missions against Arab combat forces, and in these attacks killed and damaged very few armored vehicles. Thus there was little attrition of forward Arab combat power from airstrikes. Second, the

IAF did considerably more damage to the logistical trains of the Arab armies, but the war did not last long enough for this to have had a real impact. In particular, because nearly all of the key battles of the war took place within the first 24 hours, Arab combat forces did not really have the opportunity to run out of fuel, ammunition, food and so on. There is no evidence – even anecdotal – from either the Israeli or Arab side that Arab units were hampered in combat during the first two days of the war because of logistics shortfalls. (Whereas the Israelis almost always seemed to be short of fuel and ammunition.) Thus the destruction of their logistical units was essentially irrelevant to the conduct of the war by the Arab armies. Had the war gone on for even a few days longer, the Arab armies would no doubt have begun to suffer from the damage to their logistical trains, but since the Israeli ground forces finished the job so quickly, this potential encumbrance was never realized.

The Impact of Israeli Air Power

Having discussed what the IAF did not do, it is important to turn now to what the IAF did do. Although the impact of the Israeli air effort was not at all what the legends about the war claim – that it was the decisive factor in Israeli victory and that it did so by killing lots of Arab tanks – the IAF's contribution was still very significant. Indeed, by clearing away these myths, it helps us to develop a better understanding of how air forces affect ground operations. In particular, what stands out regarding the impact of Israeli air power was its powerful psychological effects, which dwarfed the physical damage it inflicted on the Arab armies. The IAF made its presence felt primarily by sowing confusion and panic in the Arab armies, causing paralysis, demoralization and then a near-total breakdown in unit cohesion as they fled.

Turning Retreat into Rout

To a great extent, it was the constant, relentless Israeli air pressure on 7 and 8 June that turned the Egyptian and Jordanian retreats into routs. Many Arab units were badly disrupted by Israeli airstrikes that caused soldiers to scatter (with or without their vehicles), inflicted casualties and gave the survivors a serious scare. Indeed, the widespread exaggeration of the number and lethality of Israeli airstrikes among Arab accounts is, at least to some extent, attributable to the deep impression they left on Arab personnel. Although analyses of the psychological impact of airstrikes are few, those available, and numerous anecdotal accounts from every conflict since World War I, make clear that an airstrike can be a terrifying experience for ground troops, especially those with little warning or ability to fight back, as was true for all of the Arab armies in 1967.[47] In a number of cases during the Six-Day War, although Israeli airstrikes caused only minor damage to an Arab unit, they caused so many personnel to abandon their vehicles or lose their nerve that the unit was essentially incapable of further action. Arab units generally decided that retreat was the better part of valor after suffering through Israeli airstrikes not because of the physical damage they had sustained, but because of their fear of future attacks. For instance, the Jordanian 2nd Armored Regiment (battalion) of the 60th Armored Brigade was moving to try to bolster the crumbling defenses of Jerusalem during the night of 6/7 June when it was caught by the IAF and subjected to severe airstrikes.[48] Although the Jordanians lost few tanks or APCs in the raid, the unit was so thoroughly demoralized that it retreated back to the Allenby bridge rather than press on and risk further airstrikes.[49]

Isolating the Battlefield

In addition to its impact on the Egyptian and Jordanian retreats, Israeli air power also had a meaningful impact on the course of the war by preventing Arab reserves from participating in the battle. This was particularly important in the battle for the West Bank, but may also have played a role in the conquest of the Golan. Israeli airstrikes against Arab reserves moving forward to reinforce the frontlines invariably delayed, halted and occasionally either scattered or routed these units. Against Jordan, Israeli airstrikes prevented elements of the 27th Infantry Brigade, the 2nd Armored Regiment (battalion) of the 60th Armored Brigade, and the entire 6th Infantry Brigade from reaching Jerusalem in time to contribute to the defense of the city.[50] Likewise, IAF raids against the Iraqi 8th Motorized Brigade so paralyzed the unit, that it never even made it to Amman. Against Syria, there is no specific evidence that IAF airstrikes prevented particular Syrian reserves from joining the fray. However, it seems likely that the IAF did prevent the main Syrian operational reserve – the 42nd Brigade Group, which Tel Aviv was watching very carefully – from moving to the front because of air interdiction. Indeed, the Syrian high command may have pulled this formation back to Damascus during the night of 9/10 June as much to shelter it from further pounding by the IAF as to protect the capital (and the regime) from an Israeli ground attack.[51]

Paralyzing the Arab High Commands

The psychological blows inflicted by Israeli air power on the Arab armies not only hit the combat units at the bottom of the chain of command, but also struck at the generals at the very top of the hierarchies. The destruction of the Egyptian Air Force shocked Egypt's political-military leadership.[52] Nasser; Field Marshal 'Amr, the Commander-in-Chief of the Armed Forces; Defense Minister Shams Badran; Air Force Commander Sidqi Mahmud; and much of Cairo's General Staff, had never expected such a devastating attack, nor thought that they would be deprived of their air force at the start of a war with Israel. In his memoirs, Anwar as-Sadat described Marshal 'Amr's reaction on the morning of the attack:

> I went into Amer's office to find him standing in the middle of the room, looking about with wandering eyes. 'Good morning,' I said, but he didn't seem to hear me. I said 'Good morning' again, but it took him a minute to return my greeting. I immediately realized something had gone wrong. I spoke to those present and was told that our air force had been completely destroyed on the ground.[53]

It would be reasonable to expect any group of commanders to have been shocked by this sudden catastrophe; however, in many ways, this was a mercurial bunch, and their reaction may have been even more extreme. 'Amr became catatonic, Nasser panicked and attempted to convince King Hussein to blame the Americans, and few others in Cairo seem to have been able to think straight for the first hours of the war.[54] In the words of Field Marshal Gamasy (then Chief-of-Staff to the Egyptian Eastern Command controlling all forces in Sinai), 'In the face of this disastrous situation, the general command and the air force command went into a state of paralysis and were unable to think or plan a response'.[55]

All of the Arab armies had extremely rigid and centralized command and control systems, and with their highest echelons overcome by shock, they had difficulty responding to the Israeli onslaught. Although the unwillingness of Egyptian division and brigade commanders in Sinai to take the initiative and act on their own greatly contributed to Egypt's problems, it is still the case that much of the general inertia afflicting Egyptian forces can be traced to the shock and panic that gripped the highest echelons of the Egyptian command as a result of the sudden destruction of their air force. The Israeli ground offensive into Sinai was conducted with such speed and power that the Egyptians had to react very quickly to every IDF move, but the high command was in no shape to make rapid, effective decisions. As a result, little guidance came from Cairo in the crucial early hours of the war. Many Egyptian units sat in place, not even acting to implement their well-defined missions to counterattack or reinforce specific sectors without explicit orders from the high command to do so.[56] The Jordanians were equally stunned to learn that the most powerful air force in the Arab world had been reduced to ashes in a few hours. According to Mutawi, Amman's realization on 6 June that the Egyptian air force had been destroyed, 'had a devastating effect on the Jordanian command', and was one of the reasons the king and his senior military commanders decided to abandon the West Bank.[57] Although the Jordanians reacted better than the Egyptians in terms of moving reserves and launching counter-attacks, it was still the case that Jordanian reactions were rarely in time to stop the flexible, quick-moving Israelis. Given how decisively the Israelis prevailed in every tactical engagement – whether it was a frontal assault against dug-in Arab defenders, or a meeting engagement with Arab armor – there is no reason to believe that if the Egyptian and Jordanian high commands had been able to react more quickly and order the counterattacks, their subordinates would not undertake on their own that this would have changed the ultimate outcome. But again, it would have changed the course of the fighting and might have bought them some more time to conduct a more orderly withdrawal.

The Importance of Israeli Air Supremacy

Arguably the most important accomplishment of the IAF was its near perfect counter-air effort. The IAF removed the Arab air forces almost completely from the military balance. As a result, Israeli ground forces were able to move and fight with almost no interference from enemy aircraft. Although Israeli ground forces were so much more competent than their Arab counterparts – and in the limited operations they were able to conduct, the Arab air forces revealed themselves to be mostly hapless – that the Israelis likely would have prevailed even if they had had to contend with Arab air interference, the war would have looked very different. Even inaccurate airstrikes that cause little damage can be disruptive of ground operations, costing time as the ground forces must disperse and then regroup after each attack. If Israeli ground forces had had to fend off Arab airstrikes they almost certainly would have moved more slowly than had been the case. Likewise, a few well-timed air attacks (even ineffective ones) could have broken up key Israeli assaults such as those on the fortified Arab positions at Abu Ageilah, Rafah, the Jiradi pass, Radar Hill, Ammunition Hill and the northern Golan. Arab airstrikes might also have impeded the IDF's logistical operations. Many Israeli armored units did run out of fuel or ammunition at different points during the course of the fighting, and these problems could only have been worse had Israeli supply columns been subjected to Arab airstrikes. Even a near miss from a 500-lb bomb can destroy a fuel or ammunition truck, and even inaccurate airstrikes can badly slow a column of thin-skinned trucks with little ability to move off-road.

Furthermore, if the IAF had had to spend its time contesting air superiority with Arab fighters rather than striking Arab army units, the various Arab retreats would have looked very different. The constant Israeli airstrikes served as a lash, driving on the Arab armies, stoking their panic and impressing on them a constant need for haste. Absent the airstrikes, the Arabs might have been able to conduct a more orderly retreat, perhaps established and scouted routes of march, deployed more rearguards, and better coordinated and prioritized the movement of units out of the theaters of war. Less Arab equipment would have been destroyed, either directly by air-delivered ordnance, or indirectly as drivers and crews wrecked equipment in their haste to flee Israeli fighter-bombers. Similarly, far less equipment would have been abandoned by crews panicked by airstrikes.

Reassessing the Role of Air Power in Israel's Victory

Overall, the IAF played a major role in the conquest of the Sinai, West Bank and Golan Heights, but was not the decisive element in the defeat of the Egyptian, Jordanian and Syrian armies. Israeli air power eliminated the Arab air forces, and so allowed the Israeli ground forces to conduct operations without fear of air attack. Israeli air power also mauled Arab units as they retreated from the frontlines, causing a fair number of the casualties the Arabs suffered in the Six-Day War. Finally, Israeli air power was an important element in preventing the Arab armies, particularly the Jordanians, from reinforcing their forward defense lines. However, the inability of entrenched Arab units – often deployed in superb defensive terrain – to hold off equal or smaller-sized Israeli units, combined with the inability of larger and usually better-armed Arab armored forces to beat Israeli armored units in maneuver battles, ultimately were the crucial factors in Israel's victory. In these contests, the IAF essentially played only a supporting role. The carnage wrought by Israeli air power during the pursuit phases of the various campaigns contributed to the strategic outcome of the war by ensuring that the Arabs would escape with little of their military power intact, but essentially was irrelevant to the defeat of the Arab armies and the Israeli conquest of Sinai, the West Bank and the Golan.

Had the IAF not been able to participate in the campaigns against Egypt, Jordan and Syria, Israeli casualties would undoubtedly have been higher and it would have taken Israel longer to secure its conquests, but the Israelis still would have won, and probably quite handily. The tactical performance of all three Arab armies was sufficiently poor that there is little reason to believe they could have held back the Israelis even without IAF participation. At Abu Ageilah,

Khan Yunis, Rafah, Jebel Libni, Latrun, Radar Hill, throughout Jerusalem and across the Golan, Arab combat units were only capable of fighting from their fixed defenses, and once the course of battle changed, they were incapable of adapting. Arab units in every theater rarely counterattacked, mostly would not redeploy or reform their lines to meet Israeli flanking operations, and generally would not maneuver in battle against Israeli tactical formations. Thus, as soon as the Israelis had penetrated the Arab infantry lines, the fight was effectively over because the Arabs would not shift their forces or counterattack to prevent the Israelis from clearing their positions.[58] Indeed, in his memoirs, General Narkiss notes that the Israeli 10th Mechanized Brigade's attack up the ridge of the Jerusalem corridor should have been suicidal, but because Jordanian resistance was so incompetent, it turned into a rout.[59]

Arab mechanized formations did little better than the infantry: Egyptian, Jordanian and Syrian armored reserves were butchered by Israeli armor at Rafah, Abu Ageilah, B'ir Lafhan, Jebel Libni, B'ir Gifgafah, the Dotan valley, Tel al-Ful, the Janin-Tubas road and on the Golan. In many of these cases, the Arab units not only outnumbered the Israelis, but also outgunned them. Thus at Tel al-Ful, the M-48 battalion of the Jordanian 60th Armored Brigade was easily defeated by a smaller force of Israeli Super Shermans with 76-mm guns (not even 105-mm guns). Indeed, the only Arab armored formation that even held its ground against the Israelis was the Jordanian 40th Armored Brigade under the command of the exceptional Rakan al-Jazi.

Given these problems, it is difficult to believe that the absence of the IAF would have somehow completely transformed the situation. There were several instances where IAF intervention had a major *and direct* impact on the course of operations on the ground, but in each of these cases, had the IAF not intervened, there is no reason to believe the Israelis would not still have won the war – just not as quickly or easily. In particular, the IAF interdicted many Jordanian attempts to reinforce Jerusalem, preventing elements of the 60th Armored Brigade and 27th Infantry Brigade, and the entire 6th Infantry Brigade, from reaching the city on 6 June. However, had those units been able to reach the city there is nothing to suggest they would have dramatically altered the outcome of the fighting. Even if they had been able to arrive around midday on 6 June (probably the earliest possible given their start times), Jebel Mukhaber, Radar Hill, Abu Tor, Tel al-Ful and Ammunition Hill/ Shaykh Jarrah all would have fallen already, the Old City would still have been isolated on three sides and the Jordanian defense of Jerusalem would already have been undermined. Nor is there reason to believe that these units would have fought better than the Jordanian forces that actually were there, and thus been able to counterattack, drive back the Israelis and reform a coherent defense of East Jerusalem. Obviously, the additional forces would have delayed Israel's advance, if only because it might have taken the IDF a day or more to destroy them, but this probably would not have made much difference in the end. Israel consistently rejected UN and superpower pressure for a ceasefire, suggesting that an extra day or two to complete the conquest of the West Bank would not have radically altered the outcome of the war. Similarly, if the Israelis had not been able to conduct airstrikes against Qabatiyah crossroads on the morning of 7 June, the Jordanian 40th Armored Brigade might have been able either to hold its position on the ridgeline or else fall back to another position farther south. However, this almost certainly would still have proven irrelevant: one capable Jordanian brigade was not going to stop the Israeli conquest of the West Bank.

Lessons for the Employment of Air Power from the Six-Day War

It is unfortunate that air power enthusiasts continue to obsess over the need to prove that air power can be a decisive instrument of war, and mischaracterize historical events to try to catch this will-o'–the-wisp. There is no longer any substantive reason to do so. The US Air Force is well established as an independent service. Over time, it is gaining an equal share of the most senior command billets (particularly as commanders-in-chief of the Unified and Specified Commands). The Congress and the services all appear content with the longstanding division of the budgetary pie and any changes are likely to be marginal. Nor is proving or disproving this theory ever likely to affect war-fighting strategies: the US military is not going to try to win a war using just one service when it can employ two, three or all four. It will employ whatever forces are necessary to do the job. Kosovo made this clear: the US started with air power and had that failed, Washington was already prepared to move on to a ground campaign. Falsely claiming that the Six-Day War proved that air power could be decisive was not going to affect a presidential decision about whether to use ground forces if air power could not bring the Serbian regime to heel, nor will it at any time in the future. However, by distorting the history of the Six-Day War, air power enthusiasts have muddied important lessons about the impact of air forces on ground operations.

The Six-Day War highlighted four key effects of air power on land warfare.

First, air forces can create severe temporary panic and delays, disrupting and demoralizing tactical forces and leaving them highly vulnerable to ground attack. This effect has tremendous immediate influence, rendering enemy ground forces vulnerable to a friendly ground assault, but fades quickly as the psychological trauma of the air attack wears off.

Second, because of the ability of air forces to concentrate for massed operations, air power can deliver stunning psychological blows to the enemy's chain of command as a whole. The key here is to find an appropriate target – one which is both vulnerable to a short, violent air campaign, and important enough so that its destruction or incapacitation will cause widespread shock. By the same token, like the psychological dislocation of tactical forces, this kind of impact is likely to wear off quickly as the enemy recovers from the initial trauma.

Third, air forces can shut down supply flows and destroy the logistical stores carried by combat forces. Over time, this will first hinder the operational movement of motorized and mechanized forces, then the fighting power of enemy maneuver units (as ammunition and fuel is depleted) and finally their morale.

Fourth, air forces can prevent operational and theater reserves from reaching the front in time and in shape to immediately execute combat operations. In some extreme cases, air power may be able to prevent entire units from reaching the front, or even cause them to retreat, by itself.

The Six-Day War also illustrated the difficulty of a fifth effect, namely the ability of air forces to directly defeat enemy ground forces by physically destroying, or rendering 'combat ineffective', entire enemy maneuver units. This is, perhaps, the most obvious lesson of the Israeli use of air power against the Arabs in 1967: the primary effects of airstrikes are psychological, not physical.[60] As the IAF demonstrated despite its own exaggerated claims, air power works its influence on ground forces principally out of its shock effect and the panic it creates. Physical damage from an airstrike may be modest, but its psychological impact will often exceed its physical impact by orders of magnitude. Historically, airstrikes have prevented far more of an enemy's combat power from engaging friendly ground forces by demoralizing them and convincing them to retreat or abandon their weapons and flee, than by actually killing men and weapons themselves. Air power typically has caused far greater paralysis in the enemy's chain of command by creating the impression of omnipresent fighter-bombers preventing all movement, than by actually destroying physical lines of communication or command nodes.[61]

Consequently, the Six-Day War experience suggests that attriting the combat power of frontline ground forces is one of the least productive missions for an air force. As the Israelis found, airstrikes are an inefficient way to kill tanks, APCs and other major weapons systems. Throughout the twentieth century it was generally the case that, to the extent ground forces were affected by airstrikes, it was usually a result of crews freezing up or abandoning their vehicles rather than actual damage, and this effect often did not last because crews eventually calmed down and returned to their vehicles when given the time to recover and the opportunity to do so. On the other hand, airstrikes can cause considerable damage to the unarmored vehicles that form the logistical tail of modern armies. Because trucks lack the armor of tanks and APCs, aircraft can inflict considerable direct damage to supply and transport companies. Moreover, because trucks also lack the off-road movement capability of tracked vehicles, airstrikes can cause an equal or greater amount of damage from accidents and breakdowns as drivers attempt to get off the road and disperse while under attack.

Today, we are witnessing the introduction of a wide range of new air-delivered munitions that promise to solve all of these problems. New 'smart' and 'brilliant' weapons such as the Joint Stand-Off Weapon (JSOW), the CBU-97/BLU-108 Sensor-Fuzed Weapon (SFW or 'Skeet'), the Brilliant Anti-Tank weapon (BAT), JDAMs and the Joint Air-to-Surface Stand-off Missile (JASSM), may greatly increase the ability of aircraft to quickly destroy enemy armored fighting vehicles.[62] However, there is precedent for skepticism. History offers numerous lessons suggesting that these new systems will take time to live up to their promise, and may never achieve a true revolution in targeting. We have repeatedly seen the claims of wondrous new technology undermined by the simple conundrums of war: atmospheric conditions, terrain masking, improved enemy air defenses, limited information and/or the ubiquitous fog of war could all confound the new wonder weapons as they have previous technologies. In the Persian Gulf War, smart munitions proved to be far more accurate than traditional 'dumb' bombs, but far less accurate than originally claimed.[63] In that conflict, US forces using earlier generation smart weapons achieved an AFV-kill-per-sortie rate of about 0.039, which is *lower* than historical norms.[64] Even in Kosovo in 1999, when some of these new weapons were available, NATO's 78-day air campaign only achieved an AFV-kill-per-sortie rate of 0.07, still slightly *lower* than the historical average, albeit nearly twice the rate of the Gulf War.[65] In fact, Allied military leaders were somewhat embarrassed by how many Serbian tanks and APCs survived the NATO air campaign.[66] Initial reports from both Operations 'Enduring Freedom' and 'Iraqi Freedom' suggest that American strikes employing this new class of weaponry are proving to be far more accurate than in the past, but as of this writing we do not yet have reliable evidence to make a judgment one way or the other.

This is not to suggest that the US should give up trying to improve the ability of aircraft to destroy enemy AFVs, only that we should be prepared for our accomplishments to be modest. If this is the case, the lessons of the Six-Day War argue that the US would do well to focus its air operations in ground warfare on delaying, demoralizing and disrupting enemy ground formations, rather than trying to destroy them outright. Perhaps the most compelling lesson of the Six-Day War is that air power can have a psychological impact on ground forces far greater than any physical damage, and physical damage to armored fighting vehicles is inherently difficult to achieve – points confirmed by our Gulf War and Kosovo experiences, and hinted at by the preliminary accounts from the invasions of Afghanistan and Iraq. Ultimately, therefore, our first objective should be working to harness the ability of air power to render enemy ground forces vulnerable to friendly ground forces, rather than having air power try to do it all alone. It is a lesson that has been proven time and again, but it is worth repeating regularly: air and ground forces working together create a combination far more deadly than either operating independently. Thus, the real lesson of Israel's stunning victory in 1967 is the exact opposite of what most of the world learned.

ACKNOWLEDGEMENTS

I would like to thank Daniel Byman, Eliot Cohen, Michael Eisenstadt, Daryl Press, Jeremy Shapiro and Brent Sterling for their comments on this article. Of course, any mistakes in fact or judgment are mine alone.

NOTES

1 Donald Neff, *Warriors for Jerusalem* (NY: Linden Press 1984) p.246. See also, Brig. Syed Ali El-Edroos, *The Hashemite Arab Army, 1908–1979* (Amman, Jordan: The Publishing Committee 1980) p.269; James Lunt, *Hussein of Jordan* (London: Macmillan 1989) pp.99–100; Samir Mutawi, *Jordan in the 1967 War* (Cambridge: Cambridge University Press 1987) pp.128, 142.

2 Emanuel Wald, *The Wald Report: The Decline of Israeli National Security Since 1967* (Boulder, CO: Westview 1992) pp.88–91, 94–7, 107–10. See also, Anthony H. Cordesman and Abraham R. Wagner, *The Lessons of Modern War: Volume I, The Arab-Israeli Conflicts, 1973–1989* (Boulder, CO: Westview 1990) pp.52–64; Ariel Sharon with David Chanoff, *Warrior: The Autobiography of Ariel Sharon* (NY: Simon and Schuster 1989) pp.303–305.

3 For a recent example, see Lt. Col. David K. Edmonds, USAF, 'In Search of High Ground: The Airpower Trinity and the Decisive Potential of Airpower', *Airpower Journal* XII/1 (Spring 1998) p.13. Air Force historian Richard Hallion has written that the Six-Day War was the first that 'came close to being a decisive victory for air power alone'. Richard Hallion, *Storm Over Iraq: Air Power and the Gulf War* (Washington, DC: Smithsonian 1992) p.2.

4 Author's interview with General Uzi Narkiss, Sept. 1996. Narkiss was the commander of Israel's Central Command during the war, the front responsible for conquering the West Bank from Jordan.

5 The American-led invasion of Iraq relied far more heavily on ground forces for much of the fighting. Indeed, US ground operations were unquestionably the decisive element of the campaign. Consequently, those who argue that air power can be decisive in its own right have generally not employed it as evidence to support this claim.

6 For a sample of the strategic air power debate, see Bernard Brodie, *Strategy in the Missile Age* (Princeton, NJ: Princeton University Press 1959); Daniel L. Byman, Matthew C. Waxman and Eric Larson, *Air Power as a Coercive Instrument* (Santa Monica, CA: RAND 1999); Mark Clodfelter, *The Limits of Air Power: The American Bombing of North Vietnam* (NY: Free Press 1989); Eliot A. Cohen, 'The Mystique of Air Power', *Foreign Affairs* 73/1 (Jan./Feb. 1994); Malcolm Cooper, *The Birth of Independent Air Power: British Air Policy in the First World War* (London: Allen and Unwin 1986); Giulio Douhet, *Command of the Air*, translated by Dino Ferrari (NY: Coward-McCann 1942); Robert Futrell, *Ideas, Concepts, Weapons: Basic Thinking in the United States Air Force*, (Maxwell Air Force Base, AL: Air University Press 1989); Gian P. Gentile, *How Effective is Strategic Bombing?: Lessons Learned from World War II to Kosovo* (NY: NYU Press 2001); David MacIsaac, 'Voices from the Central Blue: The Air Power Theorists', in Peter Paret (ed.), *Makers of Modern Strategy: From Machiavelli to the Nuclear Age* (Princeton, NJ: Princeton University Press 1986) pp.624–47; Robert A. Pape, *Bombing to Win* (Ithaca, NY: Cornell University Press 1996); Alexander P. de Seversky, *Victory Through Air Power* (NY: Simon and Schuster 1942); Michael S. Sherry, *The Rise of American Air Power: The Creation of Armageddon* (New Haven, CT: Yale University Press 1987); Air Chief Marshall Sir John Slessor, 'Air Power and World Strategy', *Foreign Affairs* (Oct. 1954) pp.43–53; John A. Warden III, 'Employing Air Power in the Twenty-First Century', in Richard H. Schulz Jr. and Robert Pfalzgraff Jr. (eds), *The Future of Air Power in the Aftermath of the Gulf War* (Maxwell Air Force Base, AL: Air University Press 1992) pp.57–82; Edward Warner, 'Douhet, Mitchell, Seversky: Theories of Air Warfare', in Edward Mead Earle (ed.), *Makers of Modern Strategy: Military Thought from Machiavelli to Hitler* (Princeton, NJ: Princeton University Press 1943) pp.485–503; Lt. Col. Barry D. Watts, USAF, *The Foundations of U.S. Air Doctrine: The Problems of War* (Maxwell Air Force Base, AL: Air University Press 1984).

7 For other work on this topic, see, e.g., M.J. Armitage and R.A. Mason, *Air Power in the Nuclear Age* (Champaign, IL: University of Illinois Press 1983); Lt. Col. Prince T. Bingham, USAF (Ret.), 'The United States Needs to Exploit its Airpower Advantage', *Airpower Journal* VII/3 (Fall 1993) pp.62–71; Edward Luttwak, 'Air Power in U.S. Military Strategy', in Schultz and Pfaltzgraff, pp.17–38 ; MacIsaac (note 6); Pape (note 6); Reina Penington, 'A Commentary: Prophets, Heretics, and Peculiar Evils', *Airpower Journal* X/2 (Summer 1996) pp.65–7; Col. Richard Szafranski, USAF, 'Interservice Rivalry in Action: The Endless Roles and Missions Refrain', *Airpower Journal* X/2 (Summer 1996) pp.48–59; W.W. Rostow, *Pre-Invasion Bombing Strategy* (Austin, TX: University of Texas Press 1981).

8 For good accounts of the Israeli counter-air operations, see Col. Eliezer Cohen, *Israel's Best Defense*, trans. by Jonathan Cordis (NY: Orion Books 1993); Col. Trevor N. Dupuy, *Elusive Victory*, third edn. (Dubuque, IA: Kendall Hunt 1992); Eric Hammel, *Six Days in June* (NY: Scribner's 1992); Chaim Herzog, *The Arab-Israeli Wars* (NY: Random House 1982); Neff (note 1); Lon Nordeen, *Fighters over Israel* (London: Greenhill Books 1991); Lon O. Nordeen and David Nicole, *Phoenix Over the Nile: A History of Egyptian Air power, 1932–1994* (Washington, DC: Smithsonian Institution Press 1996); Edgar O'Ballance, *The Third Arab-Israeli War* (Hamden, CT: Archon Books 1972); Wald (note 2); Ehud Yonay, *No Margin for Error: The Making of the Israeli Air Force* (NY: Pantheon Books 1993).

9 Cohen, *Israel's Best Defense* (note 8) pp. 193–217; Dupuy (note 8) pp.245–7; Hammel (note 8) pp.165–71; Herzog, *The Arab-Israeli Wars* (note 8) pp.151–3; Samuel M. Katz, *Soldier Spies: Israeli Military Intelligence* (Novato, CA: Presidio 1992) p.191; Neff (note 1) pp.32, 202–203; Nordeen (note 8) p.67; Nordeen and Nicole (note 8) p.212; Yonay (note 8) pp.202–13, 231–44; author's interviews with senior IAF officers, Sept. 1996.

10 Cohen, *Israel's Best Defense* (note 8) pp.220–221; Dupuy (note 8) p.247; Hammel (note 8) p.392.

11 Cohen, *Israel's Best Defense* (note 8) pp.218–20; Dupuy (note 8) p.247; Hussein of Jordan, *My 'War' with Israel*, as told to and with additional material by Vick Vance and Pierre Lauer, trans. by June P. Wilson and Walter B. Michaels (NY: William Morrow and Co. 1969) pp.66–7; Nordeen (note 8) p.67; O'Ballance, *The Third Arab-Israeli War* (note 8) p.70; Yonay (note 8) pp.250–53.

12 Nordeen (note 8) pp.77–82, 148.

13 For good accounts of the Israeli offensive against Egypt in Sinai, see Randolph S. Chuchill and Winston S. Churchill, *The Six Day War* (London: Heinemann 1967); Moshe Dayan, *Diary of the Sinai Campaign* (NY: Harper and Row 1965); Dupuy (note 8); Abdel Magid Farid, *Nasser: The Final Years* (Reading, UK: Ithaca Press 1994); Field Marshal Abdel Ghani el-Gamasy, *The October War: Memoirs of Field Marshal El-Gamasy of Egypt* trans. by Gillian Potter, Nadra Marcos and Rosette Frances (Cairo: American University in Cairo Press 1993); George Gawrych, 'The Egyptian Military Defeat of 1967', *Journal of Contemporary History* 26 (1991); George W. Gawrych, *Key to the Sinai: The Battles for Abu Ageilah in the 1956 and 1967 Arab-Israeli Wars* (Fort Leavenworth, KS: US Army CGSC 1989); Hammel (note 8); Herzog (note 8); Israel Ministry of Defense, *The Six-Day War* (Israel :Israel Press 1967); Samuel M. Katz, *Fire and Steel: Israel's 7th Armored Brigade* (NY: Pocket Books 1996); S.L.A. Marshall, *Swift Sword: The Historical Record of Israel's Victory, June 1967* ([town of publication and US state code?] US: American Heritage Publishing Co. 1967); Nordeen and Nicole (note 8); Wald (note 2); Sharon (note 2); Shabtai Teveth, *The Tanks of Tammuz* (NY: Viking Press 1968); Various authors, in Avrahaim Shapira (ed), *The Seventh Day: Soldiers Talk About the Six-Day War* (London: Andre Deutsch 1970).

14 Gawrych, 'The Egyptian Military Defeat of 1967' (note 13) p.279; Hammel (note 8) pp.44, 145; Herzog, *The Arab-Israeli Wars* (note 8) p.152; Israel MoD (note 13) p.52; Neff, *Warriors for Jerusalem* (note 1) p.193; Nordeen and Nicole (note 8) pp.193, 199–200; O'Ballance (note 8) p.99.

15 Dupuy (note 8) pp.242–4; Hammel (note 8) pp.149–61; Wald (note 2) pp.82–3.

16 For good accounts of the Israeli campaign against Jordan on the West Bank, see Churchill and Churchill (note 13); Dupuy (note 8); El-Edroos (note 1); Lt. Gen. Mordechai Gur, *The Battle for Jerusalem* (NY: Popular Library 1974); Hammel (note 8); Herzog (note 8); Lunt, *Hussein of Jordan* (note 1); Israel MoD (note 13); Robert J. Moskin, *Among Lions: The Definitive Account of the 1967 Battle for Jerusalem* (NY: Arbor House 1982); Mutawi, *Jordan in the 1967 War* (note 1); Lt. Gen. Uzi Narkiss, *The Liberation of Jerusalem* (London: Valentine Mitchell 1983); Neff (note 1); O'Ballance (note 8); Abraham Rabinovich, *The Battle for Jerusalem* (Philadelphia, PA: The Jewish Publication Society 1987); Wald (note 2); Brig. Peter Young, *The Israeli Campaign, 1967* (London: William Kimber 1967).

17 El-Edroos (note 1) pp.353–5; 373–83; Hammel (note 8) pp.149–50; 284–7; Young (note 16) p.51.

18 Dupuy (note 8) pp.284–5; El-Edroos (note 1) pp.353–5; Narkiss (note 16) p.87; O'Ballance (note 8) p.223.

19 History Branch, Israeli Air Force, correspondence with the author, 10 Sept. 1997.

20 For good accounts of the Israeli offensive against the Syrians on the Golan, see Churchill and Churchill (note 13); Dupuy (note 8); Hammel (note 8); Herzog (note 8); Israel MoD (note 13); Neff (note 1); O'Ballance (note 8); Patrick Seale, *Asad of Syria* (London: I.B. Tauris and Co. 1988); Wald (note 2).

21 Dupuy (note 8) pp.318–9; Hammel (note 8) pp.388–9; Herzog (note 8) p.186; J.M. Moreaux, 'The Syrian Army', *Defence Update* 79 (July 1986) p.38; Seale (note 20) p.117.

22 Dupuy (note 8) p.319; Hammel (note 8) pp.394–5; Nordeen (note 8) pp.83–4; author's interviews with senior IAF officers, Sept. 1996.

23 The campaign against Syria really did not begin until 9 June; however, having crushed the much larger Egyptian armed forces and the much better-trained Jordanian armed forces in just three days, Israel's ability to defeat Syria thereafter was never really in doubt. The Israelis were very confident, and the Syrians began the war demoralized. Consequently, the decisive campaigns were those against Egypt and Jordan.

24 History Branch, Israeli Air Force, correspondence with the author, 10 Sept. 1997; Historical Evaluation and Research Organization, *A Historical Analysis of the Effectiveness of Tactical Air Operations Against, and in Support of Armored Forces,* hereafter referred to as HERO, (McLean, Va: NOVA Publications 1980) p.36.

25 Author's interview with Lt. Gen. Mordechai Hod, IAF (Ret.), Sept. 1996.

26 HERO (note 24) p.89.

27 History Branch, Israeli Air Force, correspondence with the author, 10 Sept. 1997.

28 Compare Mutawi (note 1) p.133 with Gur (note 16) pp.56, 60–155; and Narkiss (note 16) pp.158–9, 164. Elsewhere, Mutawi argues that 'Most Arab and Western commentators believe that Israeli air supremacy was the most important military factor which led to the defeat of the Arabs' (Mutawi [note 1], p.128). In the footnote to this statement, Mutawi cites two Arab authors and two Western authors, and one of the authors – Col. Trevor Dupuy – makes no such claim on the pages cited by Mutawi, or anywhere else in his book. (The pages in question are Dupuy [note 8] pp.246–7.) The other source sited is the Churchills' book *The Six-Day War* (note 13) which does claim that Israeli air supremacy was vital to Israel's victory over Jordan. However, the Churchills also note that Israeli airstrikes 'were not particularly effective against the Jordanian armour', and helped mainly by interdicting Jordanian movements through the hills of the West Bank (Churchill and Churchill [note 13] pp.144–6.] It is also worth noting that the Churchills' book is among the least reliable accounts of the fighting and is prone to considerable exaggerations.

29 Lunt (note 1) pp.99–100.

30 Dupuy (note 8) p.299; Hammel (note 8) pp.315–20, 335–6; Herzog (note 8) p.176; Moskin (note 16) pp.289–90; Narkiss (note 16) pp.192–3, 200, 211–13; O'Ballance (note 8) pp.192–3; author's interviews with senior IDF officers, Sept. 1996. For a Jordanian source claiming air power decided this battle, see El-Edroos (note 1) pp.379–80.

31 For confirmation that neither the 75-mm high velocity gun on the Israeli M-51 nor the 76-mm gun on most of the Israeli Super Shermans could effectively penetrate the frontal armor of the Jordanian Pattons, see Joint Technical Coordinating Group for Munitions Effectiveness [JTCGME], *Special Report: Survey of Combat Damage to Tanks*, 3 Volumes (Washington, DC: Defense Intelligence Agency, 1 Nov. 1970) pp.105–21. Some Israeli Super Shermans mounted a French 105-mm gun, but none of these were at Tel al-Ful.

32 Dupuy (note 8) p.299; Hammel (note 8) pp.315–20, 335–6; Herzog (note 8) p.176; Moskin (note 16) pp.289–90; Narkiss (note 16) pp.192–3, 200, 211–13; O'Ballance (note 8) pp.192–3; author's interview with Lt. Gen. Uzi Narkiss, IDF (Ret.), Sept. 1996.

33 O'Ballance, *The Third Arab-Israeli War* (note 8) pp.182–3, 190, 220.

34 S.L.A. Marshall has observed that the psychological impact of airstrikes tends to fade within ten minutes of the attack unless repeatedly reinforced. S.L.A Marshall, 'The Devil and the Sea', in Donald Robinson (ed.), *Under Fire: Israel's 20-Year Struggle for Survival* (NY: W.W. Norton and Co. 1968) p.148. For similar conclusions regarding the duration of psychological trauma from airstrikes, see Ian Gooderson, 'Allied Fighter-Bombers Versus German Armour in North-West Europe 1944–1945: Myths and Realities', *Journal of Strategic Studies* 14/2 (June 1991) pp.210–31; Ian Gooderson, 'Heavy and Medium Bombers: How Successful Were They in the Tactical Close Air Support Role During World War II?', *Journal of Strategic Studies* 15/3 (Sept. 1992) pp.367–99.

35 Churchill and Churchill (note 13) p.112; Dupuy (note 8) pp.263–4; Hammel (note 8) pp.223–6; Israel MoD (note 13) pp.67–8; O'Ballance (note 8) pp.135–6; Nadav Safran, *From War to War* (New York: Pegasus 1969) pp.342, 346; Wald (note 2) p.84; and author's interviews with IDF military personnel, Jan. 1994 and Sept. 1996.

36 HERO (note 24) pp.35–9; JTCGME (note 31) esp. pp.1–18.

37 Dupuy (note 8) pp.263–4; Hammel (note 8) pp.223–6; Israel MoD (note 13) pp.67–8; O'Ballance (note 8) pp.135–6; Safran (note 35) pp.342, 346; Wald (note 2) p.84; and author's interviews with IDF military personnel, Jan. 1994 and Sept. 1996.

38 Dupuy (note 8) pp.310–11; El-Edroos (note 1) pp.386–8; Hammel (note 8) pp.369–81; Herzog (note 8) p.179; Israel MoD (note 13) p.104; Mutawi (note 1) p.137; O'Ballance (note 8) pp.203–207; author's interviews with senior IDF officers, Sept. 1996. Jordanian accounts of the battle claim that the 40th Armored Brigade was virtually obliterated by the IAF *in place* on the ridgeline, and the Israeli ground forces merely swept its remnants off the ridge after the battle was over. I have several reasons for disregarding this version. Although it is clear that the IAF was an important element of the Jordanian defeat at Qabatiyah crossroads, the speed of the final Israeli victory suggests that air power was not the only element of Jordan's defeat. The Israelis allowed only 15–30 minutes (accounts vary) for air strikes and artillery bombardment prior to their ground assault at dawn on 7 June. Even if the entire IAF had participated in the attack, it is almost inconceivable that 100–120 Jordanian armored vehicles camouflaged and dug in along a wooded, rocky ridgeline could have been destroyed by the IAF *in the dark* in just half an hour. In the best of circumstances, against exposed Jordanian columns moving during the day along the narrow roads of the Judean hills and unable to flee or hide, Israeli airstrikes appear to have achieved an armored-fighting-vehicle-kill-per-sortie rate of no better than 0.5 (and probably closer to 0.3). It is extremely unlikely that the IAF could have achieved a similar ratio against the 40th Armored Brigade in its positions at Qabatiyah crossroads. For the sake of argument, however, let us assume the IAF achieved a kill-per-sortie ratio of 0.5 and that it only killed 60 Jordanian tanks and APCs, causing the rest to flee. To accomplish this, the Israelis would have had to have flown 120 attack sorties just against Qabatiyah crossroads. Because half an hour was too short a time even for Israeli jets to conduct an attack on dug-in armor, return to base, refuel and rearm, return to the battlefield and conduct another strike, the Israelis would have had to have committed 120 aircraft to this mission. The entire IAF at the start of the war was only 207 operational aircraft, and by 7 June it was down to about 160–170. One would expect that if the Israelis had pulled nearly all of their aircraft off other missions to make a massive attack on Qabatiyah crossroads someone might have mentioned it, but none of the accounts of the Six-Day War or the histories of the IAF do, nor did any of the senior IAF officers I interviewed remember devoting the lion's share of their assets to this target. Instead, by all accounts, the air effort against Qabatiyah, while significant, did not necessarily receive more attention than the constant Israeli air effort against Jordanian forces retreating from the West Bank or those against the Egyptian army retreating from Sinai.

Moreover, the Israelis flew only 233 air-to-ground sorties against Jordan on 7 June, and again, if the majority of these had gone against Qabatiyah crossroads, I would expect this to have been mentioned in IAF accounts of the war or stuck in someone's memory. Of course, the Israelis found after the war that, in fact, they generally had achieved an armored vehicle kill-per-sortie rate of no better than 0.2 – which is also more in tune with historical norms (see HERO [note 24] pp.36–42). There is no reason to believe that the IAF strikes on the 40th Armored Brigade achieved better than this average and, given the disposition of the Jordanians (dug-in and dispersed), the terrain (wooded hills) and the time of day (before dawn), if anything, the IAF probably did worse than average here, probably closer to 0.1 armor kills-per-sortie. Consequently, to have inflicted even 25 per cent casualties on this brigade, a far cry from the Jordanian claims, would have required the Israelis to have flown roughly 300 sorties if the more accurate kill-per-sortie rate of 0.1 is used. Of course, with only 160–170 operational aircraft, it would have been physically impossible for the IAF to have generated so many sorties in half an hour, and we know for a fact that they only flew 233 air-to-ground sorties against all targets on the West Bank all day on 7 June. If the Israelis had flown every single ground-attack sortie that went against the West Bank on 7 June against Qabatiyah crossroads, they probably would have killed no more than 20–25 Jordanian armored vehicles given their actual armored vehicle kill-per-sortie rate. Assuming that the Israelis probably flew 50–75 sorties against Qabatiyah crossroads on the morning of 7 June, we could have expected them to have destroyed 5–8 Jordanian tanks and APCs, and no more than 15 tanks and APCs even if the IAF was able to achieve a 0.2 kill-per-sortie rate. This figure is entirely in keeping with the experience of other Arab armored forces under air attack during the rest of the war, and would hardly constitute the obliteration of 40th Armored Brigade by the IAF. It also accords well with the official US survey of damaged tanks after the war, which found that less than two per cent of Arab tanks destroyed during the war were destroyed by airstrikes. Consequently, it seems highly unlikely that the IAF alone could have mauled the 40th Armored Brigade given how little evidence there was of armored vehicles being destroyed by airstrikes (see JTCGME [note 31] esp. pp.1–17). As a final note, Nordeen also argues that the Israeli ground forces played the key role in defeating the 40th Armored Brigade at Qabatiyah. Nordeen's book is generally sympathetic to the IAF, and so if he claims that the IAF's role was secondary to the ground forces at Qabatiyah, there is strong reason to believe him (see Nordeen [note 8] p.79).

39 Gamasy (note 13) p.57; Hammel (note 8) p.244. Sadat claims that Nasser was told about the destruction of the Air Force around noon (p.175), while Nordeen and Nicole (note 8) state only that the EAF commanders were 'slow to report the true extent of the defeat' (p.211).

40 Author's interviews with former senior Egyptian military officers, Dec. 1997.

41 The survey team examined 203 destroyed Israeli and Arab tanks. Of these 176 were Arab tanks. Since all three Arab armies combined probably had 500–600 tanks destroyed in the fighting, this represents roughly 30 to 40 per cent of all Arab tanks destroyed. The tanks examined were all of the Arab tanks that in 1970 (the year of the US study) remained in the condition in which they had been found at the end of the war. All other tanks that had been both destroyed and captured by the Israelis had either been repaired and put into service with the Israeli army or used for target practice and thus were no longer reliable for the survey. There is no reason to believe that this sample represented anything but a random sampling: many of the tanks examined were slated to be used for target practice or refurbishment, hence it was not the case that the Israelis left the worst damaged tanks (or the least damaged) for last. See JTCGME (note 31) pp.1–16.

42 JTCGME (note 31) p.1.

43 HERO (note 24) pp.35–9, 41–2, 56, 89.

44 See for instance, Dupuy (note 8) pp.263–78, 304–305, 325–6; El-Edroos (note 1) pp.386; Gamasy (note 13) pp.64–5; Hammel (note 8) pp.246, 260–79, 374–5, 382–3, 418–24; Herzog (note 8) pp.160–65, 188; Israel MoD (note 13) pp.138–9; John Keegan, *World Armies* (London: Macmillan 1979) p.687; Moskin (note 16) p.351; Mutawi (note 1) 139–40; O'Ballance (note 8) pp.142–65, 250–58; Safran (note 35) p.342; Sharon (note 2) pp.198–203; Wald (note 2) pp.84–5; Young (note 8) p.112; author's interviews with senior IDF officers, Sept. 1996.

45 HERO (note 24) p.36. It is worth reiterating that, at least with regard to armored vehicle kills, Israeli pilot claims were two to four times greater than actual numbers as revealed in postwar assessments.

46 If IAF claims regarding destruction of trucks was even half as badly exaggerated as their claims regarding the destruction of armored fighting vehicles, then there is reason to believe the IAF may have destroyed no more than 400–500 trucks in Sinai. Of course, IAF claims may not have been as badly exaggerated with regard to thin-skinned vehicles. One reason for the discrepancies between IAF claims and actual kills of tanks and APCs was that many of the aerial munitions, particularly the 20- and 30-mm ammunition in the cannons on IAF fighter-bombers, could not penetrate tank armor. However, they could tear up a truck or jeep. Similarly, near misses from bombs or rockets were found to have had little impact on armored fighting vehicles, but could have destroyed trucks, cars and jeeps.

47 On the psychological impact of airstrikes on ground forces, see e.g. Ian Gooderson, 'Allied Fighter-Bombers Versus German Armour in North-West Europe 1944–1945: Myths and Realities', *Journal of Strategic Studies* 14/2 (June 1991) pp.210–31; Ian Gooderson, 'Heavy and Medium Bombers: How Successful Were They in the Tactical Close Air Support Role During World War II?', *Journal of Strategic Studies* 15/3 (Sept. 1992) pp.367–99; Stephen T. Hosmer, *Psychological Effects of U.S. Air Operations in Four Wars, 1941–1991: Lessons for Commanders* (Santa Monica, CA: RAND 1996); Group Capt. P.N. Lambert, RAF, 'Shattering Impact: The Psychology of Air Attack', in Richard P. Hallion (ed.), *Air Power Confronts an Unstable World* (London: Brassey's 1997) pp.83–109; Barry D. Watts and Thomas A. Keaney (eds), *Gulf War Air Power Survey, Volume II, Part 2: Effects and Effectiveness* (Washington, DC: GPO 1993) esp. pp.202–205, 221–26.

48 Dupuy claims that the head of the IAF History Branch told him that Israel flew 300 ground attack sorties against this unit (HERO [note 24] p.89). Given that the IAF flew only 549 ground-attack sorties against Jordan in the entire war (according to the IAF History Branch), it is ludicrous to claim that the IAF would have flown over half of all its ground attack sorties against Jordan on one day against one target.

49 HERO (note 24) p.89; Lunt (note 1) p.103; Narkiss (note 16) pp.242–3. IAF accounts claim that the Jordanian 2nd Regiment was destroyed in this action. However, other evidence indicates that, as with so many Israeli airstrikes, the actual damage inflicted was far less than claimed. In this case, the 60th Armored Brigade managed to retreat back to the east bank of the Jordan with 40 intact tanks on 7 June. The vast majority of these were from the 2nd Armored Regiment, which began the war with about 40 tanks, indicating that the Israeli airstrikes during the night of 6/7 June may have been terrifying, but did not destroy many Jordanian tanks (see Lunt [note 1] p.103). Again this finding brings what otherwise would have been a significant statistical outlier back in line with the conclusions of the Israeli and American postwar assessments which found that few armored vehicles were actually destroyed in airstrikes and that actual physical damage from airstrikes had been greatly exaggerated. Note that the HERO account, drawing as it does on the official IAF historian, incorrectly claims that

it was an entire Jordanian armored brigade mauled on the Edom ascent. Because by the time this event occurred, two battalions of the 60th Armored Brigade had already been smashed at Tel al-Ful, and the 40th Armored Brigade was then slugging it out at Qabatiyah crossroads, this could only have been the last battalion of the 60th Brigade, moving to reinforce Jerusalem which it was ordered to do after the defeat of the rest of its brigade earlier in the day. The eyewitness account in Narkiss also makes clear that the force was a battalion in strength, not a brigade.

50 Hammel (note 8) p.344; HERO (note 24) p.89; Narkiss (note 16) pp.242–3.

51 This was suggested in an interview with former IAF Commander, Lt. Gen. Binyamin Peled, IAF (Ret.), in Sept. 1996.

52 Nordeen and Nicole (note 8) pp.208–212.

53 Anwar el-Sadat, *In Search of Identity* (NY: Harper and Row, 1997) pp.174–5.

54 Gamasy (note 13) pp.52–60; Sadat (note 53) pp.174–5.

55 Gamasy (note 13) p.56.

56 See in particular on this point, George Gawrych, 'The Egyptian Military Defeat of 1967' (note 13) pp.277–305.

57 Mutawi (note 1) p.139.

58 For a more thorough discussion of the tactical incompetence of all three Arab armies during the Six-Day War, see Kenneth M. Pollack, *The Arabs at War: Arab Military Effectiveness, 1948–1991* (Lincoln, NB: University of Nebraska Press, 2002).

59 Narkiss (note 16) p.113.

60 For a concurring assessment, see Hosmer (note 47).

61 This is also an important lesson for military intelligence personnel and other military experts who must try to gauge the effectiveness of air attacks: if the effect of an airstrike is measured simply by the number of tanks, APCs or artillery destroyed, the analyst will greatly underestimate the actual impact of the attack.

62 For a good argument that these new weapons could radically improve US air capabilities, see David A. Ochmanek, Edward R. Harshberger, David E. Thaler and Glenn A. Kent, *To Find and Not to Yield: How Advances in Information and Firepower Can Transform Theater Warfare* (Santa Monica, CA: RAND 1998).

63 Tim Weiner, 'Pentagon Loses Fight to Keep Report on Gulf War Secret', *The New York Times*, 29 June 1997, p.20; Tim Weiner, 'Smart Arms In Gulf War Are Found Overrated: Pentagon's Reliance On High-Tech War Questioned in Review', *International Herald Tribune*, 10 July 1996, p.1. In Kosovo, precision-guided munitions achieved an accuracy of 70 per cent hitting their aim points – see William Arkin, 'Smart Bombs, Dumb Targeting?' *The Bulletin of the Atomic Scientists* 56/3 (May/June 2000) pp.46–53.

64 During Operation 'Desert Storm', Coalition air forces flew over 41,000 ground-attack sorties against Iraqi forces in the Kuwaiti Theater of Operations (KTO): Eliot A. Cohen (ed.), *The Gulf War Air Power Survey, Volume V, Part I: Statistical Compendium*, hereafter referred to as *GWAPS* (Washington, DC: GPO 1993) p.467. The reader should note that this number, while at first blush seeming outrageously high, corresponds with the same categories as flown by the Israelis in the Six-Day War. The IAF flew 2,591 interdiction and close-air support missions in 1967, and not all of these were specifically intended for tank-killing: many blew up bridges, command and control facilities, depots, etc. Thus these numbers correspond with the 41,000 listed by GWAPS for interdiction and CAS sorties in the KTO. They also correspond with the numbers used in the HERO study of airstrikes against armored forces (HERO [note 24] esp. p.59).

These 41,000 sorties eventually may have destroyed as many as 1,000 Iraqi tanks and 600 Iraqi APCs, producing an AFV-kill-per-sortie rate of 0.039. Accurate numbers of Iraqi equipment destroyed by airstrikes during the Gulf War remain elusive. We may never know precisely how many vehicles were killed in the air campaign because the Iraqis did not keep kept accurate records themselves (debriefs of unit commanders and records captured during the Gulf War are extremely spotty: some units kept very accurate counts of destroyed equipment, others did not). I arrived at these numbers in the following manner. First, according to the most accurate assessment of Iraqi strength – a CIA equipment count using U-2 imagery of the entire theater taken immediately after the war – the Iraqis had 3,475 tanks and 3,080 APCs in theater at the start of the war (E. Cohen (ed), *The Gulf War Air Power Survey, Volume II, Part I: Operations*, p.254). Second, the CIA found that of the 2,665 tanks in the 12 heavy divisions Iraq deployed to the KTO, 1,135 (43 per cent) did not move to fight or flee during the ground war. Likewise, of the 2,624 APCs in these same 12 divisions, 827 (32 per cent) did not move during the ground war. (See Central Intelligence Agency, *Operation Desert Storm: A Snapshot of the Battlefield* [Washington, DC: GPO Sept. 1993].) These 1,135 tanks and 827 APCs represent the upper end of the number of tanks and APCs in these 12 divisions that might have been destroyed by the air campaign. However, the actual number is probably considerably lower because many of these tanks did not move not because they were destroyed by airstrikes, but because they were abandoned by frightened crews or were inoperable because of maintenance problems. Postwar inspections of Iraqi equipment found that as many as 50 per cent of tanks and APCs in some units were simply abandoned in this fashion. Debriefings of Iraqi prisoners of war revealed that large numbers of vehicle crews abandoned their tanks and APCs and surrendered or fled rather than try to stop the Coalition ground offensive. Similarly, the Iraqis have always had abysmal maintenance practices and an operational readiness rate of 65 per cent is the norm in combat units. Indeed, during the Iran–Iraq War, readiness rates of around 50 per cent for tanks and APCs was commonplace in Iraqi line formations. Consequently, the actual number of tanks and APCs in the 12 heavy divisions that were destroyed by airstrikes during the Gulf War was probably no more than half of those that the CIA study found did not move during the ground war – or 568 tanks (21 per cent) and 414 APCs (16 per cent).

In addition to the tanks and APCs in the 12 armored and mechanized divisions in the KTO, the Iraqis also deployed 810 tanks and 456 APCs in independent brigades and battalions, as well as battalions attached to some of the 39 infantry divisions deployed in the KTO (CIA, *Operation Desert Storm*). The air campaign clearly destroyed some of these too. In fact, air power probably destroyed a greater percentage of these vehicles than those in the Iraqi heavy divisions. The Coalition flew more sorties and had much longer loiter times with its deadliest tank-killing aircraft – the A-10 Thunderbolts – against the frontline infantry divisions and their supporting armor in the south of the KTO, than against the heavy divisions farther north (*GWAPS, Volume II, Part I: Operations*, pp.268–82). Moreover, both the CIA study and anecdotal accounts from Iraqi personnel captured during the war also indicate that tanks and APCs attached to frontline infantry divisions suffered more than those in the heavy divisions. See CIA, *Operation Desert Storm;* Richard Hallion, *Storm over Iraq* (Washington DC: Smithsonian Institution Press, 1992) p.217; US Department of Defense (DoD), *Conduct of the Persian Gulf War* (Washington, DC: GPO 1992) pp.158–60. Consequently, it seems reasonable to believe that these units had a greater percentage of tanks and APCs killed than the heavy divisions, perhaps as much as 50 per cent tank losses and 40 per cent APC losses, which would equate to another 405 tanks and 182 APCs. (I would not go higher than this because many of these armored vehicles were deployed along the Kuwaiti and Iraqi coasts and in the northern KTO where they did not suffer the same pounding as those units along the Iraqi frontlines.)

These rough estimates produce a total number of armored vehicles probably destroyed by the Coalition air campaign of 983 tanks (28 per cent) and 596 APCs (19 per cent). As a final note, the largest US survey of Iraqi armor captured during the war found that only 10–20 per cent had been destroyed by air attack. Although this survey examined only six per cent of all Iraqi tanks destroyed during the war, and then only those in a small part of the KTO, it nonetheless indicates that while my numbers may not be precise, they are probably not off by much and, if anything, probably *overstate* the amount of physical destruction caused by airstrikes. See, United States General Accounting Office, *Operation Desert Storm: Evaluation of the Air Campaign*, Report to the Ranking Minority Member, Committee on Commerce, House of Representatives (Washington, DC: GPO June 1997).

The tank kills-per-sortie ratio for Coalition air forces in the Gulf may seem strange given the complete air superiority and sophisticated munitions of the US Air Force. However, there were several factors that mitigated against killing Iraqi tanks. First, only a small percentage of the munitions dropped on Iraqi ground forces were precision munitions. Second, the Coalition high command largely insisted that airstrikes be conducted from medium altitudes where aircraft were not vulnerable to Iraqi anti-aircraft guns or man-portable surface-to-air missiles. Third, the Iraqis were able to disperse, camouflage, berm and dig-in their armor during the coalition air campaign. Because of the extensive Iraqi passive defenses, really only precision-guided munitions were able to destroy dug-in Iraqi armor, and even for PGMs, their accuracy was considerably degraded by the need to stay at medium altitudes. See *GWAPS, Volume II, Part II: Effects and Effectiveness*, pp.202–30.

65 During Operation 'Allied Force', NATO aircraft flew 3,400 ground attack sorties against Serb ground forces in Kosovo and claim to have destroyed 246 Serb tanks and other armored fighting vehicles. See Department of Defense, *Report to Congress: Kosovo/Operation Allied Force After Action Report* (Washington, DC: DoD 2000) p.86; Ivo H. Daalder and Michael E. O'Hanlon, *Winning Ugly: NATO's War to Save Kosovo* (Washington, DC: Brookings 2000) p.154. This works out to an AFV-kill-per-sortie rate of 0.07, slightly *lower* than the historical average found by the HERO study of 0.1 to 0.3 (HERO [note 24] pp.36–42, 59). On the other hand, this figure is nearly double the 0.039 rate Coalition forces appear to have achieved during the 1991 Gulf War. This increase probably reflects the impact of the new munitions, but may also in part be attributable to the fact that KLA pressure on the Serb army forced them to concentrate forces, making them more vulnerable to airstrike. Whereas, the vast majority of Coalition airstrikes against Iraqi armor occurred before the Coalition ground offensive, and so the Iraqis were able to remain in their bermed positions (Daalder and O'Hanlon , pp.153–4, 200–202).

66 Richard Norton-Taylor, 'How the Serb Army Escaped Nato', *The Guardian*, 9 March 2000, available at < http://www.guardian.co.uk/Kosovo/Story/0,2763,193539,00.html > (accessed July 2001).

[4]

The 'Big Lie': Inventing the Myth of British–US Involvement in the 1967 War

ELIE PODEH

During the June 1967 War the Egyptian media initiated a propaganda campaign. As part of this campaign it contended that Britain and the USA were in collusion with Israel. The swift Israeli victory in the war seemingly substantiated this allegation. As a result, all Western attempts to deny the collusion story – termed the 'Big Lie' – during and after the war fell on deaf ears. Although 'Abd al-Nasir, the Egyptian president, repudiated the allegation in 1968, it had succeeded in taking root to such an extent that it turned into a myth, appearing in Egyptian and Arab historiography, as well as in school textbooks. By arguing that the collusion myth fulfilled certain functions in Egyptian society, this article traces the political and psychological processes by which a historical fabrication turns into a myth that then becomes part of the national collective memory up to the present day.

> The United States' Role: Israel was not [fighting] on its own in the [1967] war. Hundreds of volunteers, pilots and military officers with American scientific spying equipment of the most advanced type photographed the Egyptian posts for it [Israel], jammed the Egyptian defense equipment and transmitted to it the orders of the Egyptian command.

The above quote is not a propaganda item; it is taken from an Egyptian high school history textbook published in 1999.[1] Similar statements can be found in other textbooks for lower grades. Older textbooks present even more fictitious depictions.[2] Thus, for more than 30 years a fabricated narrative of the 1967 War has been transmitted by the Egyptian education system as a fact of life. This narrative first appeared on the second day of the war, when Egypt launched a massive propaganda campaign that also included reports on false victories. Cairo Radio and the Voice of the Arabs (*Sawt al-'Arab*), under the direction of Ahmad Sa'id, had played key roles in this campaign. While most of the 'news' transmitted during the war was later repudiated by the authorities and Ahmad Sa'id himself was discharged, the story of US 'collusion' with Israel became a myth, found in Egyptian historiography and history textbooks even decades later. Though educated and informed

---

Elie Podeh is senior lecturer in the Department of Islam and Middle Eastern Studies at the Hebrew University of Jerusalem, and head of the Middle East Unit at the Harry S. Truman Institute, Jerusalem, Israel.

Egyptians may not be part of it, the collusion story is part of the Egyptian and Arab collective memory. This article traces the process through which this fabrication been turned into myth.

When war broke out in the Middle East in June 1967, both the USA and Britain had two main concerns: that the crisis would not escalate into World War III, and that they would not be accused of open collaboration with any of the warring parties. The first goal was immediately achieved; on the morning of June 5 the Hot Line for direct communication between the White House and the Kremlin was used for the first time since its installation in 1963. This direct contact between the USA and the Soviet Union articulated assurances that the two superpowers would not militarily intervene in the war. British Prime Minister Harold Wilson adopted the same position in his contacts with the Soviets and the Americans.[3] With the threat of a world war eliminated, the Western powers concentrated on attaining the second goal: they were apprehensive lest the Arab countries sever diplomatic relations with them, which would lead to the imposition of an oil embargo and the closure of the Suez Canal to shipping. These fears were particularly vivid in Britain, where memories of the 1956 collusion with Israel still haunted many politicians.

Against this backdrop, it is clear why the USA and Britain reacted with such vigor to the Egyptian allegation – made on June 6 – that their forces were involved in the war, supporting the Israeli military campaign. Yet, in spite of major attempts to refute this allegation – aptly called by an American diplomat the 'Big Lie' – it stuck in the Arab world. Thus, a piece of propaganda that originated in a certain climate and as a response to a certain need found its way into the annals of Arab history.

This article, which is based primarily on declassified archival material, has several aims. First, to explore how and to what end Nasser's Egypt invented the Big Lie. Second, to describe the US–British counter-attack and the reasons for its failure. Third, to analyze the conditions facilitating the fabrication of this story and its subsequent absorption among the masses. And, finally, to expose the process through which the story has turned into a living myth in Egypt and the Arab world.

PROPAGANDA AND ITS FUNCTIONS

Scholars not only differ in their views on what constitutes 'propaganda', they also question the validity of the term because of its pejorative and dishonest connotations. Other terms – *communication*, *information* and *persuasion* – that imply no value judgment are sometimes preferred. For the sake of simplicity we shall adopt the definition of Garth Jowett and Victoria O'Donnell: 'Propaganda is the deliberate and systematic attempt to shape the perceptions, manipulate cognitions, and direct behavior to achieve a response that furthers the desired intent of the propagandist'.[4] They distinguish between three types: *white* propaganda that 'comes from a source that is identified correctly, and the information in the message tends to be accurate'; *black* propaganda that comes from 'a false source, and it spreads lies, fabrications, and deceptions'; and *gray* propaganda that stands in the middle.[5]

Jowett and O'Donnell equate black propaganda, for example, with the Nazi technique of the Big Lie espoused by Josef Goebbels. A more sophisticated type of black propaganda does not falsify facts – as these may be confronted with counter-proofs – but deals with motivation, intentions or interpretations of facts.[6] The success of this kind of propaganda depends upon the willingness of the receiver to accept the credibility of the source and the contents of the message.[7]

Propaganda is often associated with psychological warfare conducted against an enemy with an eye to shaping and manipulating his perceptions. This study is more concerned with the social and psychological functions of propaganda within the state. In this respect, propaganda has two major roles: first, it attempts to preserve order; according to Fred MacDonald, 'in authoritarian states the principal function of propaganda is to legitimize and reinforce the power of the elite; in democratic regimes propaganda functions primarily to convince and manipulate'.[8] Second, it provides both confidence and coherence; according to Jacques Ellul, propaganda furnishes the individual with 'a complete system for explaining the world, and provides immediate incentives to action'. In his opinion, propaganda 'tries to surround man by all possible routes, in the realm of feelings as well as ideas, by playing on his will or on his needs, through his unconscious and his conscious'.[9] Since propaganda has psychological functions, the use of lies (or 'black propaganda') should not be considered a complete invention. Ellul asserts that a lie may be 'an expression of a belief, of good faith – which leads to a lie regarding intentions because the belief is only a rationalization, a veil drawn deliberately over a reality one wishes not to see'.[10]

In order to maximize effect, propaganda has to be total – that is, the propagandist must utilize all the technical means at his disposal: the press, radio, television, movies, posters, computers and so on. Even diplomacy may serve as an effective channel of propaganda.[11] Though it is often transmitted in a factual manner, propaganda is also associated with the use of emotive language and presentations.[12] To ensure success, each propaganda tool has to fit the targeted audience while relating to the prevailing mood of the times (*Zeitgeist*). Jowett and O'Donnell attach importance to the knowledge and understanding of the historical background, the deeply-held beliefs, values and myths of the targeted society because 'messages have greater impact when they are in line with existing opinions, beliefs, and dispositions'.[13] Ellul adds that the propagandist builds his techniques on the knowledge of man, his tendencies, his desires, his needs, his psychic mechanisms and his conditioning.[14]

Propaganda is more likely to have influence, and ultimately succeed, in a mass society. It is a society, according to Ellul, 'with considerable population density in which local structures and organizations are weak, currents of opinion are strongly felt, men are grouped into large and influential collectives, the individual is part of these collectives, and a certain psychological unity exists'.[15] On the whole, however, Ellul's analysis places the needs of the individual at the center of his argument. For propaganda to succeed, he asserts, it must correspond to a need on the individual's part. He further elaborates in this connection:

> The secret of propaganda success or failure is this: Has it or has it not satisfied the unconscious need of the individual whom it addressed? No propaganda can have an effect unless it is needed, though the need may not be expressed as such but remain unconscious. And if we take into consideration that propaganda exists in all 'civilized' countries...this need appears to be practically universal; it is an intrinsic part of the setting in which man finds himself in the technological society.[16]

Conceivably, the impact of propaganda would be enhanced if transmitted by a leader enjoying authority and prestige among the masses because he is considered 'a sublimation of the "ordinary man"'.[17] Ultimately, the most important effect is whether the purpose of the propaganda has been achieved. This is also connected with the techniques being used. Ted Smith offers four categories in this regard: (1) falsehoods (that include lies, historical reconstruction and symbolic fiction); (2) omissions; (3) distortions (that include errors of statistical inference, multiple standards and fallacy of impossible certainty); (4) and suggestions (such as name-calling).[18]

The use of propaganda may be associated with the dissemination of conspiracy theories. In such cases, when the propagandist arranges the information according to a certain paradigm, he attaches to the enemy – or to any other opponent – the intention of effecting a collusion aiming to achieve a prohibited goal. Whether the use of a conspiracy is out of genuine conviction or sheer manipulation, the conspiracy theory exists only in the imagination of the propagandist. Daniel Pipes sees the Middle East as fertile ground for the emergence of conspiracy theories of which the 1967 Egyptian allegation of US–British–Israeli collusion is a good illustration.[19]

PROPAGANDA AND THE EGYPTIAN MEDIA

Surprisingly, in spite of academic interest in the study of propaganda, this dimension in the Arab–Israeli conflict has largely been neglected. This observation is all the more astonishing since the use of propaganda, both by the Arab states and Israel, has always been extensive. Although each party to the conflict perceived its media campaign to be a neutral transmission of information while its adversary was engaged in propaganda warfare,[20] in fact *all* parties were engaged in an ongoing propaganda war-of-words. The number of studies dealing with this issue is meager and their academic contribution is questionable.[21]

In contrast to the scant research on the role of propaganda in the Arab–Israeli conflict, there are relatively more studies on the Egyptian media as an instrument of its foreign policy. It was the revolutionary regime, which in the 1950s transformed the Egyptian State Broadcasting (ESB) from a modest station to the strongest corporation in the Middle East with 26 Arab and foreign services and a total broadcasting time of 500 hours per week.[22] The most important and popular service was *Sawt al-'Arab* (Voice of the Arabs), which started on July 4, 1953 with a modest half-hour daily broadcast, eventually becoming a 24-hour-a-day service. Under the guidance of Ahmad Sa'id, who also served as the station's main propagator, the Voice of the Arabs became the most popular radio service in the Arab world. According to Douglas Boyd, a leading authority in the field of communication, *Sawt al-'Arab* 'was the first major propaganda radio station in the Middle East to have an impact on listeners. It provided the model for others'.[23] Acknowledging its influence, Charles Issawi was highly critical of the station, which, in his words, had no rivals in the Arab world in terms of its 'venom, vulgarity, and indifference to truth'.[24] The so-called 'Ahmad Sa'id era' ended with the 1967 War.

Under the control of the state, Nasser's regime used the radio as an instrument to further its interests in Egypt and around the Arab world. There are indications that Nasser himself, seeing the station 'as a personal mouthpiece and a key weapon in his revolution', was involved in the selection of the content of the transmissions.[25] The fact that the number of radio sets in Egypt had sharply risen in the 1950s meant that large audiences were now exposed to these broadcasts.[26] Moreover, in a highly illiterate society (around 70 percent of the population), it is little wonder that the radio had become such a useful device for transmitting and receiving information. Nasser, in one of his speeches, referred to the radio as the nation's 'general university'.[27] The fact that Egypt was considered to be the cultural center of the Arab world enabled the regime to turn the radio into an important tool in its education and foreign policies.[28]

Interestingly, although some of the Egyptian propaganda was transmitted in Hebrew and directed at Israel, the target audience of most broadcasts was the Arab masses. Its major aim was not to influence enemy (Israeli) thinking, as has often been the case with propaganda in international conflicts. Rather, by reaching as many people as possible, this propaganda was aimed at strengthening Nasser's domestic legitimacy and enhancing his prestige in the Arab world. The Egyptian collusion allegation of 1967 should also be analyzed in this context.

THE EGYPTIAN ALLEGATION AND ITS IMPACT

In the early hours of June 6, the Egyptian media began to spread the allegation that both the USA and Britain were taking part in the Israeli aggression. According to Radio Cairo, US and British aircraft carriers provided an air umbrella for Israel and played an active role in the operations. In the following days the Egyptian media constantly repeated the same argument in a variety of ways. Thus, for example, on the afternoon of June 6, Radio Cairo reported that British Canberra bombers had taken part in air strikes against Egyptian posts in Sinai, and that US aircraft had left the US airbase in Libya for Israel. The next day, Egypt's leading journal *al-Ahram*, quoted by Radio Cairo, reported that British and American pilots were flying Israeli planes under the guise of volunteers and that Israeli pilots used aerial photographs taken by US spy aircraft. The Egyptian propaganda machinery then let launch a barrage of reports on the Anglo–American aggression with Israel.[29] The aim of this sustained campaign was to establish the thesis that the current war, much like the 1956 war, was the result of collusion between Israel and its Western allies, only that now the USA had joined Britain instead of France and US President Johnson had replaced Eden as the main culprit. It should be noted the 'collusion' thesis constituted the lion's share of the Egyptian propaganda campaign conducted during and after the war.

The Egyptian allegation, whether as a result of prior coordination or simply as an act of solidarity, was strengthened by broadcasts from Jordan and Syria – the two other parties to the fighting. In the early morning of 6 June, Radio Amman broadcast that 'foreign forces...effectively support Israel', and that 'there are three aircraft carriers close inshore from which aircraft take off to bomb our forces'. The radio quoted a statement by the director of military intelligence to the effect that the radar in Ajlun detected three aircraft carriers located between 20 and 80 kilometers off Tel Aviv. According to the report, a group of 16 and later a group of 12 aircraft had flown from the carriers to the Israeli airfield at Ramat David on 5 June.[30]

Syrian broadcasts also participated in the propaganda warfare, repeating the same information transmitted by Cairo and Amman. Radio Damascus added that a captured Israeli pilot had admitted that 17 British Vulcan bombers, with a full complement of equipment and pilots, had arrived in Israel ten days before the outbreak of the war. These aircraft, in addition to other bombers coming from British bases in Cyprus, allegedly attacked Syria and Egypt on 5 June. The radio further reported that at the end of May some 3,000 British troops had arrived in Israel from Cyprus and British aircraft carriers had moved to an Israeli port on 6 June.[31]

The collusion theme was immediately picked up by broadcast stations in other Arab countries, which were not participating in the war, such as Algeria, Iraq, Kuwait and Sudan. The volume of the Egyptian broadcasts, as well as the spread of the message, succeeded in turning the US–British–Israeli collusion into a 'fact' in the Arab world within a short period of time. Still, the success of the Egyptian propaganda should not be attributed only to its efficiency but also to the credibility of the message: in light of the historical lessons of 1956, it was 'reasonable' to assume that in 1967 another collusion occurred. Belief in the collusion theme had also a psychological function for both decisions-makers and the masses, providing them with a clue to understanding the astonishing and unexplainable fact that little Israel had defeated three militarily superior Arab states. In addition, the Arab masses, which had been exposed – through historiography, school textbooks and other socialization instruments – to a master collective memory that attributes many of the calamities in the Arab world to Western imperialism, found no reason to doubt the occurrence of another collusion.

The credibility of the Egyptian allegation in Arab eyes had an immediate political impact: following Egypt, Syria, Iraq, Algeria, Sudan and the Republic of Yemen severed diplomatic relations with both the USA and Britain, while Lebanon recalled its ambassadors.[32] In addition, Arab oil-producing countries announced either an oil embargo on the USA and Britain or suspended oil exports altogether. Syria, for its part, closed the pipelines crossing its territory from Iraq and Saudi Arabia. Radio Cairo and Damascus even called on Arab workers in the Arabian Peninsula and the Gulf to sabotage the oil pipelines. In order to damage Western economic interests, Egypt decided to close the Suez Canal to shipping on June 6.[33] In addition, demonstrators vandalized American and British institutions in several Arab countries.[34] With all these measures, US and British fears were realized. In contrast to 1956, however, this time it was with no good reason.

On June 8, as Egyptian reports from the battlefield proved to be largely fabricated, the propaganda campaign regarding the 'collusion' story was intensified. Thus, the fact that Israel mistakenly attacked the US *Liberty*, a radio ship of the American Sixth Fleet near the shores of Sinai, served as 'proof' that US forces – in contrast to US denials – were located near the battlefield and therefore took part in the war. It was claimed that the ship was engaged in monitoring Egyptian operational messages and relaying them to Israel.[35]

Indirectly, the Soviet Union played an important role in disseminating the Egyptian allegation by adopting a duplicitous strategy. On the one hand, as a party to the non-intervention understanding with the USA (see above), the Soviet media in Russian, Arabic and other foreign languages avoided any reference to active Western military involvement. On the other, it quoted the reports from Radio Cairo and Damascus on the 'collusion', stating that all the evidence proved that the USA and Britain 'are playing a leading role in the aggression which Israel has launched'.[36] This double-sided policy was employed even though the Soviet Union was aware that the collusion story was a fabrication. This is evident from the fact that the Soviet foreign ministry notified Arab ambassadors in Moscow in the early days of the war that the story of Western collusion had no foundations.[37] Undoubtedly, Arabs listening to Soviet radio could be led to believe that some kind of collusion did exist. This ambiguous Soviet attitude would continue also after the cessation of hostilities.[38]

The credibility of the collusion theme was further strengthened when President Nasser himself referred to it in his 'resignation speech' on June 9, when he said:

(1) The enemy, whom we were expecting from the east and north, came from the west – a fact that clearly showed that facilities exceeding his own capacity and his calculated strength had been made available to him.
(2) The enemy covered in one go all military and civilian airfields in the UAR. This means that he was relying on some force other than his own normal strength to protect his skies against any retaliation action from our side. The enemy was also leaving other Arab fronts to be tackled with outside assistance...
(3) There is clear evidence of imperialist collusion with the enemy – an imperialist collusion, trying to benefit from the lesson of the open

> collusion of 1956... What is now established is that American and British aircraft carriers were off the shores of the enemy helping his war effort. Also, British aircraft raided, in broad daylight, positions of the Syrian and Egyptian fronts, in addition to operations by a number of American aircraft reconnoitering some of our positions... Indeed, it can be said without exaggeration that the enemy was operating with an air force three times stronger than his normal force.[39]

With the cessation of hostilities, Egypt's propaganda campaign continued. While most of the information provided during the war was proven to be falsified and fabricated (such as that relating to alleged Egyptian–Arab victories on the battlefield and the huge Israeli losses), the collusion theme had been magnified to unparalleled proportions. Reports on the nature of the collusion were broadcast by local radio stations and by Radio Cairo,[40] as well as appearing in reports and bulletins by Egyptian embassies throughout the Middle East. Indeed, the Egyptian foreign ministry played an important role in disseminating this information. Though details sometimes varied, the narrative in these reports was largely identical.

A typical report, published by the Egyptian embassy in Baghdad on June 11, stated that the 'new tripartite plot' against the Arab states was prepared in advance by the USA, Britain and Israel and drafted during Moshe Dayan's visit to South Vietnam in June 1966. The plot included a secret shipment of American and British arms to Israel and a promise by the Sixth Fleet to protect it. The statement then provided a day-by-day account of the plot, beginning on May 21 and leading to the outbreak of hostilities on June 5. By offering a twisted interpretation of various diplomatic and BBC statements, the document seemingly 'substantiated' the foundations of the collusion. It also misinterpreted various innocent diplomatic maneuvers as calculated steps on the road to collusion. To show that the war was 'an Anglo–American–Israeli attack', the document repeated all the known 'facts' broadcast during the war, including the report of the Jordanian radar, the 'confessions' of Israeli prisoners in Damascus and other pieces of 'intelligence' information about the movements of US–British forces. Finally, the document attempted to refute US–British claims that Egypt had initiated the war.[41] Here, the Egyptian allegation rested on a more solid basis since Israel actually fired the first shot (though it can fairly be claimed that Egyptian moves prior to the war constituted the first act of aggression). Consequently, it can be concluded that the soundness of the last Egyptian allegation made the whole Egyptian document appear reasonable, at least in the eyes of Arabs.

While diplomatic activity was underway, both in the international and the Arab arenas, the collusion story remained a major topic in the Egyptian media campaign. Interestingly, however, most of the Egyptian attacks were directed against the USA, with Britain treated only as an addendum. A new, religious tone appeared in the propaganda to the effect that the USA, by supporting Israel in the war, enabled it to occupy Jerusalem and thus 'sully the birthplace of Christ, the Via Dolorosa and the Church of the Holy Sepulchre'.[42] But perhaps more influential were the articles of Muhammad Heikal, Nasser's close advisor and editor of the influential *al-Ahram*. In a series of weekly articles that were simultaneously broadcast on Radio Cairo, Heikal attempted to uncover the secrets of the war. With his eloquent language he presented a blend of facts, documents and interpretations that could not but win the confidence of his readers. Most of the evidence was circumstantial at best and in many cases it was a twisted interpretation of the events. Yet, Heikal's conclusion was clear-cut: there was a secret US–Israeli collusion against Syria and Egypt. Though Britain's role was somewhat downplayed, Heikal thought it was party to the collusion as well.[43]

The collusion story was not exclusively Egyptian; it soon turned into accepted Arab rhetoric. When the leaders of Egypt, Syria, Iraq, Algeria and Sudan met in Cairo in mid-July to discuss the situation, their final communiqué referred, *inter alia*, to the 'aggression committed by Israel in which the US and British imperialist forces took part against the Arab nation'.[44]

Although no reliable statistics exist, it is reasonable to assume that the intensive Egyptian media campaign, coupled with the broadcasts of other Arab stations, affected many in the Arab world who believed that 1967 was another 'Suez' with some minor change of actors. The fact that these actors categorically denied any involvement in the collusion (see below) did not necessarily make an impressive impact. The thinking was, as a British diplomat cleverly observed, that 'just as it has taken ten years for the truth about the UK–Israel collusion at Suez to come out, so in ten years the British Prime Minister of 1967 will be shown to have been in collusion with Israel'.[45] In spite of this pessimistic forecast, the USA and Britain were not willing to surrender without a fight: after all, this time, in contrast to 1956, the collusion thesis was a Big Lie.

THE WESTERN COUNTER-ATTACK

It should be emphasized that both the USA and Britain intuitively understood the adverse potential of the Egyptian allegation. Well aware of the lessons of the 1956 collusion, the decision-makers in both countries were anxious to avoid its repetition. Knowing that the collusion story was a calumny, they were determined to fight it. But the problem was that in the light of Arab experience, a mere denial was not sufficient; a firmer response was needed but the Western powers did not fully grasp it.

On June 6, when the magnitude of the Egyptian propaganda campaign was revealed, Britain and the USA issued clear denials. The British, for example, stated that:

> Her Majesty's Government are shocked by reports emanating from the Middle East... that planes from a British aircraft carrier have been involved in the fighting. This is a malicious fabrication. There is not a grain of truth in it. It is the policy of Her Majesty's Government to avoid taking sides in this conflict and to do everything they can to bring about a cease fire as soon as possible.[46]

This statement was followed by another denial by Foreign Secretary George Brown and Prime Minister Harold Wilson in the House of Commons. At the same time, Brown sent a personal letter to all Arab ambassadors in London, in which he ridiculed Cairo's allegations and indicated that the only two British aircraft carriers in the area were located at Malta and Aden.[47] Furthermore, American and British heads of missions in the UN, Arthur Goldberg and Lord Caradon, respectively, delivered strong denials at the Security Council, describing Arab allegations as a 'complete lie'. Letters in this spirit were also sent to the President of the Security Council.[48] Both delegations, at the Americans' initiative, offered to arrange a UN investigation into the charges of US–British involvement in the war.[49]

Sensitive to the possible ramifications of the collusion story, the State Department nominated an official, Chet Cooper, to be responsible for countering Arab charges. On June 7, he met with the British foreign minister in Washington to discuss ideas on how to 'nail the big lie'. Several ideas were raised in this meeting concerning a UN investigation:

(1) That a statement be issued by Johnson and Wilson on this issue.
(2) That there be pressure on the UN to lead open the investigation.
(3) That a third party, probably Canada, propose that it be handled by members represented in UNEF.

In addition, Cooper suggested that once a ceasefire was observed an attempt would be made to 'find a prominent Arab willing to expose Egyptian mendacity for what it was'. He also offered to increase publicity about the Egyptian use of poison gas in Yemen in order to discredit Nasser in Arab eyes. Finally, it was agreed to see if it was possible to step up the transmission hours of the Arabic broadcasts of the Voice of America and the BBC.[50]

This conversation, taking place only a day after Egypt's initiation of the 'collusion campaign', served as an indication of the gravity with which both the USA and Britain viewed the issue. Yet, it seems that they still thought that they would be able to arrest the tide if proper measures were taken. The Big Lie was seen primarily as an alibi for Nasser for his defeat; it was not seen as a psychological mechanism for satisfying some basic needs of society.[51] Therefore, the British thought that a statement by the prime minister at that stage would be to 'over-egg the cake'.[52] Gradually, however, the USA and Britain realized that the Egyptian–Arab allegations were gaining ground in the Arab world in spite of their efforts to counter them. Thus, Lord Caradon, the British representative at the UN, bitterly commented on June 8: 'The Arabs [simply] do not want to believe our denials'.[53]

Indeed, reports from Arab capitals indicated that the collusion story was warmly embraced by both leaders and the masses. Attempts by the USA and by British diplomats to refute it with logic and reason were usually brushed aside. In Baghdad, an official British démarche was received with disbelief. According to the Iraqi foreign minister, it was impossible to ignore the assurances received from Jordan (he was referring to the report of the Jordanian radar, see above), as well as the statements of Egypt's Foreign Minister Mahmud Riad concerning Western involvement in the war.[54] The surprising Iraqi response led the British ambassador in Baghdad to conclude with a measure of prophecy that 'it is likely that the story of Anglo-American intervention in the hostilities...will become a myth'.[55]

The Iraqi response was typical; other Arab capitals responded similarly. Even long-time allies of the West, such as Jordan, Kuwait, Lebanon and Saudi Arabia seemed convinced. It is possible that some Arab leaders did not believe in the collusion story, but they could not swim against the tide. The Lebanese, for example, told the British that they had to demonstrate their solidarity with the Arabs and therefore any attempt to dissuade them would not be in Western interests.[56] Particularly hurting was the attitude of Jordan's King Husayn, who frequently relied on Western support for his survival but now was demanding an impartial UN investigation on reports that British Hawker Hunters and Canberras participated in the fighting.[57]

With many Arab states either severing relations or announcing an oil embargo, it became clear to both the USA and Britain that the effects of the Egyptian collusion story were more severe than initially expected. Though the main problem lay in the realm of psychology, the USA and Britain continued playing on logic and reason. They believed that if accurate and reliable information were supplied, the Egyptian lie would be openly exposed. This line of thinking dictated the nature of their operations. On June 8, with timing that might hint at some prior coordination, the Israel Defense Forces (IDF) made public a recording of a telephone conversation between Nasser and Husayn, monitored in Israel at 4.50am on June 6. Considered to be authentic, the recording proved that the collusion idea had been concocted between the two leaders.[58] The British also attempted to refute Jordan's allegation that British aircraft had been detected by Jordan's radar. Following an investigation, the Ministry of Defence came to the conclusion that no Jordanian or any other Arab radar was capable of detecting British aircraft. These findings were delivered to Jordan.[59] In addition, both the USA and Britain continued their diplomatic activity under the auspices of the UN aiming to erode the credibility of the Egyptian story.

Since most Western diplomatic activity was focused on achieving a ceasefire, it was clear that the struggle against Egyptian propaganda became a secondary matter. But when the war was over, as both the USA and Britain were in a position to tackle the issue more forcefully, it soon became evident that the Egyptian allegation had turned into a fact of life in the Arab world. An excellent report by the British representative in Jedda was a typical assessment of the situation in Arab capitals:

> President's Nasser allegation ...is firmly believed by almost the whole Arab population here who listen to the radio or read the press ... Our broadcast denials are little heard and just not believed. The denials we have issued to the broadcasting service and press have not been published. Even highly educated persons basically friendly to us seem

> convinced that the allegations are true. Senior foreign ministry officials who received my formal written and oral denials profess to believe them but nevertheless appear skeptical. I consider that this allegation has seriously damaged our reputation in the Arab world more than anything else and has caused a wave of suspicion or feeling against us which will persist in some underlying form for the foreseeable future ... Further denials or attempts at local publicity by us will not dispel this belief and may now only exacerbate local feeling since the Arabs are understandably sensitive to their defeat with a sense of humiliation and resent self justification by us who in their eyes helped their enemy to bring this about.[60]

The fact that such was the atmosphere in a country that was remote from the scene of the battle and an enemy of Egypt since 1962 (as a result of the Yemen War) meant that the collusion story had indeed taken hold in the Arab world. This report is even more telling in the light of the fact that the Saudi king himself refuted the Egyptian allegation. In a reception for the heads of diplomatic missions in Jedda, the Egyptian ambassador was at pains to persuade the participants of the existence of an Anglo–American scheme to support Israel in the war. The king, however, in front of foreign representatives, at length and in detail, ridiculed the ambassador's claims.[61] From a Western point of view, Faysal's behavior was encouraging but it was hardly surprising in light of his long-time enmity with Nasser. It meant, therefore, that the existence of some Arab leaders unwilling to succumb to Egyptian propaganda did not change the overall picture of general Arab belief in the collusion story. The story suddenly received a boost from an unexpected source on June 13, when Louis Heren reported in *The Times* on growing belief among Washington diplomats of US collusion with Israel. Such a report in a respected British journal at this critical moment could only add credibility to the Egyptian allegation. Naturally, the Arab press was quick to pick up Heren's article as final proof of collusion.[62] No wonder that the British ambassador in Washington desperately concluded the Heren affair thus: 'The harm is already done and there can be no hope of a retraction'.[63]

The Foreign Office and the State Department did not necessarily share this view. After succeeding in enlisting Saudi and Turkish support in their campaign to counter the Big Lie,[64] they attached great importance to the position of their old ally in the Arab world – King Husayn. The fact that in two interviews with Western agencies Husayn stated that Jordan had 'no physical evidence or definitive proof that there was any intervention by Britain or the United States'[65] served as an indication that he, too, had retreated from his previous negative position. These positive signs coincided with some reports from Middle East posts recommending that the USA and Britain should stop attacking the Big Lie and, instead, speak of a mistake which must be corrected so as to give the Arabs a 'way out'. Indeed, the Foreign Office now believed that a switch of emphasis was needed and therefore instructed its posts to convey the opinion that 'genuine mistakes of radar interpretation were made in

Jordan in the tense atmosphere of June 5. It is also understandable that other Arab countries were misled by the continued accusations by Cairo Radio'.[66] The adoption of this line in the face of the ongoing Egyptian propaganda campaign reflected the US and British inability to grasp the deep causes of the Egyptian success.

At the same time, both the USA and Britain renewed their call for opening an immediate UN investigation of Egyptian charges. Britain expressed readiness for UN officials to visit its aircraft carriers, as well as its airfields in Cyprus and Malta.[67] As discussions at the UN revealed that the chances for initiating such an inquiry were slim, the British searched for other alternatives. In their quest for a neutral body they approached the Swedish Institute of Peace Research, but it was hesitant to carry out such an investigation.[68]

Still eager to prove their non-complicity in the collusion, the Labour government, in response to demands from the opposition, decided on an unusual step. The Ministry of Defence was asked to deposit in the library of the House of Commons photographic copies of the flying logs and daily routine orders for the dates June 5–7 of the Royal Air Force and the Royal Navy stations within a 1,000-mile radius of a circle including Gaza, Alexandria and Cairo. The aim of this step was to demonstrate once and for all the falsity of the Egyptian charges.[69] This was done by the end of June. The British even entertained the idea of holding a similar 'exhibition' at the UN but eventually decided against it.

Important as they were, these steps were mainly intended to placate local opposition[70] and could in no way affect Arab public opinion. A good illustration of Western inability to change Arab opinion was connected with a press conference held by Rear Admiral Ashmore, Deputy Commander of HMS *Hermes,* in Singapore on June 22. He indicated that the *Hermes* was in the harbor from before the start of hostilities until June 6 and then operated within 30 miles of Aden. He also explained in detail why jet strike aircraft could not have been involved in the fighting.[71] The press conference was part of the campaign to refute Egyptian allegations, but Egyptian propaganda offered an interpretation supporting the collusion story. The broadcast on the Voice of the Arabs that day reported that Admiral Ashmore admitted that flights had taken off from the British carrier *Hermes*, thus confirming once more that Britain participated in the plot.[72]

By the end of June, the British and the Americans had to admit that their propaganda counter-campaign had succeeded only in persuading the already-persuaded and perhaps some more Western-oriented elements. The bulk of the population in the Arab world remained convinced that a US–British–Israeli collusion, in some way or another, did indeed take place. A British guidance telegram to Middle East posts concluded: 'The Arabs' reluctance to disbelieve all versions of the big lie springs in part from a need to believe that the Israelis could not have defeated them so thoroughly without outside assistance'.[73] Though most probably true, this explanation does not attempt to provide a more comprehensive response to the issue.

The reverberations of the collusion story affected British interests in both Libya and Cyprus. In the former, Britain received a formal request to evacuate land forces from its base at Benghazi (though not from the base at Tobruk).[74] In the latter, President Makarios demanded Britain evacuate its bases because British planes based in Cyprus attacked the Arab states.[75]

Yet the collusion allegation was not an obstacle to the resumption of diplomatic relations with the Arab countries that severed them during the war. Even relations with Egypt were resumed in December 1967 and the able British Ambassador Harold Beeley returned to his post in Cairo.[76] The oil embargo, too, was lifted shortly after the announcement of a ceasefire, and Egypt's propaganda against Britain was substantially reduced.

In contrast, however, Egyptian attacks against the USA barely abated, as reflected in Heikal's articles in *al-Ahram*. The USA, for its part, insisted that Nasser retract the 'US air-cover story' as a condition for resuming diplomatic relations.[77] Indeed, in March 1968 Nasser retracted his charge that the USA had given air support to Israel during the war, implying that the accusations resulted from a misunderstanding based on suspicion and faulty information. Heikal, whose writings had reflected official policy, admitted in June that charges of US–Israeli 'collusion' were unfounded. He laid the responsibility squarely on the purged military commanders who used the story as an excuse for their defeat.[78] US–Egyptian relations were resumed only in February 1974. But by now the story had turned into myth.

THE MAKING OF A MYTH

The literature on the role of myths and symbols in society is abundant. For the purpose of this article we may follow Henry Tudor's definition:

> It is a story told with a view to promoting some practical purpose, and it is successful only insofar as it is believed to be a true story. This does not, of course, exclude the possibility that, in advancing a mythical argument, the mythmaker either deceives himself or deliberately sets out to deceive his audience. A myth may well be believed to be true by those among whom it circulates; but this does not prevent it from being, in actual fact, a set of extravagant illusions.[79]

Like propaganda, myths have institutional/structural and psychological functions: in the former, they serve as effective tools for indoctrinating and mobilizing the masses with the aim of legitimizing (or strengthening the legitimacy) of the existing authority and order. In the latter, they interpret reality (usually a changing reality) in a way that dovetails with existing beliefs and norms of behavior of the common citizen. In other words, myths and mythmaking, like propaganda, are products of societal needs and they aim to alleviate the anxiety created by uncertainty and insecurity.[80] To be sure, propaganda is associated with mythmaking. According to Ellul:

> Through the myth it creates, propaganda imposes a complete range of intuitive knowledge, susceptible of only one interpretation, unique and

> one-sided, and precluding any divergence. This myth becomes so powerful that it invades every area of consciousness, leaving no faculty or motivation intact. It stimulates in the individual a feeling of exclusiveness, and produces a biased attitude. The myth has such motive force that, once accepted, it controls the whole of the individual, who becomes immune to any other influence.[81]

The process by which a piece of propaganda is transformed into a myth is highly complex and not easily traced. Arguably, the role of the media, education system, literature and historiography may be considered paramount in this process. Though a systematic analysis of mythmaking according to those criteria is beyond the scope of this article, what will follow is an attempt to illuminate some of the signposts along the process that led to the emergence of the 1967 collusion myth.[82]

A salient characteristic of President Nasser was his ability to mobilize the masses, which was explained in terms of his charisma. This charisma was associated, *inter alia*, with a remarkable mythmaking ability, which meant that Nasser's performance-message often fitted the crisis situation, the cultural ethos of society and the deeply-felt needs of the individual.[83] When Nasser began spreading the collusion story he was not engaged in mythmaking but attempting to offer an explanation to a chaotic situation. Whether the story was a deliberate attempt to manipulate gullible masses or emanated from a sincere belief cannot be established with certainty. Daniel Pipes tells us that 'the larger and more pervasive an alleged conspiracy, the more likely it is put forward sincerely'.[84] Such an assertion suggests that Nasser's belief in the collusion may have been sincere. Moreover, the fact that Nasser came to power through conspiracy perhaps led him to perceive others' activity through the prism of conspiratorial behavior.[85] Since Nasser did not leave behind any personal memoirs, we should rely on his associates in order to understand his motives.

The major testimony, as expected, comes from Nasser's close advisor and *al-Ahram* editor, Muhammad Heikal, who stated that Nasser 'genuinely believed' the collusion story.[86] Yet Heikal's close association with Nasser may stain the credibility of his version; therefore, we cannot establish with certainty whether the president did sincerely believe the collusion story, which had its own logic, or whether it was part of a wider propaganda campaign. What we can conclude, is that when Heikal was making this statement, ten years after the repudiation of the collusion story, *he* was, in fact, engaged in mythmaking. In his 1990 book on the war, Heikal continued this trend, providing more details on the 'collusion' between Israel and the Johnson administration on the basis of a twisted interpretation of selected US archival material.[87] Thus, the importance of his testimony derives from his knowledge and close involvement with Nasser and his work as a respected historian of the Arab–Israeli conflict. His credibility, therefore, has not often been questioned.[88]

Even more forceful testimony appears in the memoirs of Mahmud Riad, the Egyptian foreign minister during the war. He claims that the USA had played a major role in the 1967 aggression in three forms: first, it provided Israel with intelligence information on Egyptian moves. Second, President Johnson deceived Nasser by pressing him not to make the first move while giving a 'green light' to Israel to strike a preemptive blow. Riad arrived at this conclusion on the basis of his interpretation of the diplomatic contacts between Nasser and Johnson prior to the war. Third, the vessel *Liberty* was sent to monitor and jam Egyptian communications. To emphasize the US role, Riad uses the term 'perfidy' (*'amaliyyat khida'*).[89] However, Richard Parker, who served as a diplomat in the US embassy in Cairo during the war, refuted all these allegations, based on his experience and academic research.[90]

The conspiratorial nature of US–Israeli relations was also apparent in the testimony of Salah Bassiouny and Tahsin Basheer, who played minor roles in the Egyptian foreign ministry during the war. In a conference held on the twenty-fifth anniversary of the war, both expressed some cautious belief in the existence of some form of US–Israeli conspiracy.[91] In should be emphasized, however, that the collusion story had not pervaded all Egyptian historiography of the war.[92]

But perhaps the most important role in entrenching the collusion story in the Egyptian collective memory should be attributed to the education system and its textbooks. All post-1967 history school textbooks repeated the thesis that Israel had launched the war, aiming to establish Greater Israel from the Nile to the Euphrates, with the support of Britain and the USA. The narrative also established a direct link between the 1967 War and former imperialist attempts to control the Arab world, thus portraying Israel as an imperialist stooge.[93] This narrative, as indicated in the opening quote of the article, is still found in current history textbooks. The appearance of this fabricated story, with only minor variations, in all history school textbooks means that all Egyptian schoolchildren have been exposed to, and inculcated by, the collusion story. Since school textbooks epistemologically are considered authoritative sources of information, their influence cannot be overstated. Undoubtedly, the repetition of the story in all history textbooks transformed it into a living myth.[94]

Support for the Egyptian collusion thesis came again from an unexpected direction in 1984. Based on an unidentified source participating in the operation, an American journalist, Stephen Green, claimed that several US aircraft flew daily reconnaissance missions for Israel over Egypt during the war. In fact, Green made a more serious accusation: based on a selective reading of several US messages, he came to the conclusion that the USA had sought to resolve the dispute peacefully while covertly 'unleashing' Israel against Egypt.[95] No evidence has been found to support Green's allegation. Parker claimed in response that 'it is difficult to believe that an operation...would not have left a trail among the bureaucrats and that some of the details would not have leaked authoritatively by now'. And, since the book is generally marked by carelessness, Parker concludes, this information should be treated with reserve.[96]

Among the Arab masses the collusion story was, as we have seen, wholeheartedly embraced. Three reasons may account for this. First, its frequent repetition and transmission via various communication channels (radio, press, television, diplomacy, and later the education system and historiography) made the story seem truthful. A foreign diplomat once remarked in a different context: 'Tell a lie long enough and often enough and inevitably you start believing it'.[97] This well fits the Egyptian case study.

The second reason was that the ground was well prepared for the acceptance of the collusion theory. During the first half of 1967, following an anti-US speech by Nasser on February 22, Heikal published a series of eight articles in *al-Ahram*, entitled 'We and the US'. He blamed the USA for adopting a neo-colonialist policy in the Middle East, involving 'the economic and psychological warfare, the hatching of plots and assassinations, and a basic and fundamental reliance on secret activities'. The last in this series of aggressive articles was published on May 12, less than a month before the outbreak of the war.[98] In addition, there were indications that in late 1966 Nasser became convinced that the USA was bent on assassinating him. The successful military coup in Greece and other coups in Third World countries led him to believe that the USA had chosen him as the next target.[99] No wonder, therefore, that the collusion story was seen as a natural continuation of the charges against the Johnson administration and the fulfillment of Nasser's dreaded fears. After the war, Heikal continued his 'crusade' against the USA with another series of four articles.[100]

The third reason for the absorption of the story was psychological. Propaganda and myths are both created as a response to a certain emotional need. On June 6, there was an urgent need to explain a chaotic situation to perplexed and largely gullible masses. Moreover, Ellul asserted that 'a person listens to a particular propaganda because it reflects his deepest unconscious convictions without expressing them directly'. In his opinion, 'propaganda must always choose the trend that normally will triumph because it agrees with the great myths of the time, common to all men'.[101] In the Egyptian context, it meant that not only the story attempted to explain reality; it was conceived as truthful because it dovetailed with the conspiratorial nature of Middle Eastern history, according to the Arab narrative. In other words, there was no need to 'sell' or 'market' the myth to the public because it was like an old wine in a new bottle. Thus, for example, Tahsin Basheer, a perceptive Egyptian diplomat, claims that though he did not believe in conspiracy theories, 'the modern Middle East started with the [1916] Sykes–Picot agreement – a huge conspiracy hatched in the British and French embassies in Cairo'.[102] It is only natural, therefore, that Arab thinking would associate the collusion story with previous conspiracies, some of which did actually happen (the last and most vivid was the 1956 Suez affair). In such a way, the target audience was not only responding to existing myths; it was a party to the formation of a new one: the myth of the 1967 collusion.

CONCLUSIONS

The fabricated story of the collusion between Israel, the USA and Britain in 1967 is a case study of the role of propaganda, as much as it is a case study of the process of mythmaking. When Nasser and his media initiated the propaganda campaign they were not concerned with mythmaking; rather, facing a humiliating military defeat, they were preoccupied with shoring up their waning domestic legitimacy. Whether they believed in the story or invented the Big Lie to manipulate public opinion is difficult to establish. But since the course of Middle Eastern history suggested that Western collusion and conspiracies do occur (though how often is a matter of interpretation) and as US–Egyptian relations had been deteriorating since the mid-1960s, it could very well be that Nasser sincerely believed in the story.

Both the USA and Britain invested considerable efforts to repudiate the collusion charge. Since at stake were vital Western interests, such as oil, communication and transportation, they attempted – unsuccessfully, we should emphasize – to prevent the Arab states from severing relations with them and imposing an oil embargo. The Western struggle was conducted in a 'rational' way: by providing data and evidence refuting the charge, the USA and Britain hoped to expose Arab duplicity. Yet, as the belief in the collusion stemmed mainly from psychological reasons, this kind of response could hardly affect Arab thinking. The result was that whatever evidence was provided by the USA and Britain against the allegation, Arab public opinion always interpreted it in a way that substantiated deeply ingrained beliefs.

When Nasser and Heikal, who were both responsible for disseminating the Big Lie, repudiated the collusion story in the course of 1968, it could have been expected that the issue would die out. Instead, the story gained momentum, turning into a myth in Egypt and throughout the Arab world. Undoubtedly, there were several 'objective' variables that contributed to the entrenchment of this myth, such as the place of Western collusion and conspiracies in the Arab collective memory and its inclusion in Egyptian history school textbooks in the post-1967 period. In addition, perhaps some Arabs, familiar with the 'Protocols of the Elders of Zion', associated the collusion story with this fabricated book on a worldwide Jewish conspiracy.[103] All these factors tended to validate the 'truthfulness' of the story.

Yet, beyond these factors it seems that the belief in the collusion story filled an important psychological need for a society that was suffering from a pervasive inferiority complex vis-à-vis the West. Vatikiotis claimed that Nasser 'took upon himself the whole burden of the national feeling of inferiority'.[104] If he did not take it upon himself, at least he personified it. From an Arab perspective, the notion that little Israel could have defeated three Arab states was unbearable and shameful; on the other hand, the idea that Western involvement played a crucial role in the defeat was somewhat comforting. In 1956 there was no expectation that Nasser would win the war. By portraying the 1967 War as another collusion, Arab leaders could throw the blame squarely on the Western powers and thus shun responsibility for the defeat. Such a depiction also places the Arabs as victims of the West and the war as another episode in the long chain of Arab–Islamic victimization that started with the Christian Crusaders.[105] This kind of thinking, however flawed, contained a grain of optimism – namely, that once the might of the Western superpowers was neutralized, it would be possible to fight on a par with Israel. Perhaps such thinking was needed in order to bring the Arab states back to the battlefield, only six years after the disaster of 1967.

History suggests that myths usually do not disappear; rather, as the anthropologist Claude Lévi-Strauss concludes, they wither away or exhaust themselves.[106] Though this will probably be the case with the 1967 collusion myth, it is still too early to predict when that will happen.

NOTES

I would like to thank the Harry S. Truman Institute for the Advancement of Peace at the Hebrew University of Jerusalem for the financial support of this project.

1. 'Abdallah Ahmad Hamid al-Qusi, *Al-Wisam fi al-Ta'arikh*, history textbook for high school/stage three [twelfth grade] (Cairo: al- Mu'asasa al-'Azabiyya al-Hadith, 1999), p. 284. See also the narrative in 'Abd al-Mun'im Gharib, *al-Muhit fi al-Dirasat al-Ijtima'iyya: Gugrafiyat al-'Alam wa-Ta'arikh Misr al-Hadith* (Cairo: al- Mu'asasa al-'Azabiyya al-Hadith, 1999), p. 238; *al-Mu'llam fi al-Ta'arikh*, for high school (Cairo: al- Mu'asasa al-'Azabiyya al-Hadith, 1999), p. 427 [Arabic].
2. See the list of sources in note 93.
3. *Middle East Record* (MER), Vol. 3, 1967 (Jerusalem: Israeli Universities Press, 1971), p. 235.
4. Garth Jowett and Victoria O'Donnell, *Propaganda and Persuasion*, 2nd edn. (London: Sage, 1992), p. 4. For an overview of older definitions of propaganda, see John C. Clews, *Communist Propaganda Techniques* (London: Methuen, 1964). For some further useful sources, see J.A.C. Brown, *Techniques of Persuasion: From Propaganda to Brainwashing* (London: Penguin Books, 1963); Oliver Thomson, *Easily Led: A History of Propaganda* (London: Sutton Publishing, 1999); Anthony R. Pratkanis and Elliot Aronson, *Age of Propaganda: The Everyday Use and Abuse of Persuasion* (New York: W. H. Freeman and Company, 1991); Robert W. Chandler, *War of Ideas: The U.S. Propaganda Campaign in Vietnam* (Boulder, CO: Westview, 1981); Allison B. Gilmore, *You Can't Fight Tanks with Bayonets: Psychological Warfare against the Japanese Army in the Southwest Pacific* (Lincoln: University of Nebraska Press, 1998). Many studies were published on the propaganda aspects of the Gulf War. See, for example, Philip M. Taylor, *War and the Media: Propaganda and Persuasion in the Gulf War* (Manchester: Manchester University Press, 1992); Robert E. Denton, Jr. (ed.), *The Media and the Persian Gulf War* (Westport, CT: Praeger, 1993); John R. MacArthur, *Second Front: Censorship and Propaganda in the Gulf War* (Berkeley, CA: University of California Press, 1993); Leonard W. Doob, 'Propaganda', in Erik Barnouw et.al (eds.), *International Encyclopedia of Communications* (New York: Oxford University Press, 1989), pp. 374–8.
5. For more on this distinction, see Jowett and O'Donnell, *Propaganda and Persuasion*, pp. 8–15. See also 'Propaganda as a Form of Communication', in Ted J. Smith III (ed.), *Propaganda: A Pluralistic Perspective* (New York: Praeger, 1989), pp. 49–64; Michael Kunczik, *Images of Nations and International Public Relations* (Mahwah, NJ: Lawrence Erlbaum Associates, 1997), pp. 15–16.
6. Jowett and O'Donnell, *Propaganda and Persuasion*, p. 56.
7. Ibid., pp. 8–10.
8. J. Fred MacDonald, 'Propaganda and Order in Modern Society', in Smith, *Propaganda*, p. 24.
9. Jacques Ellul, *Propaganda: The Formation of Men's Attitudes*, translated from the French by Konrad Kellen and Jean Lerner (New York: Alfred A. Knopf, 1966), p. 11.
10. Ibid., p. 60.
11. Ibid., p. 13; Jowett and O'Donnell, *Propaganda and Persuasion*, p. 219. On propaganda as an instrument of policy, see K. J. Holsti, *International Politics: A Framework for Analysis*, seventh edn. (Englewood Cliffs, NJ: Prentice Hall, 1995), Ch. 7.
12. Jowett and O'Donnell, *Propaganda and Persuasion*, p. 225.
13. Ibid., p. 221. See also pp. 214–15. They assert that a myth is not merely a fantasy or a lie, but rather a model for social action (p. 215).
14. Ellul, *Propaganda*, p. 4.
15. Ibid., p. 93.
16. Ibid., pp. 138–9.

17. Ibid., p. 96. See also Jowett and O'Donnell, *Propaganda and Persuasion*, p. 222.
18. Ted J. Smith III, 'Propaganda and the Techniques of Deception', in Smith, *Propaganda*, pp. 85–90.
19. Daniel Pipes, *The Hidden Hand: Middle East Fears of Conspiracy* (New York: St. Martin's Press, 1996), pp. 9–10, 16–19, 251.
20. The Israeli information service has euphemistically been called *hasbara* (literally 'explaining'). The Arab side has used the term *i'lam* (literally 'information'). In a book entitled 'Arab Information Strategy' there is a section on 'Zionist Propaganda', see al-Sayyid 'Aliwa, *Istrategiyat al-I'lam al-'Arabi* (Cairo: al-hai'a al-Misziyya al-'Amma lid-kitab, 1978) [Arabic].
21. See, for example, Yonah Alexander, *The Role of Communications in the Middle East Conflict: Ideological and Religious Aspects* (New York: Preager, 1982); Eli Tavin and Yonah Alexander (eds.), *Psychological Warfare and Propaganda: Irgun Documentation* (Wilmington, DL: Scholarly Resources Inc., 1982). For a study that deals with propaganda during the 1956 War but from the British angle, see Tony Shaw, *Eden, Suez and the Mass Media: Propaganda and Persuasion during the Suez Crisis* (London: I.B. Tauris, 1996). For a study that deals with Soviet propaganda role in the conflict, see Baruch A. Hazan, *Soviet Propaganda: A Case Study of the Middle East Conflict* (Jerusalem: Keter, 1976). For a study that deals with some aspects of the relation between language and propaganda, see Adel Safty, *From Camp David to the Gulf: Negotiations, Language and Propaganda, and War* (New York: Black Rose Books, 1992).
22. For more information, see Rais Ahmad Khan, 'Radio Cairo and Egyptian Foreign Policy, 1956–1959', unpublished Ph.D. thesis, University of Michigan, 1967, pp. 21–5; Douglas A. Boyd, *Broadcasting in the Arab World: A Survey of the Electronic Media in the Middle East* (Ames, IA: Iowa State University Press, 1999), pp. 15–40; Sydney W. Head (ed.), *Broadcasting in Africa: A Continental Survey of Radio and Television* (Philadelphia, PA: Temple University Press, 1974), pp. 15–29.
23. Boyd, *Broadcasting in the Arab World*, p. 28. For more on *Sawt al-'Arab*, see Khan, 'Radio Cairo', pp. 26–7; Douglas A. Boyd, 'Development of Egypt's Radio: 'Voice of the Arabs' under Nasser,' *Journalism Quarterly*, Vol. 52 (1975), pp. 645–53; A. Loya, 'Radio Propaganda of the United Arab Republic – An Analysis', *Middle Eastern Affairs*, Vol. XIII, No. 4 (April 1962), pp. 98–110.
24. Charles Issawi, *Egypt in Revolution: An Economic Analysis* (London: Oxford University Press, 1964), p. 217.
25. Julian Hale, *Radio Power: Propaganda and International Broadcasting* (London: Paul Elek, 1975), p. 72. See also the testimony of Mustafa Amin in *Akhbar al-Yawm*, Aug. 12, 1974, as quoted in Boyd, 'Development of Radio Cairo', p. 653.
26. On data of radio sets in Egypt and other Arab countries, see Khan, 'Radio Cairo', p. 28; Daniel Lerner, *The Passing of Traditional Society: Modernizing the Middle East* (London: Glencoe, 1958), pp. 232–4, 253–4.
27. Quoted in Shaw, *Eden, Suez and the Mass Media*, p. 5.
28. A. I. Dawisha, *Egypt in the Arab World: The Elements of Foreign Policy* (London: Macmillan, 1976), pp. 162–72; Hale, *Radio Power*, pp. 72–3.
29. For some examples, see Cairo Radio, June 6 (*BBC*, ME/2485/A, June 7, 1967); June 7 (*BBC*, ME/ 2486/A, June 8, 1967); June 8 (*BBC*, ME/2487/A, June 9, 1967).
30. For this and other Jordanian reports, see *MER*, p. 242; Public Record Office, Kew Gardens, London [hereafter: PRO], FCO 17/598, Adams (Amman) to FCO, Tel. 527, June 5, 1967; Minutes by Brooke-Turner, June 8, 1967.
31. For this and other Syrian reports, see *MER*, pp. 242-43; PRO, FCO 17/598, Evans (Damascus) to FCO, Tel. 376, June 6, 1967.
32. *MER*, 1967, p. 244. In fact, Egypt's relations with Britain were cut off already in Dec. 1965.
33. Ibid., pp. 244–5.
34. Michael Bar-Zohar, *The Longest Month* (Tel Aviv: Levin-Epstein, 1969), p. 232 [Hebrew].
35. See, for example, a commentary by Ahmad Sa'id, Radio Cairo, June 8 (*BBC*, 2487/A2, June 10, 1967). See in this regard, Stephen Green, *Taking Sides: America's Secret Relations with a Militant Israel* (New York: William Morrow and Company, 1984), pp. 212–42.
36. PRO, FCO 17/598, Dean (Washington) to FCO, Tel. 1959, June 7, 1967. See also *MER*, 1967, p. 243.
37. PRO, FCO 17/599, Harrison (Moscow) to FCO, Tel. 1050, June 16, 1967. See also FCO 17/598, Arthur (Kuwait) to FCO, Tel. 222, June 6, 1967.
38. PRO, FCO 17/599, Fall (Moscow) to Eastern Department, June 14, 1967.
39. Cairo Radio, June 9 (*BBC*, ME/2488/A/2, June 11, 1967). The 'collusion' was also mentioned

by Syrian President Nur al-Din al-Atasi, Radio Damascus, June 9 (*BBC*, ME/2488/A11, June 11, 1967); and by Algerian President Hawwari Bumadyan, Radio Algiers, June 10 (*BBC*, ME/2489/A11, June 12, 1967). Significantly, it was not mentioned in King Husayn's statement, Radio Amman, June 11 (*BBC*, ME/2489/A10, June 12, 1967). For an explanation, see below.

40. It is most probable that television was used as well as an instrument of persuasion, but in the absence of any concrete evidence, the role of this medium will not be analyzed.
41. For the text, see Radio Baghdad, June 11 (*BBC*, ME/2489/A6, June 12, 1967). For a similar statement issued by the Egyptian embassy in Nicosia on June 13, see PRO, FCO 17/599, Sir Costar (Nicosia) to FCO, Tel. 555, June 17, 1967. For a slightly different document issued by the Egyptian embassy in Morocco, see PRO, FCO 17/599, Guidance Tel. 143, June 14, 1967.
42. See, for example, commentary by Ahmad Sa'id, Radio Cairo, June 12 (*BBC*, ME/2490/A3, June 13, 1967).
43. See, in particular, Radio Cairo, June 16 (*DR*, B1-B9, June 16, 1967); and 22 June (*DR*, B2-B11, 23 June 1967).
44. PRO, FCO 17/600, Moberly's Minute, July 17, 1967. Though the final communiqué of the Khartoum summit did not specifically referred to the 'collusion,' one of the decisions adopted was to encourage the evacuation of all foreign bases in the Arab world.
45. PRO, FCO 17/598, Adams to FCO, Tel. 573, June 7, 1967.
46. For the full text, see PRO, FCO 17/598, FO Guidance Tel. 128, June 6, 1967.
47. PRO, FCO 17/598, Brown to Sheikh Abd al-Rahman al-Helaissi, Ambassador of Saudi Arabia, June 6, 1967. A similar letter was sent to all Arab ambassadors, except that of Egypt, which had not had diplomatic relations with Britain since Dec. 1965.
48. PRO, FCO 17/598, Lord Caradon (UK Mission in New York) to FCO, Tel. 1240, June 6, 1967; Tel. 1249, June 7, 1967.
49. PRO, FCO 17/492, Caradon to FCO, Tel. 1244, June 6, 1967; FCO 17/599, Caradon to UN Security Council, June 7, 1967.
50. PRO, FCO 17/598, Dean (Washington) to FCO, Tel. 1958, June 7, 1967.
51. See, for example, the attitude of Dean Rusk, the American Secretary of State, PRO, FCO 17/598, Dean to FCO, Tel. 1949, June 6, 1967.
52. PRO, FCO 17/598, FCO to Washington, Tel. 6273, June 8, 1967.
53. PRO, FCO 17/598, Caradon to FCO, Tel. 1262, June 8, 1967.
54. PRO, FCO 17/598, Beaumont (Baghdad) to FCO, Tel. 704, June, 6; Tel. 706, June 6; Tel. 711, June 6; Tel. 707, June 6, 1967.
55. PRO, FCO 17/598, Beaumont to FCO, Tel. 776, June 10, 1967.
56. PRO, FCO 17/598, Riches (Beirut) to FCO, Tel. 536, June 6, 1967. But it seems that even in Lebanon many believed the collusion story, see Tel. 555, June 7, 1967.
57. PRO, FCO 17/598, Moberly's Minute, June 11, 1967.
58. For the full text, see PRO, FCO 17/598, Hadow (Tel Aviv) to FCO, Tel. 546, June 8, 1967. See also *MER*, 1967, p. 242.
59. PRO, FCO 17/598, Brooke-Turner's Minutes, June 8, 1967.
60. PRO, FCO 17/599, Cranston (Jiddah) to FCO, Tel. 393, June 11, 1967. For a similar assessment from Iraq, see FCO 17/598, Vincent (Baghdad) to FCO, Tel. 801, June 12, 1967. For a more cautious attitude displayed by King Husayn, see FO 17/599, Adams to FCO, Tel. 665, June 14, 1967.
61. PRO, FCO 17/599, Cranston to FCO, Tel. 446, June 20, 1967.
62. See, for example, the report in the Kuwaiti press, PRO, FCO 17/599, Arthur (Kuwait) to FCO, Tel. 319, June 15, 1967.
63. PRO, FCO 17/599, Dean to FCO, Tel. 2078, June 17, 1967. See also FCO 17/598, FCO to Washington, Tel. 6684, June 15, 1967.
64. See, for example, PRO, FCO 17/599, Cranston to FCO, Tel. 434, June 17, 1967.
65. PRO, FCO 17/599, Adams to FCO, Tel. 722, June 21, 1967; Wright (New York) to FCO, Tel. 104, June 27, 1967. In all fairness it should be noted that Husayn did not say that there was no collusion but only that no evidence was found.
66. PRO, FCO 17/599, Guidance Tel. 156, June 21, 1967.
67. For the text, see PRO, FCO 17/599, Caradon to President of the Security Council, June 17, 1967; Caradon to FCO, Tel. 1395, June 16, 1967.
68. PRO, FCO 17/599, FCO to Stockholm, Tel. 537, June 15; Mackenzie to FCO, Tel. 308, June 16; Tel. 312, June 19; Tel. 315, June 19, 1967.
69. PRO, FCO 17/600, Extract from *Hansard*, June 15, 1967. When the Ministry of Defence was preparing the material for inspection, it revealed that the carrier logs contained some embarrassing features. First, one carrier arrived in Malta only during the morning of June 5 and

was not actually tied up in the harbor as stated by Lord Caradon at the UN. Second, that another carrier left Aden on June 6 and that two planes took off for training missions. Third, that there were a number of Canberra and Vulcan flights in the Mediterranean area on June 5–7. All these planes could theoretically have reached the area of the conflict but they were engaged in a variety of routine training exercises far from the battle scene. All these details were marginal, but the British were afraid that the inaccuracies would lead Egypt and its allies to take them as evidence of British duplicity. See PRO, FCO 17/599, Brooke-Turner's Minutes, June 23, 1967; FCO to Kuwait, Tel. 454, June 27, 1967. For an explanation of how such a mistake occurred, see FCO 17/600, Glass to Lambert, July 3, 1967; Lambert to Glass, July 25, 1967.

70. See, for example, the letter of the foreign secretary to MP the Earl of Dalkeith, which, in response to the latter's assertion that Britain was 'simply not being believed,' listed all the steps taken to refute the Egyptian allegation: PRO, FCO 17/600, June 29, 1967.
71. PRO, FCO 17/599, Singapore to FCO, Tel. 226, June 22, 1967.
72. Voice of the Arabs, June 22, 1967 (*DR*, June 23, 1967, b2); PRO, FCO, 17/599, Guidance Tel. 158, June 23, 1967.
73. PRO, FCO 17/600, Guidance Tel. 168, June 30, 1967.
74. *MER*, 1967, pp. 59-60.
75. PRO, FCO 17/600, Tesh (Nicosia) to FCO, Tel. 676, July 8, 1967.
76. *MER*, 1967, p. 61.
77. Ibid., p. 51. For a summary of Heikal's articles against the USA, see p. 49.
78. *MER*, Vol. 4, 1968 (Jerusalem: Israel Universities Press, 1973), p. 77.
79. Henry Tudor, *Political Myth* (London: Macmillan, 1972), pp. 132–3. See also Jean Holm (ed.), *Myth and History* (London: Pinter Publishers, 1994), pp. 2–3; David Adams Leeming, *The World of Myth* (New York: Oxford University Press, 1990), pp. 3–8.
80. Emmanuel Sivan, *Arab Political Myths* (Tel Aviv: Am Oved, 1988), p. 10 [Hebrew]; Tudor, *Political Myth*, pp. 48–51; Haggay Ram, *Myth and Mobilization in Revolutionary Iran: The Use of the Friday Congregational Sermon* (Washington, DC: The American University Press, 1994), pp. 8–9.
81. Ellul, *Propaganda*, p. 11.
82. For analysis of another myth, see Nachman Ben-Yehuda, *The Masada Myth: Collective Memory and Mythmaking in Israel* (Madison, WI: The University of Wisconsin Press, 1995). See also Yael Zerubavel, *Recovered Roots: Collective Memory and the Making of Israeli National Tradition* (Chicago, IL: Chicago University Press, 1995).
83. R. Hrair Dekmejian, *Egypt under Nasir: A Study in Political Dynamics* (Albany, NY: State University of New York Press, 1971), pp. 4–5, 56–7.
84. Pipes, *The Hidden Hand*, p. 246.
85. On Nasser's conspiratorial character, see P. J. Vatikiotis, *Nasser and His Generation* (New York: St. Martin's Press, 1978), pp. 306, 351.
86. Mohamed Heikal, *The Sphinx and the Commissar: The Rise and Fall of Soviet Influence in the Middle East* (New York: Harper and Row, 1978), p. 181.
87. Idem, *1967: Al-Infijar* (Cairo: al-Ahram Center, 1990), pp. 715–23 [Arabic].
88. See, for example, Ahmad 'Abd al-Rahim Mustafa, *al-Hilal*, Vol. 98, No. 10 (Oct. 1990); Gamal al-Shelby, *Muhammad Hasanayn Heikal: Istimrariyya am Tahawwul* (Beirut, 1999), pp. 170–84 [Arabic].
89. *Mudhakkirat Mahmud Riad: Amrika wa-al-'Arab* (Cairo: Dar al-Mustaqbal al-'Arabi, 1986), pp. 40–42 [Arabic.]. Unfortunately, Abdel-Kader Hatem, the Director of Information Service and one-time Minister of Information during the Nasserite period, evaded the issue in his memoirs, though from the way he defined propaganda one may deduce that Nasser was manipulating the masses. Hatem, however, did blame Israel for 'cynical use of lies and deception' during the 1967 War. See his *Information and the Arab Cause* (London: Longmans, 1974), p. 228.
90. Richard Parker, *The Politics of Miscalculation in the Middle East* (Bloomington, IN: Indiana University Press, 1993), pp. 109–10.
91. A full session was devoted to 'conspiracy theories', see Richard Parker (ed.), *The Six-Day War: A Retrospective* (Gainesville, FL: University Press of Florida, 1996), pp. 237–88. See, in particular, pp. 244–52; 276–7. It should be noted that several Western observers supported the conspiracy theory.
92. See, for example, 'Abd al-'Azim Ramadan, *Tahtim al-Aliha: Qissat Harb Yunyu 1967* (Cairo: Madbuli, 1988) [Arabic].
93. See, for example, Abu al-Ftuh Radwan, *Usul al-'Alam al-Hadith*, lil-Saff al-Awal al-Thanawi [Tenth Grade] (Cairo, 1969), p. 221; Muhammad 'Abd al-Ghani Saudi, *Ta'arikh al-'Arab al-*

Hadith wa-al-Mu'asir, lil-Saff al-Thalith al-Thanawi [Twelfth Grade] (Cairo, 1973), pp. 142–3; Yusuf Ahmad Mahmud al-Qusy, *al-Wahda fi Ta'arikh al-'Arab al-Hadith wa-al-Mu'asir* (Cairo, 1975), pp. 75–6; Muhammad Ibrahim al-Dasuqi, *al-Dirasat al-Ijtima'iyya Misr wa-al-'Alam*, lil-Saff al-Thalith al-I'dadi [Ninth Grade] (Cairo, 1991), pp. 290 [Arabic].

94. Elie Podeh, *The Arab-Israeli Conflict in Israeli History Textbooks, 1948–2000* (Westport, CT: Bergin and Garvey, 2002), Ch. 1.
95. Stephen Green, *Taking Sides: America's Secret Relations with a Militant Israel* (New York: William Morrow and Company, 1984), p. 198. For a description of the whole episode, see pp. 195–211.
96. Richard Parker, *The Politics of Miscalculation in the Middle East* (Bloomington, IN: Indiana University Press, 1993), p. 108.
97. Quoted in Pipes, *The Hidden Hand*, p. 236. It was said with regard to Saddam's propaganda during the Gulf crisis in 1990.
98. The articles were published on Feb. 24; March 3, 10, 17, 31; April 21; May 5, 12, 1967. For a summary in English, see *MER*, 1967, pp. 49–50.
99. Pipes, *The Hidden Hand*, p. 17. See, in particular, the report of Parker in *The Politics of Miscalculation*, p. 106.
100. Published on August 4, 10, 17, 24, 1967. For a summary, see *MER*, 1967, p. 50.
101. Ellul, *Propaganda*, pp. 41–2.
102. Quoted in Parker, *The Six-Day War*, p. 244.
103. See in this connection Hadassa Ben-Itto, *The Lie That Wouldn't Die* (Tel Aviv: Dvir, 1998) [Hebrew].
104. Vatikiotis, *Nasser and His Generation*, p. 322.
105. See, for example, Sivan, *Arab Political Myths*, p. 42.
106. 'How Myths Die', in Claude Lévi-Strauss, *Structural Anthropology*, Vol. 2, translated from the French by Monique Layton (Chicago, IL: the University of Chicago Press, 1976), p. 268.

The 1969–1970 War of Attrition

[5]

Israel and the War of Attrition

Robert S. Bolia
U.S. Air Force Research Laboratory
Wright-Patterson Air Force Base, Ohio

SINCE IT BECAME an independent nation in 1948, Israel has fought six wars against its Arab neighbors. The 1948-1949 conflict, called the War for Independence, demonstrated that the new nation could stand on its own despite violent opposition to its existence. In the 1956 war in the Sinai, Israel proved that a preemptive strike could delay an enemy's preparation for war for years.

The Israeli Defense Force (IDF) reached its apogee during the Six-Day War of June 1967 and demonstrated the value of intelligence and planning. The Yom Kippur War of 1973 revealed the dangers of Israeli overconfidence. Israel's invasion of Lebanon 9 years later produced internal dissent and conflict between military and political objectives. But Israel's longest war, the War of Attrition, fought between Israel and Egypt from 1967 to 1970, is hardly remembered at all.[1] When people do remember it, they usually remember it only as being a prelude to the Yom Kippur War.[2]

The Six-Day War resulted in Israel's occupying the Sinai Peninsula. Egypt sought to force Israel to withdraw from the territories conquered in the Six-Day War; Israel sought to retain its foothold on the Sinai Peninsula to prevent an Egyptian or pan-Arab offensive and to achieve a regional cease-fire. Combat operations were generally limited to cross-border shellings, raids, ambushes, naval and air strikes, terror, and sabotage.[3]

The War of Attrition was Egypt's first attempt to force Israel to recognize that its continued occupation of the Sinai Peninsula was not in its best interests. The war was also a testing ground for Egypt and Israel to gauge the effectiveness of weapons that they would use again in 1973.[4] The war profoundly affected Egypt's and Israel's perceptions of each other's combat effectiveness and deterrent power, which in turn, had far-reaching effects in the next struggle.

Nasser's strategy was to inflict a level of casualties that would be unacceptable to the Israeli people or to escalate the conflict so Israel would have to mobilize for an extended period of time. Nasser assumed that political and economic considerations would compel Israel to withdraw its armed forces behind its pre-1967 borders.

Egyptian President Gamal Abd el-Nasser's purpose in initiating the War of Attrition was to compel Israel to withdraw from the east bank of the Suez Canal and, eventually, from the Sinai Peninsula. Nasser based his decision to begin hostilities on an analysis of Israeli strengths and weaknesses. In his view, Israel's one notable weakness was a small population relative to Egypt's. Because of this, Nasser concluded that Israel could absorb fewer casualties than Egypt could. He also knew that Israel did not have a large professional army but relied largely on citizen-soldiers. Not only would casualties significantly affect Israel's economy, so would mobilization for war.

Nasser's strategy was to inflict a level of casualties that would be unacceptable to the Israeli people or to escalate the conflict so Israel would have to

An Israeli soldier approaches a branch of the Suez Canal.

Embassy of Israel

Nasser aimed to take armor out of the equation by striking the Israeli front along the Suez Canal with massive artillery bombardments and commando raids. He hoped the physical barrier of the Canal would limit Israel's response to these attacks. If Israel wanted to strike back, it would have to cross the waterway, risking heavy casualties, possibly provoking Soviet intervention.

mobilize for an extended period of time. Nasser assumed that political and economic considerations would compel Israel to withdraw its armed forces behind its pre-1967 borders.[5]

Nasser also proposed to turn the IDF's major strength—its doctrine of flexible mobility—against itself by forcing Israel to use unfamiliar tactics. Israel's successes in 1956 and 1967 largely resulted from Israel's ability to rapidly bring decisive force to bear on enemy centers of gravity, relying heavily on the use of armor.

During the War of Attrition, Nasser aimed to take armor out of the equation by striking the Israeli front along the Suez Canal with massive artillery bombardments and commando raids. He hoped the physical barrier of the Canal would limit Israel's response to these attacks. If Israel wanted to strike back, it would have to cross the waterway, risking heavy casualties, possibly provoking Soviet intervention, and further lengthening its already overextended lines of communication.[6]

Anticipating Egypt's strategy, Israel changed its strategy. Following the Six-Day War, the Israeli high command devised a way to defend its newly occupied territory in Sinai against an Egyptian crossing of the Canal. Major General (MG) Israel Tal and MG Ariel Sharon proposed a system of defense in depth. Sharon wrote, "[M]aintaining an Israeli presence at the western edge of the Sinai did not mean we had to sit down along the entire length of the Canal. We could carefully choose one or two locations, on the Great Bitter Lake, for example, where we would not be directly under their guns. . . . I proposed that we should base our defense on the natural line of hills and dunes that runs parallel to the Canal [5 to 8] miles to the east and dominates the Canal plain. A second line with our mobile reserves should be established [15 to 20] miles from the Canal, where the mountains begin and the Mitla and Gidi passes cut toward the interior. Between the first line and the Canal we should run mobile patrols, keeping on the move constantly and unpredictably so that we would not be sitting ducks for ambushes, snipers, and artillery."[7]

The Bar-Lev Line

In spite of these and similar arguments, MG Yeshayahu Gavish and IDF Chief of Staff Lieutenant General Chaim Bar-Lev overruled Sharon and Tal. Gavish and Bar-Lev advocated building a long series of static fortifications, subsequently known as the Bar-Lev Line, on the east bank of the Canal. Most Israeli soldiers were not familiar with this type of defense and had never used barbed wire, mines, or sandbags. The defense also nullified Israel's advantage in mobility and made the War of Attrition possible.[8]

Nasser's strategy appeared sound, but it did not work. Thousands of Egypt's artillery shells impacted Israeli positions on the Bar-Lev Line, but even as Israeli casualties mounted, no general cry for an end to the war went up in Israel. Nasser had underestimated Israel's will to fight. Although a group of high school students wrote to Prime Minister Golda Meir suggesting that the government was too content with the idea of keeping the nation in a state of war, and that it was difficult to reconcile their upcoming mandatory military service with the notion of "ein breira" (Hebrew for "no choice"), Israel's students proved willing to endure the War of Attrition. The conflict's costs never became a political issue.[9]

E-Systems

The Mitla West Watch Station in the Sinai.

New Weapons

The War of Attrition was an opportunity for both countries to try out their newest weapons. Egypt had received hundreds of Soviet T-54 and T-55 tanks to replace the T-34s and T-54s lost during the Six-Day War, but the war was not to be one of large-scale tank battles. As it happened, the T-55's most noteworthy appearance occurred when the Israelis crossed the Canal with a unit of six captured T-55s, thus attacking Egyptian positions with Egyptian tanks.[10]

[Major General Sharon wrote that] "[M]aintaining an Israeli presence at the western edge of the Sinai did not mean we had to sit down along the entire length of the Canal. . . . We should base our defense on the natural line of hills and dunes that runs parallel to the Canal [5 to 8] miles to the east and dominates the Canal plain. A second line with our mobile reserves should be established [15 to 20] miles from the Canal, where the mountains begin and the Mitla and Gidi passes cut toward the interior."

Naval technology played only a minor role in the conflict, although it did have major repercussions for the future of warfare. On 21 October 1967, two Egyptian missile boats, anchored inside Port Said harbor, launched three Soviet-supplied Styx surface-to-surface missiles at the Israeli destroyer *Eilat*, the flagship of the Israeli Navy. *Eilat* was patrolling off the Sinai coast when all three missiles struck and sank it, killing 47 and wounding 90. This was an important event in naval history; it was the first time a surface-launched missile sank a ship. The attack prefigured the missile boat battles of the Yom Kippur War and the Exocet attacks on British shipping in the Falklands War a decade later.[11]

The War of Attrition also marked the first use of drones or unmanned aerial vehicles (UAVs) for reconnaissance. An Israeli major serving in IDF intelligence suggested mounting cameras on remote-controlled toy aircraft, which were too small to hit with antiaircraft fire, and sending them over the Canal as reconnaissance aircraft. Israel purchased three such drones from the United States for $850, and trial runs conducted over Israeli positions demonstrated the concept's validity. When the first UAV flew over Egyptian positions, the Egyptians did not even fire at it, and it returned with excellent photographs. Advanced versions of these vehicles have since become important to the intelligence-gathering process.[12]

Evaluating Technology on the Battlefield

Egyptian and Israeli air forces used the War of Attrition to conduct major evaluations of technology. The Egyptian Air Force (EAF) accepted over 100 MiG-21s and hundreds of other aircraft from the Soviet Union to replace Egyptian aircraft that Israel had destroyed on the ground in its preemptive strike

The crew of an SA-3 Goa surface to air missile rushes to their station during a training exercise.

Egyptian Embassy

IAF pilots flew in below the level of Egyptian radar and took out the SA-2s along with the Egyptian artillery positions. . . . Nasser begged the Soviets for help. Reluctantly the Soviets sent additional SA-2s, batteries of the newer SA-3s, and radar-guided antiaircraft batteries, which Soviet soldiers networked and operated.

on Egypt at the outset of the Six-Day War. To counter this, the Israeli Air Force (IAF) arranged to buy A-4 Skyhawks and F-4 Phantoms from the United States.[13]

Initially, the IAF's role in the War of Attrition was relatively minor. The IAF sent Mirages to intercept Egyptian MiGs only when they crossed the Canal to attack positions on the Bar-Lev Line and at Sharm el-Sheikh. Eventually, though, IAF commander MG Mordechai "Moti" Hod sent teams of Mirages into Egypt to attack MiG patrols and to bait the EAF into dogfights, the majority of which the Israelis won. At first the EAF responded by increasing the number of MiGs sent to tangle with the Mirages, but the IAF countered the escalation. Eventually Nasser conceded air superiority to the IAF and simply stopped sending planes.[14] But this did not mean an end to the war. Nasser still had artillery, and the Egyptian guns continued to pound Israeli positions along the Canal.

Having gained air superiority, the IAF now supported the Army by mounting a massive strike on Egyptian positions, dropping 159 tons of bombs and 72 canisters of napalm during a 2-hour period. The Egyptians had erected a networked air defense system that included Soviet-supplied SA-2 surface-to-air missiles, but the SA-2 was only effective at hitting targets above 3,000 feet, as the Israelis were well aware. IAF pilots flew in below the level of Egyptian radar and took out the SA-2s along with the Egyptian artillery positions.[15]

Still, neither the Egyptians nor the Israelis gave up. In fact, both sides escalated the strikes. Nasser begged the Soviets for help. Reluctantly the Soviets sent additional SA-2s, batteries of the newer SA-3s, and radar-guided antiaircraft batteries, which Soviet soldiers networked and operated. The batteries were effective against targets flying at any altitude and were sufficiently dispersed so that Israel could not easily attack them from the air. When Israel sent its new F-4 Phantoms to neutralize these batteries, Egypt shot two of the Phantoms down in a single day.[16]

The United States reacted by sending Israel more Phantoms and Skyhawks as well as jamming pods for the Phantoms. The pods, designed to confuse the SA-2s' radar, worked—but only against the SA-2s. During the first raid in which Israel used the jamming pods, SA-3s hit one Phantom and shot down another. Still, Israel destroyed 4 of the 10 batteries it targeted.[17]

Throughout the war, the Soviets blamed Egyptian losses on operator cowardice or failure to understand Soviet training. They chided the Egyptians for having lost numerous pieces of high-technology military equipment to the Israelis, including T-55 and T-62 tanks and a P-12 radar. In July 1970, the Soviet Union decided to teach Israel a lesson by patrolling the Canal Zone with MiG-21s. The Israelis responded by shooting down five Soviet MiGs on 30 July. Three of the aircraft that scored MiG kills were older Mirages. Israeli planes were not better than Russian aircraft, but IAF pilots were better than their Russian counterparts.[18]

Neither side could afford continued escalation. Israel had humiliated the Soviets, but the Soviets could not afford to raise the stakes against one of the United States's major allies. Similarly, Israel had gained a tactical victory but could not afford to pressure the Soviets further. Egypt and Israel accepted a cease-fire, which went into effect on 8 August 1970.[19]

A Hollow Victory

Both Egypt and Israel claimed victory, and there were arguments to be made for both sides' claims. Although Nasser's purpose in prosecuting the war had been to compel Israel to withdraw from the Canal Zone, at war's end, Israelis remained on the east bank, secure in the knowledge that they had suffered far fewer casualties than had the Egyptians. Still, despite the high cost in human life, Egypt felt it had won at least a moral victory.

Unlike in 1967, when Egyptian troops had fought disgracefully, the Egyptian Army and Air Force had stood their ground against the superbly trained IAF, and the average Egyptian regained some lost pride. This was thought to have been an important factor when the Egyptian Army invaded Sinai 3 years later.

Almost everyone in Israel claimed victory, but the real problem for the Israelis was what they had failed to learn. For example, they clung to the concept of static defense, despite the fact that static defense had proven costly and had not stopped Egyptian shelling. This fault in Israeli military doctrine was further exposed 3 years later when the Bar-Lev Line failed to prevent a massive Egyptian crossing along the entire length of the Canal.

Also, when the war ended, the IAF had less command of the air than when the war had begun. The Egyptians and their Soviet allies had built a defensive missile umbrella. After the war ended, they moved the umbrella to the edge of the Canal where it would be more effective against the IAF. The umbrella disrupted the strategic bombing and interdiction that were normal parts of Israel's doctrine of taking the war to the enemy as quickly as possible, it also impeded the progress of Israeli ground forces because the IDF favored close air support rather than artillery support.

Israel believed it had won the war and assumed the Egyptians knew they had lost it. Israel believed Egypt would not fight again until it had an air force equal to Israel's. This, of course, was not the case. Israel's mistaken conclusions led it to become complacent, which allowed Egypt to surprise Israel with an attack in October 1973. Former Commander of the IAF Ezer Weizman wrote, "It is no more than foolishness to claim that we won the War of Attrition. On the contrary, for all their casualties, it was the Egyptians who got the best of it. . . . We, with our own hands, smoothed Israel's path to the Yom Kippur War."[20] ***MR***

MiG wreckage near the Suez Canal.

Embassy of Israel

Throughout the war, the Soviets blamed Egyptian losses on operator cowardice or failure to understand Soviet training. . . . In July 1970, the Soviet Union decided to teach Israel a lesson by patrolling the Canal Zone with MiG-21s. The Israelis responded by shooting down five Soviet MiGs on 30 July.

NOTES

1. Martin van Creveld, *The Sword and the Olive: A Critical History of the Israeli Defense Force* (New York: PublicAffairs, 2002), and Chaim Herzog, *The Arab-Israeli Wars* (New York: Vintage Books, 1984), provide overviews of Israel's six wars. In *The War of Atonement: The Inside Story of the Yom Kippur War* (London: Greenhill Books, 2003), 7, Herzog's says that the War of Attrition is regarded in Israel as a "passing event" In *No Margin for Error: The Making of the Israeli Air Force* (New York: Pantheon Books, 1993), 265, Ehud Yonay says the IAF has never issued a War of Attrition campaign ribbon. The official dates of the war are March 1969–August 1970, although many historians also include the time between the end of the Six-Day War and March 1969.
2. Herzog, *The War of Atonement*.
3. For more information, see on-line at <www.inter.net.il/~eizen/english.html#attrition>, accessed 9 February 2004, and <www.palestinefacts.org/pf_1967to1991_warofattrition.php>, accessed 9 February 2004.
4. Nasser bestowed the name "The War of Attrition" on the conflict.
5. Benny Morris, *Righteous Victims: A History of the Zionist-Arab Conflict, 1881-2001* (New York: Vintage Books, 2001), 348; Avi Shlaim, *The Iron Wall: Israel and the Arab World* (New York: W.W. Norton, 2001), 289; John Laffin, *The Israeli Army in the Middle East Wars, 1948-73* (London: Osprey Publishing, 1982), 4.
6. Herzog, *The Arab-Israeli Wars*, 207; Morris, 348.
7. Ariel Sharon with David Chanoff, *Warrior: An Autobiography* (New York: Touchstone, 2001), 219-20.
8. Ibid., 219-21, 229-31; David Eshel, *Chariots of the Desert: The Story of the Israeli Armoured Corps* (London: Brassey's, 1989), 89-90; Laffin, 21.
9. Sharon with Chanoff, 231-32; Shlaim, 294-95.
10. Walter J. Boyne, *The Two O'Clock War: The 1973 Yom Kippur Conflict and the Airlift that Saved Israel* (New York: St. Martin's Press, 2002), 297-98; Eshel, 90-91; Herzog, *The Arab-Israeli Wars*, 199.
11. Samuel M. Katz, *Arab Armies of the Middle East Wars 2* (London: Osprey Publishing, 1988), 4; Herzog, *The Arab-Israeli Wars*, 197-98.
12. Ian Black and Benny Morris, *Israel's Secret Wars: A History of Israel's Intelligence Services* (New York: Grove Press, 1991), 283.
13. Lon Nordeen, *Air Warfare in the Missile Age* (Washington, DC: Smithsonian Institution Press, 2002), 105-106.
14. Yonay, 269-73.
15. Ibid., 273-76.
16. Ibid., 281-88.
17. Ibid., 292-97; Nordeen, 118.
18. Nordeen, 118-19; Herzog, *The Arab-Israeli Wars*, 218.
19. Herzog, *The Arab-Israeli Wars*, 218-19.
20. Van Creveld, 215; Shlaim, 296-97; Morris, 262-63.

[6]

'The Russians are not coming': Israel's Intelligence Failure and Soviet Military Intervention in the 'War of Attrition'

DIMA ADAMSKY and URI BAR-JOSEPH

This paper describes and analyzes Israel's intelligence failure to warn against Soviet intentions and preparations to intervene in the Egyptian–Israeli War of Attrition in 1969–1970. Drawing mostly on primary sources, the paper describes (1) how the thesis that the Soviets will avoid intervention in the Arab–Israeli conflict was born and became dominant; (2) the information that refuted this thesis, which had become available to Israel's intelligence community since summer 1969 (when the Soviet decision to intervene was made), and how it was ignored; (3) the political and military outcomes of this intelligence failure, and (4) its causes.

INTRODUCTION

Soviet military intervention in the Egyptian–Israeli War of Attrition – a war that took place between March 1969 and August 1970 – is a remarkable event in the history of the 'cold war'. For the first time since 1945, Soviet combat units were sent to fight outside the borders of the Warsaw Pact countries and far from the borders of the Soviet Union. The decision to deploy an air-defense division in Egypt – 'Operation Kavkaz', as it was dubbed in the USSR – was taken in August 1969.[1] It was the Soviet response to Egypt's increasing difficulties in confronting the Israeli Air Force (IAF) which from July 1969 was used as 'flying artillery' to compensate for the weaknesses in the Israeli Defense Forces' (IDF) ground artillery during this static war along the Suez Canal.

The Soviet division was composed of air defense units from the Air Defense military districts of Dnieperpetrovsk, Byelorussia, Moscow and Leningrad. Its core component consisted of SA-3 medium altitude surface-to-air missile batteries (NATO codename 'Goa'), backed by two squadrons of MiG-21 interception planes (70 planes, 102 pilots), ZSU 23X4 anti-aircraft guns ('Gundish') and SA-7 personal anti-aircraft missiles ('Strela 2'). In early September 1969, the two MiG-21 squadrons began training near the Black and Caspian Sea. In December the 'Kavkaz' task force's first jets landed in Egypt. At approximately the same time, the division's air defense units were sent to the Ushuluk training area in Kazakhstan. Their shipment from the port of Nikolayev in the Black Sea to Alexandria began three months later. By 5 March 1970 the first units arrived in Egypt and by 15 March the division became operational.

During this first stage, the Soviet division was used for the defense of Egypt's rear – which included the Cairo, Alexandria, and Aswan areas. Once this target was secured and the IAF ceased flying in the Egyptian hinterland, the Soviet deployment started moving eastwards, in order to provide anti-aircraft defense for the Egyptian ground forces in the Canal front zone. This led to bitter clashes with the IAF resulting in relatively heavy casualties on both sides.[2]

'Operation Kavkaz' surprised Israel. Until March 1970, when elements of a Soviet air-defense division were pinpointed in Egypt, Israel's Military Intelligence (AMAN), the nation's sole intelligence estimator, assessed that, for a number of reasons which will be discussed below, the Kremlin would avoid such a move. Although in many respects this intelligence failure constituted a dress rehearsal for Israel's more devastating October 1973 debacle, it had barely received any attention, even when compared to the moderate discussion of the American intelligence performance in this case.[3] This article aims at bridging the gap between the significance of the Israeli failure on the one hand, and, on the other, the lack of information about how it actually took place.

This introduction will be followed by a discussion of how Israel's conception of Soviet behavior in the Arab–Israeli conflict – a conception that ruled out the possibility of Soviet intervention – was born during the May–June 1967 crisis that preceded the Six Day War and how it was subsequently shaped. The following section will describe the indications of Soviet involvement that became available to Israel after the Kremlin made its decision to intervene and how they were misread in AMAN's Research Department – the main analytical organ of Israel's intelligence community until 1974 – and by Foreign Office analysts, who played second fiddle in shaping the national intelligence estimate. Finally, we will show the impact of these erroneous intelligence estimates on Israel's policymaking process during the war and provide several explanations for this intelligence fiasco.

One note concerning the sources: With the exception of Israel's Foreign Office documents that have been recently released, most of the intelligence documentation of this episode is not yet accessible and is unlikely to become available until the year 2020. Consequently, the account below relies to a great extent on interviews conducted with the relevant intelligence officers, in addition to Foreign Office documents which only provide a partial description of the story.

'THE RUSSIANS ARE NOT COMING' 3

THE CONCEPTION AND ITS ORIGINS

Despite the traditional apprehension of Israeli leaders such as David Ben-Gurion towards Soviet anti-Zionist policy and the fact that since 1955 the USSR had become the main supplier of arms for the Arabs, it was not until the mid-1960s that Israel's intelligence community had any independent collection and analytical capabilities vis-à-vis the Soviet Union. In 1966 a Soviet section was created within the framework of a new branch – Branch 3 (the 'Superpower' branch) – of AMAN's Research Department. However, the ability of the new section to properly assess Soviet conduct was rather limited since its analysts did not speak Russian, possessed no academic or practical background in Soviet affairs and had no independent sources of information about the Kremlin's policies. Instead, they primarily relied on Western press reports, the translation of articles from the Soviet media into English, and occasional intelligence support from counterpart services.[4]

When the May–June 1967 crisis started and the possibility of war loomed, Israeli policymakers became concerned about the probable risk of the Soviet military intervening in the conflict in order to rescue their Arab clients from a devastating defeat. Based upon these circumstances, the inexperienced experts of Branch 3 were urged to present their assessment of the Kremlin's mode of operation. Since they had no independent ability to properly assess this issue, they found a solution in the form of oral evidence given by Yizhak (Berger) Barzilai – one of the Jewish leaders of the Palestinian Communist Party in the 1920s, who returned to the USSR in the 1930s, was expelled to Siberia, survived and, allegedly, was in touch with various segments of the Soviet leadership from the mid-1950s. He returned to Israel in the early 1960s. In a forum of senior IDF and intelligence officers that was convened during the May 1967 crisis in the 'pit' (the IDF underground headquarters in Tel Aviv), he estimated that although the USSR would use propaganda and diplomacy to support its clients, it would avoid any direct military intervention.[5] Probably on the basis of this assessment, the Director of Military Intelligence (DMI), Maj. Gen. Aharon Yariv, estimated that if Israel were to launch war the Soviets would make do with a warning at first stage but might intervene militarily later.[6] The war's events confirmed this thesis – the USSR threatened but eventually did not intervene. Thus the foundations of Israel's intelligence conception with regard to Soviet intervention were laid.[7]

The post-war period witnessed a massive shipment of Soviet arms by air and sea to their regional clients, a substantial increase in the number of their military advisors in the Arab armies, and an accelerated growth of their independent presence – in the form of Soviet naval and aviation bases – in Egypt and Syria. In late 1967, in order to meet the emerging challenge, AMAN had established a Soviet-designated Sigint unit titled 'Masrega' ('knitting-needle'). 'Masrega' was made up of Russian-speaking officers, most of whom immigrated to Israel in their youth during the 1950s. They were called 'Grechkoes' after the then Soviet Minister of Defense Andrei Grechko. Shortly after its inauguration, 'Masrega' started to yield high quality information about the Soviet advisors and their coordination of activities in Egypt and Syria (including any friction between the Soviets and their clients), the structure of their advisory contingents in both countries, and the activities of the Soviet Mediterranean fleet and the naval aviation units based in Egypt.[8]

AMAN's improved collection abilities regarding Soviet activities were not matched by a similar upgrading of its analytical know-how. The poor professional background of the Soviet analysts of Branch 3 remained the same as it was before the 1967 war and so did their main sources of information. The newly built 'Masrega' provided them with an important tool for the collection of Soviet military regional conduct but the understanding of Soviet politics and decision-making at the strategic level remained a weak spot.[9]

AMAN's conception of Soviet intervention in the Arab–Israeli conflict had been crystallized and occasionally tested since the war of 1967. In a nutshell, it said that since the Soviets had avoided intervening directly in conflicts outside the Warsaw Pact perimeter, and since this cautiousness was mainly the outcome of the Kremlin's fundamental interest in avoiding a direct confrontation with the US, the likelihood of such a move in the Middle East was very low. More specifically, the conception relied on three assumptions: *First*, the USSR's prime concern is the United States. Hence, it regards the Arab–Israeli conflict as a confrontation between a US proxy (Israel) and the Soviet ones – first and foremost Egypt and Syria. Consequently, the future of the conflict will not be decided by the local actors but by the two superpowers.[10] *Second*, the Soviets' most vital interest was to avoid a direct confrontation with the United States. Hence, firm US conduct is the best guarantee against Soviet expansion. AMAN's experts assessed that if faced by a Soviet challenge the United States would take all necessary measures to meet and deter a Soviet offensive move.[11] Finally, the *third* assumption is derived from the second: A firm Israeli stand against the Soviets is likely to reduce the probability of Soviet regional expansion.[12]

It is important to note that this conception and its practical implications constituted a sharp contrast to the traditional way in which senior leaders of the Jewish state – such as David Ben-Gurion, Moshe Sharet, Levi Eshkol, and Golda Meir – perceived the Soviet threat. Many of them were born in Tsarist Russia, experienced anti-Semitism and pogroms first-hand, and opted for Zionism rather than Communism as the way to solve the Jewish problem. They all feared the might of Communist Russia and the brute resolve of its leaders to use that power at all costs. Consequently, adopting a view that confronting the Soviet threat rather than avoiding it was the proper approach was a very difficult task for Golda Meir, who was the last national leader of this older generation. This position was more acceptable to the politicians of the younger generation, and among those who led the IDF to its stunning victory in 1967.[13]

Until the summer of 1969 the conception was not put to a test. Following the Six Day War, the Egyptians rejected the new status quo and attempted to change it by force. Israel's actions were mostly responsive and none of the IDF's acts were considered to be provocative enough to trigger a direct Soviet response. In July 1969, however, Israel began using the IAF as 'flying artillery' to compensate for its relative weakness in fire power in the static war along the Suez Canal. Despite AMAN's assessment that the Soviets would avoid direct intervention in order to assist their clients, the IDF Chief of Staff, Lieutenant General Haim Bar Lev issued an order to avoid attacking Surface to Air Missiles (SAM) sites and other military installations where Soviet advisors were present. Consequently, AMAN was requested to provide accurate information about the deployment of Soviet personal in the Egyptian Army – particularly within Egyptian anti aircraft artillery deployments which were a prime target of the IAF's attacks.[14]

As a result of the war's escalation, the threat of Soviet intervention had now become more realistic. AMAN's experts continued to argue in favor of taking a firm stand as a means to deter the Kremlin from stepping into the arena. If their advice was ignored, they regarded it as a grave mistake. Thus, following an aerial raid on the naval facilities of Port Said, in which a Soviet ship was damaged, Israel used secret diplomatic channels in order to make it clear to the USSR that the IDF was instructed to avoid targeting nonbelligerent personnel,[15] implying that the damage to the Soviets was collateral and unintended. AMAN's Soviet experts estimated that a far tougher message, one that would put the Soviets on notice that their personnel in the war zone were not immune, would have been a more suitable measure to deter them from expanding their involvement in the conflict.[16] Within a short period of time the dynamics of the conflict appeared to have confirmed this stand. Once again, the IAF attacked targets in Port Said and a number of Soviet sailors from a ship docked in the port were wounded. The Kremlin delivered a protest but it was drafted in a rather cautious tone. The Americans estimated that this was 'the least they [the Soviets] could do'.[17] Backed by the American assessment, AMAN recommended refraining from responding to the Soviet note and Golda Meir accepted the advice. As AMAN's analysts forecasted, the Soviets avoided making any additional protests and shortly afterwards their ships left the port.[18]

As is the case in many intelligence blunders, what at the time appeared to be a confirmation of the conception of Soviet unwillingness to take active measures against Israel enhanced the belief of AMAN's analysts in the validity of their thesis. It also boosted the confidence of the intelligence clients in the agency's ability to understand the Soviet *modus operandi*. As we now know, when the Soviet ships left Port Said, the Soviets were already preparing for the dispatch of significant military forces to the region. In other words, by the summer of 1969 AMAN's conception was already invalid.

In the autumn of 1969, Israel's policymakers started to contemplate the idea of using the IAF to destroy Egypt's air defense system in order to bring the war to the Egyptian hinterland. The main proponent of this escalation was Yitzhak Rabin – the IDF Chief of Staff during the 1967 war and the current Israeli Ambassador to Washington – who maintained close relations with President Nixon's national security advisor, Henry Kissinger. Deputy Prime Minister Yigal Alon, Israel's most brilliant commander in the 1948 war, who now had aspirations to replace Golda Meir upon her expected retirement, was Rabin's main supporter. Defense Minister Dayan – Alon's primary rival in the race to the office of Prime Minister – held a far more reserved view for fear of Soviet intervention. Abba Eban, the Minister of Foreign Affairs, objected to the deep-penetration raids mainly out of concern that it might damage the sale of American weapons systems, first and foremost additional Phantom F-4 planes, to Israel. Prime Minister Meir had no firm stand in favor of or against the operation. Her main concern was the Soviet reaction and she was about to make her decision based upon this standard.[19]

AMAN's estimate was presented in October 1969. It assessed that the USSR was unlikely to intervene directly in the conflict, primarily because of the risk of a direct confrontation with the US.[20] Nevertheless, even the orthodox supporters of this thesis did not completely rule out Soviet intervention if the IAF's attacks created a grave threat to Egypt. Some analysts in AMAN and the Foreign Office even questioned the United States' firm stand vis-à-vis the USSR in light of the lukewarm US response to aggressive communist moves, such as the Pueblo incident in January 1968, the invasion of Prague in August 1968, and Soviet encouragement of communist and Palestinian cooperation in Lebanon.[21]

AMAN's analysts determined that if the Kremlin was to take more drastic measures, it would be done gradually. That is to say that prior to any commitment of Soviet troops to fight in the region, Moscow would provide Egypt with more advanced weapons – perhaps the most advanced systems available in the Soviet arsenal.[22] To some extent, such an assessment was supported by reliable information. For example, AMAN knew that in early December 1969 an Egyptian delegation headed by Sadat arrived in Moscow in order to demand Soviet help and that the Soviets promised to supply SA3 batteries. AMAN was also aware of the fact that Egyptian crews had already started their SA3 training in the USSR and that they were likely to complete their course in July 1970.[23] On the one hand, such information confirmed the fundamental thesis, since it said nothing about Soviet intervention. But, on the other hand, Egypt's inability to operate the systems was also a well-known fact that AMAN's analysts ignored. Based upon the conception of cautious Soviet conduct, they also ruled out the possibility that the Soviet crews would be dispatched to bridge the time gap until July 1970 – a period that would allow Israel to exploit Egypt's vulnerability.

While AMAN's analysts avoided describing scenarios under which their conception might cease to be valid, a Foreign Office paper written by the head of Department Europe 3, Yohanan Cohen, attempted to raise a few situations under which Soviet intervention might take place. Although Cohen repeated the known arguments regarding why such a move was highly unlikely, he nevertheless developed, under the title of 'A speculative exercise', three scenarios of intervention. The first was a gradual escalation of the conflict with possible damage to strategic Egyptian and Soviet assets such as the Aswan Dam and Soviet Naval or Air Force installations. Such circumstances might trigger the enforcement of the Egyptian Air Force by Soviet pilots, allegedly in response to the assistance of American pilots in the IAF. Further escalation of the Israeli offensive might lead to the expansion of an independent Soviet presence in Egypt. A second scenario involved the breakout of an unintentional new war, in which Egypt would find itself in a dire situation. Under such a circumstance, the Soviets might send 'volunteers' in order to put an end to the fighting and promote a political solution which would call for the reopening of the Suez Canal. The third scenario involved an Israeli attack against Egypt and/or Syria. Here the paper foresaw an immediate Soviet intervention since the Kremlin could not afford another humiliation and would probably assess that 'progressive public opinion' in the US might hamper a strong American reaction.[24]

This paper, however, was an exception; the dominant view continued to regard Soviet intervention as highly unlikely. It was not only supported by the CIA's estimates but also by the assessments of other senior officials in the US administration, as well as in US academia. Thus, for example, in a talk with an Israeli diplomat in mid-September Professor Nadav Safran of Harvard University estimated that the USSR would, under all circumstances, avoid any direct intervention.[25] Helmut Sonnenfeld from the Operations Staff of the NSA assessed in late October that in light of the Israeli deterrent factor and due to insufficient logistical means and fear of a confrontation with the US, the Kremlin was unlikely 'to operate a military force' in the region.[26] The head of the USSR Department in the State Department assessed in mid-January 1970 that the Soviets did not seem to be too worried about the situation in the Middle East, since they were not rushing to find a diplomatic solution.[27] Two weeks later, despite the information about Nasser making a secret visit to Moscow on 22–26 January, his chief expert for Soviet affairs still estimated that the Kremlin wanted to avoid any expression of support for a military solution. Instead, the Soviets would try to deal with Israel through diplomatic negotiations with the US, hoping that this would lead to American pressure on Israel to restrain its attacks against Egypt.[28] Professor Zbigniew Brzezinski of Yale University forecast that the Kremlin would adhere to a prudent policy and would not cross the line between involvement and intervention because of the American deterrent posture.[29]

AMAN's estimate regarding Soviet reaction to the deep-penetration raids became crucial as soon as the Israeli Security Cabinet's approval for Operation 'Blossom' was requested in late 1969. AMAN presented a type of interrelated assessment that explained the casual links between three main factors: Soviet reaction to the Israeli raids, the firm US stand vis-à-vis the Soviets and the probability of obtaining a desirable outcome on the Egyptian front. The Director of Military Intelligence, Major General Aharon Yariv, assessed on the basis of the common intelligence conception that the expected volume of the bombings would not prompt the Soviets to send fighting forces to Egypt.[30] Moreover, according to AMAN's experts, the US was expected to take a firm posture to prevent the Soviets from making an aggressive move, thus reducing the likelihood even more. AMAN also estimated that the deep-penetration raids might have a devastating effect on Egypt's morale and stamina and should pressure Nasser to end his War of Attrition[31] or should even cause him to lose power.[32] Rabin, the ambassador in Washington, provided the Cabinet with a detailed explanation as to why the White House would support the increase in Israeli pressure on Egypt,[33] and the IDF commanders, encouraged by the newly appointed Transportation Minister, Major General (Res.) Ezer Weitzman, who was the former legendary commander of the IAF, assured the Cabinet that the new Phantom F4 attack planes were capable of conducting this mission.[34] Notwithstanding continued concerns and debates over the US reaction to such an escalation, but relying solely upon AMAN's assessment of the low probability of Soviet intervention, the Cabinet approved the IAF to launch Operation 'Blossom' on 6 January 1970.[35]

THE INFORMATION

When the decision was made to start the deep-penetration raids in Egypt, AMAN (and as far as is known the CIA also) had no concrete information that Soviet preparations to intervene in the conflict were underway. And yet an ample quantity of indirect but valuable indications that if interpreted properly might have changed the estimate of Soviet behavior was available to AMAN. This, however, was only recognized *post factum*. The closer the arrival of the Soviet military units was, the more hard intelligence indications were accumulated in AMAN. But as will be shown below, for a period of almost five months the incoming information failed to convince AMAN's analysts that their thesis – that the Soviets would not intervene – had lost its solid ground. It only collapsed in early March 1970 when the presence of Soviet military units on Egyptian soil was positively identified.

Below are the indications about the coming Soviet intervention which AMAN received between the summer of 1969 and March 1970. These are organized according to their means of collection.

The Soviet Media

First indications of a possible shift in the Kremlin's traditional policy appeared in the Soviet press shortly after Operation 'Boxer' – the massive use of the IAF as 'flying artillery' – began. In August 1969, the Soviet media started an intensive anti-Zionist campaign.[36] On 31 October 1969 a TASS statement accused the US of 'offering Israel an opportunity not only to receive military aircraft, but also to have the use of American pilots and personnel for technical maintenance'. The statement warned that the American decision 'could lead to serious complications in the Middle East'.[37] Such accusations, which continued to appear in the Soviet media, especially in *Pravda* and *Krasnaya Zvezda*, were aimed at preparing local public opinion for a similar move by the Soviet military.[38] Following the TASS statement, the anti-Zionist campaign intensified. In addition to comparing Israel to Nazi Germany, it also repeated accusations regarding the alleged participation of American pilots in the IAF's bombing campaign in Egypt.[39]

The November 1969 issue of *Communist*, the journal of the Central Committee of the Communist Party of the Soviet Union (CPSU), carried an editorial written by the chairman of the International Department of the Central Committee, V. Rumiantsev, about the USSR's relationship with Egypt since 1955. It emphasized the military aspects of the relationship and the way they developed after the 1967 war.[40] Given that *Communist* was a major organ of the Communist Party it can be assumed that the editorial aimed at preparing party members for a major shift in the Kremlin's Middle Eastern policy. Another indication of this potential shift was a firm statement about the Arab–Israeli conflict that was released at the end of the Prague summit of the Eastern bloc (excluding Romania).[41] The anti-Zionist campaign reached its peak around March 1970 and was at a level similar to that of the anti-Israeli campaign waged in the Soviet media during the 1967 war.[42] But by this stage belief in the Soviet non-intervention policy was already shaken.

The media campaign failed to change the conclusions drawn by AMAN's analysts for three reasons: First, since they did not read Russian they had to rely on the western press, primarily on the *New York Times*,[43] as the main source of information about the content of the Soviet press campaign. This resulted in their tendency to neglect this source of information. Second, such campaigns were not unprecedented. During the 1967 war Moscow conducted a similar anti-Zionist campaign without intervening in the conflict, and allegations about the American Jewish pilots in the IAF had surfaced before. And, finally, since their American counterparts did not estimate that this campaign was aimed to promate Soviet intervention, AMAN's analysts, who were highly dependent on their assessment, chose to ignore any information that pointed in a different direction.

Diplomatic Signals

Kosygin's note. The most important means used by the Kremlin to signal its intention to take drastic measures was a note from Prime Minister Alexei Kosygin delivered to President Richard Nixon of the USA, President Pompidou of France, and British Prime Minister Harold Wilson on 31 January 1970. Together with the USSR, these three western nations combined to form a central forum to advance peace in the Middle East. At the center of the note stood the following warning:

> We would like to tell you in all frankness that if Israel continues its adventurism to bomb the territory of the UAR and of other Arab states, the Soviet Union will be forced to see to it that the Arab states have means at their disposal, with the help of which a due rebuff to the arrogant aggressor could be made.[44]

Compared to previous Soviet warnings, this one was phrased in a relatively mild tone and avoided threatening the use of Soviet force. This is certainly the case when compared to the warning which was delivered during the last hours of the 1967 war and was still fresh in the memory of both the Israelis and the Americans. Demanding that Israel accept a ceasefire and stop advancing towards Damascus, Kosygin's note of 10 June 1967 did not mince words and simply warned that unless these demands were met, the USSR would take necessary steps, including military. The threat unambiguously implied the Kremlin's readiness to use force to stop Israel advancing any further.[45]

Given the lack of a blatant threat of intervention, there is little wonder that the present note was interpreted in Washington, as well as in London, Paris and Jerusalem as an indication of the Kremlin's reluctance to send forces to the region. Joseph Sisco, the State Department Undersecretary for Political Affairs, in a closed meeting with congressmen estimated that the note was the result of high-level Egyptian pressure on the USSR to take firmer measures, and reflected the Soviet interest in advancing diplomatic negotiations in order to find a way out of a difficult situation. The only threat that it carried was to supply the Arabs with more advanced weapon systems[46] – a move regarded as being very likely for quite some time. The British Foreign Office spokesman responded to media reports that the note carried an intervention threaten by saying that, 'It was completely untrue that... [it] contained at any point a threat of military intervention in the Middle East conflict'.[47] The French estimated that the Soviets probably had no intention of intervening directly in the conflict and therefore they 'would not send pilots or other volunteers'.[48] The recipients of Kosygin's note were not the only ones to underestimate its implications. A Romanian diplomat in the embassy in Paris told his Israeli colleague that although the Kremlin wanted to prevent Nasser's fall, they 'would not go beyond threats to ship more modern aircraft and surface to surface missiles. They are unlikely to send volunteers'.[49]

Despite this generally calming approach, some officials with whom the Israelis spoke were more concerned. Helmut Sonnenfeld, who four months earlier regarded a possibility of Soviet intervention as highly unlikely, estimated in mid-February that the Kremlin faced two options in response to what it regarded as an Israeli provocation: Either to pressure Egypt to agree to a ceasefire – a move that was unacceptable for both Egypt and the USSR since it would be interpreted as an American–Israeli victory, or to take more effective measures to help Egypt, which actually meant the 'deployment of Soviet personnel'. He added that such a move constituted an escalation, which the Soviets did not like.[50] Most notably, Foreign Minister Eban, who objected to the deep-penetration raids, regarded Sonnenfeld's estimate as sufficiently disturbing to demand that Prime Minister Meir re-open a discussion of the issue in the government.[51]

The Soviet note was discussed in the Israeli government on 8 February. The IDF Chief of Staff and AMAN's director participated in the meeting and influenced its atmosphere, as was reflected in a report in the daily *Maariv* – a report that was based on confidential information.[52] According to this report, the ministers heard that:

> The analysis of President Nixon's response [to Kosygin's note] indicates that Washington decided not to retreat... and that it sees the proposals of [Secretary of State William] Rodgers as the final border of concessions. If, indeed, the USSR will realize that there was no chance to blackmail additional concessions from the US, it could recommend that the Egyptian President be more flexible with regard to the Rodgers plan. Summing it up [the ministers] agreed that the Soviet move did not reach its goals.[53]

While AMAN's written estimates are not yet accessible to the public, the available Foreign Office intelligence assessment (which appears to reflect the consensus within the Israeli intelligence community) summarized the situation as follows:

1. The Soviet note is the outcome of the pressure yielded by the IAF raids in Egypt's hinterland, the Egyptian inability to prevent it, and a lack of a clear decision within the Soviet leadership as to how to act in the near future.
2. At this stage, the USSR is not inclined to respond to Nasser's demands and is not prepared to expand its involvement and the level of its intervention.
3. The present situation presents [the Soviet] leaders with a dilemma for which they cannot provide a complete answer. The developments in the Middle East are not according to their forecast and the horse on which they gambled is unsatisfactory – on the one hand they cannot abandon Nasser, but on the other there are limits to their support of Egypt.
4. Kosygin's notes served as an immediate outlet. But it is possible that the USSR will not be content in the future with such means and will look, in light of a new estimate of the situation, for new ones.[54]

When attempting to predict future Soviet policy if Operation 'Blossom' continued, the document assessed that the Kremlin would gradually increase its support of Egypt by supplying advanced weapon systems such as the Mig-23 and Sukhoy-9 aircraft, in addition to increasing the number of advisors, instructors, and technicians necessary for the manning of the more complicated systems. The paper also anticipated a quantitative projection of Soviet independent presence in sensitive sites in Egypt and Syria in order to deter Israel from attacking these targets. Summarizing its forecast, the Director of Europe 3 noted that:

1. The USSR does not want war and will do everything to avoid a direct intervention.
2. The USSR will do its best... to strengthen its regional base and bring the Arab nations a political solution.
3. Nevertheless, the possibility that the USSR would see no other outlet except for a very direct intervention (and on this I wrote separately) cannot be ruled out.[55]

In mid-February AMAN was continuing to reject any notion that the Soviets might intervene in the conflict. To some extent they could rely on American estimates. Two weeks after Kosygin's note was delivered the Americans, according to an Israeli Foreign Office document, assessed that the Soviets would attempt to avoid escalation and 'would limit themselves to strengthening Egypt's anti-aircraft capability by 85mm and 100mm anti-aircraft guns, SA2 missiles and also perhaps SA3 (*the maximum the Americans foresee is a Soviet manning of the SA3*); they would avoid supplying surface to surface missiles'.[56]

Additional signals. While Kosygin's note provided no clear warning that the USSR might intervene in the conflict, a number of Soviet officials hinted that this might be the case. The first known message to this effect came from a Soviet diplomat in London, who leaked to the press that the USSR will intervene if a war breaks out. It did not cause any concern in Jerusalem and was regarded as part of the Soviet campaign to wage psychological warfare that was aimed at intimidating Israel.[57] A few days later, the second secretary in the Soviet embassy in Washington, Boris Sedov, invited the American Jewish journalist Milton Friedman, who was known to have good connections with Israel, to lunch. Sedov spoke about the Israeli hijacking in late December 1969 of a Soviet-made P-12 RADAR from Egypt (IDF code-name Operation 'Rooster 53'). He added that some in the USSR 'regarded this act as a very serious business. In the future we will ensure that modern and costly equipment will be guarded by us and the Israelis will have to think twice'. The Russian diplomat repeated this warning a few times and Friedman interpreted it as a Soviet message that was to be passed on to Israel.[58]

The Americans received additional warnings but it is unclear whether they delivered them to Israel. On 6 February a junior Soviet official initiated a conversation with the Pentagon correspondent of the *New York Times* in order to emphasize the gravity of Kosygin's note and to warn that if the demand to stop supplying American weapons to Israel was ignored, the Soviets would deliver 'offensive weapons' to their clients. King Hussein of Jordan told the American ambassador in Amman that the most recent Soviet warning was more serious than before. The American representative in Cairo heard a similar evaluation from the Finish ambassador stationed there. According to another source, French journalist, Eric Rouleau, Egyptian officers on the Suez front interpreted Kosygin's note as an indication of impending Soviet intervention.[59]

It is important to note that Kosygin's message and the above-mentioned information were only interpreted *post facto* as indications for the incoming Soviet intervention. In fact none of these pieces conveyed any clear message about the Kremlin's decision to intervene. Moreover, recently released information shows that the Soviets not only tried to hide their military preparations but also attempted to mislead the Americans regarding the reality of the looming threat. Thus, for example, following a *Pravda* article and talks with a Soviet representative in the embassy in Washington on the eve of Kosygin's note, State Department officials concluded that the Kremlin had no interest in committing itself to an action that might be interpreted as support for an Egyptian military solution.[60] Later in February an American journalist was approached during the opening ceremony for a Soviet exhibition in Washington by two Soviet officers – Lieutenant Colonel Kuzma Meleshkov and Commander Ivan Sakul'kin – who served as assistants to the Soviet Military Attaché in the capital. The two told him that in light of Nasser's difficult situation he now insisted on 'unconventional weapons' – a demand that Moscow fully rejects. Instead, they assessed that the Egyptians were likely to 'receive more artillery and aircraft, so that they'll have additional equipment to lose'. The State Department interpreted this incident as a deliberate message to the US – that they should not be worried about a possible supply of weapons that could hit Israel's population centers.[61] In retrospect, it can also be interpreted as being aimed at calming American fears (emerging by this stage) of a drastic Soviet move. The Israeli reaction to this incident affirms this possibility: The Director of Department Europe 3 in the Foreign Office assessed that through this initiative the Soviets tried also to 'underestimate the value of the military support they provide Egypt and its effect'.[62]

Given the vague character of the Soviet signals, it is not difficult to understand why American and Israeli analysts failed to comprehend their true meaning. One veteran State Department analyst concluded that the American failure stemmed from two main causes: the low-level channels through which the Soviet warnings were delivered (e.g., from low-ranking Soviet officials to American journalists) instead of the Dobrynin–Kissinger channel; and the conception that was shared by all about Moscow's prudent Middle Eastern policy and its supreme interest in avoiding a confrontation with the US.[63] The Israelis maintained a similar approach because of the ambiguity of the Soviet messages. After all, none of these vague signals specified that direct intervention was a possible course of action.

Information from Intelligence

History shows us that it is very difficult to hide preparations for a large scale deployment of new missile systems. British intelligence located the German 'ski jump' launchers of the V1 in the area of Pa de Calet months before the first ones were launched in the second week of June 1944. Similarly, air photographs of SA2 sites, and later on of ground preparations for sites of the SS4 constituted the clearest indication of the positioning of offensive missiles in Cuba in September–October 1962.[64] In both cases, photographic evidence combined with additional intelligence information helped to properly assess the nature of the developing threat. The case of the deployment of the Soviet-manned SA3 batteries in Egypt in 1970 could be said to be a similar situation, but eventually it was shown to be an exception.

The T dugouts. In mid-January 1970 Israeli interpreters of aerial photographs started to identify trenches in the shape of the letter T that had been excavated in several districts of Egypt.[65] In early February it became clear that the gigantic building project was gaining momentum and T-shaped dugout sites were now located near Cairo and in a strip of a 15 kilometer-wide section of the central sector of the Canal front, which was around 15–30 kilometers from the frontline. AMAN's analysts, who had never seen such dugouts, could not explain what stood behind them.[66] But even without realizing exactly what their role was, in early February the IAF started to bomb them. The Egyptians intensified their efforts to build them up and in the process paid a heavy price. In the months that followed, thousands of Egyptian workers were killed in the IAF raids that succeeded in delaying the completion of the project.[67]

Israel's inability to correctly assess what stood behind the intensive Egyptian building project seems odd even when viewed with a 35-year perspective. AMAN's analysts estimated that the dugouts were prepared for the positioning of new weapons systems, perhaps anti-aircraft ones, but did not specifically indicate that they were the SA3 missiles. Instead, they assumed that they were being prepared for the positioning of decoys for the SA2 systems, or surface to surface missiles.[68] This was strange since the shape of the SA3 dugouts in the USSR was no secret to their American counterparts who had uncovered them through satellite photography in the military districts of the USSR and the Warsaw Pact countries since the early 1960s.[69] And yet, despite their acquaintance with these systems and the close intelligence cooperation between the two states, Israel did not receive any indications from the US that the newly built dugouts were sites for SA3 missiles.[70] Since the Israelis themselves had no familiarity with modern Soviet weapon systems which were not present in the local arena, they had to depend on American expertise in this area. Consequently, the failure of the American analysts to link the build-up project in Egypt to the already familiar SA3 dugouts in the Soviet Block hindered the Israeli ability to properly grasp their meaning.[71]

SIGINT information. Unlike the officers in AMAN's Research Department who believed that the USSR had no intention of intervening in the conflict, the officers in AMAN's SIGINT unit responsible for monitoring Soviet regional activity ('Masrega') were highly concerned by such a possibility. Since the establishment of 'Masrega', one of its main tasks was to collect indicators for a possible Soviet intervention. And although the most likely scenarios were an intervention by airborne or marine forces, the unit's staff always kept their eyes wide open for any indication of other forms of intervention as well.[72]

Since late 1969 and early 1970, 'Masrega' had begun identifying indicators for irregular activity in Soviet behavior in Egypt. While the precise content of these indicators is still not accessible, the available information shows that it included the following information:

- *A marked increase in the number of Soviet VIP flights between the USSR and Egypt.* This started in late December 1969 and since the officers of 'Masrega' were already able to differentiate between civilian and military flights, they concluded that most of the Soviet visitors to Egypt were high-ranking military officers.[73]
- *The buildup and activation of new communication channels between the USSR and Egypt.* Given that the number of Soviet military personnel in Egypt was to quadruple from 3,000 advisors and instructors in February 1970 to more than 12,000 in mid-March, a new communication infrastructure had to be built to support this radical change. Since the beginning of 1970, 'Masrega's' staff had begun identifying new communications traffic from Egypt. The patterns of the new traffic differed from that of advisors and instructors. The traffic was sporadic and, as became known at a later date, it reflected the activation of the new communication lines to the USSR.[74]
- *The appearance, from early 1970, of electronic intelligence signals that indicated the introduction of new weapon systems, including Radars.*[75]
- *A sharp increase, from late January, in the number of Soviet cargo flights to Egypt.* No information was available about the nature of the cargo, but it was estimated that the planes carried equipment for anti-aircraft systems (missiles, guns, and Radar and command and control equipment).[76]
- *Verbal communication.* In some cases 'Masrega' intercepted verbal communiqués that indicated the possible arrival of Soviet combat units in Egypt. The most important piece of information, from a highly reliable Egyptian source with good access, discussed operational parameters of the arrival of Soviet fighter planes implicitly with their pilots. The interpretation of not only this communiqué but of others, of a lower quality, had become a bone of contention between the commander of 'Masrega', Tuvia Fineman, and his chief, the head of the communication intelligence branch in unit 848, Colonel Yoel Ben Porat, on the one hand, and the analysts of AMAN's Research Department, who refused to give up their conception that the Soviets would not intervene, on the other. Eventually, it was AMAN's analysts and not the COMMINT experts who determined Israel's national assessment of this issue.[77]

SURPRISE AND ITS CONSEQUENCES

In mid-February Israeli officials began to realize that the American analysts were no longer certain that the Soviets would continue to refrain from becoming actively involved in the conflict. The first sign, as recalled, was Sonnenfeld's warning of 13 February, which led Foreign Minister Eban to request that Prime Minister Golda Meir raise the possibility in the upcoming government meeting of an unexpected Soviet move.[78] A few days later Mark Sedan, a senior analyst from the Bureau of Intelligence and Research (INR) in the State Department, told Israeli diplomats that he, personally, did not 'exclude the possibility that the USSR will provide ground teams to operate new weapons, at least for the first period'.[79]

The only concrete information regarding possible shipments of SA-3 batteries to Egypt came from French diplomatic sources. Already, on 10 February, they made inquiries as to whether Israel had any information concerning massive deliveries of Soviet anti-aircraft equipment, including SA-3 batteries, to Egypt.[80] After checking with the relevant agencies (i.e., AMAN), the Israeli answer was negative.[81] Five weeks later, based upon British information, this same French source reported that the shipment of SA-3 batteries, and the Soviet teams necessary to operate them, had occurred. The source had doubts about the validity of this information in light of the fact that the French diplomats in Cairo did not report such an event.[82] But this time the Israelis did not need external help to understand that for more than half a year their strategy in the conflict with Egypt had rested on the basis of an erroneous estimate.

Most Israeli intelligence officers involved in the assessment of Soviet policy during the War of Attrition agree that the Soviet decision to intervene militarily in Egypt in 1970 surprised Israel, although they disagree about the magnitude of the surprise and its gravity.[83] David Korn, then a political officer in US Embassy in Tel Aviv, reinforces this view, stating that Israeli officials were shocked by the Soviet move.[84] The CIA memorandum issued at that time declared that Israel had apparently not anticipated that its air attacks on potentially sensitive targets deep inside Egypt would result in such a rapid and impressive Soviet reaction, and that Israel was uncertain as to what the presence of the Soviet forces signified, not only in the short term but for the years ahead.[85]

Israeli policymakers clearly expressed their concerns regarding the unexpected developments. On Friday 20 March 1970, Defense Minister Moshe Dayan faced the nation to inform the Israelis about the dramatic change in the situation. He also used this opportunity to propose new rules of engagement to the Soviets. Dayan admitted that Israel's situation had not only worsened since the IAF now had to face a more sophisticated anti-aircraft system, but also because of the introduction of Soviet forces *per se* to the theatre. He minced no words in signaling to the Kremlin that Israel had no interest in engaging Soviet soldiers in combat, since it had had no capability to confront the USSR. His analysis of the situation was rather straightforward: 'if ultimately the USSR would decide to enter this war in full force and the US will not restrain it and help Israel, then we will certainly be in a difficult situation'.[86] Consequently, he made it clear that the IAF would stop its raids on Egypt's rear which was now defended by the Soviets. But Dayan also insisted that Israel would do all it can to prevent the introduction of Soviet-manned weapons systems and Soviet pilots to the Canal front. At this stage, it was still unknown that two Soviet fighter squadrons had already been deployed in Egypt for approximately two months. Dayan assessed that the Kremlin would avoid such a move.[87]

Dayan's attempt to set the rules of engagement with the Soviets failed. They did not respond to his public statements and, in contradiction to American assessments that they would avoid reinforcing their units in Egypt with fighter planes[88] and would not advance the SA-3 batteries' layout to the front,[89] the Kremlin chose to escalate the situation. By early April 1970 the Soviet deployments started moving eastward to the front, using the T-shaped dugouts that the Egyptians continued to build throughout this period. On 18 April 1970 Mig-21s flown by Soviet pilots who made no effort to disguise their identity intercepted two IAF Phantom jets on a reconnaissance mission south of Cairo. According to the rules suggested by Dayan a month earlier, the IAF started bombings the T-shaped positions in order to prevent the Soviet SA-3 batteries from reaching the front line. At the same time, in a more defensive response, Israel limited its activity in Egypt's rear.

The following months saw the emergence of a direct Soviet–Israeli confrontation. In early May 1970 the Egyptians intensified their efforts to construct positions for the SAM layout on a 15-kilometer-wide strip along the central sector of the front, approximately 15 kilometers from the Canal. The intensive IAF bombings put an end to this attempt and led to thousands of Egyptian casualties, of mainly civilian workers. Both the Soviets and the Egyptians drew the proper conclusions from this failure: Instead of trying to build a comprehensive anti-aircraft layout, they shifted to a system of 'missile ambushes' by sneaking, under the cover of darkness, a number of SA-3 batteries to the front and surprising the IAF attack planes by shooting at them from behind. In June and July 1970 the confrontation escalated. Soviet-manned SA-3 batteries shot down five Phantom Jets and killed some of Israel's best pilots. Moreover, in a number of cases, Soviet pilots intercepted the attacking Israeli planes. The Israelis decided to signal to the Kremlin the possible cost of this escalation. On 30 July 1970 the IAF ambushed Soviet interceptors and a large-scale Soviet–Israeli air battle took place. Five Soviet

Mig-21s were shot down and three Soviet pilots died without the Israelis suffering any casualties. A week later, under heavy American and Soviet pressure, Israel and Egypt signed a 90-day ceasefire agreement that ended the War of Attrition and launched a more than three-year period of calm along the Suez Canal. But unlike the situation in January 1970, when Operation 'Blossom' started and the Egyptian ground forces lacked any effective air defense, by late August 1970 the Egyptians, with considerable Soviet assistance, had built a large-scale anti-aircraft layout west of the Canal. Its effectiveness was fully realized only in October 1973 more than a year after the Soviet Air Defense division had left Egypt.

THE CAUSES OF ISRAEL'S ASSESSMENT FAILURES

The available information about the dynamics that led to AMAN's failure to correctly assess, at least by mid-February 1970, the likelihood of a Soviet military intervention, is limited. Nevertheless, a number of explanations can be proposed to explain this intelligence blunder. The most plausible one is the fact that the move made by the USSR in 1970 had no precedents. The Soviet Army was sent to Budapest in the autumn of 1956 to crush the Hungarian uprising and in August 1968 it invaded Czechoslovakia in order to put an end to the Prague Spring. Soviet and Chinese soldiers also clashed along the Soviet–Chinese border in 1969. But a direct military intervention in a conflict far from the Soviet borders and outside of the Warsaw Pact parameter was a move that until 1970 the Soviets had avoided, even when they threatened to take it, as was the case in both the Suez campaign of 1956 or the Arab–Israeli war of 1967. Certainly, the Soviets took grave risks by directly confronting a very close and militarily capable ally of the United States far from their borders. But the Israelis failed to realize how risky, from Moscow's perspective, the alternative was. After the 1967 war the Soviets made an unprecedented effort to rebuild and to reshape the Egyptian military. Therefore, an Egyptian defeat in 1970 could have been interpreted as a defeat of the Soviet Army itself. The false realization of Soviet intentions also stemmed from an insufficient understanding of the gravity of Egypt's situation. In retrospect, Israeli intelligence analysts admitted that they did not comprehend the devastating effect that the deep-penetration raids had on Egypt.[90] Consequently, they could not properly assess the Soviets' interest in assisting the Egyptians – an interest that led the Kremlin to break from its policy of avoiding direct military intervention far from the USSR's borders.

Another source of the blunder involves the failure of both American and Israeli intelligence experts to correctly realize the extent of influence that the Soviet hawks had amassed in the Kremlin of the late 1960s, and their proclivity for risk-taking behavior. In addition, Soviet experts in both countries failed to comprehend the global view that the Kremlin held during this period. In 1969, the Soviet leaders estimated that the US was too preoccupied with the war in Vietnam and felt that, for that reason, it would not react militarily to an escalating Soviet move in the Middle East.[91] And the Kremlin was correct in its analysis. As Kissinger had put it in his memoirs, 'President Nixon has not abandoned [in 1969] some vague notion of a trade-off with the USSR between the Middle East and Vietnam'.[92]

A third explanation involves the relatively calming nature of Soviet signals prior to the intervention. Kosygin's note could not be interpreted as laying the ground for a drastic move and the Kremlin avoided making any gestures that might indicate that it was ready to commit forces to the Egyptian–Israeli conflict. To some extent, the opposite was true. The obvious interpretation of Kosygin's warning, that the continuation of the IAF bombings would lead the USSR to 'see that the Arab states have means at their disposal to rebuff the arrogant aggressor' was that the Soviets would provide their clients with more advanced weapons. In this sense, one cannot escape drawing the conclusion that the prime consideration once the decision to intervene was made, was to bring the Soviet forces in Egypt to an operational status as quickly as possible, in order to minimize the window of vulnerability when the IAF could hit them before they were ready to defend themselves. Consequently, the Kremlin opted to disguise the nature of its move rather than to provide a more blatant warning that could have generated stronger American pressure on Israel to cease the bombings in Egypt.

A fourth explanation involves the lack of sufficient experience of AMAN's analysts in Soviet affairs. As previously noted, AMAN's Soviet analysts did not speak Russian and had no academic background in Soviet studies. Moreover, AMAN had only recently begun allocating resources to this field, and had no institutionalized memory of Soviet politics and Soviet foreign policy conduct. As a result, the agency lacked sufficient expertise to independently assess the Kremlin's policy, especially when it came to such an unusual move as the intervention decision. Hence, in making its judgment AMAN had to rely excessively on foreign assessments, primarily on the American estimate of the likelihood of Soviet intervention. And, despite their well-established professional expertise and their far better means of intelligence collection regarding Soviet global and regional conduct, the Americans erred here as well.

The best explanation for the American misdiagnosis of Soviet policy was offered by Richard Parker, who in 1970 was already an experienced diplomat in Middle Eastern affairs and was involved in the State Department's assessment of the Egyptian–Israeli conflict and the Soviet conduct. According to Parker, the American mistake was largely the outcome of a lack of imagination that came into expression by placing too much reliance on past Soviet conduct as the basis for their assessment of any future behavior. It also stemmed from an insufficient understanding of the Egyptian suffering caused by Israel's overwhelming air superiority. American Soviet experts were not familiar enough with the situation in the Middle East and this increased their difficulty in grasping the urgent Soviet need to assist the Egyptians. In addition to these professional mistakes, Parker maintains that in the atmosphere that dominated the American intelligence community at the end of the 1960s expressing sympathy with any Soviet difficulties could have been viewed as being anti-American. Consequently, the intelligence analysts were reluctant to put themselves in the Soviets' shoes.

Finally, it should be noted that one group of intelligence experts did suggest that the Soviets were likely to intervene. This group was made up of the 'Masrega' officers who in early 1970 started warning that the Soviets were coming. But since their role was collection and not analysis, their warnings were ignored. The fact that the collection officers assessed the situation correctly, while the analysts in AMAN's Superpowers Branch did not, reaffirms the common wisdom that adhering to intelligence information rather than indulging in sophisticated conceptions is the correct way to formulate good intelligence estimates. Unfortunately for them, the Israelis did not learn this lesson in 1970. Less than four years later, on 6 October 1973, they paid a very dear price for a similar failure.

NOTES

1 The USSR considered dispatching Air Defense combat components to Egypt as early as 1968, following the permanent deployment of both Soviet Navy and Air Force units to Soviet installations based in Egypt in 1967–68. In light of Egypt's inability to provide adequate air defense for these bases, the Kremlin considered sending in its own air defense units. See Anatoly Egorin, *Egypet Nashego Vremeny* (Moskva: Institut Vostokovedenia RAN 1998) p.185; Alexandr Vasiliev, *Rossia na Blizhnem I Srednem Vostoke: ot Messianstva k Pragmatizmu* (Moksva: Izdatelstvo Nauka I Vostochnaia Literatura 1993) pp.94–7; Igor Hmelnov, *Rossiiskii Flot: Doblest I Nisheta – Zapiski Admirala* (Moskva: Ast Press Kniga 2003).

2 For Soviet perspectives of 'Operation Kavkaz', see Vladimir Il'in, *Istrebitel Phantom F4* (Moskva: OOO Izdatelstvo Astrel 2001); Valerii Safonov (ed.) *Grif Sekretno Sniat. Kniga ob uchastii sovetskih voennosluzhashih v Arabo Israelskom konflikte* (Moskva: Sovet Veteranov Boevih Deistvii v Egypte 1998); Evgeny Pirlin, *Trudnii I dolgii put' k miru; Vzgliad iz Moskvi na problemy blizhnevostochnogo uregulirovanija* (Moskva: ROSSPEN 2002); Mihail Zhirohov, *Krilia Vozmezdia: Kratkaia istoria VVS Israelia* (Moskva: ACT Harvest 2001); V. Zolotariov and V. Vartanov (eds.) *Rukopozhatie Cherez Chetvert' Veka 1973–1998; Materiali Nauchno-Prakticheskoi Konferencii* (Moskva: Institut Voennoi istorii Ministestva Oboroni Rossiiskoi Federacii 1999); V. Zolotariov, *Rossia (SSSR) v Lokalnik Voinah I Voennih Konfliktah Vtoroi Polovini XX veka* (Moskva: Institut Voennoi Istorii Ministerstva Oboroni RF 2000); Vladimir Vinogradov, *Diplomatija: Liudi I Sobitija. Iz Zapisok Posla* (Moskva: ROSSPEN 1998); Vadim Kirpichenko, *Iz Archiva Razvedchika* (Moskva: Mezhdunarodnie Otnoshenia 1993); Ruslan Bedoev, 'Voevali tri tovarisha', *Krasnaia Zvezda*, 16 August 2003; V. Buligin, 'Voina v Nebe Egypta: K 30 Letiu uchastia sovetskoi

aviacii v boevih deistviah na Blizhnem Vostoke v 1969–1971 godah', *Tehnika I Vooruzhenia* (March 2000); A. Dokuchaev, 'Tot Egypetskii god', *Krasnaia Zvezda*, 25 March 1989; Valerii Elchanin, 'Dan Prikaz Emu v Egypet!', *Soldat Udachi* (February 2001); Anatoly Usikov and Valery Jaremenko, 'Flot kak Instrument Politiki', *Nezavisimoe Voennoe Obozrenie*, 29 August 2003; Lieutenant A. Klemenko, 'Vstrechia Spustia Destiletija', *Vo Slavu Rodini*, 15 March 2003; Igor Kulikov, 'Kremnii Prevratilsia v Pesok', *Nizhegorodskiy Rabochiy*, 11 July 2000; A. Lialkina, 'Dni Sadata', *Zerkalo Nedely*, 13–19 October 2001; Dmitry Makarov, 'Neizvestnaja Voina na Vostoke', *Argumenty I Fakty*, 25–31 August 1990; E.I. Malashenko, 'Linija Fronta – Suezky Kanal', *Voenno – Istoricheskiy Zhurnal* (March and April 1994) No.3 pp.36–42; No.4 pp.53–60; L. Nikolaev, 'Arabskii As Nomer Odin', *Eho Stolici*, 6 August 1999; idem, 'Znoinoe Leto Semidesiatogo', *Eho Stolici*, 6 September 1999; Ratmir Orestov, 'Operacia Kavkaz', *Eho Planeti* (February–March 1994); Andrei Pochtariov, 'Moskva I Kair Vspominajut', *Krasnaia Zvezda*, 31 January 2003; idem, 'Kak podrezali Krilia Fantomam – Interview with General Colonel Aleksei Smirnov', *Krasnaia Zvezda*, 14 January 2000; idem, 'Blizhnevostochnii Triumf Letuchih lisic', *Krasnaia Zvezda*, 17 August 2002; Col. I. Probilov, 'Mi tozhe voevali', *Krasnaia Zvezda*, 8 March 1989; Vladimir Siniutin, 'Zharkie Dni v Strane Piramid', *Voennie Znaniia* (September 1989); Igor' Timofeev, 'Voina v Sudnii Den', *Novoe Vremia* 45 (1988).

3 Richard Parker, *The Politics of Miscalculation in the Middle East* (Bloomington: Indiana University Press 1993) pp.148–64; Henry Kissinger, *White House Years* (Boston: Little Brown 1979) pp.574–89.

4 Interviews conducted with Abraham Leef (The first head of Branch 3 in the Research Department), Ramat Ha Sharon, 6 January 2004, and Haim Yaabetz (The first head of the Soviet section in the Research Department), Kiryat Atidim, 8 February 2004.

5 Excerpts taken from Leef's January 2004 interview.

6 Ami Gluska, *Eshkol, Give the Order* (Hebrew) (Tel Aviv: Maarachot 2004) p.287.

7 Leef's January 2004 interview.

8 Uri Dromi, 'The Man Behind "Masrega"', *Haaretz*, 2 February 2002; Interviews with Nehemia Burgin (head of the translations desk in 'Masrega'), Holon, 11 December 2003; Moshe Noiman (officer in the political section of the Egyptian desk and then the first head of the Soviet military desk in the Research Department), Tel Aviv, 1 January 2004; Ephraim Lapid (officer in AMAN's SIGINT Unit 848 – later Unit 8200), Jerusalem, 28 March 2004; and Gideon Gera (deputy director, in charge of political assessment in AMAN's Research Department), Tel Aviv, 29 December 2003.

9 Interviews with, Gideon Gera; Nehemia Burgin; Arie Idan (head of the American desk in Branch 3 in the Research Department), Tel Aviv, 26 January 2004; Arie Shalev (the director of the AMAN's Research Department), Tel Aviv, 28 November 2003; Yoal Ben Porat (head of the COMMINT branch in AMAN's SIGINT Unit) Herzliya, 3 March 2004.

10 Interviews conducted with Arie Idan and Haim Yaabetz.

11 In order to make this issue clear, Israeli officials repeatedly emphasized the threat of a Soviet southern outflanking of American interests in Asia and Europe to their American counterparts. Needless to say, a firm American stand, taken through the backing of the Israeli proxy, was the primary means to avoid this threat. See, for example, from the Israeli Envoy in Washington DC, Shlomo Argov, to the deputy general manager of the Foreign Office, Moshe Bitan, 'Soviet Expansion in the Middle East – Providing the USA with a Proper Cognition', 2 September 1969, Box 1448/4156, File No.3 (103.1), *Israel State Archives*, Jerusalem (hereafter referred to as *ISA*).

12 Interview conducted with Haim Yaabetz.

13 The only major exception among them was Defense Minister Moshe Dayan – Israel's supreme authority on security affairs between 1967 and 1973. Dayan was known to have a 'Soviet phobia' as was expressed, for example, by his objection to the conquest of the Syrian Golan Heights towards the end of the 1967 war, due to fear of Soviet intervention in order to help Damascus. In an IDF GHQ meeting held in November 1968 Dayan warned that if Egypt succeeded in occupying Israeli-held territory east of the Canal in a sudden, limited scale attack, the UN Security Council may call for an immediate ceasefire that would prevent the IDF from regaining its losses. Under such circumstances, he added: 'The Russians will say that if the

Israelis continue their attack to take "their [the Egyptian's] land" ... then we [the Soviets] intervene and we mean it. This means to position ships in the Mediterranean and shoot at Tel Aviv' (the Secretariat of the Supreme Command, the Office of the Chief of Staff. 'Operational Discussion – Our Border with Egypt', 21 November 1968, p.6. *Private collection.*

14 Interview conducted with Moshe Noiman.

15 Moshe Zak, *40 Years Dialogue with Moscow* (Tel Aviv: Maariv 1989) p.173.

16 Interview conducted with Haim Yaabetz.

17 From the Israeli Envoy in Washington DC, Shlomo Argov, to the Deputy General Manager of the Foreign Office, Moshe Bitan, 'Soviet Note on Port Said', 15 August 1969, Box No.1448/4156, File No.3 (103.1), *ISA*.

18 Zak (note 15) p.163.

19 Yaacov Bar-Siman-Tov, *The Israeli–Egyptian War of Attrition, 1969–1970: A Case Study of Limited Local War* (New York: Columbia 1980) p.45; Dan Margalit, *Message from the White House* (Hebrew) (Tel Aviv: Otpaz 1971) pp.40–44.

20 Interview conducted with Ben-Porat.

21 From the Director of Europe 3 Department (East European bloc and the USSR) to the Director General, 'The Soviets and the Apprehension of a Confrontation with the USA', 5 November 1969, Box No.3401/4221, File No.19 (110.1), *ISA*.

22 Interview conducted with Moshe Noiman.

23 Interviews conducted with Gideon Gera and Abraham Leef.

24 From the Director of Department Europe 3 to the General Manager of the Foreign Office, 'Is there a Threat of Soviet Intervention in the Region', 17 November 1969, Box No.3401/4221, File No.19 (110.1), *ISA*.

25 From the Consul in Boston to the Department of North America in the Foreign Office, 'The Soviet Stand', 16 September 1969, Box No.1448/4156-3 (103.1), *ISA*.

26 From Shlomo Argov, the envoy in Washington DC, to Moshe Bitan, Deputy Director General, the Foreign Office, 29 October 1969, Box No.1448/4156-3 (103.1), *ISA*.

27 From Nir Baruch, the Israeli Embassy in Washington DC, to Moshe Bitan Deputy Director General, 14 January 1970, Box No.3306/4548, File No.8 (103.3), *ISA*.

28 From Nir Baruch, to Moshe Bitan, 'A Talk with Norman Anderson', 29 January 1970, Box No.3306/4548, File No.8 (103.3), *ISA*.

29 From the Minister in Washington DC, to the Director of the Department of North America, 'A Talk with Prof. Brzezinski', 11 February 1970, Box No.3306/4548, File No.8 (103.3), *ISA*.

30 Parker (note 3) p.153; Bar Siman Tov (note 19), 129.

31 Taken from an interview with Shimshon Yitzhaki (head of the Egyptian Political Section in the Research Department), Herzlia, 26 December 2003.

32 Bar Siman Tov (note 30) p.122; Abba Eban, *An Autobiography* (Hebrew) (Tel Aviv: Maariv 1978), p.459; Avi Shlaim and Raymond Tanter, 'Decision Process, Choice and Consequences: Israel's Deep Penetration Bombing in Egypt, 1970', *World Politics* 4 (July 1978) pp.483–516, at 492.

33 Yitzhak Rabin, *The Rabin Memoirs* (Hebrew) (Tel Aviv: Maariv 1979) pp.253–4.

34 Shlaim and Tanter (note 32) pp.492–495.

35 Benny Morris, *Righteous Victims* (Tel Aviv: Am Oved 2003) p.335.

36 Jonathan Frankel, 'The Anti-Zionist Press Campaign in the USSR 1969–1971: An Internal Dialogue', *Soviet Jewish Affairs* 3 (May 1972) pp.5–6.

37 TASS statement, Moscow, 31 October 1969.

38 Yaacov Roi, *From Encroachment to Involvement; A Documentary Study of Soviet Policy in the Middle East, 1945–1973* (New York: John Wiley & Sons 1974). 514–15; Ilana Kass, *Soviet Involvement in the Middle East: Policy Formulation, 1966–1973* (Boulder, CO: Westview 1978) pp.155–70.

39 Frankel (note 36) pp.5–6.

40 V. Rumiantsev, 'Arabskii vostok na novom puti', *Kommunist* 16 (November 1969) pp.90–101, cited in Alvin Rubinstein, *Red Star on the Nile; Soviet – Egyptian Influence Relationship since the June War* (Princeton: Princeton University Press 1977) pp.103–5.

41 The Head of Europe 3 to the Director General, 'The Statement of Six Parties and East European Governments', 30 November 1969, Box No.3401/4221, File No.19 (110.1), *ISA*.

42 Yeshayahu Nir, *The Israeli–Arab Conflict in Soviet Caricatures 1967–1973, A Research Monograph in Visual Communication* (Jerusalem: Shamgar Press 1976).
43 Interviews conducted with Ben Porat and Burgin.
44 Parker (note 3) p.145.
45 Uri Bar-Joseph and John P. Hannah, 'Intervention Threats in Short Arab–Israeli Wars: An Analysis of Soviet Crises Behavior', *Journal of Strategic Studies* 11/4 (December 1988) pp.436–67, at 455–6.
46 From Amos Eran, Washington DC to the North American Department, the Foreign Office, 'Sisco in a closed-door meeting with members of the Congress', 4 February 1970, Box No.3306/4548, File No.8 (103.3), *ISA*.
47 From London to the Foreign Office, 'An un-attributed response of the Foreign Office spokesman to the writers of Reuters and J.T. with regard to the Soviet note', 3 February 1970, Box No.5243/4604-5 (103.1), *ISA*.
48 From Paris to the Foreign Office, 'The Kosygin note', 10 February 1970, Box No.5243/4604, File No.5 (103.1), *ISA*.
49 From Paris to the Foreign Office, 'A talk with the Romanian attaché in charge of Middle Eastern affairs', 6 February 1970, Box No.5243/4604, File No.5(103.1), *ISA*.
50 From Washington DC, to the Foreign Office, 'A talk with Sonnenfeld', 13 February 1970, Box No.3321/4549, File No.1 (103.1), *ISA*.
51 From the Foreign Minister to the Prime Minister, 'The estimate of an American expert on the USSR', 14 February 1970, Box No.3321/4549, File No.1 (103.1), *ISA*.
52 A day later, the General Manager of the Foreign Office, Gideon Raphael, wrote a note to the General Director of the Prime Minister's Office, informing him that *Maariv* had published, almost in its entirety, the text of a cable from the ambassador in Washington to the Foreign Office regarding the secret American assessment of the situation. The article also included the cabinet members' evaluation of the Soviet note. From Gideon Raphael to Simha Dinitz, 'The article by Mr. Harif in *Maariv'*, 10 February 1970, Box No.5252/4605, File No.2 (104.1), *ISA*. Paradoxically, this note is an indication that the article accurately described the outcome of the meeting. In general, the protocols of Israeli government meetings are considered to be a state secret for at least 40 years.
53 Yoseph Harif, 'Nixon in his answer to Kosygin: Egypt is responsible for breaking the cease fire', *Maariv*, 9 February 1970.
54 From the Director of Europe 3 [Soviet Bloc department in the Foreign Office] to the Director Manager, 'Kosygin's notes and their role in the Soviet Middle Eastern policy', 11 February 1970, Box No.5252/4605, File No.2 (104.2), *ISA*.
55 Ibid. For the document concerning a possible Soviet intervention, see note 22.
56 From the Manager of Europe 3 to the General Manager, 'The Soviet options in the eyes of American speakers', 15 February 1970 (emphasis added), Box No.5252/4605, File No.2 (104.2), *ISA*.
57 From the Director of Europe 3 to the Director General, 'Kosygin's notes and their role in the Soviet Middle Eastern policy', 11 February 1970, Box No.5252/4605, File No.2 (104.2), *ISA*.
58 From Washington to the Foreign Office, 'A talk between the journalist Milton Friedman and Boris Sedov', 6 January 1970, Box No.5252/4605, File No.2 (104.2), *ISA*.
59 Parker (note 3) pp.161–2.
60 From Washington DC, to the Foreign Office, 'A talk with Norman Anderson, an expert on the Middle East in the department for USSR', 29 January 1970, Box No.3306/4548, File No.7 (103.3), *ISA*.
61 From Washington DC, to the Foreign Office, 'The journalist Leon Volkov's talk with two Soviet military attaches', 25 February 1970, Box No.3306/4548, File No.7 (103.3), *ISA*.
62 From the Director of Department Europe 3 to Washington DC, 'Volkov's talk with Soviet military attaches', 11 March 1970, Box No.3306/4548, File No.7 (103.3), *ISA*.
63 Parker (note 3) pp.145–6.
64 See, for example: F.H. Hinsly with E.E. Thomas, C.F.G. Ransom and R.C. Knight, *British Intelligence in the Second World War: Its Influence on Strategy and Operations*, Vol.3, Part I (London: HMSO 1984) pp.47, 403–4; Christopher Andrew, *For the President's Eyes Only* (New York: Harper Perennial 1995) pp.286–7.

65 Bar Siman Tov (note 30) p.147; Interview with Yehuda Porat (Head of the Air Force Intelligence Research Department), Ramat Ha Sharon, 17 February 2004.
66 Interviews with Yehuda Porat; and Yoav Dayagi (Officer in the Air Force Intelligence Research Department), Tel Aviv, 4 February 2004.
67 Mohamed Heikal, *The Road to Ramadan* (Collins St. James Place: London 1975) p.77.
68 Interviews conducted with Yehuda Porat; Yoav Dayagi; Shimshon Yitzhaki; and Hanan Alon (Assistant to the IDF Chief of Staff Bureau Head), Sharon, 26 February 2004.
69 National Intelligence Estimate NIE 05-01-69, 'Soviet Guided Missiles, Surface to Air and ABM Missiles, SA-3 Goa', p.34, 5 January 1969.
70 Interviews conducted with Dayagy, Porat, Burgin, and Noiman.
71 Interview conducted with Dayagy.
72 Interviews with Lapid, Ben Porat, Burgin, Dayagi, and Noiman.
73 Excerpts taken from Burgin's interview. Today we know that the reason for the frequent VIP flights at the end of December 1969 was to prepare the groundwork for the deployment of the Soviet air-defense division which had already started by that date.
74 Burgin's interview. Ariel Vitan, *Soviet Military Presence in Egypt 1967–1972: Objects and Achievements* (MA thesis, Tel Aviv University 1991) pp.107–10; see also Ariel Vitan, 'The Soviet Military Presence in Egypt 1967–1972: A New Perspective', *Journal of Slavic Military Studies* 8/3 (September 1995) pp.547–65.
75 Interview conducted with Nehemia Burgin.
76 Dan Shueftan, *Attrition* (Hebrew; Tel Aviv: Ministry of Defense 1989) p.261.
77 Interviews conducted with Gideon Gera, Yoel Ben Porat, and Nehemia Burgin.
78 From Washington DC, to the Foreign Office, 'A talk with Sonnenfeld', 13 February, 1970; From the Foreign Minister to the Prime Minister, 'The estimate of an American expert on the USSR', 14 February 1970, Box No.5243/4604, File No.5 (103.1), *ISA*.
79 From Washington DC, to the Foreign Office, 17 February, 1970, Box No.5254/4605, File No.4 (104.2), *ISA*.
80 From Paris to the Foreign Office, 10 February 1970, Box No.5254/4605, File No.4 (104.2), *ISA*.
81 From the Foreign Office to Paris, 11 February 1970, Box No.5254/4605, File No.4 (104.2), *ISA*.
82 From Paris to the Foreign Office, 18 March 1970, Box No.5254/4605, File No.4 (104.2), *ISA*.
83 All the Israeli interviewers cited above.
84 David A. Korn, *Stalemate: The War of Attrition and Great Power Diplomacy in the Middle* p.194.
85 Special National Intelligence Estimate No.30–70, 'The USSR and the Egyptian Israeli Confrontation', 14 May 1970, *National Archives*, Washington, D.C.
86 *Maariv*, 21 March 1970.
87 Ibid.
88 From Washington to the Foreign Office, 'A talk with the Director of INR in the State Department, Ray S. Cline', 25 March 1970, Box No.3306/4548, File No.7 (103.3), *ISA*.
89 From Washington to the Foreign Office, 'A talk with Kenneth Kerst, Deputy Director for Soviet analysis in the State Department', 16 April 1970, Box No.3306/4548, File No.7 (103.3), *ISA*. The CIA experts were split on this issue. Some, like Mel Goodman and Bob Gates estimated that the Soviets would move eastward in an attempt to provide anti-aircraft defense for the Egyptian forces. Others, primarily the chief of SOVA (the Soviet Affairs Division), believed that the Kremlin would avoid such a risky move (from private email correspondence with Mel Goodman).
90 Interview with Yehuda Porat and Arie Shalev; Parker (note 3) p.153.
91 Memorandum of a conversation between A.F. Dobrynin, the Ambassador of the USSR to the USA, and Henry Kissinger, Aide to President Nixon. A. Gromyko to members and candidate members of the Politburo, CC CPSU, Secret. Document #5, Cold War History Project: available at http://cwihp.si.edu/cwihplib.nsf/4f5f68a9959c15b48525649e0083a3a6/bd00a5b4edfeae9d852564bf006878f8?OpenDocument or see excerpts from the Memorandum: http://www.turnerlearning.com/cnn/coldwar/detente/dete_re1.html
92 Kissinger (note 3) p.559.

The October 1973 Yom Kippur War

[7]

The Yom Kippur War: Indications and Warnings

Lieutenant Youssef H. Aboul-Enein, U.S. Navy

The 1973 Yom Kippur War ranks high in the annals of intelligence failures. Although the Israelis scored a tactical victory against the Syrians and the Egyptians, the victory came at a high cost in men and materiel. Syrian forces penetrated the Golan Heights and came within 10 kilometers of securing a key bridge that would have left northern Israel vulnerable to attack. On the southern front, Egyptian forces broke through and overwhelmed the Bar-Lev Line. This surprise attack brought down the government of Israeli Prime Minister Golda Meir and severely dented the reputation of Defense Minister General Moshé Dayan.

Israeli Security Doctrine

Despite winning three wars prior to 1973, the Israelis had to cope with how to address security needs with their small population. Compared to Egypt and Syria, the Israelis could not economically field a huge standing army on its borders and could not sustain a protracted war on four fronts (Egypt, Jordan, Lebanon, and Syria). Thus, the Israeli concept of security was based on deterrence, early warning, and air supremacy. The rapid mobilization of its reserve forces depended on early warning, taking the battle to the enemy, and reaching a rapid decision on the political front.[1]

Egyptian Security Doctrine

The Egyptians took great pains to study Israeli doctrine. Soon after the 1967 Six-Day War, they began to restructure their offensive and defensive techniques. Their objective was to fortify the western Suez and lay out plans for capturing the Sinai. Unlike previous wars in which national objectives were not outlined and weapons systems did not match doctrine or the education of the troops, the Egyptians began to clearly define the ways, means, and ends for the eventual liberation of the Suez Canal and parts of the Sinai. They broke their objectives down into three phases: defiance, active defense, and war of attrition. A fourth phase resulted from the cease-fire brokered by U.S. Secretary of State William Rogers.

The defiance phase (June 1967-August 1968) was to provide politically favorable conditions for the reconstruction of the armed forces and defense of the western side of the Suez Canal. The active defense phase (September 1968-February 1969) consisted of harassing Israeli forces and delaying their fortification of the east-

---

All Arabic sources were translated by the author and represent his translation.—Editor

ern side of the Suez. Egyptian troops began to probe the Sinai and to draw plans for crossing the Suez and penetrating Israeli fortifications. The war of attrition phase (March 1969-August 1970) projected day and night raids into the Sinai that would eventually reach company strength. This was augmented with constant exchanges of artillery fire across the canal and imbued the Egyptian fighting soldier with a sense of confidence. Harassment tactics also included Egyptian frogmen who were to sink transports at the port of Eilat. (This action occurred in November 1969 and again in February 1970 and was the catalyst for using high-pressured water to breech the Bar-Lev Line.) As the result of the cease-fire, a fourth phase was developed and labeled "No War, No Peace" (August 1970-October 1973).

In spite of the cease-fire, the Egyptians continued to collect information and plan for a massive campaign, should political negotiations fail to return the Sinai. In November 1972, Egyptian President Anwar Sadat made the official decision to go to war.[2]

Egyptian Deception

The Egyptian deception plan can be divided into three components: economic, political, and military. A flood of communiqués and reports on Egypt's economic instability and its inability to afford another war were made public. The reports stressed the importance of a political solution to return the Sinai. Political deception stressed the status quo through the "No War, No Peace" slogan. The Egyptians highlighted their acceptance of the Rogers Plan and publicly expelled Soviet advisers. Many believed Sadat was working toward Western rapprochement. An element of the Rogers Plan stipulated that the United States would not look kindly at an Israeli first strike. This would cloud the Israeli decision to react in spite of growing intelligence indicating an attack. Military deception, the final component, involved a series of military exercises designed to act as faints for the real attack.[3]

The concept of surprise occupied a large portion of the Egyptian general command's planning. Coordination with Syria occurred 6 months before D-Day (6 October). Militarily, the plan was to deceive Israel as to the intention of launching an offensive operation. The Egyptians also had to be concerned with concealing its main assault's timing, size, and direction. An emphasis on the defense was undertaken as part of the deception. After enduring four wars, the Israelis had become accustomed to the Egyptians and Syrians fortifying and conducting defensive operations. Thus, preparations for defensive operations continued as normal and were even heightened because the Arabs knew that Israel expected this. This defensive strategy was heavily emphasized in military radio traffic. False reports of negligent standards, faulty missile systems, and the difficulty of absorbing tons of Soviet equipment were exchanged on open radio to deceive Israeli signals intelligence operatives.

The Egyptian military staged exercises with different force structures and sizes along the Suez so as to hide the true order of battle for the Suez Canal crossing. They assembled troop concentrations for the actual attack over a 4-month period, with crack units being moved three weeks before D-Day under the pretense of massive engineering projects for defensive fortifications of the western side of the canal. Crossing equipment was brought from the rear to the front and back again, along with incidental moves of combat engineer units, to deceive the Israelis into believing this was simply movement training. They conducted well-practiced mobilization of reserves in a way in which the maximum number of forces would be ready for zero hour. Forty-eight hours before H-Hour, 20,000 reservists were demobilized as Israeli monitors watched.[4]

Perhaps the simplest, yet most effective, deception plan was the use of a company of the most undisciplined soldiers in the Egyptian Army one day before D-Day. They were to further reinforce Israeli contempt of Egyptian forces by washing their clothes along the canal and loitering about in an undisciplined fashion. Israeli reports indicated that they were pitiful in their appearance. They were eating oranges, swimming, and sucking on sugar cane stalks. Other deceptions included a public announcement by the War Ministry accepting applications from armed forces personnel wishing to make a mini-pilgrimage (Umrah) to Mecca during the holy month of Ramadan.[5]

An Egyptian destroyer squadron deployed in August 1973 under the pretence of an overhaul in India and Pakistan with port visits scheduled for Sudan, Yemen, and Somalia. Their operations order, delivered on 1 October, directed the squadron to blockade the Bab-el-Mandab Straits that connect the Indian Ocean to the Red Sea.[6] The squadron's objective was to deprive Israel of needed petroleum products being shipped from Iran.

Who knew and when? The Arabs practiced extraordinary operations security. In Egypt, only Sadat, War Minister Field Marshal Ismail Ali, and a dozen generals of the General Command knew the plan. In Syria, not more than 10 people were told of the plans. Egyptian and Syrian divisional commanders were told of the war on 1 October. Brigade and battalion commanders of both sides were told of the war on 5 October while most troops and officers were informed no more than 2 hours before H-Hour.[7]

What failed and why? The seeds of Israeli intelligence failure were sown in the tactical success of the Six-Day War. Israeli military intelligence developed a concept (the

Lieutenant Youssef H. Aboul-Enein is a student at the Joint Military Intelligence College. He received a BBA from the University of Mississippi, and an MBA and MHSA from the University of Arkansas, Little Rock. He is also a graduate of the U.S. Naval War College and the U.S. Marine Corps University Amphibious Warfare School. He is a prolific writer and has a regular column about the Middle East in the Foreign Area Officer Journal. *He is a frequent contributor to* Military Review.

The following two review essays are also by Lieutenant Aboul-Enein.

"Conceptzia"), which stipulated that an Arab-Israeli war would occur only if certain conditions were met. A combined Arab attack would not occur unless Arab forces possessed the means to simultaneously attack all Israeli airfields. Hence, 1975 was deemed the year of attack. Egypt was acquiring long-range Soviet fighter-bombers and would have adequate pilots and aircraft for the attack by 1975. Here, the Israelis practiced mirror imaging, a cardinal intelligence sin.

Another aspect of the concept was that the Suez Canal, with the formidable Bar-Lev Line, would be a sufficient barrier to give the Israelis enough time (48-hours) to mobilize. They did not expect an attack during Yom Kippur, the Jewish High Holiday, when only a skeletal force was deployed and mobilization was difficult. Finally, it was believed that the Arabs would not attack unless they possessed enormous stockpiles of weapons and equipment and that they were only capable of guerrilla warfare, not conventional attacks, as evidenced by their performance in three wars. This concept was adopted fully by Israel's chief of military intelligence, and he squashed any indicators that violated these "rules."[8]

Israeli intelligence is divided into five branches, with a focus on Aman, which according to the Agranat Commission that looked into the Israeli military's failures, was responsible for the national intelligence estimate and bore responsibility for the intelligence failure. Israeli intelligence was vulnerable to the Conceptzia.

Since 1963, reorganization led to Aman being arranged along strict military lines. Officers were assigned at Aman for 6 to 7 years. By the early 1970s, no outside thinking took place, and no contradiction of analysis was made within this strict military hierarchy.[9] A searing example of this is the story of Lieutenant Benjamin Simon-Tov, who 5 days before the war argued that Egyptian exercises and deployments were a camouflage for a real assault. His report sat on the desk of his commander, Lieutenant Colonel Gedaliah.[10]

What did not help was Sadat's declaration that 1971 would be the Year of Decision. The year came and went with Sadat being seen as crying wolf. His threats in 1972 and 1973 were not taken seriously.

Indicators and Distractions

Israeli and U.S. estimates analyzed Egyptian and Syrian exercises over the course of several years, and each year the exercises grew larger, finally involving division-size maneuvers. Landlines were installed between Cairo and the Suez, which negated the need for radio traffic. There were civil defense exercises and the stockpiling of war materiel. In addition, elite Egyptian commando units were detected along the front. Even with such indicators, Aman was still distracted.

From 1969 on, Israeli intelligence distracters included a new emphasis—terrorism. The Palestinian Liberation Organization was active in 1972. The Lod Airport was attacked, the Munich Olympics was disrupted, an Israeli diplomat was killed in London with a letter bomb, the Israeli Naval attaché was gunned down in front of his home in Maryland, and a train carrying Soviet Jews on their way to Israel through Austria was hijacked. Many intelligence specialists were drained from other sources to form a new antiterrorism cell, which affected Israel's ability to concentrate on Syrian and Egyptian maneuvers.[11]

In May 1973, at a cost of $10 million, the constant deployment of Egyptian and Syrian forces caused a mass mobilization of Israeli reserves. The attack never came, and there was criticism within the government regarding frivolous deployments. The most intriguing warning came from King Hussein of Jordan, who personally warned Meir that a combined Egyptian-Syrian attack was imminent.[12] Hussein had an interest in keeping Damascus weak through Israeli clashes. His message went unheeded.

The results of the war that Israel failed to imagine was 2,700 Israeli dead, the majority of whom died in the first two days of combat. Israeli losses in tanks and armor were so enormous that Israel beseeched the United States for an immediate airlift. The Egyptians, used to advancing under a timetable, stopped and did not advance toward the Giddi and Mitla Passes. This allowed the Israelis to mobilize an effective counterattack and surround the Egyptian Third Army.

Aman is guilty of several violations of intelligence analysis, including building an impregnable psychological barrier through adherence to strict concepts. Instead of influencing policymakers, Aman was influenced by disinformation.

The Israeli military establishment lacked private strategic think tanks to provide checks and balances on intelligence appraisals. The Israelis had no contingency plans for a surprise attack. They relied solely on the competence of their early warning apparatus. All force mobilizations were based on the success of their intelligence organizations. Israeli defense forces' mobilization plans were based on having 48-hour warning, not the 10 hours that occurred during this war.[13]

When B. Lidell Hart visited Israel in March 1960, he stressed to Israeli officers that the Israeli Defense Force's greatest danger lay in its success; victorious armies become overconfident.[14] Hart's prophetic warning became reality 13 years later. ***MR***

NOTES

1. Colonel Michael Herzog in Chaim Herzog, *The War of Atonement: The Inside Story of the Yom Kippur War, 1973* (London: Lionel Leventhal, 1998), xiii.
2. Hassan el-Badri, *The Ramadan War, 1973* (Fairfax, VA: Dupuy Associates, 1978), 9-14.
3. Manaal Nour-al-Din, *Butulaat Waraa al-Sittar* [Heroism Behind the Lines] (Cairo: Egyptian Writer's Association, 1997), 29-30.
4. Badri, 46-47.
5. Abd-al-Ghanny Al-Gamassy, *Muzakiraat Al-Gamassy: Harb Oktoober Thallathah wa Sabeen* [Memoirs of Al-Gamassy: The October War 1973 (Cairo: Egyptian Writers Association, 1998), 287.
6. Al-Gamassy, 90.
7. Doron Geller, "Israeli Intelligence and the Yom Kippur War," *The Department for Jewish Zionist Education*, Lecture Series (April 1999), online at <www.jajz.ed.orl.il/juice/ service/ week11.html>, accessed 29 January 2002, 2.
8. Steven Stewart, *The Spymasters of Israel* (New York: MacMillan, 1980), 295.
9. Insight team, London *Sunday Times*, 94.
10. Herzog, 46-47.
11. Lester A. Sobel, *Israel and the Arabs: The October 1973 War* (New York: Facts on File, 1974), 8-39.
12. Geller, 5.
13. Zeev Schiff, *October Earthquake: Yom Kippur 1973* (Tel Aviv: University Publishing Projects, 1974), 29-37.
14. Ibid., 29.

[8]

Strategic Surprise or Fundamental Flaws? The Sources of Israel's Military Defeat at the Beginning of the 1973 War

☆

Uri Bar-Joseph

Abstract

During the decade after the 1973 War of Yom Kippur, the consensus was that Israel's military defeat in the war's first stage was caused by the failure of intelligence to provide a warning prior to the Arab attack, but many experts maintained later that it reflected improper preparations for war. Using recently released evidence, this article analyzes Israel's inadequate war deployment when firing commenced and its impact on the failure to repel the attack. It concludes that since this deficient deployment resulted from the absence of a sufficient intelligence warning, the intelligence failure was at the root of the Israeli failure at the war's start.

Introduction

The first stage of the 1973 War of Yom Kippur (the War of Ramadan or the October War in Arab terminology) constitutes the worst defeat in Israel's military history. The Syrian-Egyptian attack, which started on 6 October at 2:00 p.m., came as an almost-complete surprise to the Israelis. In the south, five Egyptian mechanized divisions crossed the Suez Canal with only minor losses. They occupied some forts of the Bar-Lev defense line and encircled the rest, thus creating a solid bridgehead in the Israeli-held Sinai Peninsula. On the morning of 7 October, the Southern Command of the Israel Defense Forces (IDF) was left with

Uri Bar-Joseph teaches in the Division of International Relations in the School of Political Science at Haifa University. In addition to more than thirty refereed journal articles on national security issues, intelligence, and the Arab-Israeli conflict, he is the author of four books, the most recent of which is *The Watchman Fell Asleep: The Surprise of Yom Kippur and Its Sources* (SUNY Press, 2005).

fewer than 110 tanks out of the approximately 300 with which they had started the war. They had to block an Egyptian army of close to 100,000 soldiers, supported by hundreds of tanks, until the arrival of Israel's reserve army.[1] In the north, three Syrian infantry divisions, supported by 600 tanks and about 80 artillery batteries, attacked Israel's defense line in the Golan on 6 October. By the early hours of the next day, the southern sector of the defense line had collapsed, and there were no ground forces to stop the Syrian tanks from invading Israel at its pre-1967 borders.[2] In the air, two major operations carried out on the morning of 7 October to destroy the Arab antiaircraft layouts – Operations "Tagar 4" (Quarrel 4) in the south and "Dugman 5" (Model 5) in the north – completely failed at a high cost. Their failure prevented the Israeli Air Force (IAF) from effectively participating in the ground warfare throughout most of the war. Given that the IAF had been allocated about 50 percent of the IDF's annual budget during the period of 1967-73 and that it had carried much of the burden during the War of Attrition in 1969-70, its inability to provide the expected results in October 1973 was a major setback to Israel's military effort.

The dire situation on the battlefield had a considerable effect on the Israeli leadership. Following a visit to the IDF command post in the north, about sixteen hours after the war started, Defense Minister Moshe Dayan concluded that this might be the end of "the Third Temple," meaning the end of the Israeli state. According to a general's evidence, Dayan wept when he said it.[3] A few hours later, during a visit to the war room in the south, Dayan suggested making a deep withdrawal in the Sinai. If accepted by the IDF command, such a move could have led to a complete defeat on this front.[4] Upon returning to Tel Aviv, Dayan described the grim situation on both fronts to Prime Minister Golda Meir. By some accounts, following this discussion she started contemplating suicide.[5] In the aftermath of his talk with the Prime Minister, he offered his resignation. Golda Meir rejected his offer.[6]

1. Amiram Ezov, "'Ministerial Advice' – the Southern Command in the Yom Kippur War, October 7, from Hold-Defense to a Counter-Offensive," in *Yom Kippur War Studies* [in Hebrew], ed. Haggai Golan and Shaul Shai (Tel Aviv: Maarachot, 2003), 205-58, 210. Ezov's study of the war in the Southern Command was conducted within the framework of the IDF's History Department and is considered as the most detailed and authoritative account of its kind.

2. Hanoch Bartov, *Daddo, 48 Years and 20 more days* [in Hebrew] (Or Yehuda: Dvir, 2002), 408; Lt. Col. (Res.) Elhanan Oren, *The History of the Yom Kippur War* [in Hebrew] (Tel Aviv: IDF History Department, 2004), 81. Oren's study, which is the official IDF history of the war, was completed in 1978 and was released to the public in late 2004.

3. Abraham Rabinovich, *The Yom Kippur War: The Epic Encounter that Transformed the Middle East* (New York: Schocken Books, 2004), 209.

4. Bartov, *Daddo*, 420-21.

5. Shlomo Nakdimon, *Low Probability* [in Hebrew] (Tel Aviv: Revivim, 1982), 143.

6. Golda Meir, *My Life: The Autobiography of Golda Meir* (London: Futura, 1976), 361.

The speedy arrival of the Israeli reserve army to the fronts on 7 October brought this stage of the war to its end. By the early evening hours of 7 October, the situation in the Golan and in the Sinai had been stabilized, and the IDF was planning counteroffensives on both fronts for the next day. The attack in the north succeeded, and by 10 October the IDF had regained almost all of its territorial losses in the Golan. Although the offensive in the south was a complete failure, the IDF had accumulated sufficient forces on this front to block any additional Egyptian offensive, even after its 8 October defeat.

The Debate

Until the early 1990s, the consensus among most students of this dark chapter in Israel's military history was that the main cause of the failure was the element of strategic surprise. This surprise was the outcome of the failure by Military Intelligence (AMAN), Israel's sole national intelligence estimator in 1973, to provide, on the eve of the war, a warning that war was highly likely. Despite the fact that AMAN had an almost-perfect picture of the deployment of the Egyptian and Syrian armies along the border with Israel, as well as numerous other war indications and information obtained from reliable Arab sources (including King Hussein of Jordan) that Egypt and Syria planned an attack, AMAN's experts estimated twenty-four hours before the outbreak of the war that "the probability that the Egyptians intend to renew fighting is low . . . . [and] the probability of a Syrian independent (without Egypt) action remains low."[7] On the basis of this estimate the Chief of Staff avoided demanding the mobilization of the reserve army and additional moves. He changed his mind only ten hours before firing commenced, after he received a warning from the head of the Institute for Intelligence and Special Tasks (the Mossad), Major General (Res.) Zvi Zamir, who had been informed in a meeting with a top Egyptian source that war would start on 6 October.[8] Zamir's source was Dr. Ashraf Marwan, a close advisor of President Anwar Sadat and a son-in-law of the late President Gamal Abdel Nasser.

That surprise was the main cause of Israel's military failures at the beginning of the war was the conclusion of the Agranat Commission, the only official investigation into the war.[9] The larger share of senior IDF officers agreed. The Chief of Staff in 1973, Lieutenant General David Elazar, assessed

7. Uri Bar-Joseph, *The Watchman Fell Asleep: The Surprise of Yom Kippur and Its Sources* (Albany: State University of New York Press, 2005), 170. For a detailed analysis of the causes for the Israeli surprise in October 1973, see pages 81-188, 235-51.

8. Ibid., 185-89.

9. Agranat Commission, *The Agranat Report* [in Hebrew] (Tel Aviv: Am Oved, 1975).

that a warning twenty hours in advance would have been sufficient to prevent the catastrophe.[10] His predecessor, Lieutenant General (Res.) Haim Bar-Lev, who was nominated during the war to command the Suez front, maintained that "it was all the outcome of the surprise."[11] Likewise, Major General (Res.) Avraham Adan, a division commander on the southern front, was firm in his conviction that "had we received an early warning . . . . the picture would have been completely different."[12] And Major General Yitzhak Hofi, the commander of the Northern Command, estimated that "if a mobilization of the reserves had taken place in time, at least of one tank brigade, the whole picture of the fighting in the first two days would have been completely different."[13] Academic scholars tended to agree as well.[14]

Confronting this thesis, more and more studies of recent years maintain that the main source for Israel's initial defeats was not strategic surprise but fundamental flaws in the IDF's preparations for the next war, mainly in its operational plans, force structure, and the poor quality of staff work. Major General Eli Zeira, who served as the Director of Military Intelligence (AMAN) in 1973, maintained that overconfidence led the IDF to make unrealistic plans for the coming war by depending on regular forces alone to prevent any Arab achievement, in both the Suez and the Golan, without the support of the IAF, which was to destroy the Arab antiaircraft systems during the first forty-eight hours of the war.[15] Mordechai Gazit, the General Manager of the Prime Minister's office in 1973, regarded the source of the problem not as insufficient warning, but as a mistaken military assertion that the post-1967 borders would enable Israel to repulse any coordinated, simultaneous surprise attack by Egypt and Syria.[16] In a similar manner, Eliot Cohen and John Gooch emphasized the impact of the great victory in 1967, which had prevented the Israelis from properly assessing the true capabilities of the Egyptian and Syrian armies and led them to rely on

10. Evidence of Elazar before the Agranat Commission, quoted in Yoel Ben-Porat, *Neila, Locked-on* [in Hebrew] (Tel Aviv: Idanim, 1991), 103; Bartov, *Daddo*, 342-43.

11. Gai Carmit, *Bar-Lev: A Biography* [in Hebrew] (Tel Aviv: Am Oved, 1998), 256.

12. Avraham Adan, *On Both Banks of the Suez* [in Hebrew] (Jerusalem: Idanim, 1979), 72.

13. Yitzhak Hofi, "The Northern Front," in *The War of Yom Kippur: A New Look* [in Hebrew], ed. Ofaz Haim and Bar-Siman-Tov Yaacov (Jerusalem: Davis Institute of the Hebrew University, 1999), 89-99, 98.

14. Martin van Creveld, *The Sword and the Olive: A Critical History of the Israeli Defense Force* (New York: Public Affairs, 2002), 239-40; Avi Shlaim, *The Iron Wall: Israel and the Arab World* (London: Penguin, 2001), 318-19; Benny Morris, *Righteous Victims: A History of Zionist-Arab Conflict, 1881-2001* [in Hebrew] (Tel Aviv: Am Oved, 2003), 379-90.

15. Eli Zeira, *Myths versus Reality: The October 1973 War – Failures and Lessons* [in Hebrew] (Tel Aviv: Miskal, 2004), 79-80.

16. Mordechai Gazit, "Could the War be Avoided?" in Ofaz and Bar-Siman-Tov, *War of Yom Kippur*, 9-21, especially 12.

an unrealistic balance of forces on the front prior to the outbreak of the war.[17] And Zeev Maoz concluded that had strategic warning been provided in time to allow a complete war deployment of the IDF in the Suez and the Golan, the situation could have been even worse: In the north, the Syrian failure to occupy the entire Golan Heights had less to do with the Israeli surprise and more to do with fundamental weaknesses of the Syrian army, and in the south, more Israeli tanks could be hit by Egyptian antitank missiles without blocking the crossing forces.[18]

Other analysts emphasized the improvements that had taken place between 1967 and 1973 in Egyptian and Syrian weaponry and command and control.[19] Arab accounts of the war tend to emphasize this point as well. Egypt's leader, Sadat, regarded the Egyptian air strike at the outset of the war as the key to Egypt's military success during the war's first stage. A semi-official history of the Egyptian Air Force (which quotes Sadat) paints this air strike in very bright colors as well.[20] The Egyptian army's chief of staff in 1973, Lieutenant General Saad el Shazly, maintained that the key to Egypt's military achievement was the excellent preparation of his army for war, as well as the fact that the Egyptian strategy combined limited territorial goals with maximal employment of force. His operations officer, who planned the war and replaced him in its aftermath, regarded the war's prudent and detailed planning as the main cause for the Egyptian success.[21]

An in-depth analysis of the course of the fighting led many observers to pinpoint other weaknesses of the IDF. Most of them agree that its unbalanced force structure was a major source of its initial failure. Following the war of 1967, the IDF allocated most of its resources to air power and tank forces at the expense of other elements, primarily the infantry, artillery, and combat engineering. The result was an army in which the ability of the tank to meet all types of challenges in ground warfare and of the IAF to provide close air support and to decide the war was overestimated. As the war showed, the tanks were surprised, mainly on

17. Eliot Cohen and John Gooch, *Military Misfortunes: The Anatomy of Failures in War* (New York: Vintage Books, 1991), 124-35, 130.

18. Zeev Maoz, *Defending the Holy Land: A Critical Analysis of Israel's Security & Foreign Policy* (Ann Arbor: University of Michigan Press, 2006), 148-51.

19. Ze'ev Schiff, *A History of the Israeli Army* (London: Sidgwick and Jackson, 1987), 226; Chaim Herzog, *The War of Atonement* (Jerusalem: Steimatzky, 1975), 273-75.

20. Anwar el-Sadat, *In Search of Identity* (New York: Harper and Row, 1978), 249; Lon Nordeen and David Nicolle, *Phoenix Over the Nile: A History of Egyptian Air Power, 1932-1994* (Washington: Smithsonian Institution, 1996), 278-79. The Israelis were not too impressed by the Egyptian air attack (Oren, *History of the Yom Kippur War*, 75), and other observers concur (e.g., Kenneth M. Pollack, *Arabs at War: Military Effectiveness*, 1948-1991 (Lincoln: University of Nebraska Press, 2002), 108.

21. Saad el Shazly, *The Crossing of the Suez* (San Francisco, Calif.: American Mideast Research, 1980), 24-89; Mohamed Abdel Ghani el Gamasy, *The October War: Memoirs of Field Marshal El-Gamasy of Egypt* (Cairo, Egypt: American University of Cairo Press, 1993), 134-38.

the Egyptian front, by the massive use of personal antitank weapons, and the IAF was unable to provide effective air support because it had failed to destroy the Arab antiaircraft layouts and had to perform in a zone defended by ground-to-air missiles throughout most of the war.[22] Other observers emphasize different elements. The IAF commander in the war, Major General Benny Peled, blamed the ineffective air support on the lack of cooperation between Military Intelligence, which was to provide the operational intelligence, and the IAF, which was supposed to launch the attacks.[23] The poor quality of staff work and operational discipline, which manifested itself on the eve of the war and during the first days of fighting, is also cited as a major cause for its poor outcome during the initial stage.[24]

In order to systematically address this debate, I will identify six Israeli decisions – two concerning the Suez front, two in the Golan, and two in the use of the IAF – which led to the military catastrophe. By analyzing the nature of these mistakes, their context, and their sources, I will show that despite certain deficiencies in war preparations, the primary source of the IDF's failure at the outset of the war was indeed the element of strategic surprise.

The Mistakes Responsible for the Military Failure

The Suez Front

The grave situation on the Suez front, which became apparent in the early morning hours of 7 October, was the outcome of two developments. First was the ease with which the Egyptian divisions crossed the canal and reached their operational goals. Second was the fact that only a third of the IDF force that had been tasked to defend the Sinai under a worst-case surprise attack scenario was actually operational. This was the result of two errors.

The first error was that the forts of the Bar Lev line were mostly manned by second-rate reserve soldiers rather than the high-quality regular service paratroopers specified by plans in case of war. The soldiers assigned to the central and northern sectors of the line belonged to Battalion 68 of Brigade 16 (the Jerusa-

22. Herzog, *The War of Atonement*, 270-79; Avner Yaniv, *Deterrence Without the Bomb: The Politics of Israeli Strategy* (Lexington, Mass.: Lexington Books, 1987), 132-34; Avi Kober, *Military Decision in the Arab-Israeli Wars 1948-1982* [in Hebrew] (Tel Aviv: Maarachot, 1995), 334-35; van Creveld, *Sword and the Olive*, 241-42; Emanuel Wald, *The Wald Report: The Decline of Israeli National Security Since 1967* [in Hebrew] (Tel Aviv: ShoKen, 1987), 99-104; Maj. Gen. (Res.) Yaacov Amidror, "The Failure of Warning and the Outcomes of the Yom Kippur War," in *The Yom Kippur War and Its Lessons* [in Hebrew], ed. Pinhas Yehezkely (Tel Aviv: Ministry of Defence, Broadcast University Series, 2005), 18-27, 19-20; Col. (Res.) Yaacov Hisdai, "The Holding Defense Battle in the Yom Kippur War," in Yehezkely, ed., *Yom Kippur War*, 47-56, 50-51.

23. Gen. (Res.) Benjamin Peled, *Days of Reckoning* [in Hebrew] (Tel Aviv: Modan, 2004), 332-33.

24. Amidror, "The Failure of Warning," 21.

lem territorial brigade). This battalion had not previously been deployed to the Bar Lev line. Many of its soldiers were relatively old, poorly trained, and lacking in combat experience. Moreover, since their service was to take place during the Jewish high holidays, the command of the brigade had selected for this particular mission many reservists who had escaped service on previous occasions, as a form of what was termed "educational punishment."[25]

Making matters worse was the fact that some of the forts were under-staffed, and fourteen out of thirty were not manned at all. None of them were prepared for war, and information that war would break out reached the soldiers there only minutes before firing commenced.[26] Thus, rather than serving as a means to interrupt the Egyptian crossing, the forts turned into a liability. Lack of professional training and preparation for the coming events led the reserve soldiers to demand assistance or to be rescued immediately after fire commenced. Henceforth, assisting and rescuing them rather than fighting the Egyptian forces became the main task of Division 252 during the war's first night.

The way in which the Bar Lev line fell led a number of experts to maintain that the professional concept behind it was mistaken from the start.[27] Some even maintained that its collapse in 1973 was inevitable.[28] Yet, one has to remember that a similar concept of defense was employed on the Golan Heights, where none of the strongholds were taken, with the exception of the isolated Hermon fort. Therefore, it seems that the variance in the fate between the two defense lines has less to do with the fundamental concept itself and more to do with the way they were prepared for war: In the north, the line was ready for war and the strongholds were manned by high-quality regular service infantry soldiers. They were staffed, moreover, with forward observation officers and surgeons. Consequently, even when encircled by the Syrian army, they continued participating effectively in battle, reporting back to central command, and repulsing Syrian attacks.[29] None of these essential elements were in place in the Bar Lev line, and this was the reason for its rapid collapse.

The second mistake that decided the fate of the battle on the southern front was the decision, made by the Command's head, Major General Shmuel Gonen, to delay the deployment of the main bulk of his available force, the 300

25. Ronen Bergman and Gil Meltzer, *The Yom Kippur War – Moment of Truth* [in Hebrew] (Tel Aviv: Yedioth Ahronoth, 2003), 24; Moti Shkenazy, *A War Will Break Out at Six p.m.* [in Hebrew] (Tel Aviv: Hakibbutz Hameuchad, 2003), 16.

26. Bartov, *Daddo*, 392.

27. Yaniv, *Deterrence Without the Bomb*, 135-36; van Creveld, *Sword and the Olive*, 211-12; Israel Tal, *National Security: The Few Against the Many* [in Hebrew] (Tel Aviv: Dvir, 1996), 152-56.

28. Yaacov Hisdai, who was one of the military investigators of the Agranat Commission, asserted that "it was impossible to avoid the success of the Egyptian crossing, even if Israel was not surprised"; Hisdai, "Holding Defense Battle in the Yom Kippur War," 51.

29. Eitan Haber and Ze'ev Schiff, *Yom Kippur War Lexicon* [in Hebrew] (Or Yehuda: Zmora-Bitan-Dvir, 2003), 258-62.

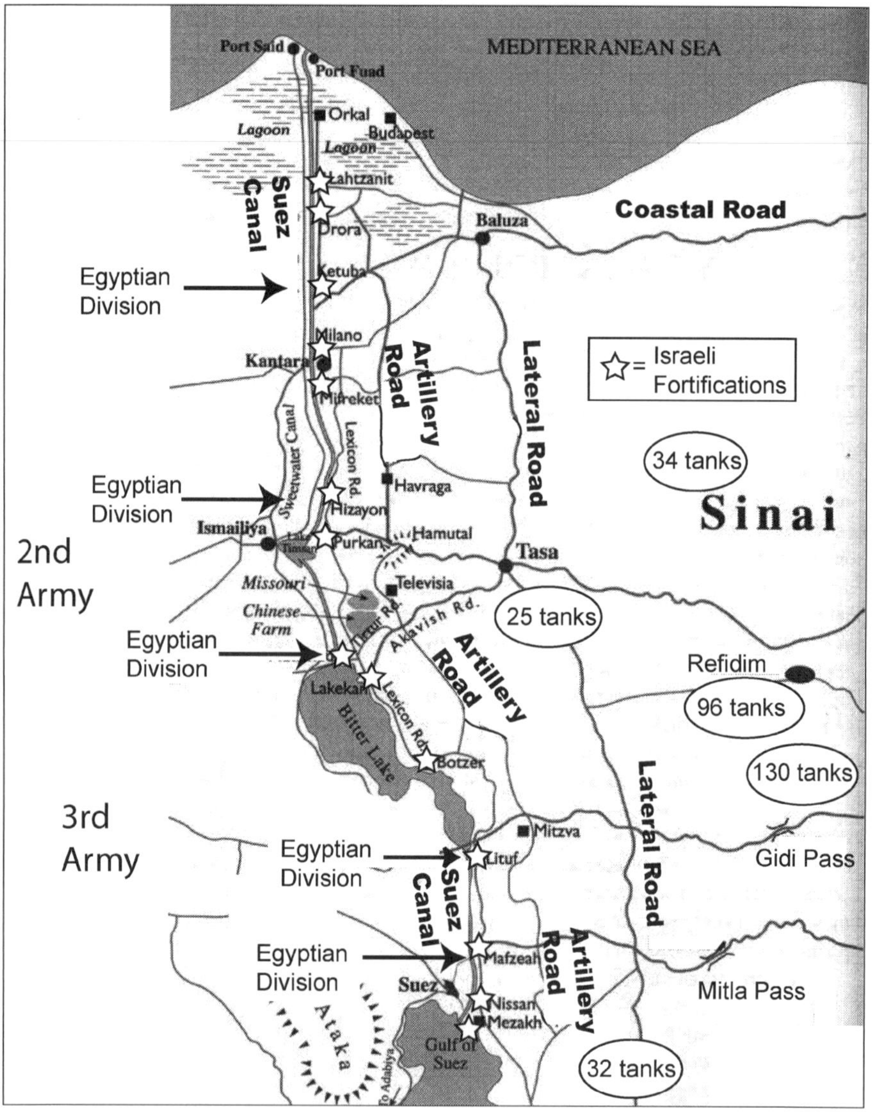

The Suez Front, 6 October 1973 (Map: James R. Arnold)

tanks of Division 252, until the afternoon hours of 6 October. Gonen gave this order despite the fact that the Chief of Staff had instructed him personally at 6:00 a.m. to prepare for the war that was expected to start at 6:00 p.m.[30] By his own account, he did so because he was not certain that Egypt would start a war at all and because he wanted to avoid a too-early large-scale deployment that might provoke the Egyptians to start firing. Since the expected "H" hour was 6:00 p.m., he instructed the commander of Division 252 to avoid breaking routine activity until 4:00 p.m.[31]

The emergency plan for war in the south, Shovach Yonim (Dovecote), called for the disposition of two tank brigades (220 tanks) between the water line and the Artillery Road, about 8 to 10 kilometers east of the canal. A third brigade (110 tanks) was to be deployed between the Artillery Road and the Lateral Road, about 25 kilometers from the canal. In addition, 21 batteries of 155mm and 175mm self-propelled guns, capable of covering the canal crossing zones by laying down fire, were to be deployed as well.[32] Had Gonen ordered at 7:00 a.m. that this deployment plan be carried out, most of the forces could have been ready for war when the firing started. As a result of his decision, however, only three tanks were in their positions at 2:00 p.m., and most of the force was still hours away from its deployment areas. Consequently, the first waves of the Egyptian crossing forces met no Israeli resistance at all, precisely at the stage when they were most vulnerable.

Each of the five Egyptian divisions used a regiment of 144 rubber boats (720 boats altogether) for crossing. During the first hours of the war, they transferred 32,000 soldiers to the Israeli side of the canal in twelve waves without meeting any resistance.[33] What would have been the outcome had the boats crossed the canal under fire from at least 100 tank guns as well as artillery, machine gun fire from the forts, and air bombardment? Under such circumstances, it is quite probable that hundreds of boats would have been hit and that the rate of the Egyptian crossing operation would have been far slower. Had the Israelis received a three- to five-day warning, there would have been seventy-two

30. According to the detailed Egyptian attack plan that was available to Israel's Military Intelligence more than a year before the war and reconfirmed by Israel's best human intelligence source in Egypt on the night of 5 October, the war's "H" hour was to be 6:00 p.m. However, in negotiations with the Syrians, the Egyptians had agreed on 3 October to change "H" hour to 2:00 p.m. There is no doubt that this last-minute change, which was not known to the Israelis, confused them further (Bar-Joseph, *The Watchman Fell Asleep*, 119-21, 185-86).

31. Oren, *History of the Yom Kippur War*, 64; Investigation Committee, *The Yom Kippur War, An Additional Partial Report: Reasoning and Completion to the Partial Report of April 1, 1974* [The Agranat Report] [in Hebrew], 7 vols. (Jerusalem, 1974), 205-8.

32. Oren, *History of the Yom Kippur War*, 20-21, 63; Map 7: "'Dovecote' deployment in the south."

33. "The Instruction of the Chief of Staff of the Armed Forces, No. 41: Infantry Division in the Crossing of a Water Obstacle," The War Office (Egypt), March 20, 1973, in *Enemy documents-Egypt*, The Library of the Center for the Memorialization of the Intelligence Community (MALAM), Glilot, Israel. This document summarizes the detailed Egyptian crossing plan.

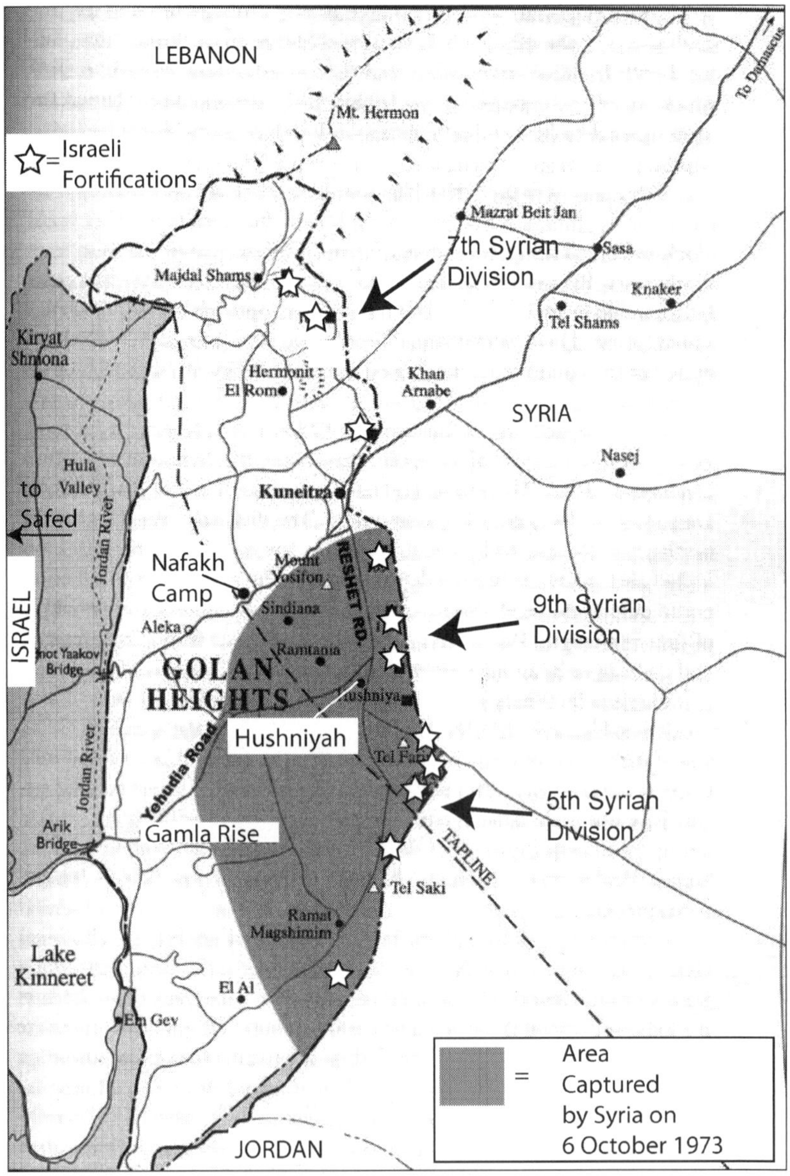

The Golan Front, 6 October 1973 (Map: James R. Arnold)

artillery batteries in place on the front,[34] and the results could have been even more devastating for the Egyptians.[35] In addition, since the firing positions of the Israeli tanks were empty when the first waves of the crossing took place, the Egyptian infantry occupied them. Using personal antitank guided missiles ("Sagger") and RPG-7 rocket launchers, they hit the Israeli tanks in the midst of approaching their positions, causing them heavy casualties.

Altogether, the combination of the completely unprepared Bar Lev line and the unnecessary delay in the deployment of Gonen's available forces led to the fall of Israel's first defense line on the Suez front, with hardly any disruption to the Egyptian crossing, while two-thirds of the 300-tank force of Division 252 were lost (as compared with 122 tanks lost in the occupation of the Sinai Peninsula in the 1967 war[36]).

The Golan Heights

The severity of the situation in the Golan became clear about eight hours after the war began, when reports of Syrian tanks in the Hushniya area started arriving. Two hours later, at midnight, it became evident that Syrian tanks had advanced about eight kilometers into the Golan, and by 2:00 a.m. on 7 October, the IDF force on the southern sector of the front was left with only half of its tanks – about thirty-seven. By 4:00 a.m., Syrian tanks had arrived at the Gamla ascent over the Sea of Galilee. At this stage, when at least four Syrian armored and mechanized brigades (43, 46, 51, and 132) broke into the southern sector of the Golan, no viable Israeli force was available to block their way into the Jordan Valley.[37]

The collapse of the southern defense sector in the Golan was the outcome of two fundamental mistakes. The first involved the way in which the IDF was deployed when the war began. The second took place shortly after firing commenced.

The IDF had three deployment plans for this front. The first two were to meet local challenges aimed at, for example, a Syrian attempt to capture a civilian settlement or a fort or two. In order to meet such threats, Plan Hol Yam (Sea Sand), which was solely based on regular forces, called for the deployment, in preplanned positions, of one tank brigade (100 tanks), ten artillery batteries, and

34. Oren, *History of the Yom Kippur War*, 21; Map 9: "'Desert Cat ' plan."

35. The director of the IAF intelligence department in 1973, Brig. Gen. (Res.) Rafi Harlev, estimated, for example, that had the IDF deployed for war according to plan, "there would simply be no crossing"; "Sadat Just Planned to Cross the Canal and Stop," Interview with Brig. Gen. (Res.) Rafi Harlev, *IAF Journal* 123, October 1998.

36. Trevor N. Dupuy, *Elusive Victory: The Arab-Israeli Wars, 1947-1974* (New York: Harper and Row, 1978), 333.

37. Oren, *History of the Yom Kippur War*, 79-81; Dani Asher, "The Syrian Plan for the Yom Kippur War and the Course of the Fighting," in Golan and Shai, *Yom Kippur War Studies*, 287-310, 296.

three high-quality infantry battalions within twenty-four hours from warning. Plan Gir (Chalk), which called for the deployment of an additional reserve tank brigade (altogether 222 tanks) and five more artillery batteries, necessitated a warning time of thirty-six hours.[38] The IDF's plan for an all-out war with Syria (in parallel to an all-out war with Egypt) was Plan Sela (Rock), which required a warning time of forty-eight hours. This plan involved the deployment of 693 tanks and 258 artillery pieces to face 1,300 Syrian tanks, as well as 700 artillery pieces on the front line, out of a Syrian arsenal of 1,650 tanks and 1,600 artillery pieces.[39]

Since the Syrian buildup for war started in late August and since Israel lacked strategic depth in the Golan, which was populated by civilian settlements, Israeli awareness of the looming threat in the north was higher than in the south. Yet, in light of AMAN's insistence that Egypt did not intend to launch a war and that Syria would not initiate a war without Egypt, the main concern of Defense Minister Dayan and the IDF high command was a small-scale Syrian attack.[40] This was the logic behind the gradual process of reinforcement in the Golan that started on 26 September. Altogether, on the morning of 6 October, the IDF order of battle (OB) there amounted to 176 tanks (77 from Brigade 188 and the rest from Brigade 7); eleven artillery batteries (two 160mm heavy mortars, seven 155mm medium artillery, and two 175mm heavy artillery); two high-quality regular service infantry battalions (Battalion 13 of the Golani Brigade and Battalion 50 of the Paratroopers Brigade) manning the line, including the forts; and one mechanized infantry battalion.[41] Although it was far from the planned OB for an all-out war, it could suffice, under certain circumstances, to prevent a catastrophe until the reserve army arrived.

Here, however, the first mistake came into play. At 6:00 a.m., the Chief of Staff ordered the commander of the Northern Command, Major General Yitzhak Hofi, to prepare for war, which was expected to start at 6:00 p.m. At 10:00 a.m., Hofi met the commanders of the three brigades in his command (188, 7, and territorial brigade 820) and informed them that the war would start at 6:00 p.m. However, he did not order them to deploy for war and returned to Tel Aviv for an additional meeting at general headquarters (GHQ). Thus, when firing commenced, his command was still deployed for a local clash rather than for a full-scale war.

According to Plan Gir (the defense of the Golan by two tank brigades), the front was to be divided into two regimental sectors. In each, one battalion was to be deployed at the front and two others in the rear. Each brigade commander was to filter the information from his sector to the rear in order to enable the Command's war room to properly digest the vast quantities of operational

38. Oren, *History of the Yom Kippur War*, 25; Map 12: "'Chalk' plan."

39. Ibid., 25; Map 14 "'Prairie wolf' plan."

40. For a discussion of the intelligence collection concerning the Syrian deployment, the way it was processed, and the military and political conclusions, see Bar-Joseph, *The Watchman Fell Asleep*, 81-199.

41. Oren, *History of the Yom Kippur War*, 60-61.

information and to conduct a high-quality decision-making process. In addition, two separate advance supply centers were to be established in order to efficiently provide the two brigades with the necessary logistics for fighting.

However, the Gir order was not given, and therefore the front was not divided into two regimental sectors. Most of the tanks of Brigade 188 that were deployed on the front were not in their positions when firing commenced, and the tanks of Brigade 7 were still in battalion concentrations at the rear. The Command's war room was set up in the Nafakh camp, at the center of the Golan, together with the war rooms of Brigades 188, 7, and 820. Only one advance supply center was in existence, creating a cumbersome and risky situation in which a Syrian breakthrough in the front's center could have destroyed the ability of the whole force in the Golan to continue fighting.[42]

The main outcomes of this deployment became evident shortly after the war started. Since there were no regimental war rooms that could filter the flow of information to the Northern Command's war room, the front's command and control system almost collapsed when firing commenced. The commander of Brigade 7, Colonel Avigdor Ben-Gal, described the chaotic situation there about forty minutes after the Syrians opened fire:

> The war room bunker was an unbelievable mess. I will not exaggerate if I say that there were there about 200 persons – commanders, soldiers, women soldiers, and all sorts of functionaries. The war rooms of the command, of the territorial brigade of the 188 Brigade and the first elements of Division 36 were located in the same hall [of about 50 sq. meters]. Ben Shoham, the commander of 188, ran the operations as if it was a regular local incident.[43]

The order to divide the front into two sectors was given only at 4:00 p.m., when Hofi returned to his command post in Nafah. By then, however, it was too late. At 2:45 p.m., the second mistaken decision had been made in the battle for the defense of the Golan.

Israel had been in possession of the plan for a Syrian offensive since April 1973. An updated version of the plan was delivered during a secret meeting between King Hussein of Jordan and Prime Minister Golda Meir on the evening of 25 September. On 30 September, the U.S. Central Intelligence Agency provided Israel with the same plan. It was distributed by AMAN on 2 October, together with an estimate that war was unlikely.[44]

42. Ibid., 60-61; Zvi Ofer, *Barak in the War of Yom Kippur, Brigade 188 in the Defense Battle in the Golan Heights in the War of Yom Kippur – an Event Analysis* [in Hebrew] (Tel Aviv: IDF History Department, April 1996), 371-73; interview with Zvi Ofer (22 August 2005).

43. Uri Milstein, *The Lessons of Collapse* [in Hebrew] (Kiron: Sridut, 1993), 211.

44. Bar-Joseph, *The Watchman Fell Asleep*, 87-89, 96-97, 114-15.

URI BAR-JOSEPH

According to the plan, the main effort of the Syrian army was to take place in the Rapid area, about fifteen kilometers south of Kuneitra, in the southern sector of the Israeli defense line. However, the officers of the IDF Northern Command were not aware of this document. Instead, they expected that the main Syrian effort would be staged in the Kuneitra open land area in the northern sector of the front.[45] Acting on this assumption, the Command's Assistant Chief of Staff G3 (operations), Lieutenant Colonel Uri Simhoni, authorized the dispatch of Brigade 7, the Command's reserve force, to the northern sector of the front in order to block what he deemed to be the center of gravity of the Syrian attack. He did so despite the fact that prior to 3:30 p.m., when the first Syrian tanks crossed the border, it was not yet possible to identify the offensive's main thrust.[46]

The Nevo report, which was prepared as part of the official investigation of the war by the Agranat Commission, assessed this decision as "perhaps the most decisive decision in the battle of the Golan Heights."[47] Its gravity became clear as the evening hours approached, when battlefield reports indicated that the pressure of the offensive was increasing, despite earlier expectations that the Syrians were not prepared for night fighting. Even more importantly, it became apparent that the main Syrian effort was not concentrated in the north, where Brigade 7 faced no considerable pressure that night, but rather in the south, where 360 out of 600 Syrian tanks were deployed. The southern sector was held by Brigade 188, which had only 76 tanks, as compared with more than 100 in Brigade 7.

Consequently, the situation in the southern sector deteriorated rapidly. By 10 p.m., the Syrians had penetrated the line of defense in the Hushniya area, and by midnight the tanks of Brigade 46 had already advanced eight kilometers into Israeli-held territory and had taken positions along the Tapline road, about ten kilometers south of the central command post at Nafakh. Mechanized Brigade 132 penetrated the line a few kilometers to the south. At 1:20 a.m., the commander of the Northern Command and his aides left the war room in Nafakh and returned to Command's central command post in Mount Canaan near the town of Safed in the Galilee. By 2:00 a.m., Brigade 188 was left with fewer than forty tanks.

In order to prevent a Syrian breakthrough into Israel proper, preparations were made for the demolition of the six bridges over the Jordan River. Shortly afterwards, orders were issued to destroy secret documents in the Golan's military camps. At 4:00 a.m., Syrian forces arrived at the Gamla ascent over the Sea of Galilee. No IDF unit stood between them and the Jordan Valley,[48] but requests

45. Ibid., 211-13.

46. Ofer, *Barak in the War of Yom Kippur*, 386.

47. The Nevo Report, 42, quoted in Ofer, *Barak in the War of Yom Kippur*, 386.

48. Oren, *History of the Yom Kippur War*, 79-81; Bar-Joseph, *The Watchman Fell Asleep*, 217; Rabinovich, *The Yom Kippur War*, 168-69.

by Syrian commanders to advance westward were denied by their commanders.[49] They opted instead for an attack northward, toward the IDF's central command post in Nafakh – a move that, if successful, could have led to the collapse of the Israeli defense in the Golan.

By 3:00 a.m., the gravity of the situation in the north began to be clear to the IDF GHQ in Tel Aviv. The fear of a Syrian breakthrough into Israel prompted Defense Minister Moshe Dayan to go to the central command post in Mount Canaan in order to receive first-hand reports of the state of affairs. Upon his arrival, at about 6:00 a.m., he heard from General Hofi that the situation in the Golan was very grave and that the bridges over the Jordan River had been prepared for demolition. Another senior and experienced officer told him that "the fighting in the Heights' southern sector has ended and we lost. We don't have a force to stop them." A third officer added that there was nothing to stop the Syrians if they were to reach the Jordan Valley. Consequently, Dayan sent him to organize the defense there.[50]

In reality, by the early morning hours the situation had slightly improved since the first reserve forces had started arriving at the front. But due to the thick fog of war and insufficient information about Syrian intentions, Dayan estimated that only massive air support would enable the IDF to hold its positions in the Golan Heights. Accordingly, he ordered the commander of the IAF, Major General Benny Peled, to allocate all its resources to that front. This order led to the last two mistakes that doomed the ability of the IAF to quickly decide the fate of the war.

The Air Force

Unlike the IDF on the southern and northern fronts, the IAF was well prepared for the war before it started. The IAF had been gradually increasing its state of alert since mid-September and needed only a small-scale mobilization because it functioned mainly on the basis of regular manpower. Given its critical role in the defense of the country, the Chief of Staff authorized mobilization of the manpower required to bring the IAF to a full state of alert about twenty-four hours before the outbreak of the war.[51]

49. Rabinovich, *The Yom Kippur War*, 186. According to the Syrian war plans, which Israel had prior to the war, the Syrian territorial goals did not include the crossing of the Jordan River into Israel at its pre-1967 borders. According to one Syrian official, in a rather rare Syrian statement on this issue, President Hafez al Asad estimated that he could occupy the Golan within forty-eight hours, but feared an Israeli nuclear response. Consequently, he gave the order to stop the progress of his forces on 7 October so as to signal Jerusalem that there was no reason for panic and that there was no need to let the nuclear genie out of the bottle (Richard B. Parker, ed., *The October War – A Retrospective* [Gainesville: University of Florida Press, 2001], 102-3, 119).

50. Bartov, *Daddo*, 408.

51. Bar-Joseph, *The Watchman Fell Asleep*, 175.

The IAF started the war with 311 fighter and attack planes in operational status out of a total of 391 in service. They were deployed at five main airfields (Ramat David, Tel Nof, Hazor, Hazerim, and Etzion) and two secondary ones (Bir Gafgafa and Sharm al Sheik in the Sinai).[52] Upon learning at about 4:30 a.m. on 6 October that war would start the same day, the Chief of Staff ordered the IAF to prepare for a preemptive strike. However, Operation Dugman, the code name for the air strike against the Syrian antiaircraft layout, which was planned for 10:00 to 12:00 a.m., was called off due to bad weather conditions. It was replaced by a planned strike against the Syrian Air Force (code name Negiha or Butt), but Moshe Dayan and Golda Meir vetoed this act for fear of American reaction. Consequently, when shooting commenced, many IAF planes were in the process of changing their ammunition from an attack to a defense mission.[53] Nevertheless, some IAF fighters succeeded in taking off on time and interrupting the Arab aerial attack. By the evening hours, they had shot down 22 Egyptian and Syrian planes and helicopters. In addition, the IAF took part in the defensive battle, primarily on the Suez front, though it proved to be quite ineffective since it had to operate in darkness and within the zone defended by Egyptian ground-to-air missiles.[54]

During the night hours, the IAF made final preparations for Operation Tagar – the destruction of the Egyptian antiaircraft layout west of the Suez Canal – which had been carefully prepared since the end of the War of Attrition in the summer of 1970. The operation combined a massive use of ground and airborne means of electronic warfare, with the use of ground artillery for shooting chaff and shelling nearby batteries, in order to support the main effort – an all-out attack by more than 100 Phantom F-4s, flying at low altitude at about 550 knots and simultaneously dropping bombs to destroy the missile sites. The preparations for Operation Tagar took thirty-six hours and were dependent, among other things, on good weather and updated information about the location of the surface-to-air missile (SAM) batteries.[55] The operation was to last six hours and to be carried out in four phases. The first flight – preparation – was aimed at suppressing the antiaircraft guns and isolating the battlefield from enemy planes. The second and main stage was the bombing and the destruction of the missile sites. The aim of the third flight was to destroy the surviving batteries. The last one was to take out missile reinforcements, which were expected by this stage to start advancing from the rear.[56] Repeated exercises and detailed planning led the IAF command-

52. Oren, *History of the Yom Kippur War*, 14.

53. Bar-Joseph, *The Watchman Fell Asleep*, 189-90, 218-19.

54. Oren, *History of the Yom Kippur War*, 75-76.

55. Peled, *Days of Reckoning*, 346-47.

56. Lecture by Yosef Abudy, *The IAF in the Yom Kippur War,* Ramat Efal, 4 May 2000. Lt. Col. (Res.) Abudy was the head of the IAF History Division and extensively studied the role of the IAF in the Yom Kippur War; Ran Ronen, *Hawk in the Sky* [in Hebrew] (Tel Aviv: Hemed, 2002), 335.

ers to conclude that although the operation might incur significant casualties, its chances of success were high, and it would give the IAF freedom of action over the combat zone. In spring 1973, following the buildup of a Syrian antiaircraft layout on the Golan front, the IAF prepared a similar operation (Dugman) for the destruction of this layout as well.[57]

The main challenge for these operations was the introduction in 1972 of the mobile SA-6 batteries in Egypt and Syria. Previously, a reconnaissance mission that revealed the location of the fixed SA-2 and SA-3 batteries had enabled an immediate attack within hours. Now, the localization of the sites prior to attack was of little use, since the SA-6 batteries had the capability to start moving within ten minutes from order and to start firing within twelve minutes after arriving at their destination.[58] In addition, the SA-6 batteries could hit targets flying at 300 feet and above.[59] Until the introduction of these systems, IAF pilots used very low flight as the main tactic for evading missiles. The deployment of the SA-3 in Egypt in the spring of 1970 had already limited their freedom of action significantly. The introduction of the SA-6 limited it further.[60] The challenge posed by the SA-6 batteries was even greater in the north. The Syrian layout was made up of fifteen SA-6 batteries and sixteen fixed SA-2 and SA-3 batteries. The Egyptian layout, west of the canal, was made up of fifty-five batteries, only ten of which were SA-6.[61] For this reason, the IAF planners believed that Operation Tagar had quite a high chance of success.[62]

The first stage of Tagar 4 began on 7 October at 6:45 a.m., when eighty Skyhawk A-4s attacked antiaircraft gun batteries and other targets. Half an hour later, approximately forty Phantom F-4s attacked seven Egyptian airfields in order to prevent Egyptian fighters from taking off while the main stage of Tagar was being carried out. The attacks were successful and achieved their operational goals with only minor losses.[63]

57. Interview with Brig. Gen. Rafi Harlev, the IAF's intelligence officer in 1973 (Ramat Hasharon, 29 July 1999); interview with Lt. Col. Yehuda Porat, the head of the Research Division of the IAF intelligence department in 1973 (Shfayim, 25 July 1999); Ronen, *Hawk in the Sky*, 339.

58. Abudy lecture.

59. *Jane's Weapon Systems* (New York: Jane's Information Group, 1976), 92. According to this source, the use of optical means could enable the SA-6 to hit targets flying at 150 feet.

60. Harlev interview, 29 July 1999.

61. Oren, *History of the Yom Kippur War*, 11.

62. Interview with Brig. Gen. (Res.) Giora Furman, (Ma'abarot, 11 August 2005). In 1973, Furman served as Chief of the Operations Department of the IAF.

63. Oren, *History of the Yom Kippur War*, 97-98.

At this stage, however, the operation had already been called off. Under the impression that the IAF was the only means by which the Syrian tanks could be blocked on their way into Israel, Dayan, who failed to get in touch with the Chief of Staff, called General Peled directly at about 6:40 a.m. He described the desperate situation in vivid terms, referring to a threat to "the Third Temple," and pressed Peled to use all his aircraft to save the situation. The Chief of Staff, who heard about the exchange, was alarmed by Dayan's pessimistic estimate of the situation. Together with the IAF commander he decided, at about 7:00 a.m., to discontinue Operation Tagar and to redirect the main effort of the IAF from the south to the north.[64]

The decision to stop Tagar was a major mistake – and this is more than just a post factum assessment. A number of senior IAF officers who were present in the IAF war room when it was made regarded it then as such, but none of them objected when Peled called off the second stage of the operation.[65] There were two considerations that favored the continuation of Tagar 4. First5, there was a good chance that it would succeed. Although the conditions under which it was launched were not ideal (for example, artillery support was not available), it still had a high chance of success given the operational preparations that had been made the night before and the fact that most of the Egyptian SA batteries were fixed. The successful completion of the operation's first stage had already indicated that strategy to be a good one.[66]

Second, during the critical hours of the morning of 7 October, the conditions on the Golan front prevented the IAF from effectively deciding the fate of the battle. The small area of terrain in which the battle took place and the fact that the Israeli planes had to operate within the range of the Syrian antiaircraft layout did not allow a massive use of aircraft. In addition, two squadrons of A-4s had already been participating in the battle since the early morning hours, so there was no need to stop Tagar 4 for this purpose.

At this juncture, the second mistake took place. Since the doctrine of the IAF called for destruction of the enemy's antiaircraft defense prior to its engagement in ground warfare, Peled ordered the IAF to be prepared to carry out Operation Dugman 5 at 11:30 a.m. – at precisely the same time that he ordered the discontinuation of Tagar. As the IDF's official history of the war describes the situation:

> the preparations and the briefings [for Operation Dugman 5] were done under time pressure, especially because of a change of plans following the movement of the mobile SAM batteries and antiaircraft

64. Ibid., 87; Peled, *Days of Reckoning*, 363; Bar-Joseph, *The Watchman Fell Asleep*, 220.

65. Furman interview; Ehud Yonai, *No Margin for Error: The Making of the Israeli Air Force* [in Hebrew] (Jerusalem: Keter, 1995), 255.

66. Harlev interview, 29 July 1999; Furman interview. Harlev's estimate was that Tagar had a 70 to 80 percent chance of success.

> batteries on the front. Nevertheless, the IAF started the operation at 11:30, but without updated air-photographs and without the planned artillery preparation that the artillery corps could not provide under the conditions of the battle. The loss of the Hermon [Israel's eyes, ears, and a major platform for electronic warfare in the Golan] and the inability to transfer electronic warfare helicopters from the south to the north have made the mission even more difficult.[67]

As could be expected, Operation Dugman 5 was an almost complete failure. Since the IAF had no updated information concerning the location of the fifteen Syrian SA-6 batteries, it bombed the sites in which they had been located about twenty-four hours earlier. Consequently, only one battery was located and destroyed, and another one was damaged. The rest of the bombs fell on empty sites. The immediate cost of this failure was the shooting down of six F-4s. In the longer run, the failure shattered the IAF's belief in its ability to destroy the SAM layouts in Egypt and Syria.

Was it necessary to improvise the most critical and dangerous IAF operation in the north and to carry it out in the absence of so many components that were deemed necessary for its success? In retrospect, the answer is negative. The commander of the Northern Command testified that, with the exception of one case, the ground forces did not feel the aerial support even after the main effort of the IAF was diverted from the south to the north.[68] Israel's official history notes that "the effective air support was provided . . . . by the planes that attacked the Syrian armor since dawn . . . . These sorties were stopped when operation 'Dugman' was underway – and this was during the most difficult hours in the fight for Nafakh."[69] Indeed, immediately after Dugman 5 ended, the IAF was called again to divert its forces to the south, thus denying effective air support for the ground forces in the Golan.[70] Nevertheless, the ground forces succeeded in containing any additional Syrian advancement, and by the evening hours, following the failure of a Syrian attempt to occupy the Nafakh camp, the initiative taken by the IDF appeared to be sufficient.

According to the official history of the war, the way in which the IAF was used on 7 October was indeed a mistaken one, though specific accusations are avoided:

> After the war, a widespread opinion was that the enemy succeeded to provide a cover for its ground forces by a dense "umbrella" of missiles and anti-aircraft guns with which the IAF planes were crushed. But the picture will not be a complete one unless it is remembered that the

67. Oren, *History of the Yom Kippur War*, 98.

68. Maj. Gen. Yitzhak Hofi, "The Supreme Command in the War of Yom Kippur," *A lecture in the framework of a seminar on the Yom Kippur War*, Ramat Efal, 2 December 1999.

69. Oren, *History of the Yom Kippur War*, 99.

70. Ibid.

> SAM layout did not face an organized attack, since the IAF did not have the chance to act in a concentrated manner against the missiles at the beginning of the war. In the south, the operation was stopped after the preparation stage and the temporary paralysis of the Egyptian airfields. And in the north, the IAF acted under time pressure and under operational conditions that caused its failure.[71]

Conclusions

This article posed the question as to what the main source of Israel's military defeats was in the first stage of the Yom Kippur War: strategic surprise or – as many now claim – inherited deficiencies in the IDF's preparation for the war.

Any attempt to answer this question involves a certain amount of counter-factual analysis and should therefore be phrased in probabilistic rather than definitive terms. In order to answer the question, this article identified six major mistakes that doomed the initial phase of the war. Two took place on the Suez front: the first was the manning of the Bar Lev line by second-rate forces rather than by high-quality soldiers, as was planned earlier; and the second was the decision to delay the deployment of Division 252 until the last moment. Two took place in the north: the first was the failure to deploy the available regular forces according to Plan Gir, despite the fact that the IDF had sufficient time to do so; and the second was the decision to allocate the Command's stronger force (Brigade 7) to the defense of the front's northern sector before locating the Syrian main effort. Finally, two mistakes took place in the air: the first was the decision to call off Operation Tagar 4 despite its high chance of success, and the second was the decision to launch Operation Dugman 5 despite its very high chance of failure.

An analysis of the context within which these mistakes were made shows that their main source, and therefore the source of Israel's military failures at the beginning of the war, was the fact that until eight hours before it broke out, war was not expected, certainly not with such a short warning time. Adding to the high level of strategic surprise was the fact that firing commenced four hours earlier than predicted.

The way in which the IDF's ground forces were deployed on the Suez front when the war started was primarily the direct outcome of the intelligence failure. All operational plans in the south assumed that even a twenty-four-hour warning would enable the replacement of the reservists in the Bar Lev line forts by high-quality regular infantry soldiers and the deployment of the tanks and the artillery of Division 252 in their pre-prepared positions. The fact that force replacement at the Bar Lev line did not take place was solely the outcome of surprise. The fact that Division 252 was not deployed for war in time was largely attributable to the same cause, as well as to the poor judgment of the Southern

71. Ibid., 99-100.

Command commander, who, by his own testimony, avoided immediate deployment of his available forces based on the belief that such a move might trigger an unintentional war. Obviously, that belief was the direct outcome of insufficient warning. This scenario is somewhat reminiscent of Joseph Stalin's belief, during the first hours of the German invasion of the U.S.S.R. on 22 June 1941, that the German action did not necessarily mean war and that a prudent Soviet reaction might prevent it.[72] Gonen's decision also reflected his poor level of military professionalism, as he did not act in accordance with the direct orders given to him by the Chief of Staff.

In the Golan, insufficient warning was the only cause for the the IDF's failed order of battle on 6 October. The IDF plan for war called for the deployment of two divisions, with close to 700 tanks and over 250 artillery pieces on this front. The course of the fighting in the Golan during its first hours indicates that even with the minimal available forces, the IDF had a chance to contain the Syrian advancement until the arrival of the reserve army, had the order to allocate the larger share of the scarce available forces to the northern sector not been given. The fact that the weaker Brigade 188 had to block the Syrian main effort while the stronger Brigade 7 faced a lesser Syrian pressure is the key event that led to the collapse of the southern sector. The mistaken deployment order that was given less than an hour after the outbreak of the war was not the direct outcome of surprise, but was influenced by the fast pace of unexpected events and the chaotic atmosphere in the unprepared war room. Moreover, it should be remembered that had sufficient forces been available to the Command from the start, there would have been no need to make such a critical decision. In this sense, the second mistake in the Golan reflected, perhaps, a poor level of professional judgment, but was also highly influenced by the element of strategic surprise.

Finally, the two mistaken decisions regarding the use of the IAF were indirectly the outcome of surprise. Both were made in light of the dire situation that was developing in the southern sector of the Golan front and the fear that, with hardly any ground forces to block them, the Syrian tanks were likely to invade Israel at its pre-1967 borders. This grim situation was, as concluded above, primarily the outcome of insufficient intelligence warning for the impending outbreak of war. Certainly, when Moshe Dayan made his demand to allocate the whole might of the IAF to the north, there were other available alternatives to this decision, and when the IAF commander complied with it, other officers around him regarded it as a mistake. As such, the decision to discontinue Operation Tagar and to initiate Operation Dugman was not the only, and ultimately not the best, approach to the difficult situation. But the fact that both operations failed probably has less to do with fundamental mistakes in the preparations of the IAF for war than with the way it was used to destroy the antiaircraft layouts in Egypt and Syria.

72. David E. Murphy, *What Stalin Knew: The Enigma of Barbarossa* (New Haven, Conn.: Yale University Press, 2005), 212, 216.

This article has shown that the main cause for Israel's military defeats in the first stage of the 1973 war was the element of strategic surprise. Nevertheless, it is important to note other sources as well: certain weaknesses in the IDF war planning (overreliance on intelligence warning and ineffective air-ground coordination); force structure (overreliance on aircraft and tanks); and, in some cases, poor quality of staff work and military professionalism. But all these weaknesses were a secondary cause for the military defeats, and in most cases were manifested under pressure, which by itself was the direct outcome of surprise. It is quite probable that if surprise had been averted and sufficient IDF forces had been deployed when the war started, the Egyptian and Syrian offensives would have been repelled from the start, at a rather low cost to the defending forces. Under such circumstances, it is quite possible that no one would have faulted the IDF for any mistakes in its preparation for the war. Above all, it must be emphasized that the final outcome of the war was an Israeli military victory – one that was achieved despite the difficult conditions under which the IDF had to function once the first stage of the war was over. This constitutes the best proof that strategic surprise rather than fundamental weaknesses was the independent variable that doomed the outcome of the war in its initial stage. Improper preparations for war served, and only in some respects, as an intervening variable that facilitated this outcome.

The 1982 Lebanon War

[9]

Personal Whim or Strategic Imperative?

The Israeli Invasion of Lebanon

Avner Yaniv and Robert J. Lieber

The 1982–1983 crisis in Lebanon has again demonstrated the persistent tendency of observers to identify Israel with the idiosyncracies of specific individuals. In the 1950s, Israel's conduct was instinctively associated with the virtues and vices of David Ben Gurion; in the 1960s, especially after the Six-Day War, Israel was often reduced to the attributes of Moshe Dayan; and, in the first part of the 1970s, it became a kind of Golda Meir writ large. Since 1977, it has become indistinguishable from the personality of Menachem Begin or, worse still, the vices of Begin and Ariel Sharon. When Israel's leaders have been international favorites, the country and its diverse and complicated society have appeared upright, enlightened, and virtuous. When its elected leaders have been disliked abroad, it has been widely reproached. The result has been the trivialization of complex strategic issues.

The events of last year, for example, have been reduced, as in the title of a recent article, to a matter of "Begin's Rhetoric and Sharon's Tactics."[1] But, does such an approach do justice to the subject? Or does it make more political sense to depict the Israeli invasion of Lebanon as the culmination of a complex strategic sequence? Surely the *personification de pouvoir* is as sterile for scholarship and statecraft as it is appealing in its dramatic simplicity. Therefore, tracing these events to the imperatives of security rather than the idiosyncracies of individuals provides a far sounder basis for policy analysis.

Specifically, the invasion must be evaluated against the backdrop of the past three decades. For twenty years after 1948, a succession of Israeli Cabinets had maintained a continuity of policy toward Lebanon. But this pattern was broken nearly a decade before Begin's victory in the 1977 elections. A

Avner Yaniv is Director of The Institute of Middle Eastern Studies, Haifa University, Israel. He was educated at the Hebrew University, the London School of Economics, and Oxford. In 1982–83 he was Visiting Professor in the Department of Government, Georgetown University. His writings have been published in Israel, Europe, and the United States. Currently he is preparing a book on the making of Israel's national security policy. Robert J. Lieber is Professor of Government, Georgetown University. His latest book is The Oil Decade: Conflict and Cooperation in the West *(New York: Praeger, 1983). He is also contributing author and co-editor of* Eagle Defiant: U.S. Foreign Policy in the 1980s *(Boston: Little, Brown, 1983).*

---

1. Amos Perlmutter, "Begin's Rhetoric and Sharon's Tactics," *Foreign Affairs*, Vol. 61, No. 1 (Fall 1982), pp. 67–83.

new pattern was initiated under Prime Minister Levi Eshkol, a moderate by all accounts, and was maintained under Golda Meir, Itzhak Rabin, and Menachem Begin. The source of the change in policy was not a replacement of one individual with another. Rather it was the result of the encounter between Israel and the Palestine Liberation Organization (PLO) in which the latter had to sustain its military pressure and the former could not forgo countermeasures.

Israel's options were inaction, limited action, or massive action. For nearly fifteen years, it repeatedly opted for the intermediate—limited—alternative. Even Begin initially followed this pattern. But, increasingly, the limited action alternative brought diminishing returns. By the beginning of the 1980s, it was becoming apparent that the number of Israel's alternatives had been reduced to either inaction or massive action. Begin and Sharon may have moved earlier and more ferociously than other Israeli policymakers would have done, but, it will be argued here, the nature of the problem was such that, under any leadership, Israel eventually would have reacted in much the same way.

Cooperation to Conflict: Israel, Lebanon, and the Advent of the PLO

Stability along the Israeli–Lebanese border during the two decades after 1948 can be explained by the situation in Lebanon as much as by Israeli decisions. Early in 1954 there had been suggestions that Israel should turn Christian–Moslem friction in Lebanon to its own advantage. Dayan and Ben Gurion, who supported this thesis, approached Prime Minister Moshe Sharett with the idea that Israel should find itself a supporter among Lebanon's Maronite Christian army officers. These officers would move with Israeli backing to force Lebanon towards a new foreign policy orientation based on an alignment with Israel and, as a result, the encirclement of Israel would be broken. But Sharett, who had grown up in Palestine, spoke excellent Arabic, and took great pride in his knowledge of the Middle East, preferred a diplomacy of accommodation. He therefore opposed the idea. Convinced that minorities in general could not be reliable allies, he argued that, in the final analysis, Israel would have to settle its differences with the Middle Eastern (Sunni Arab) majority. He doubted whether a truncated Christian Lebanon, cut off from its Moslem hinterland, could be economically solvent, and he strenuously opposed what he called a "policy of setting the Middle East ablaze." Hence while authorizing the cautious cultivation of covert contacts with friendly elements among Lebanon's Maronites, he rejected the plan as a "mirage."[2]

It appears that Sharett's decision shaped Israel's strategy vis-à-vis Lebanon until the aftermath of the 1967 war. However, another, not entirely implausible, possibility is that Israel's resourceful intelligence community did try to go beyond Sharett's guidelines but failed to elicit from the Lebanese themselves any enthusiasm for an explicit alliance. When Sharett was eventually forced to resign after the disastrous Lavon Affair, Ben Gurion returned to both the Premiership and the Ministry of Defense and, in this capacity, worked actively towards the establishment of a network of alliances with non-Arab or non-Muslim states and minorities such as Turkey, Iran, Ethiopia, the Kurds in Iraq, and the Christians in the Sudan. The ideological underpinning of this policy was that the Middle East was not a homogeneous Sunni ocean waiting eagerly to be reorganized by Pan-Arabism but rather a mosaic of cultures, religions, and races. In such an area a Jewish state had a legitimate place, whereas in the Pan-Arab alternative it appeared an alien intruder. Moreover, strategically, the cultivation of such relations held the promise of friction between Israel's enemies and other minorities and nations. In turn, there was evidently some hope that Arab armies could be pinned down to theaters of operation other than Israel's borders, thus alleviating Israel's pressing security situation.[3]

In view of these calculations, it seems likely that a dormant link with certain elements in Lebanon was established and maintained throughout the 1950s and 1960s. It was therefore available when the events which followed the 1967 war augmented its value from the point of view of the Lebanese in question. Most important was the rise of the PLO and its attempt to turn Lebanon into its main base of operations. To be sure, this in itself does not explain every aspect of the Lebanese problem. Had it not been for the weakness of the Lebanese polity, as well as the coincidence of religious, ethnic, geographic, social, and economic cleavages in Lebanon's society, the appearance of the PLO might not have had the consequences it did. But given the intensity of strife among Lebanon's competing groups, the weak-

---

2. Moshe Sharett, *Yoman Medini* [Political Diary] (Tel Aviv: Sifriyat Poalim, 1978), Vol. 2, pp. 334, 377, 677; Vol. 3, p. 639.
3. Nadav Safran, *Israel: The Embattled Ally* (Cambridge, Mass.: Harvard University Press, Belknap Press, 1978), pp. 379–380; Shimon Peres, *David's Sling* (London: Weidenfeld & Nicolson, 1970), pp. 285–286.

ness of political institutions, the absence of effective means of conflict resolution, and the accelerating disintegration of the Lebanese state under the strain of rapid economic growth, the appearance of the PLO and its activities against Israel were critical.[4]

From the Israeli point of view the arrival of the PLO created a dual security problem. One aspect was the use of Lebanese territory by PLO personnel for hijacking operations and for training and equipping members of other guerrilla organizations (Baader-Meinhof, Tupamaros, the Japanese Red Army, the IRA, and others). These groups could offer the PLO in return a kind of infrastructure for operations against Jewish and Israeli targets in different parts of the world. Meanwhile, Israel's air connections with the rest of the world, its trade, and tourism were in jeopardy while Jewish communities in Europe became exposed to violent attacks.[5] The second aspect of the problem related to security along Israel's border with Lebanon. For two decades this border had been relatively quiet. Occasionally the Lebanese gendarmerie would allow small numbers of civilians to cross quietly into Israel. At other times they would stop them from doing so.[6] But all this never led to any military engagements. After 1968, however, the atmosphere was disrupted by an increasing number of PLO attacks against Israeli communities along the border, most of the casualties from which were non-combatants. To be sure, the total number of casualties was not more than a few hundred over a period of several years.[7] But the Israeli population along this border was composed primarily of new immigrants from North Africa whose low ideological motivation, weak social fabric, and shaky morale increased their vulnerability. Thus, to apprehensions about Israel's air travel links and the safety of Jewish communities and Israelis in different parts of the world was added a fear of "falling social dominoes" inside Israel itself—i.e., of panic among settlers inside Israel's boundaries gradually spreading into other parts of the country.

4. Michael C. Hudson, *Arab Politics: The Search for Legitimacy* (New Haven: Yale University Press, 1977), pp. 280–294.

5. Peter J. Snow and David Phillips, *The Arab Hijack War* (London: Pan, 1970).

6. John K. Cooley, "The Palestinians," in P. Edward Haley and Lewis W. Snider, eds., *Lebanon in Crisis* (Syracuse: Syracuse University Press, 1979), p. 22; Sharett, *Yoman Medini*, Vol. 2, p. 477.

7. Bard E. O'Neill, *Revolutionary Warfare in the Middle East: The Israelis vs. the Fedayeen* (Boulder, Colo.: Paladin Press, 1974). Between 1967 and 1970 the PLO killed 317 Israelis and wounded 1674. IDF Spokesman, *Basic Facts: The IDF and the Security of Israel* (Tel Aviv: IDF, April 1982), p. 40.

Against this background, a new strategy gradually evolved. In similar circumstances involving other Arab states (Egypt before the 1956 Sinai war, Jordan and Syria at the same time, and then again between 1967 and 1970), Israel had resorted to a policy of "reprisals" which sought to compel neighboring governments to stop these activities by force. The logic of this method was simple. No Arab government could be relied upon to act merely in response to Israeli appeals; nor would the United Nations restrain terrorism against Israel. Hence, the Arab governments in question had to be made to pay a high price—in damage to their own citizens as well as in prestige—through punitive Israeli action. Whenever Israel acted in such a fashion its decision-makers took severe U.N. condemnation for granted—but, in view of the alternatives, this was a small price to pay.[8] Moreover, Israeli policymakers were sensitive to the possibility of escalation. But, again, escalation leading to a showdown in which Israel was very likely to emerge victorious seemed a far better outcome than continued deterioration. Indeed, viewing every show of force as an index of the balance of power, they imputed to these measures a larger virtue, namely the gradual aggregation of what came to be known in Israeli official parlance as "deterrence power."[9]

Insofar as Jordan was concerned this policy worked. In the 1950s, it led the Jordanian Government to deploy large formations along its Israeli border with specific orders to shoot potential infiltrators in the back. In the case of Egypt it led to severe escalation and ultimately to the Sinai campaign; but, after this 1956 conflict, the Israeli–Egyptian border was quiet for eleven years—a span of time which in the Israeli perception justified the price. Syria proved more difficult to deal with, yet after the 1967 war it too seemed to observe restraint.

Lebanon, however, posed a different problem. For whereas the other governments were stable and clearly capable of disciplining the PLO if they wished, Lebanon had a government which could not, even if it desired to. The simultaneous rise of the PLO's strength in Lebanon—especially following its expulsion from Jordan in September 1970—and the weakness of Lebanon's legitimate government led the various private armies in the country to undertake a sustained effort to increase their own capabilities. Thus Israeli

8. The most lucid explanation of the rationale of reprisals was offered by its chief architect, Moshe Dayan, in "Peace-Time Military Operations," *Ma'arachot*, No. 119 (May 1959).
9. Oz Chen, "Reflections on Israeli Deterrence," *The Jerusalem Quarterly*, No. 24 (Summer 1982), pp. 26–40.

policymakers could reason—and evidently did—that if the reprisal technique were applied in Lebanon it would sooner or later force the Christian and anti-PLO forces of the Phalange and the Chamounites to back an official Lebanese drive against the PLO or alternatively to take on the PLO by themselves. In either case, Israel could expect significant benefits. The least that could happen would be the increased preoccupation of the PLO with its own security. If everything went well, it might even be driven out of Lebanon or subdued into insignificance. If not, Israel could at least anticipate expansion of its own dormant links with the Christian forces. And even if such a strategy led to a major escalation involving Israeli forces on a large scale, it would still be far better than to sit back and watch the PLO increase its strength while Israel was being bled in its economy, credibility, and morale.

Of course, Israeli policymakers could not possibly have foreseen everything which would unfold in the course of the 1970s. Nor were they geared to detailed and comprehensive planning of moves which involved more than military action.[10] In a sense, they were merely reacting to new circumstances by relying on deeply ingrained rules of thumb whose origins lay in the experience of the 1950s. Yet given Israel's real security problem, the strategy which they applied was anything but irrational.

Escalation and Civil War

A closer look at some of the events between 1968 and the outbreak of the civil war in 1975 seems to prove this point. The reaction to the hijacking operations of the PLO included a variety of defensive and preventative measures (such as arming El Al personnel and tightening security arrangements in all airports used by the Israeli airline and by other airlines flying to Israel). But it culminated with the December 28, 1968 attack on the Beirut airport. The brunt of the damage was done to Lebanese Christian-owned property and it could therefore be argued that Israel was harming its own allies. Yet set against the background of Christian reluctance to cooperate with Israel, there was logic in the Israeli reprisal. This suggested that only by harming Christian interests could the Israelis force these groups into attempting to control the PLO.

10. Gerald E. Caiden, *Israel's Administrative Culture* (Berkeley: University of California Press, 1970).

Moreover, the same logic can be extended to Israel's reprisals in southern Lebanon. In the short run these actions led to panic and to a massive flight of villagers—most of them Christians, Shi'ites, and Druzes—to the seemingly safer north and especially to the Beirut area. Once again, Israel's potential allies bore the brunt of its punitive action. In the long run, however, the pressure generated inside Lebanon led the various Christian militias to build up their military capabilities, to increase their pressure on the Lebanese government to take punitive action against the PLO and thus, paradoxically, to an intensification of the links between them and Israel—their major source of supplies and advice.

To be sure, it sometimes appeared as if Lebanon were struggling to accommodate the PLO. Thus, after sporadic fighting between the PLO and Christian militias in the fall of 1969, President Nasser of Egypt mediated an accord which was signed in Cairo on November 2, 1969. The terms of the agreement could be interpreted as a PLO success, yet they could also be seen as a Lebanese maneuver designed to weaken the PLO. For while compromising Lebanon's sovereignty through a signed agreement allowing the PLO to operate against Israel from within Lebanon, the accords also limited the area from which they could operate to the slopes of Mount Hermon—a place known as the Arqub or later as Fatahland—and thus pointed out to Israel where to aim its reprisals. In other words while seeming to go along with the Arab world's demand to allow the PLO to operate, the Lebanese authorities were also making it easier for Israel to deal with the PLO. Thus, if it held, the Cairo Accord would have dealt the PLO a severe blow. The PLO soon understood this aspect no less than the Lebanese and the accords were ultimately abortive.

The cycle of PLO action, Israeli reprisals, and anti-PLO reactions by the Christian militias therefore continued. On May 2, 1970, a PLO unit operating from Fatahland ambushed an Israeli school bus. Israel reacted by moving into the Arqub, occupying parts of it, and paving a road on the mountains above the PLO positions. The PLO was thus weakened. As a result, the Lebanese authorities introduced additional restrictions on PLO freedom to operate through new regulations published on May 20, 1970.

The defeat of the PLO in Jordan the following September concentrated its presence in Lebanon still further. By being squeezed into small parts of Lebanon, the PLO became more vulnerable to Israeli attacks and Christian restrictions. This also drove its more militant components to step up their campaign of international terrorism. There were more attacks against Israelis

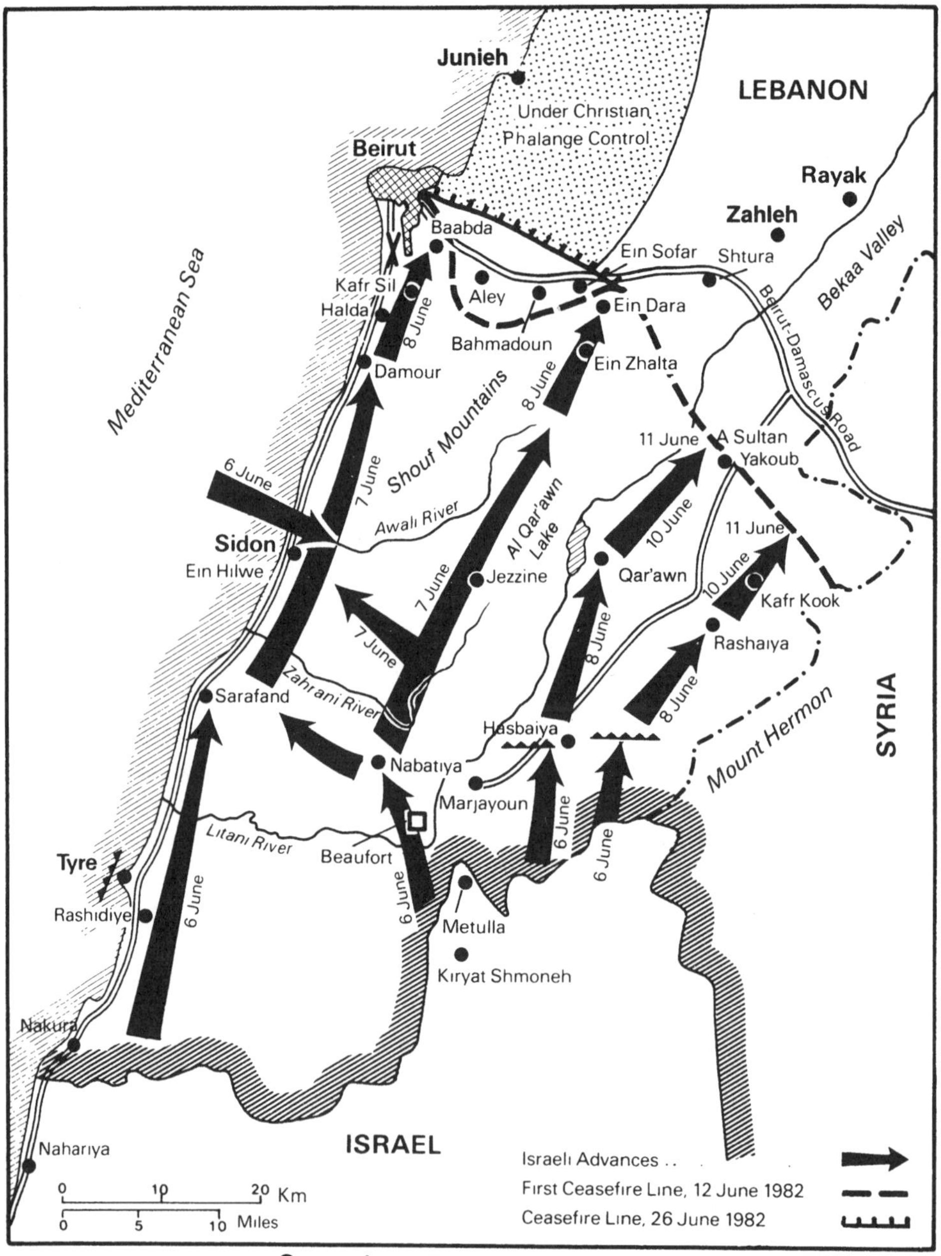

Operation "Peace for Galilee"

Reprinted with permission from *The Arab-Israeli Wars: War and Peace in the Middle East* by Chaim Herzog (New York: Random House, 1982; London: Arms & Armor Press, Lionel Leventhal Limited, 1982).

abroad and therefore more Israeli reprisals against PLO strongholds in Lebanon. On April 10, 1973, a new crisis was generated when Israeli soldiers (evidently with Christian Lebanese assistance) attacked leading PLO personalities in their Beirut apartments. The Lebanese Prime Minister Saeb Salem (a Moslem) resigned, blaming the Chief of Staff of the Lebanese Army (a Christian) for failing to defend the country. Tensions in Beirut increased and large-scale exchanges of fire took place among the militias and some units of the Lebanese army. Israeli policy had thus nudged the Lebanese one step closer to a showdown with the PLO.

The 1973 war brought a certain lull in this cycle. By 1974 the PLO had gained recognition from the Arab League, and indirectly from the United Nations, as the sole legitimate representative of the Palestinian people. In turn there was an attempt by Al Fatah to reduce border raids as well as to bring overseas operations to a nearly complete stop. But since the Popular Front for the Liberation of Palestine (PFLP) and other militant organizations within the PLO refused to follow this line, the pattern of rising tensions continued. By 1975, this created a situation so tense that a large scale showdown inside Lebanon became virtually inevitable.

The outbreak of the Civil War soon generated a new series of Israeli moves in response to events across the border in Lebanon. Broadly speaking, the Civil War proceeded in three stages. In the first stage, lasting roughly from April 1975 through the first week of October, the Christian militias took the offensive. But they did not succeed in achieving major changes in the territorial status quo and the fighting was largely confined to static but deadly exchanges of fire in which the Christians were on the whole more successful than their Moslem adversaries. Up to this point the PLO had attempted to remain neutral, while the Christians concentrated their fire on areas in which there were few Palestinians.

A second stage in the war began in mid-October 1975, when the Christians expanded the fighting to Palestinian concentrations in the Beirut area. As a result, the PLO joined the fighting, becoming the mainstay of the anti-Christian coalition. The tide of battle turned sharply against the Christians, and by the end of March 1976 the combined forces of the Moslem Left had gained control of close to 80 percent of Lebanese territory.

Throughout the winter of 1975–1976, the Syrians attempted to bring the Lebanese to a compromise. This also suited the Syrians' own attempt to increase their control over the country. Looked at from this perspective, the deployment of a Palestine Liberation Army unit in the Bekaa Valley was designed to reassure the PLO and at the same time to secure its removal from the fighting. Moreover, the Syrians succeeded in having their own ally, Elias Sarkis, elected to the Presidency. They also brought all the parties to agreement on a new national accord which reflected more equitably the demographic changes which had taken place since 1943. The fact that the Christians were prepared to sign this "Damacus Pact" reflected their weakness. But the accord dissatisfied the Moslem Left, who apparently felt that they were being denied a decisive victory and resented letting Lebanon become too closely controlled by Syria.

The Moslem Left's continued military effort following the Damascus Pact met, at first, with Syrian threats. Ultimately, in mid-April 1976, it led to a full-scale Syrian intervention on the side of the Christians, thereby starting a third stage in the Civil War. The Syrian intervention force was on the whole less successful than could have been expected, considering the disparity in forces between it and the Left–PLO coalition. Nevertheless, within less than three months, the Syrians succeeded in bringing an end to the fighting in all the areas which came under their control. And, as a result, the Christians secured a solid autonomous triangle of territory between Beirut and Junieh on the Mediterranean and the mountains some 40 kilometers to the east of Beirut. The Syrians took control of most of the Bekaa Valley as well as a large part of northern Lebanon along their border. The Druze component of the Left was restricted to its traditional base in the Shouf Mountains southeast of Beirut. There were Syrian contingents at the gates of Sidon and along the Beirut–Damascus road up to East Beirut. The PLO suffered greatly and was confined to parts of West Beirut and a number of pockets along the main road from Beirut southward to Tyre.[11]

Israeli Responses to the Civil War

Initially, the fighting must have appeared to Israeli planners as a salutary development. At last there was a showdown in Lebanon. It tied the PLO down to the internal Lebanese scene and deflected their threats to Israel. Above all, it was bound to keep Syria preoccupied with Lebanon. This occurred at a time when there were serious fears that Syria was trying to

11. For a far more detailed discussion of the Civil War and the events up to the Litani Operation, see Haley and Snider, *Lebanon in Crisis*.

form an eastern front with Iraq and Jordan for the purpose of launching another round of fighting against Israel. Moreover, the concentration of the fighting in and around Beirut caused a movement of PLO forces away from Israel's border as well as a movement of Lebanese who had fled to the north in previous years back to the south. The latter turned to Israel for first aid and food. Within a few weeks Israel could open the barbed wire fence along the border and turn it into what Shimon Peres, Minister of Defense, called the "good fence." Gradually building on that basis, Israel developed a sphere of influence based on a series of non-continguous enclaves along this border.

As the Civil War moved into its second phase, however, the Rabin Cabinet began to face acute dilemmas. These revolved around the question of what should be done if the Christians were defeated. Should Israel intervene in order to thwart a Moslem Left–PLO victory and if so how and to what extent? The immediate answer to that was partial involvement. The assistance program for Christians in northern Lebanon was expanded. In addition, the Haddad forces along Israel's border were assisted directly and encouraged to expand their enclaves—in cooperation with the Shi'ite al-Amal group and with some direct Israeli support—at the expense of the PLO. However, U.S. pressure indicated that Israeli involvement of greater proportions might lead to a serious rift, at a time when relations between Israel and the U.S. were already strained because of differences concerning the Sinai II accords. Accordingly, Israel could not go beyond material assistance to the Christians in the north. If the latter's defenses were to collapse, Israel would confront a very difficult choice: to intervene in order to save the Christians and face a major dispute with the United States, or to sacrifice the Christians on the altar of American–Israeli relations.

In the event, the dilemma was "resolved" as Syria became increasingly involved in the fighting. Yet this posed an immediate and still more aggravating dilemma: if Israel now failed to resist the Syrian intervention, it would abandon a two-decade-old commitment to oppose Syrian involvement in Lebanon. On the other hand, if Israel moved to deter Syria from intervening on the side of the Christians, it would in fact save the PLO–Left coalition and abandon its own Christian allies.

The importance of this strategic dilemma was enormous. Suddenly Israel had to choose between what had been two cardinal principles of its security policy: deter the Syrians from entering Lebanon or fight the PLO. The frame of reference was a search for deterrence. Since the time of Ben Gurion, Israeli policymakers had taken for granted an unmitigated Arab hostility and a fundamental disparity in resources. They thus sought to deflect the Arabs from engaging Israel militarily. The recipe for achieving this was complex. It involved a search for allies, both tacit and public, inside the Middle East and beyond it. It also required a strenuous effort to build adequate military capabilities. The deterrence strategy led to a continuous articulation of threats, including a list of clear *casi belli.* It also shaped Israel's attitude to the occupied territories more significantly than did any ideological commitment to retain them. And, finally, it affected in a most fundamental way Israel's style and methods of force employment.[12]

The dilemma for Israel was clear. On the one hand, if the Syrians were allowed into Lebanon, Israel could be interpreted either to have abandoned one of its most repeated commitments or to have its credibility called into question. Yet if the Syrians were deterred from entering Lebanon, a major adversary—the PLO—would be greatly served, while the Christians—an important ally whose attachment had been cultivated for two decades—would be abandoned. And this would occur at a time when the collapse of the Israeli connection with the Kurds and with Ethiopia—two other important regional allies—had just become public knowledge.

Torn by this dilemma, the Rabin–Peres team opted for a compromise. Syria was not deterred from entering Lebanon altogether. Rather it was permitted to maintain its positions in the Bekaa Valley and in the northern part of the country as well as its detachments along the Beirut–Damascus Road and north of Sidon. However, Syria was warned not to deploy its surface-to-air missiles (SAMs) inside Lebanon, which would eventually provide it with an option for further advance towards the Beirut–Junieh area in the center of Lebanon. Nor were Syrian troops to move within artillery range of the Lebanese–Israeli border or to interfere with Israeli Air Force flights over Lebanese territory.[13] At the same time Israel consolidated its own belt of influence in the south where Major Saad Haddad's militia was gradually being groomed as a major force, along with the Shi'ite Moslems. All this was worked out in the form of a package whose details were apparently set down in writing and exchanged through the U.S. intermediary, L. Dean Brown. Hence Israeli policymakers could at the time rationalize that they had com-

12. These aspects of Israel's security policy are discussed at greater length in Avner Yaniv, "Deterrence and Crisis in a Protracted Conflict: The Case of Israel," Paper presented to the XIIth International Political Science Association World Conference, Rio de Janeiro, August 1982.

13. This was disclosed by former Israeli Prime Minister Itzhak Rabin in an article published several years later. See Rabin, *The War in Lebanon* (Tel Aviv: Am Oved, 1983), pp. 55–63 (Hebrew).

promised the principle of deterring Syria from direct involvement in Lebanon but gained what might turn out to be the nucleus of an agreement with Syria from which further advances towards a more comprehensive peace could someday be made. It did not take long, however, before it became clear that this was not to be and that still another strategic dilemma was developing.

Prelude to Direct Intervention

The U.S.-mediated Syrian–Israeli understanding contained a specific line in south Lebanon (stretching between the southern tip of the Al Qar'awn Lake in the east and the Zahrani river south of Sidon in the west) beyond which Syrian forces would not be permitted to move. The reasoning behind this stipulation was that Israel could not possibly tolerate the establishment of a Syrian presence on its own northern border which would enable Syrian artillery to shell a densely populated part of Israel and even some day launch a full-scale armored offensive from both the Golan and Lebanon. In essence, Israel had accepted the stabilizing role of the Syrians in Lebanon, but it would not acquiesce in the proposition that this role be extended to that part of Lebanon closest to its own border. Indeed, from the Israeli point of view, if Lebanon was incapable of running its own affairs without foreign intervention and if the Syrians' role as a stabilizer was accepted by Lebanon, the Arab world, and the United States, then an Israeli request for a similar but far more limited role in the part of Lebanon along its own boundaries would be equally legitimate.

The non-Christian parties in the Lebanese imbroglio did not see things in the same way and argued that Lebanon's own army should be called in to police the south. But the fragility of that army and the fear that it could be penetrated by a wide array of other influences rendered this notion quite unacceptable from the Israeli point of view. Thus when Lebanese army units tried to enter the area, they were forcibly prevented from doing so. The upshot, however, was that Israel was creating inadvertently by its own policy an unstable no-man's-land along the boundary with Lebanon. Haddad's rag-tag units were incapable of effectively policing the area, and Israel itself was warned by the United States not to intervene directly. Nor was the Shi'ite al-Amal militia (which maintained liaison with Israel and Haddad) up to the task. And finally, because of their understanding with Israel, the Syrians were deterred from approaching Nabatiyeh on the northern bank of the Litani River. The area between the Litani and the Israeli border, from which

Israel had wished to expel the PLO, thus became, paradoxically, wide open for PLO operations. To be sure, the PLO met with some resistance from Haddad's forces and, of course, from Israel itself. But all this amounted to an ever-escalating chain of engagements which rather than solving the problem made it worse. By 1978 this state of affairs led Begin's Cabinet to authorize a move into southern Lebanon in force.

The Litani operation of March 1978 set out to drive the PLO beyond the Litani River, to create out of Haddad's isolated chain of enclaves a solid contiguous buffer comparable to the Syrian buffers in the Bekaa Valley and in the north of Lebanon and thus to stabilize this troubled area. Its results, however, were, from the Israeli point of view, far from satisfactory. Operationally, it evolved in a cumbersome manner leaving the PLO ample time to withdraw without incurring serious losses. Politically, it further tarnished Israel's image and added yet another source of friction with the United States. Moreover, Israel was forced to withdraw and to accept the deployment of the United Nations Interim Forces in Lebanon (UNIFIL). This might have appeared to be a reasonable solution to outside observers. But it was a rather poor one from the Israeli point of view because UNIFIL could not use force against PLO forays across its lines. In the future, Israel would have no alternative but further use of force, yet it would have to bypass the UNIFIL contingents in order to do so. UNIFIL, in short, constrained Israel's ability to take care of its own security but could not at the same time prevent Israel's adversary from mounting further attacks.

Moreover, following the Litani operation the PLO proceeded to purchase military equipment which would facilitate even larger operations against Israel. From 1978 to 1981, it acquired a great number of artillery pieces, especially multiple rocket launchers and long-range 180 mm guns which could reach the Israeli side of the border over the heads of the UNIFIL (and Haddad) forces.[14] In addition it attempted to use gliders and it improved its capability to reach the Israeli side of the border through the sea. Exchanges of fire between Israel and the Haddad forces on the one hand and the PLO on the other hand therefore continued after the Litani operation and in an ever-increasing ferocity. By July 1981 this situation led to what later became known as the "mini-attrition." PLO artillery launched a series of simultaneous attacks on 33 Israeli communities from Nahariya on the Mediterranean

14. According to *Ha'Aretz* of July 26, 1981, by that time the PLO had accumulated some 30–40 long-range 130 mm guns, 70 tanks, and 50 Katyusha multiple rocket launchers.

to Kiryat Shmoneh at the foot of Mount Hermon. Israel responded with large-scale fire, including air power. But panic in its northern settlements reached unprecedented proportions. Under pressure from this population the Government agreed to accept a U.S.-mediated cease-fire with the PLO. Thus the principle of not recognizing the PLO was compromised, security along the border had manifestly deteriorated, and a thorough reassessment of Israel's entire policy vis-à-vis Lebanon and the PLO was clearly called for. The result was the last and possibly the most important strategic decision in the chain which has been surveyed here, namely Operation Peace for Galilee.

The combination of Lebanon's internal fragility and an ever-escalating Israeli–Palestinian struggle presented Israel with a dilemma from which it could not escape. If it were to forgo retaliation against the PLO, the price in terms of domestic repercussions as well as in Israel's ability to maintain a viable deterrence posture vis-à-vis its Arab adversaries would be unacceptable. If on the other hand it were to react with force there would be a heavy price to pay. A succession of Israeli Cabinets, including Begin's, had tried to walk the tightrope between these two grim choices by reacting incrementally and—from the Israeli point of view—with relative prudence. Yet, even cautious policies had led, step by step, to a point at which the basic dilemma had to be resolved. Israel would have to choose between two costly alternatives: acquiescence in the suffering of its Galilee population and erosion of its strategic position in the north, or a massive intervention with its predictable human, material, and political costs.

The Calculus of Invasion

The decision to move into Lebanon by force and seek a fundamental improvement in the situation was apparently initiated by Prime Minister Begin, Defense Minister Sharon, Foreign Minister Shamir, and Chief of Staff Eitan and approved in a piecemeal manner by the whole of the Cabinet in a series of special sessions. Israel Defense Force (IDF) planning concerning such a move had been in existence for years and only minor revisions were necessary. The main issues at hand were the timing, the scale, and the military and political objectives of the operation. Sharon may have entertained far-reaching political objectives such as ejecting the PLO from Lebanon, restoring stability in that country under the leadership of Bashir Gemayel, and throwing a much weakened PLO into a struggle with the Hashemite Kingdom of Jordan which would neutralize both as negotiating partners in the short run and eliminate one of them altogether in the long run. He may have also believed that such a process would assist in subduing the West Bank population. But few, if any, of his colleagues were convinced of the validity of these arguments and, in the final analysis, the Cabinet addressed itself primarily to the strategic-military aspect of the problem.

Differently stated, consensus was neither sought nor obtained on wide-ranging political objectives. Rather, it was focused on the narrower military objectives of the operation at every stage of its evolution. To the extent that can be judged, Sharon, Eitan, Begin, and other members of the Cabinet who supported the notion that an operation was inevitable succeeded in persuading the rest of the Cabinet on one important point: the limited reprisals of the past decade had exhausted themselves as an instrument of policy and it made no sense to authorize a mere replay of the 1978 Litani operation. The PLO had succeeded in turning the population of Israel's north into virtual hostages. If it were to act against Israeli targets in the Middle East or beyond and Israel then carried out reprisals against its bases in Lebanon, the Galilee population could be made to pay and the IDF would be unable to prevent this. Hence, a drastic action was called for—one which would, at the very least, push the PLO beyond the Awali River (some 65 kilometers north of the Israeli border along the coast and some 40 kilometers northwest of Metulla and Kiryat Shmoneh).

The basic elements of this consensus were apparently as follows. First, the crux of the problem was PLO artillery and rocket capability with an effective range of 42 kilometers against "soft" civilian targets on Israel's side of the border. Hence, it made no sense to launch any operation which would clear an area less than 40 kilometers in depth. Second, security of this new zone was unlikely to be obtained without a serious clash with Syrian forces deployed there and if engagement with the Syrians were avoided, the fighting would end with the Syrians in control of the high ground; hence, Israel's negotiating position would then be weak whereas Syria's would be strong.

Third, given the structure of the PLO and its constant need for international visibility and at least a semblance of diplomatic momentum, any lengthy delay of a major Israeli move posed serious risks. Specifically, by leaving the initiative to the PLO, Israel faced the near certainty of yet another attrition on the July 1981 scale—or larger—but initiated by the Palestinians at a time of their own choosing.

Fourth, since July 1977, the PLO had increasingly mended its fences with Syria. Hence the rationale which had made Syrian involvement in Lebanon palatable in 1976 had lost its validity altogether. By 1981, Syria had completely reversed its alignment policy in Lebanon—a fact which was manifest militarily by its attempt to dislodge the Christians from their mountain positions between the Bekaa Valley and Beirut. The Syrians also established a forward deployment of SAMs in the Bekaa Valley, so that an eventual move of major proportions against the Christians could sooner or later take place. At the same time, the PLO had been given a free hand to establish itself solidly with depots, training grounds, and a kind of tax and welfare system in the coastal area between Beirut and Tyre.

Fifth, under the pressure of the escalating conflict between Israel and the PLO, the latter was increasingly becoming a conventional force both in its organizational structure and in its equipment and deployment.[15] In the short run this was an advantage from the Israeli point of view: it gradually stripped the PLO of the advantages of guerrilla warfare and thus increasingly exposed it to the type of combat in which Israel had a clear edge. In the longer term, however, the PLO could gradually evolve into a two-to-three division army group which, given its location, would present the same danger which the Syrian army might have posed had it not been deterred from crossing the "Red Line" during the events of 1976. To be sure, unlike Syria, the PLO was not perceived in Israel as a serious military threat to Israel's very existence. But, with every passing month PLO forces would become more difficult to dislodge, not merely from Lebanon's south but from every other part of the country in which it enjoyed autonomy of action.

Sixth, as long as the peace process with Egypt had not been completed, an Israeli campaign in Lebanon with a view to driving the PLO out of that country, or even merely out of range of Israel's north, might have been too great a provocation to Egypt, especially under a new and not yet fully established president such as Mubarak. Some of Begin's colleagues who had grave misgivings about the terms of the Camp David Accords argued that

15. By 1979 the PLO anticipated a large-scale Israeli attack and was chafing under the pressure of Israeli bombing. This was disclosed by Yasir Arafat in a discussion with Soviet Foreign Minister Andrei Gromyko on November 13, 1979. The full minutes of the discussion were captured by Israeli intelligence. The PLO armament drive during 1978–1982 was largely a response to this. According to *The New York Times* of October 12, 1982, the *Ha'Aretz* estimates quoted in the previous footnote grossly underestimated the quantities of arms which had by then been procured by the PLO. For example, instead of 80–90 artillery pieces (including rocket launchers), the PLO had closer to 500, including 150 anti-aircraft guns and, in addition, 1193 anti-tank weapons. In 1980, PLO officers participated in conventional warfare courses in the Soviet Union.

for precisely this reason a campaign in Lebanon should be undertaken before April 1982 in order to test the limits of Egypt's commitment to peace and, if Egypt were provoked, salvage for Israel the most critical sector of the Sinai, namely its easternmost part stretching from the settlements around Yamit in the north all the way to Sharm el Sheikh and the Strait of Tiran in the south. But the majority of Begin's Cabinet was convinced that the opposite course would be more prudent: Israel should not do anything which might provoke Egypt into disrupting the peace process. Hence, waiting beyond April 1982 became an important constraint on the planned invasion of Lebanon.

Seventh, with Egypt neutralized by the peace treaty and with Iraq bogged down in its war with Iran, Syria would be facing Israel alone. Jordan would not join at a time when its relations with Syria were at a very low point. Jordan could also be relied upon to sit back and watch Israel deal with the PLO—its own mortal enemy. Beyond that, the experience of the 1973 war suggested that Jordan would not join any war against Israel unless and until it could be reasonably certain of an Arab victory. Thus if Syria could be deterred from opening a second front on the Golan, through Israel's deployment there of a large military force, the PLO was likely to be deserted by all its potential Arab allies irrespective of what they might choose to say.

Eighth, the fact that Syria was likely to avoid a full-scale war suggested that a test of the Soviet commitment was also unlikely. The Soviets had enough troubles in Afghanistan, in Poland, and as a result of the Iran–Iraq war. Nor would they want to prod a United States led by a seemingly hard-line administration. Beyond that, Israeli assumptions about Soviet behavior in the Middle East impute to the Soviets very cynical calculations insofar as the PLO is concerned. The Soviets, in the Israeli view, might espouse the cause of the Palestinians and offer them training, equipment, and diplomatic backing rather generously, but they have never been expected by the Israelis to take any great risks on behalf of the PLO.

Finally, the Reagan Administration had seemed to be far more favorably disposed towards Israel than the Carter Administration. It did not appear to be in a great hurry to nudge Arabs and Israelis towards a comprehensive settlement. It was not nearly so inclined to embrace the PLO, it seemed on the whole far more prone to the use of power, and its reactions to Israeli hints of a possible military operation in Lebanon suggested approval.[16] Rea-

16. Zeev Schiff, "Green Light, Lebanon," *Foreign Policy*, No. 50 (Spring 1983), pp. 73–85.

gan himself was viewed as a friend. Secretary of State Haig was regarded as a trusted ally. And the only major U.S. policymaker not seen as sympathetic was Secretary of Defense Weinberger. There was awareness of the mounting tension between Haig and the rest of the Administration, but the implication must have been that the big move should be undertaken earlier rather than later—before any replacement of Haig.

In sum, the choice of 1982 as the year in which to act appeared rational. And since Lebanon has an inauspicious terrain, especially in winter, the choice of the early summer was logical. There remained, however, the question of pretext. Israeli policymakers, like most other policymakers, take the public presentation of their cases very seriously. But faced with a choice between the requirements of vindication and those of actual policy, the former is considered only an instrument of policy rather than its determinant. Over the years of Israel's existence, its leaders had become accustomed to referring to most military actions, including preemptive and preventive strikes, as "reprisals." Thus in the same way that Israel's 1956 Sinai operation was presented as a large reprisal when in fact it was a case of preemption, Begin's Cabinet needed a reasonably acceptable pretext for moving into Lebanon. They found the solution by, in effect, laying a trap for the PLO. In highly publicized statements, Israeli officials repeatedly presented the July 1981 cease-fire as a matter of linkage. Either the PLO respected the cease-fire on *all* fronts or the cease-fire was null and void. Thus not only was the absence of hostilities along Israel's border with Lebanon important; the cease-fire also had to be respected in the occupied territories, inside Israel proper, and vis-à-vis Israeli and Jewish property and personnel the world over.

There was a key underlying assumption in such an all-encompassing interpretation of the cease-fire. It was that the PLO's nature as an umbrella organization, including diverse factions, made it incapable of keeping the cease-fire. Yasir Arafat and Al Fatah might have wanted to respect the cease-fire in order to foster the PLO's "moderate" image. But from the perspective of the Begin Government, a PLO with a respectable image was a more dangerous opponent than an apparently militant PLO. In any case Arafat was not in control of the activities of George Habbash's PFLP, of Abu Nidal's Group, nor, for that matter, of Ahmed Jabril's Popular Front–General Command. One of these organizations was bound to break the cease-fire and offer a pretext for action, and one of them eventually did. Nevertheless, even after the attempted assassinations of the Israeli Ambassador in London on June 3, 1982, Israel hesitated. Again, the experience of previous years offered a solution: massive retaliation on the PLO's centers in Beirut would make it impossible for Yasir Arafat to overrule the pressure of his colleagues for counteraction on a large scale. This, in turn, would provide Israel with an adequate pretext.

Nor was the question of pretext merely related to public opinion in general. In all likelihood, the Israeli Cabinet had two important targets in mind: one, the Israeli public; the other, the American Administration together with its attentive public including the Jewish community. On the domestic front the Begin–Sharon team found itself under pressure from two quarters. On the one hand there were the Labour Alignment and other opposition groups, who increasingly charged the Likud with warmongering. On the other hand there was pressure from representatives of the Galilee communities who would support an all-out campaign but feared another war of attrition of which they might become the chief victims.

Meanwhile, the Begin Government's view of the United States was undoubtedly affected by an experience stretching back to the establishment of the state of Israel. The lessons of this long experience were the following. First, the United States under any administration would be reflexively averse to the use of force. This might be the result of a fear of Arab reaction, but beyond that it was fed by a fear lest events get out of hand and force the United States to intervene in order to bail out Israel. Moreover, the United States could not possibly be consulted in advance since, owing to this fundamental caution, it was certain that its response would be negative. To invade Lebanon after explicit U.S. objection to such a move would be worse than to preempt the United States, move fast, and achieve results which would ultimately be beneficial from the American point of view. Nor did such a thought originate with Begin. Israel's very creation was an act of preemption directed primarily vis-à-vis the United States. During the period from November 1947 to May 1948, the Truman Administration seemed to be withdrawing from the U.N. partition plan towards "Trusteeship" and it therefore had to be preempted by a declaration of independence barely ten minutes after the formal termination of the British Mandate. This established a lasting pattern which was manifested in the 1956 and 1967 Israeli preemptions as well as in a multitude of less historic events.

As a consequence, it was worth signalling to both the Israeli public and to the United States that a major move was in the offing, yet ultimately both would have to be preempted. A pretext could be helpful, but there was no guarantee of its being sufficient. In fact Begin's Government entered the operation conscious of a double risk: domestic criticism and a rift with the

United States. Hence a clear statement of purpose was impossible. Rather, the IDF had to plan and prepare for a comprehensive campaign which might even include a link-up with the Christians on the Beirut–Damascus Road and an entry into West Beirut, whereas the Government would only approve the plan and announce it in a step-by-step fashion.

In turn, the Israeli Government, or at least that part of it involved in the initiation of the operation, had to take account of two further risks: a grave compromise of sound operational principles and a serious problem of credibility. The declared purpose was to remove the danger of PLO artillery from Israel's north, but it was not at all inevitable that Syria would have to be challenged too. Yet, without engaging the Syrian forces in the southern part of the Bekaa Valley and of the Shouf Mountains, the 40–45 kilometer security zone would not be fully secured since in these areas Syrian forces were deployed within range of the Israeli border. But, if that was the case, it made more sense from the strategic point of view—as General Adam, an eventual casualty of the war, pointed out—to start the campaign with a massive preemption of the Syrians and then proceed to deal with the far less important PLO. However, since engaging the Syrians meant expanding the entire operation far beyond limits which the United States and Israel's own population would tolerate, it could only be done later in the hope that the blame could then be placed on the Syrians. In turn, a serious loss of surprise, concentration of force, and hence life had to be anticipated.

Moreover, the gap between what could be done and what could be announced not only perverted the operational planning but posed a grave problem of credibility. The IDF, especially when it includes large reserves, is an integral part of Israeli society. One result of this unique situation is that keeping much of what it does secret is well-nigh impossible for any length of time. If the planned operation in Lebanon were to be completed quickly and with a minimum of casualties, the IDF had to throw into battle very large quantities of manpower and equipment. In turn, the proportion of reserves to regulars would have to tilt very heavily in favor of the reserves. Consequently, keeping the truth of what was happening secret became especially difficult.

The War and Its Dilemmas

During the first week of the war, from the Israeli perspective, events went well. While the Government was announcing limited objectives which suggested an expanded version of the 1978 Litani operation, the IDF was in fact pushing rapidly towards Beirut, leaving behind it pockets of PLO forces which could be dealt with later. The United States Government seemed, to Israel's genuine surprise, receptive and allowed the operation to proceed without undue criticism and evidently without applying any real pressure to stop.

At the end of the first week, however, the operation was complicated by a number of factors. In the first place, the Syrian forces did not intervene. This posed a difficult choice for Israel: should it let them remain in place with their missiles, with their ability to offer effective sanctuary to PLO forces, and with their ability to threaten the eastern flank of the IDF from the commanding heights of the Shouf Mountains, and thus retain a critical measure of influence on the reconstruction of Lebanon after the war, or should the Syrians be forced into battle. The answer was to engage them in battle on a scale sufficient for Israel to achieve its objectives—driving the Syrians up the Bekaa Valley north of Al Qarawn Lake, down from the Shouf, and destroying their SAMs.

Having achieved this, Israel faced the toughest problem in the entire operation, that of Beirut. If the IDF were to halt the assault on the southern bank of the Awali River and establish a line from there to the Shouf Mountains, Israel would face the following political and military problems. The PLO would maintain its autonomous center in Beirut and its command structure would remain intact. Syria would preserve its access to Beirut and, as a result, its ability to back up the PLO and retain control over it. Syria would also maintain its solid positions on the commanding heights of the Shouf, overlooking the IDF's eastern (right-hand) flank. The Christian militias in the Beirut–Junieh–Zahleh triangle would remain essentially isolated, exposed to Syrian and PLO pressures, and unable to benefit from Israel's incursion.

In such circumstances, Israel would be entangled in a troublesome military situation when the international pressure to retreat (which was assumed as a matter of course by Israeli policymakers on the basis of past experience) began to mount. If Israel were to yield to such pressure, it would have gained nothing from the entire operation since the PLO would simply return to the area between the Awali and the Litani rivers as soon as the IDF withdrew. The possibility of a friendly Christian force moving into this area would not exist, while expanding the UNIFIL zone northward would simply mean resumption of the troubles of the 1978–1982 period with this hopelessly inadequate force.

Alternatively, Israel could order the IDF to race for the Beirut–Damascus road. Such a thrust would drive a wedge between the Syrians in the east and the PLO (along with a smaller Syrian contingent) in Beirut and facilitate a linkup of the IDF with the Lebanese forces of Bashir Gemayel. A Christian–IDF corridor such as this across the Beirut–Damascus Highway would have to be, however, wide enough to be defended against a possible Syrian–PLO attempt to destroy it through a coordinated attack from east and west. Differently stated, Israel and the Lebanese forces would have to establish a corridor of at least 20–30 kilometers. In turn a virtual blockade would be established around West Beirut.

Thus, quite apart from any political motivation for dislodging the PLO from Beirut, there was a compelling strategic-military logic for, in effect, besieging West Beirut. To ignore it would amount to a major compromise of the military objectives of the war. To follow it, on the other hand, would mean a major new dilemma. If West Beirut were besieged without being taken, the PLO could turn its defeat into a major propaganda event. Images of Stalingrad would soon be conjured and the pressure on Israel to lift the siege would become intolerable. If under such pressure Israel were to abandon the siege, its corridor across the Beirut–Damascus Highway would be in jeopardy and subsequently its entire operation. Moreover, lifting the siege under international pressure would give the PLO an unacceptable (from Israel's view) political victory. In 1956 Nasser had thus turned a humiliating military rout into a tremendous political triumph. In 1968 the PLO had acted in a similar fashion. Two unmanned Israeli centurion tanks—which had been incapacitated by the Jordanian army and not by the PLO—were left behind during the battle of Karameh. Karameh had been a minor Israeli reprisal operation, but the PLO had turned it into a watershed of mythical proportions in its history. Israeli planners did not need a great deal of imagination to predict what the PLO would do after the IDF retreat from Beirut if the PLO centers in the city were to remain intact. In a word, tactical military logic made it difficult to avoid a virtual encirclement of West Beirut. Once there, a broader political and strategic logic made it imperative to eject the PLO from the city.

The measures necessary to capture West Beirut were studied and debated exhaustively in advance. Militarily, the operation was far more feasible than most commentators have argued, especially if the IDF were to break straight into the center from different directions while leaving ample opportunities for PLO combatants to escape rather than fight for their lives. Yet when it came to this point, a convergence of domestic criticism and U.S. pressures caused the Begin Government to stall.

Ironically, a quick move into Beirut might then have cost more Israeli lives but almost certainly fewer Palestinian and Lebanese deaths than eventually occurred. What prompted objections by the Israeli public was the understandable fear of heavy casualties. In addition, the Israeli public began to react to the fact that throughout the previous week it had been deceived by its Government. In 1956 a similar deception for comparable purposes did not generate much criticism because the operation was quick and decisive. This time there were fears of a long and costly affair, which would demand the lives of many more Israeli soldiers.

Initially, the Reagan Administration appeared sympathetic to the idea of driving the PLO out of Lebanon. But whereas Secretary of State Haig thought that Israel should be allowed to proceed, Vice President Bush and Secretary of Defense Weinberger were concerned to avoid U.S. complicity with Israeli entry into an Arab capital. The departure of Haig left the Administration more strongly opposed to an Israeli occupation of Beirut. This did not necessarily imply an objection to driving the PLO out. But it did suggest the result would have to be sought through diplomacy and specifically through the Habib mission.

The diplomatic route made sense within the Administration, but much less so in terms of the real problem at hand. For Yasir Arafat the choice was between dying as a hero inside Beirut or—if he accepted the U.S. request to depart from Beirut quickly—possibly losing his leadership (if not his life) as a traitor to his cause. He was unlikely to respond favorably unless he were offered in return virtual U.S. recognition of the PLO. But the Administration was not prepared to go that far. Thus the American special envoy, Philip Habib, had no chance for success unless and until his mission was backed up by further military pressure. Yet Israel, having resigned itself to the prospect of a diplomatic attempt to solve the problem, could not break into Beirut as long as Habib had not admitted failure. Arafat must have understood that. He therefore played for time and for the camera—which improved his international status and blemished Israel's. And since the Israeli public and the American Jewish community were becoming increasingly split, Arafat might have reasoned that he could ultimately save the PLO position in Beirut.

Israeli policymakers understood this process and hence, after a few weeks in which Habib was given (from their point of view) a fair chance, they authorized heavy military pressure. This took the form of massive bombing and cutting off of supplies to the beleaguered part of the city. Here too the United States had to be preempted in a series of cease-fires which were broken directly or indirectly by Israel. In contrast with the first week of the war, which amounted to a clear-cut success from the Israeli point of view, the siege of Beirut became—at least in terms of Israel's international standing—an unmitigated embarrassment. Yet despite the high price, the same ineluctable logic which led to the invasion also worked against other alternatives. Indeed, the short-term costs notwithstanding, the ouster of the PLO from Beirut, as well as from its other autonomous bases in Lebanon, appeared to offset these costs.

The Begin Cabinet cannot have welcomed the prospect of further tensions with a sympathetic U.S. Administration. It is also difficult to imagine that Israeli decision-makers welcomed the fissures in the American Jewish community and the deepening schisms within Israel itself. Yet, had Begin and his colleagues—or any other Israeli Government for that matter—opted to defer action, security problems were bound to have accumulated over a longer period of time. The result would ultimately have been even more painful dilemmas of strategic choice, accompanied by still worse international and domestic repercussions, given that the PLO was incapable—ideologically and operationally—of forgoing military actions against the Jewish state. Israel could only postpone, but in the long run not avoid, the Lebanese invasion, nor could Israel limit its scope. And, once having exhausted the utility of limited reprisals and then having made the decision to act, it could not allow the PLO to salvage even a symbolic victory on Lebanese soil.

Conclusion

Israel, then, has followed a long and tortuous path from 1968, when it began reprisals in Lebanon, to its decision to invade in 1982. The experience suggests characteristic Israeli patterns of response, not merely to the PLO in Lebanon, but toward other neighboring states. Since Israel achieved statehood in 1948, it has encountered constant threats which in many ways paralleled the PLO challenge in Lebanon. This was particularly evident in the period 1953–1956 vis-à-vis Egypt and 1957–1967 in regard to Syria. In both periods, as later in Lebanon between 1968 and 1982, Israel could choose a policy of passivity, limited reprisals, or large-scale action. In each instance, it opted at first for limited reprisal. But, as in Lebanon, that policy succeeded only in the intermediate term. Inevitably, it led to an escalating pattern of action, reaction, and counteraction, which would ultimately be resolved by full-scale war. In the first of the two earlier major cases (Egypt), the eventual result was peace. In the second (Syria), it may yet take a long time to resolve the conflict through a peace agreement. Indeed, the probability of a major flare-up between Syria and Israel in the foreseeable future is very high. Nonetheless, from the perspective of 35 years, it appears that a major structural change in Israeli–Syrian relations has taken place: instead of continuous small-scale friction, the emerging pattern is of increasingly infrequent major conflagrations, separated by long periods of armed stability.

The PLO in its present configuration, however, presents a problem which does not fit into either of these alternatives: while the Israeli invasion of Lebanon may have dealt effectively with an immediate security challenge, it remains to be seen whether or not the outcome will lead to stabilization or, ultimately, to peace. At the moment, neither outcome seems likely, though in the longer term neither should be too readily dismissed. Yet, even had Israeli policymakers assumed the more pessimistic scenario—and they may well have—the weight of the analysis presented here is that strategic imperatives left them little alternative in their decision to act. To be sure, they may have stretched the strategic logic too far. Conceivably, they could have stopped at the Awali line and thus avoided the carnage of Beirut, cut short the war, denied Syria a virtual veto over a Lebanese settlement, and possibly reduced the IDF's subsequent casualties from guerrilla attacks. But, as the postwar debate in Israel reveals, even the Government's strongest critics do not argue that the invasion could have been avoided altogether. In the end, strategic logic rather than mere personal whim made a major Israeli operation in southern Lebanon virtually inevitable.

[10]

Palestinian Military Performance in the 1982 War

YEZID SAYIGH*

With the evacuation of the PLO from Beirut, and the dispersal of its forces among eight Arab countries, an entire historical phase in its existence is over. After 17 years of armed action, the PLO organizations face radically changed political, and military conditions, affecting their future choice of methods and programs. As the continuing showdown in Lebanon indicates, the military option is still very open to all parties in the conflict; but what shape it can take for the PLO is unclear. What is clear, however, is that there are many lessons to be drawn from the PLO's military experience during the 1970s in general, and in the 1982 war in particular. Over the years, there has been very little in Western or Palestinian literature on Palestinian military experience; so in tackling the issue it is more rewarding to deal with those aspects that offer material for discussion. More specifically, a choice has been made to discuss operational aspects rather than the strategic or tactical ones.[1] The choice of

*Yezid Sayigh is a researcher and writer on Middle Eastern Affairs.

1 "Strategy," in military terms, means the general method chosen to confront a particular enemy, and the management of allocated resources at the national level. "Tactics" refers to how individual soldiers or units fight in the field. The operational level lies in the middle, it oversees implementation of the general strategy in the battlefield.

the operational level, rather than the strategic or tactical is due to several reasons. First, despite the PLO's general military strategy in terms of attacking targets inside Israel's current borders, or enhancing its military strength in Lebanon in response to internal Lebanese developments and external diplomatic moves, the PLO had not formulated or even improvised a strategy to confront an anticipated or actual Israeli invasion. Second, Palestinian military strategy before the war was mainly defensive, designed to confront the sort of military operations mounted against Palestinian targets by Israeli forces since 1978: air strikes, commando raids, company or battalion size assaults, or extended confrontations such as the 1978 invasion up to the Litani River and the artillery duels in the summer of 1981. Third, the fragmented nature of the forces of the PLO's component organizations and the Lebanese National Movement meant that there was never a common military strategy. Fourth, at the tactical level, there was extremely little standardization among Palestinian units, and many lessons were learned (both negative and positive) from operational performance.

There are other important factors which make an operational evaluation more relevant, given the absence of any significant strategic planning by the Palestinian command. 1. By the summer of 1982, the size and armament of Palestinian forces had reached proportions that required increased proficiency in handling larger units in several battle zones simultaneously. 2. The peculiar situation of being in control of large sections of the country and its population demanded the ability to concentrate military strength when needed. 3. These factors made classic guerrilla small-band tactics impracticable, so operational capabilities had to step in as a substitute for the non-existent overall strategy. 4. The quick disintegration of the command network and the rapid advance of the Israeli forces left large numbers of Palestinian fighters in the field; too many in number and too used to large units to revert effectively to guerrilla warfare, but bereft of sectoral or operational command. Actual military strength and geo-political considerations had required the PLO to acquire the intermediate level between its general military strategy and its traditional military tactics: in other words, operational command.

Operational performance involves direction of combat, the size and armament of combat units, logistic and support services, the size and role of particular arms such as artillery and armor, and "battlefield tactics." Grasp of such requirements implies successful use of available human and material resources within the chosen strategy or "mode of operation," and proper choice of armament, training, organization, and deployment of forces, proper planning, and efficient management of forces during combat.

The Palestinian Context: Methodological Overview

Any evaluation of Palestinian military performance encounters several complexities. One is the scarcity of reliable, detailed and comprehensive data. Israeli and Palestinian sources tend to exaggerate or belittle, or to give partial presentations. Adequate and appropriate criteria need to be developed to measure military performance in the specific context of the Israeli-Palestinian confrontation. There has been extraordinarily little critical, analytical writing on both historical and contemporary Palestinian military experience.[2] Western literature tends to underestimate Palestinian action, whereas Arab or Palestinian writing generally avoids penetrating, critical analysis, often under the pretext of "maintaining security." Both Western and Arab writers tend to err methodologically by using rigid models and criteria, measuring Palestinian military performance against either regular armies or dogmatic interpretations of guerrilla warfare or "people's" war. The military (as well as economic, social, political and cultural) criteria and models which have evolved from the experience of other countries or liberation movements are not necessarily applicable to the Palestinian situation. The Vietnamese, Chinese, Soviet or Western military experiences instruct the Palestinians, but cannot be used as a rigid measure. Universally applicable military principles, such as economy of force, definition and maintenance of objectives, and attainment of numerical superiority, also need to be translated into specific principles of local relevance, governed by the specific military, political, psychological, economic and historical features of the country or movement concerned.

[2]Recent examples (articles): Brig. Saad Sayel, "The Military Action of the Palestinian Revolution and Prospects for Development," *Shu'un Filastiniya*, no. 105, August 1982; Col. Hasan Abu Lubdah, "The PLA is a Shield and Sword for the Revolution," *Shu'un Filastinya*, no. 114, May 1981; Maj. Wasif 'Urayqat, "Palestinian Artillery," *Shu'un Filastinya*, no. 115, June 1982; (books): Bard O'Neil, *Armed Struggle in Palestine* (Boulder, CO: Westview Press, 1980); Youssef Rajab al-Ruda'i, *The Palestinian Revolution of 1936* (Beirut: Arab Studies Institute, 1983), in Arabic. Older writings are: Abu Hamam, *The Resistance in Military Terms* (Beirut: Dar Al-Tali'ah, 1969), in Arabic; Naji Alloush, *Concerning the General Strategic Line of Our Movement and Revolution* (Beirut: Dar al-Tali'ah, 1974), and *Towards a New Palestinian Revolution* (Beirut: Dar al-Tali'ah, 1974), in Arabic; Munir Shafiq, *Some Military Principles of the Palestinian Revolution* (n.p., n.d.), in Arabic.

There would seem to be little need to prove that "Operation Peace for Galilee" meets all the standards of a *war*. Yet some Israelis and Westerners prefer to refer to it as a "large-scale operation," largely for political reasons: Israel was only "punishing" undisciplined "bands of terrorists," not seeking to destroy the infrastructure of a nation. Had the Israeli forces been able to reach Beirut without serious resistance, its description as a "search-and-destroy" operation might hold. However, the quantitative military commitment on both sides, and the scope of Israel's political objectives substantiate the use of the term war.[3] "Operation Peace for Galilee" had specific, far-reaching political aims that went well beyond any immediate military gains. Israel's political goal was the destruction of the PLO, including its social, cultural and economic institutions, and Palestinian national aspirations. The cumulative size of Israeli forces sent into action against the PLO (even accounting for those units earmarked for offensive or preventive action against the Syrian forces) was greater than any "large-scale operation" would justify. The direct and indirect costs to Israel's economy show a high-level of sustained commitment. The Israeli invasion force reached a total size of 120,000, with 1,600 tanks, 1,600 armored personnel carriers, 600 guns or multiple rocket launchers (including at least 500 self-propelled howitzers or field guns), with massive air and naval support. Israeli casualties, estimated at 322 dead and 1,900 wounded through mid-August 1982, point to a war effort comparable to that of 1956 and 1967.[4] Direct and indirect costs to Israel's economy are estimated at $2.5-4 billion. Israel had to employ a significant percentage of its standing army and reserves to fight a militarily feeble and politically contemptible Palestinian enemy. The battlefield covered the main areas of Palestinian military and population concentration in Lebanon. For the Palestinian forces, and the Lebanese and Palestinian population in the south and Beirut, the invasion tapped their full military capabilities, medical, social and other institutional resources and capabilities.

In the 1982 war, both sides declared a victory, raising the important question of the standards of victory and defeat (strategically), or success and failure (tactically), in the context of the Israeli-Palestinian military

[3]Hirsh Goodman, "How Strong Was the PLO?," *Jerusalem Post*, July 9, 1982, p.4.

[4]According to one source, Israel lost 983 killed out of 250,000 men participating in the 1967 war, and 189 killed out of 100,000 in the 1956 war. Col. T.N. Dupuy, *The Elusive Victory* (London: Macdonald and Jane's, 1978), pp. 212, 333 and 337.

confrontation. Results are often measured in terms of geographic areas occupied or retained, and the extent of human and material losses. How do such standards apply to the Palestinian case? First, such standards are useful in specific, narrow cases; for example, in minor clashes between an Israeli raiding party and a Palestinian base. If one or another unit suffers heavy losses, or the base is overrun, there is a specific instance of failure or success. If a particular strategy is based on continuous application of such tactics, then a string of tactical successes implies a strategic victory, as in Israel's ability to halt Palestinian cross-border action since 1969-70. If, for example, Israeli forces suffer unacceptably high losses (relative to Palestinian losses and Israeli domestic standards), as in the current Lebanese-Palestinian guerrilla campaign in South Lebanon, the balance shifts to the Palestinians, even if the overall balance of casualties or terrain remains in Israel's favor.[5]

Second, when an obviously superior force (in terms of numbers, armament, training, mobility, organization, logistics and technology), such as the Israel Defense Forces (IDF), confronts a Palestinian enemy of extremely limited means and capabilities, such as the PLO, there is little question as to who will gain the upper hand on the battlefield.[6] However, the militarily superior force will have failed to achieve its goal if the militarily inferior force turns a battlefield defeat into a political victory (given the balance of forces, mere survival or regrowth is an achievement). An example is the Israeli invasion of South Lebanon in March 1978. Although the IDF was able to overrun PLO bases and occupy the area south of the Litani River and east of Tyre, the PLO forces put up a fierce resistance and slowed the Israeli advance, particularly at the defensive strongholds of Bint Jbeil, Taybeh and Khiyam. The PLO inflicted heavy Israeli casualties, preserved its combat units and withdrew in relative order, losing a minimal number of fighters. The primary Palestinian measure for success is the extent and intensity of combat with Israeli

[5] R.D.M. Furlong, "Israel Lashes Out," *International Defence Review*, Vol. 15, No. 8 (August 1982), p. 1001. According to Furlong, the main Palestinian guerrilla and troop concentrations, apart from those in Beirut and Damour, were located within 40km of the Israeli-Lebanese frontier.

[6] Such classic tenets of guerrilla or people's war are extensively discussed in such sources as Robert Asprey, *War in the Shadows* (London: Macdonald and Jane's, 1975); and Robert Taber, *The War of the Flea* (London: Paladin, 1974).

forces, followed by Palestinian combat skills and casualties.[7] The Israeli measure is political; do the Palestinians admit defeat, is their political will broken? Israel's main failure in March 1978 and in the July 1981 artillery duels in the South and the bombing of Beirut was its inability to impose a unilateral Palestinian cease-fire and to alter the political position of the PLO.

Finally, one crucial aspect affecting an evaluation of the Palestinian military is to situate it technically. By June 1982, the Palestinian military had not evolved fully from guerrilla units into regular forces using classical modes of operation, despite the considerable development of its armament and structure in that direction. Israeli Chief of Staff Rafael Eitan expressed his satisfaction that the PLO was "going regular," since that gave Israel a better chance to isolate and destroy it.[8] The Palestinian forces had lost the guerrilla's advantages of mobility, flexibility, and relative invisibility, without gaining the advantages of a regular army.[9] The PLO found itself fighting with medium and heavy weapons, mounted on or towed by assorted vehicles, without the necessary levels of firepower, air defense, training, organization, and management required by regular units when fighting a technologically and numerically superior enemy.

Palestinian Military in Lebanon Before June 1982

Most developments in Palestinian armament, organization, and combat doctrine took place after Lebanon became the main PLO base, particularly since the 1975-76 civil war. In the early 1970s, the PLO absorbed large numbers of regulars from the Jordanian Army, and acquired a few light artillery pieces and rocket launchers, such as 76mm and 85mm anti-tank guns for shelling, and small jeep-mounted 107mm or 122mm multiple rocket launchers. The process of "regularization" proceeded slowly, with military ranks being introduced in mid-1971, and small independent artillery units being set up after 1973. The Lebanese

[7]Martin van Creveld comments on a decline in the performance of the Israel Defense Forces against the Arabs, tracing it to the tendency to concentrate on technological solutions at the expense of tactical originality. See his article, "The War: A Questioning Look," *Jerusalem Post*, December 12, 1982, p. 12.

[8]Van Creveld writes that the campaign "was no walkover," observing that the casualties roughly equalled those on the Egyptian front during the June 1967 war, *Ibid.*, p. 12.

[9]One Israeli opinion along these lines is that of Hirsh Goodman, *op.cit.*, p. 4.

civil war prompted much more rapid and radical transformations: heavy artillery and some armored fighting vehicles were acquired and Palestinian forces began fighting in large units stretched out along fixed frontlines, often against regular forces of superior size and armament. By 1982, the PLO had acquired large numbers of heavy weapons, including obsolete T-34 tanks, 122mm, 130mm, and 155mm howitzers, BM-21 mobile 30- or 40-tube multiple rocket launchers, BRDM-2 scout cars, BTR-152 personnel carriers, SA-7 and SA-9 missile launchers, and ZSU-23-4 mobile radar-guided anti-aircraft guns. Most of these developments in armament and organization were a response to internal Lebanese developments, although Palestinian arms procurement policy was also reactive in confronting Israel. That is, a specific weapon was sought because the enemy had come to use it, and not because it was seen to be otherwise particularly suitable to Palestinian requirements.

An essential feature of Palestinian military development since 1975 is a strong functional dualism. Palestinian forces had to fight on two fronts: external, against Israel, and internal, against enemies in Lebanon. They worked simultaneously within two military balances, each with distinct priorities and considerations. PLO forces enjoyed effective parity in the internal balance, a fact which was partly due to, and partly encouraged, the tendency towards larger formations and heavier weapons. The aging T-34 tanks, for example, had an impact on internal enemies they could never enjoy against Israeli forces. The practical implications of this dualism were that the Palestinian forces in Lebanon had to arm, train, organize, plan, and deploy in two distinct, often contradictory, ways in order to confront two separate enemies. Palestinian forces needed high flexibility when facing Israeli forces—fighting in small, lightly-armed bands according to guerrilla tactics—and high concentration in men and equipment when confronting internal enemies. In addition, the Palestinian forces had to take over security tasks in large areas of Lebanon, especially after the Syrian Arab Deterrent Force withdrew from the entire coastal strip south of Beirut. In June 1982, a large part of available Palestinian fighters in the countryside were tied down and scattered in villages to prevent internal Lebanese clashes, not due to conscious planning for the Israeli invasion.

* * *

Evaluation of Palestinian military performance at the operational level means examining the following areas: planning and prediction; armament, training, organization, and deployment of forces; and management of forces during combat. Related areas are: the role and performance of specific weapon categories; the effectiveness of response to specific threats; levels of administrative, logistic, engineering, and medical services; and the degree of flexibility, mobility, and initiative. The analysis is directly relevant to the command structure at its higher levels (battalion and up), although it has a bearing on the lower levels of military formations (companies and platoons).

Planning and Prediction

In early 1982, the Palestinian leadership spoke constantly of the imminence of large-scale Israeli military attack, possibly reaching Beirut to link up with the right-wing Lebanese Forces in an "accordion" movement. The leadership was apparently quite sincere in its warnings, but did the general expectation result in any specific, practical predictions or planning?

The answer appears to be both yes and no. The deployment of Palestinian forces was not changed, and no studied contingency plans were made by the combat forces or their sectoral commands. This suggests a continuation of the reactive character of Palestinian military action since the early 1970s. What was required, and was still lacking when the war broke out, was a careful examination of possible Israeli tactical and strategic military objectives, of probable Israeli methods for achieving these objectives, of axes of advance or landing zones, and consequently of the available human and material means for defense according to a definition of Palestinian political, geographic, and military priorities.

There was some prediction and planning, inasmuch as Palestinian forces were already deployed in many sectors to confront Israeli attacks. The bulk of available manpower was in and around the refugee camps, cities, and other densely populated areas, mainly in response to perceived internal threats. Since total Palestinian fighting power was insufficient to make a stand against Israeli main units in open countryside, and since Palestinian forces were already overstretched and unable to cover all possible axes of advance, PLO deployment in the more urban, defensible areas made military sense.[10] When the IDF actually invaded, however, the

[10]Both Palestinian and Israeli sources present this assessment, but van Creveld advances the most balanced evaluation: "The PLO (despite official Israeli attempts to prove the contrary) possessed very few of the heavy weapons crucial to the conduct of modern war and hardly any of the logistic and technical infrastructure required to maintain and deploy them," *op.cit.*, p. 12.

effect of inadequate prediction and planning became all too clear. PLO forces were scattered and surrounded, their communication routes cut, and their command structures non-existent. This meant less effective resistance to the first blow, and a subsequent inability to regroup and mount any widespread military attacks behind Israeli lines. Those Palestinian units that had made some individual planning effort made a better showing, whereas those that were taken completely by surprise lost all cohesion.[11]

The PLO should also have devised contingency plans to facilitate the movement and regrouping of friendly forces should the IDF attain its immediate objectives in the South. The scattered Palestinian units would not have been neutralized so effectively had they immediately resorted to predetermined alternative battle plans. Instead, most units or individual fighters made their way to the nearest city or zone of friendly control where they were either surrounded by or isolated from the IDF.

Armament, Training, Organization, and Deployment of Forces

Armament

A. Light Arms. The light arms used by the Palestinian forces were generally effective and suitable, consisting mainly of AK-47 assault rifles, Belgian FN and West German G-3 assault rifles of 7.62mm caliber, squad and platoon machine guns of 7.62mm and 12.7mm caliber, 82mm mortars, and RPG-7 anti-tank rocket launchers. These weapons and their ammunition were available in large quantities, and individual proficiency in their use was acceptable. The use of heavier crew-served weapons requires more detailed examination.

B. Anti-tank weapons. The individually-fired RPG-7 anti-tank rocket has been mentioned frequently in Israeli accounts of the fighting.[12] Despite basic improvements in the armor of Israeli tanks, this weapon probably

[11] This is also the assessment of American military writer Anthony Cordesman, "The Sixth Arab-Israeli Conflict," *Armed Forces Journal* (August 1982), p. 32.

[12] There were several positive Israeli assessments of Palestinian use of the RPG-7 anti-tank rocket launchers and of anti-tank tactics quoted in "Israeli Evaluations of Arab Combat Capabilities in the Lebanon War," *Strategic Review*, Vol. 3, No. 16 (September 9, 1982), p. 12 (in Arabic).

caused the most Israeli losses. Indeed, the RPG-7 is widely issued to Israeli infantry as well. Anti-tank mines were of some effectiveness against Israeli armor, but only where they had been prepositioned in the South before the war or later around Beirut. Mines were not used extensively in the South to block routes of advance not covered by PLO units, mainly due to civilian use of the road network. Foresight would have enabled greater use of rese.ve stocks of mines to hinder the IDF's movements once it was in Lebanon. The other anti-tank weapons available were recoilless rifles, anti-tank guns, guided missiles, and T-34 tanks. Of these, only recoilless rifles were used to any significant degree. Palestinian forces had Western-designed 75mm and 106mm and Soviet-made B-10 (82mm), B-11 (107mm), and SPG-9 (73mm) recoilless rifles in considerable numbers. These weapons were not used efficiently, however, due to inadequate training, poor positioning, easy detection and destruction by enemy fire, and immobility, as even vehicle-mounted guns were immobilized when the vehicles ran out of fuel or the roads were cut. The 76mm, 85mm, and 100mm anti-tank guns were not used in their anti-tank role, but as light artillery, which was probably effective in view of their weight and size. The PLO did not use effectively the Sagger anti-tank and guided missiles (although several were fired in the Khaldeh battle by Syrian soldiers) due both to poor technical and tactical training, and to the absence of specialized anti-tank guided missile squads or platoons. Finally, the T-34 tanks, which could have been used as mobile anti-tank guns, were unable to fulfill that role in the open countryside because of the limited range of their 85mm guns and their extreme exposure to Israeli aircraft. A few T-34s managed to fight and survive in Beirut, though they had no anti-tank role even then.

C. Anti-aircraft weapons. Palestinian forces possessed a wide range of anti-aircraft guns, as well as numbers of individually-launched SA-7 anti-aircraft missiles and several vehicle-mounted SA-9 systems. The guns ranged in caliber from 12.7mm, through 14.5mm, 20mm, 23mm, 37mm, 57mm, 85mm, and 100mm. In addition, the PLO had recently received several self-propelled, radar-guided quadruple-barrelled 23mm "Shilka" guns (ZSU-23-4). These weapons wre moderately effective, despite the major flaws of their non-integration into air defense systems, and poor technical and tactical training. Anti-aircraft guns, especially those of 14.5mm to 23mm caliber, bore the brunt of air defense tasks in all sectors. In fact, these guns were probably responsible for shooting down the two helicopters Israel admitted losing,[13] and for "crippling" several other aircraft.[14] The A-4 Skyhawk plane which Israel admits was lost to Palestinian fire, was probably shot down by a SA-7 missile.[15] Palestinian forces achieved appreciable results firing their anti-aircraft guns against Israeli armor, particularly APCs.

Palestinian use of anti-aircraft weapons exposed two major shortcomings. One was tactical: most of these weapons were mounted on jeeps or trucks which soon ran out of fuel and usable roads, were constantly exposed to aerial observation and vulnerable to anti-personnel munitions, and posed a considerable logistic and administrative strain on already stretched company or battalion commands. Some PLO units, with tremendous efforts, extricated their vehicles and sent them to "safe" areas, only to lose them when whole zones were cut off by Israeli landings. The second shortcoming was more fundamental: the hundreds of anti-aircraft weapons employed by the PLO were totally uncoordinated, even at a local level. Thus combat units diverted men to act as gun crews that had no hope of putting up effective anti-aircraft resistance because there were too few guns per unit to be effective. On the other hand, cities and camps that needed a heavy concentration of such weapons were dotted with gun emplacements that worked individually and separately, without any fire plan or control. Consequently, the second aim (or first in some opinions) of anti-aircraft defense was lost, namely forcing enemy aircraft to take evasive action at high altitudes and thus lose effectiveness in attacking ground targets. This was compounded by the loss of hundreds of able-bodied fighters to ineffectual roles. The presence of guided weapons did not change the overall picture, partly because the Israeli Air Force had long effected counter measures, and partly because too few of these weapons were available even if the requisite comprehensive system of deployment and fire control had been in force.

[13] A senior Israeli Air Force officer (probably the then Air Force Commander David Ivri) mentioned the vulnerability of helicopters during the 1982 war to small arms fire. "Bekaa Valley Combat," *Flight International*, October 16, 1982, p. 1111.

[14] Cordesman suggests that the Israeli Air Force lost one or two aircraft during the 1982 war due to poor maintenance, pilot error, or poor support. If that is the case, then total Israeli aircraft losses during the war are five, not three, aircraft and two helicopters, *op.cit.*, p. 29. Pentagon sources are convinced that Israel lost 11 or 12 aircraft shot down or crippled, including one F-16, the rest being Kfir, A-4 Skyhawk, and F-4 Phantom aircraft. See Clarence Robinson, "Surveillance Integration Pivotal in Israel's Success," *Aviation Week*, July 5, 1982, p. 17.

[15] According to Clifford Wright, "The Israeli War Machine in Lebanon," *Journal of Palestine Studies*, Vol. 12, No. 2 (Spring 1983), p. 48.

D. Armored vehicles. The PLO's armored vehicles included 60 T-34 tanks, a few dozen light-skinned BRDM-2 and BTR-152 carriers, and some home-made armored cars with machine guns. The precise number of T-34 tanks is not known: Israeli figures range from 50 to 500, but Israeli military correspondent Ze'ev Schiff confirmed that 38 T-34s were captured, as well as 46 T-55s (all or most of which belonged to the Syrian forces)[16] The PLO's purpose in acquiring armored vehicles is particularly unclear, as they could not be assigned a direct role in tank warfare, and their presence in the South did not add defensive weight against constant Israeli commando raids. They could neither concentrate to face a threat, nor influence the battle when dispersed. These vehicles suffered from poor maintenance, primitive wireless communications, and lack of air cover. The war found a large number of tanks, APCs, jeeps and trucks immobilized by breakdowns: some were in garages, others were left as static gun emplacements.

E. Artillery. The PLO began to develop artillery with the influx of regular army gunners and artillery officers in the early 1970s. The first significant action of light artillery came during the October 1973 war, across the Lebanese border, although artillery rockets were used in far greater numbers. By 1976, with the acquisition of heavier calibers, Palestinian artillery became more important and effective, as evidenced by the curtain of fire defending Tel al-Zaatar camp when it was besieged by the Lebanese Forces. Artillery also held its own during the March 1978 fighting and in the cross-border duels of July 1981.[17] In those encounters, Palestinian guns sustained a high rate of fire and minimal losses, despite enemy air activity.

In the 1982 war, however, Palestinian artillery lost a major advantage: previously it had not come face-to-face with Israeli ground units, as the slow Israeli advance in March 1978 allowed the guns to be withdrawn, and in July 1981 there was no ground movement. In Summer 1982, Palestinian artillery units were isolated, blocked, and suffered direct assaults which obliged the gunners to take to their personal side arms in self-defense. The central commands and field artillery observers or combat units also lost contact and cohesion, effectively paralyzing the

[16] On these and other PLO material losses, see Zeev Schiff, quoted in Michael Jansen, *Battle of Beirut* (London: Zed Press, 1982), pp. 5-6.

[17] Artillery was the main effective weapon used by the PLO during the July 1982 fighting. See "Aspects of the Palestinian-Israeli Conflict in the Coming Phase," *Strategic Review*, Vol. 3, No. 19 (October 21, 1982) p. 2, Appendix.

artillery. Palestinian rocket artillery was more flexible in this situation, due to its mobility in contrast to the towed artillery. The light artillery (76mm, 85mm, and 100mm), and heavy artillery (122mm, 130mm, and 155mm) were the most vulnerable and expensive to use because of their size and weight, and their dependence on tow vehicles, ammunition supply trucks, and large crews. Mortars of 82mm, 120mm, and 160mm caliber were rarely used, although they were in fact better suited to the battle conditions.[18] Rocket artillery, mainly 107mm and 122mm mounted on jeeps, trucks, or special mobile platforms (such as the BM-21), provided heavy concentration of fire which was particularly effective due to the rapidity of fire (which allowed quick movement back into hiding), and to its "carpet" effect, which partially made up for the lack of sighting and observation. Palestinian artillery was at its most effective during the siege of Beirut where it benefitted from both the relative protection and a measure of central command.

Training

Among Palestinian forces there was a wide discrepancy in the levels of technical training related to specific weapons, and combat training related to methods and doctrines of fighting. Technical training was glaringly deficient with regard to guided anti-tank or anti-aircraft missiles, tanks, and certain anti-aircraft guns. While some fighters were very proficient, there was no systematic attempt to standardize and generalize such abilities. Training on such weapons entails much more than combat use; both combat units and support services should acquire care and maintenance skills. Poor maintenance and ignorance of various technical limitations meant, for example, that many guided missiles failed to fire. Routine checks and better storage would have avoided such situations.

Training in artillery and rocket artillery weapons was generally better. The proportion of weapons in actual service was higher, and combat training allowed gun crews to unlimber guns, fire several shells, limber up, and move away within five minutes (i.e., before the enemy could respond with air raids or counter battery fire). Training on individual weapons such as RPG-7 anti-tank rocket launchers was also generally good, especially as both launchers and rockets were in large supply which allowed lavish consumption in training.

[18]Both Schiff and Furlong estimate that the PLO had just over 300 various artillery pieces (including rocket artillery). See Schiff *op. cit.*, p. 6; and Furlong, *op. cit.*, p. 1003.

Possibly of greater importance in combat than individual skill in weapon-handling, is tactical training. Training, both at individual and at squad or platoon level, was carried out by Palestinian instructors, often within combat units, and was generally good. Such training concentrated on small unit tactics which allowed greater mobility, flexibility, and effectiveness once the larger units had broken up. Training at battalion level, or in combined operations (of infantry, anti-tank weapons, heavy machine guns, armored vehicles, and artillery), was instituted in the late 1970s. It proved effective against Major Saad Haddad's forces, or against company- or battalion-sized Israeli raids, but totally unsuitable against a major Israeli attack.[19]

Tactical training was deficient in two other respects: on specific weapons (such as the tanks), it was often unsuited to Palestinian battlefield conditions; and training on a particular weapon was not complemented by tactical training on its use within a military unit, against a specific threat, according to physical conditions. Thus, much effort was wasted on learning inapplicable skills, and much individual skill was wasted because the handler did not know how to put his training and weapon to their best use. Training provided in Palestinian camps was more relevant than training in friendly countries, due partly to the Palestinian command's lack of clarity in what it required of external training, and partly to its added failure to evaluate, standardize, and generalize the lessons of such training.

Organization

Normally, the size and form of military units are determined by the number of fighters, type of weapons, nature of combat, and enemy qualitative and quantitative strength. In the Palestinian case, there was an added element of facing two distinct enemies simultaneously. It was important, for political, psychological and security reasons, to maintain a front in South Lebanon against the Haddad forces and minor Israeli attacks. In contrast to classic guerrilla theory, it was also important not to melt away immediately whenever Israeli forces advanced in strength. Much of the Arab and international recognition accorded the PLO was bought by staying to fight and paying a high price in human losses.

[19] This contrasts with the Israeli emphasis on training, doctrine, and original military thought, including surprise and deception, to use van Creveld's expressions. See van Creveld, *op.cit.*, p. 12. On Israeli training generally, see Edgar O'Ballance, "Training of Israeli Officer," *Arab Strategic Thought*, No. 6/7 (January/May 1983), pp. 107-120.

Palestinian forces in Lebanon were divided into fixed formations theoretically ranging from platoon level, through companies and battalions, to regiments, but most units were far below strength. Thus, they became accustomed to thinking in terms of large formations, although their units did not have the numbers and punch to undertake large-formation tasks. This was especially true of infantry units, while artillery units generally fielded only as many guns as they could man.

Possibly the only advantage of this anomalous situation was that breaking up into small bands was made easier. The disadvantages were more significant: such quick and easy change in unit size was an instinctive reaction, not one envisaged in planning or training; different unit strengths meant discrepancies in fighting capabilities of theoretically equivalent units; and some units were more dependent on formal command hierarchies, which hastened unit paralysis when command structure broke down. An added disadvantage, in the event, was the loss of fighters to various independent armor, artillery, rocket, and anti-aircraft units with separate organizational status. Once Israeli forces were in control of the countryside and in position around the coastal cities, Palestinian forces ceased to exist or operate as large formations, even when present in large numbers. In Sidon, for example, many small groups fought within unconnected sectors under virtually no localized central command.

Deployment

Palestinian deployment was concentrated around the coastal camps and cities from Rashidiyeh south of Tyre, to Damour-Naameh and Beirut. One difficulty in estimating actual strength is that many fighters were not "regulars," but part of local militias. Israeli estimates vary greatly: 6-14,000 PLO fighters in the South, and 12-18,000 in the whole of Lebanon.[20] The PLO probably had no more than 2,000 full-time fighters in the South, including the Sidon region, and another 2,000, including allied organizations, for local support. These numbers were concentrated in the Tyre region, Nabatiyeh-Rihan, the lower end of the Bekaa Valley-Arqoub, along the coast south and north of Sidon (noticeably at Zahrani, Awwali, and Damour-Saadiyat), and in Sidon itself. The actual numbers deployed in the countryside were minimal: Israeli sources claimed initially

[20]Figures varied from 6,000 PLO fighters in Lebanon (*Jerusalem Post*, June 7, 1982), to 8,000 (according to Cordesman, *op. cit.*, p. 29), or 30,000 (according to van Creveld, *op. cit.*, p. 12). Furlong estimates that there were 10,000 PLO fighters in the South alone (*op. cit.*, p. 1002).

that there were 1,000 PLO men in the Nabatiyeh command,[21] roughly eight to ten times the actual number.

There were probably no more than 2,000 PLO regulars in the Beirut area, including some who had withdrawn ahead of the Israeli forces in the South.[22] Larger population figures and the presence of PLO offices and leaders in Beirut meant, however, that the number of part-time fighters was quite large, reaching around 4-5,000. Just before the war, they were deployed within Beirut, or just south of it; during the war they defended Khalde and the Aley mountain before withdrawing into the Beirut perimeter. Finally, other PLO forces in the country (Bekaa and Tripoli) probably did not exceed 1,000 regulars and 2,000 militia (including affiliated local organizations). These forces were concentrated mainly in the central Bekaa area (including Baalbeck), and in the northern city of Tripoli or in the Beddawi and Nahr el-Bared camps nearby, where they remained during the war.

Concentration in the urban areas gave the Palestinian forces added defensive strength, a primitive form of "force mulitiplication." In the countryside, and along rural axes of advance, the Israeli forces enjoyed relatively uninhibited freedom of movement, partly because there were no local PLO reserves and once the war had started there was no possiblity of moving central reserves. More PLO fighters in the countryside would not have changed the overall picture to any significant degree, but better planning and preparation would have resulted in greater Israeli casualties and higher Palestinian survival rates.

Management of Forces

The Palestinian military leadership was not able to reappraise its deployments and plans before the Israeli forces effectively cut it off from its fighters and encircled Beirut. From the outset, therefore, the central command lost its centrality and became a local battlefield command for the Beirut zone. In evaluating the performance of battlefield commands at all levels, the most prominent feature is the rapid disintegration of effective command in all but the Beirut theaters of operation. PLO forces based in the countryside may have been forced to discard command structures from company level upwards once the Israeli forces had broken

21 The *Jerusalem Post* of June 7, 1982 also estimated that there were 1,000 PLO fighters in the Nabatiyeh area.

22 These figures exclude around 1,000 Syrian soldiers and 3,000 PLA men under Syrian control.

through, but the loss of such cohesion among units defending the cities led to a marked decrease in combat effectiveness. Thus, Israeli forces were allowed to land at the Awwali river north of Sidon despite the presence of PLO forces still capable of fighting in company strength. Later, the IDF was able to carve up Sidon as it had done in the Tyre area into convenient, isolated sub-sectors to be subdued at relative ease, mainly because there was no PLO central command in Sidon, and the defending units dissolved into a disordered mass of small groups. In such situations a battalion command should have taken over the role of central command, and company commands could have provided leadership for local sector defense.

In the defense of Beirut, the PLO displayed better use of its available forces and command structures, partly because of the presence of the PLO's main leadership and headquarters. The delay of the Israeli forces south of Beirut was also crucial for by the time the city was encircled, the defense perimeter had been established and the PLO forces had been assigned sectors and responsibilities according to their organizational establishment. The result was significantly increased combat effectiveness, as evidenced in the IDF's failure to capture Beirut Airport, despite repeated attempts, until August 1. The two cases of Sidon and Beirut demonstrate how influential the presence or absence of effective command can be, although combat in the countryside also showed how important it is that such command be flexible or even dispensable.

Other Areas of Operational Performance

Support Services

These include administrative, logistic, engineering, and medical services. Before the war, administrative services were provided at battalion or regimental level and covered matters of pay, leave, rank, and posting. Much of this information was stored by the central military administrations of the separate organizations within the PLO, in the absence of a unified military structure.

The functioning of the other services is more crucial to combat performance. Logistic supply of ammunition, food and fuel broke down almost instantly in the South because of near-total interdiction by Israeli aircraft of the roads and direct attacks on the supply and administrative centers. The defending forces had to rely on local stocks, though the loss of heavy weapons and vehicles and the dissolution into smaller units generally reduced logistic requirements. In the eastern sector of the Bekaa-Arqoub, communication routes led back to Syrian-held territory and PLO forces withdrew before being encircled or destroyed. In Beirut, a stable defense perimeter and huge stockpiles of food, weapons and ammunition allowed the supply services to operate regularly. The continued existence of recognizable military formations also greatly facilitated the determination of needs and distribution.

Before the war, the engineering services concentrated on fortifying military positions and building shelters in civilian areas. In Nabatiyeh and Rihan, the Israeli forces found extensive trenches, tunnels, and shelters. In other sectors there was very little engineering work. Although there was a large number of civilian shelters, many were either unusable or inadequate due to poor design and construction. In the South, there had been no organized mine-laying work, but this shortcoming was rectified in Beirut, where extensive minefields were laid, tunnels and trenches dug, and earth barriers thrown up within four or five days as the Israeli invasion force fought for the southern and eastern approaches to the city.

Throughout the war, Palestinian medical services were the most active in all sectors. PLO casualties were still being evacuated from the battlefield as Israeli advance units closed in around Tyre, Nabatiyeh, and Sidon. The Palestine Red Crescent Society and the medical units attached to the military forces functioned even after the IDF was in physical control of these areas, and suffered a disproportionately high number of casualties or prisoners. In Beirut, Palestinian medical services managed to function long after those in the South had been eliminated, due to the stability of the frontline and the existence of appreciable stocks of medical supplies and equipment. Many new forward casualty evacuation centers and rear hospitals were set up, which treated over 12,000 wounded.

Mobility, Flexibility, and Initiative

In these aspects of performance, there was a wide discrepancy from unit to unit. In the South, road mobility was lost in stages between the first and third day of the land invasion, while the movement of fighters in the countryside allowed large numbers of them to filter north and east to safe areas. In Beirut, the protection afforded by buildings allowed continued use of the roads despite intensive shelling and bombing. In the strictest sense, however, mobility was non-existent due to constant Israeli control by land or from the air of all roads in the general theater of operations. Mobility on foot became a basic survival tactic, but was not a planned asset, so not enough use was made of small, mobile units.

The flexibility to change into small groups did not develop according to a plan but was reactive. In Sidon, for example, many small groups ended up defending the same sector, but did not coalesce smoothly into larger, better-adapted formations. The lack of flexibility was evident in planning and improvisation to face new conditions, as reflected in the general unadaptability of combat units to new tasks required by the radically changing situation.

Initiative was low, as evidenced by the lack of aggressive military action behind Israeli lines and by the absence of junior officers stepping into the command vacuum to reorganize the defense. Battalion and company commanders did adapt themselves personally to lead whatever forces they had left, even if they could not bring under their control other forces occupying the same sector. This failure was due to the fragmented nature of the various PLO organizations and their military wings, the absence of real coordination between organizations and combat units before the war, and the lack of standardization in training and military skills. Many Palestinian units displayed great tenacity during the fighting, but not initiative. In some cases holding a defensive line against huge Israeli forces precluded offensive action, in others military training did not promote immediate offensive instincts.

The War of Attrition

Since the PLO's evacuation from Beirut, there have been daily Palestinian—Lebanese guerrilla attacks against Israeli military targets inside Lebanon.[23] Israeli troops are now concentrated in major encampments and bivouacs, with extensive fortifications, and support facilities, such as airstrips in Damour, near Aley, Ansar, and Metulla, and helicopter pads.

There are several striking features in the current Lebanese-Palestinian campaign.[24] Operations range widely over occupied areas with attacks on Israeli convoys on the coastal road, others in the Tyre, Nabatiyeh, and

23The reputable Lebanese daily, *Al-Nahar*, mentioned on May 27, 1983, for example, that while the Israeli military spokesman had announced that an explosive charge had killed one Israeli soldier and wounded 14 in eastern Lebanon the day before, another unannounced explosion had wounded 15 others in Sidon that same day, and a third charge was dismantled before detonating.

24For interesting comments on the current situation, see: "Concerning the Resistance to Israeli Occupation of Lebanon," *Strategic Review*, Vol. 4, No. 7 (May 5, 1982), p. 2.

Zahrani districts, and in central and eastern Lebanon. The methods of attack vary greatly, from remote-controlled car bombs to grenade attacks and direct automatic arms ambushes. These operations indicate better training, planning, execution, and greater confidence, such as in the effective use of wireless controlled bombs, and ambushes against sizeable Israeli units in broad daylight. Israeli military spokesmen have admitted to nearly one operation a day since September 1982—and to over 150 Israeli deaths, while resistance sources consistently report other operations at about double Israeli admissions. Palestinian and Lebanese casualties have been minimal, not exceeding 15 killed or captured. For the first time in the history of the Arab-Israeli conflict, Israeli forces are suffering worse than one-to-one casualties.

That the Lebanese and Palestinians have been able to sustain such a high level of resistance since September 1982 indicates a total change in their approach to military action. Patrols into Israeli-occupied territory are small, lightly-equipped, and mobile. This implies good reconnaissance and knowledge of terrain, local support, and improved planning and management. Israeli and Western assessments that many different political groups are undertaking the attacks with little or no coordination have two positive implications: first, that it is more difficult to uncover resistance networks, and second, that local commands (whether inside Israeli-occupied areas or outside them) can take the initiative within a general strategy. Whether such advantages, and the success they bring, will promote a fundamental reevaluation of past strategy remains to be seen.

Conclusion

The shortcomings in Palestinian military performance can be traced to three inter-related phenomena that have distinguished Palestinian armed action, at both conceptual and technical levels, since the early 1970s.

1. The lack of a military theory. There has been virtually no reappraisal of Palestinian military action since the late 1960s, when there were a few individual efforts to evaluate and criticize theory and practice.[25] As a result, the PLO went through the upheavals of the 1970-71 Jordanian civil war, and the politico-military vicissitudes of Lebanon during the 1970s

[25] See footnote 2; there is a brief critique of a general nature by Gerard Chaliand and Abu Iyad in Gerard Chaliand (ed.), *Guerrilla Strategies* (Berkeley, CA: University of California Press, 1982).

and early 1980s without any form of military theory. This in turn meant that strategy and tactics received no concentrated attention.

Fateh abandoned its short-lived flirtation with the attempt to draw the Arab armies into a final, cataclysmic showdown with Israel after the June 1967 war dispelled all its illusions. Then, the various armed Palestinian organizations espoused guerrilla war, developing into a full-fledged people's war. Up to 1970, the people's war theory meant wide-scale recruitment of guerrilla volunteers and intense action against Israeli targets on or behind the (mainly) Jordanian-Israeli cease-fire lines. With the rise in numbers, Palestinian operations became more numerous and adventurous, often involving strikes against several targets along an extended front. The withdrawal from Jordan to Lebanon ended such freedom of action.

In the early 1970s, the PLO reached a historical low in its political fortunes and near-paralysis in its military activity. It did not study the military lessons of the past, or formulate a political or military strategy that could translate the theory of people's war into the Palestinians' particular situation.

2.Regularization. In the context of this vacuum, certain events came to determine subsequent Palestinian military development: the influx of ex-Jordanian army regulars, the October 1973 war, and the 1975-76 Lebanese civil war. The net result of these events was to accelerate the introduction of heavy weapons and large formations into PLO forces. The process of "regularization" (*tajyeesh*) and the static nature of both frontlines and combat in South Lebanon led to a divorce between the official ideology of people's war and the form and role of Palestinian military structures.

Ranks were instituted in PLO forces in mid-1971, and have remained ever since. The "regularization" of PLO forces meant the adoption of traditional army forms, structures, armaments and combat doctrine. Guerrilla "sectors" (*qita'*) were redesignated as battalions, light artillery and 6-wheel trucks were introduced, and the ranking system became more complex and widespread which affected pay-scales. During the Lebanese civil war, this trend accelerated in terms of armament, structure and operation (fighting in large units or along static frontlines) greatly assisted by the influx of Palestine Liberation Army (PLA) units previously outside PLO politico-military and geographic control.

Since 1976, "regularization" had become an accepted fact: PLO forces were already "regular" but needed more heavy weapons and trappings to be accepted as such. Officers were sent abroad for training as battalion and brigade commanders, and as staff, naval, air and armor officers. Every battalion or "brigade" (*qiyadat quwwat*) had a staff of administrative and operations officers. Combat units were issued with helmets, anti-aircraft weapons, vehicles, and 120mm mortars. By Summer 1982, the PLO was fielding tanks, APCs, heavy artillery, and hundreds of crew-served weapons. It had reportedly requested SA-6 anti-aircraft missile systems and Frog surface-to-surface heavy bombardment rockets.

3. Arms procurement policy. The rationale for acquisition of particular weapons was reactive and had little to do with real needs or capabilities. Developing armament seemed desirable, given the absence of a clear military strategy derived from a particular military theory, the false security of the UN buffer zone in the South, and the search for increased international legitimacy by portraying the image of a state-in-formation. Thus, heavy artillery was introduced during the Lebanese civil war in response to the shelling of West Beirut; heavy multiple rocket launchers and guided anti-aircraft weapon systems were acquired to provide better defense and a more intense artillery response to Israeli artillery firepower following the post-Camp David shellings and the July 1981 aerial bombing of Beirut. In all these cases, development in weaponry did not follow on effective absorption and use of existing weapon systems, nor did it take into account actual Palestinian technical and human capabilities.

The PLOs "regularization" and arms procurement policies grew out of an initial vacuum, then became an alternative military practice that still claimed people's war as its inspiration. "Regularization" and the acquisition of heavy weaponry are not, by definition, inimical to the concept of people's war—the Vietnamese National Liberation Front certainly used large formations and heavy weapons—but the aims of achieving concentration of men and firepower did not require, indeed were harmed by, the PLO's methods. The PLO's political and diplomatic objectives were more influential in this matter than purely military considerations. Otherwise, the Palestinian version of marrying regular forms with revolutionary aims can only be the result of bad judgment of needs, circumstances and capabilities. The use of large formations and heavy weapons requires a much broader human base; in the absence of this, elite units built through better planning and training would be more appropriate. The present successful Palestinian-Lebanese guerrilla campaign against Israeli forces in Lebanon holds positive lessons for the future.

Part II
Gulf Wars

The Iran–Iraq War, 1980–1988

[11]

Iraq's Decisions to Go to War, 1980 and 1990

F. Gregory Gause, III

The driving force behind Saddam Husayn's decisions to go to war in 1980 and 1990 was his belief that foreign forces were working to destabilize the Ba'th Party regime in Iraq. Saddam believed that Iraq had a good chance of winning those wars. However, the circumstances that encouraged that belief were not new or unique to the time in which those decisions were made. What changed in the period before those war decisions was Saddam's perception of foreign meddling against him in Iraq itself.

On July 17, 1997, in a speech celebrating the twenty-ninth anniversary of the establishment of the Ba'th party regime in Iraq, President Saddam Husayn drew an extensive parallel between the behavior of Iraq's two most recent military opponents — the United States and Iran. "There is no difference," he said, "between American behavior and Iranian behavior." In the end, the Iranians had to acknowledge their defeat and end their hostilities with Iraq, and, Saddam concluded, the Americans would shortly be forced to do the same.[1] How could any leader claim victory in these two wars, one a terribly costly draw and the other a decisive defeat?

One could simply write off Saddam's comments as self-delusion or "big lie" propaganda. However, this speech, whose content is consistent with Iraqi public statements since well before the invasion of Kuwait on August 2, 1990, should be seen as a window onto the real motivations behind both the invasion of Kuwait and the attack on Iran. Later on in the speech, Saddam referred to the United States' failure "to conquer the will of the Iraqis and the glories of Iraq, the true companions and family of Saddam Husayn." He also said that the Iranians used the rationale of saving Iraqi Shi'is from his rule — that is, bringing down his government — as a pretext to continue the war against Iraq. "But," he reminded both the Iranians and his listeners, "I am here, and I am speaking about history to recall its lessons and expose similar foreign deceits against our country."[2] Both wars were victories because the regime of Saddam Husayn re-

F. Gregory Gause, III is an associate professor of political science at the University of Vermont. The author would like to thank the following for comments on earlier versions of the article: Raad Alkadiri, Amatzia Baram, Mark Gasiorowski, Lori Gronich, Fred Lawson, Phebe Marr, Martin Malin. Andrew Parasiliti, David Pervin, David Priess, Albadr Abu-Bakr Al-Shateri, and David Walder.

1. *Al-Hayat*, July 18, 1997, pp. 1, 6.
2. *Al-Hayat,* July 18, 1997, pp. 1,6.

mained in power.

This article develops the thesis that Iraq's decisions to go to war in 1980 and 1990 are best explained by Saddam Husayn's perception that his regime was under serious and concerted pressure from a combination of international and domestic actors and circumstances. Failure to change the trajectory of events would mean a much higher probability that his own rule in Iraq would not survive.[3] Facing the likelihood that doing nothing would mean a decline in his regime's fortunes, he chose in both cases the risky option of international war to break out of the box he perceived to be forming around him.[4]

Locating the cause of wars within the domestic political system of the initiator has a long, if somewhat checkered, history. In his review of the literature on the topic, Jack Levy points out that statistical studies of the relationship between domestic politics and war initiation show very little correlation between the two, while case study and historical accounts of particular war decisions frequently emphasize domestic causes. He convincingly argues that, in large measure, this discrepancy arises from the lack of specification of just exactly how domestic politics leads to war decisions.[5] My argument about the link between domestic regime security and war initiation in Iraqi decision-making starts from the assumption that the first goal of any leader is to stay in power. In most of the "Third World," where leaders' hold on power is not buttressed by popular election or strong traditions of constitutional order, that goal becomes even more central. The importance of regime stability considerations in the foreign policy decisions of Third World states is a central insight of much of the recent scholarship on Third World foreign policy.[6]

I do not argue, however, that the Iraqi decisions of 1980 and 1990 were directed

---

3. The logic suggested here is similar to the logic suggested in two important recent theoretical treatments of the causes of war. Stephen Van Evera, *Causes of War*, (Ithaca: Cornell University Press, 1999), Chapter 4; Dale Copeland, *The Origins of Major War*, (Ithaca: Cornell University Press, 2000). The difference is that Van Evera and Copeland see decline exclusively in terms of power relations between states, where my focus is on domestic "decline."

4. I am not the first to suggest this explanation. On the Gulf War, see Janice Gross Stein, "Deterrence and Compellence in the Gulf, 1990-91: A Failed or Impossible Task?" *International Security*, Vol. 17, No. 2 (1992); Shibley Telhami, "Between Theory and Fact: Explaining American Behavior in the Gulf War," *Security Studies*, Vol. 2, No. 1 (1992); Efraim Karsh and Inari Rautsi, "Why Saddam Hussein Invaded Kuwait," *Survival*, Vol. 33, No. 1 (1991). On the Iran-Iraq War, see Shahram Chubin and Charles Tripp, *Iran and Iraq at War* , (Boulder: Westview Press, 1988), Chapter 2; W. Thom Workman, *The Social Origins of the Iran-Iraq War*, (Boulder: Lynne Rienner, 1994); Efraim Karsh and Inari Rautsi, *Saddam Hussein: A Political Biography*, (New York: Free Press, 1991).

5. Jack Levy, "The Diversionary Theory of War: A Critique," in Manus I. Midlarsky (ed.), *Handbook of War Studies*, (Boston: Unwin Hyman, 1989).

6. Steven David, "Explaining Third World Alignment," *World Politics*, Vol. 43, No. 2 (1991); Mohammed Ayoob, *The Third World Security Predicament*, (Boulder: Lynne Rienner, 1995); Edward Azar and Chung-In Moon (eds.), *National Security in the Third World*, (Aldershot, UK: Edward Elgar Publishing, 1988); Brian L. Job (ed.), *The Insecurity Dilemma: National Security of Third World States*, (Boulder: Lynne Rienner, 1992).

at improving the standing of the regime among the Iraqi public in general or important interest groups within the Iraqi political system. I am not offering a classic "scapegoat" or "diversionary war" explanation, the core of which is that cynical leaders attempt to displace popular discontent upon a foreign enemy that has little if anything to do with the sources of that discontent. The Ba'th regime hardly rested upon the consent of the governed, and no interest group in Iraq, the military included, had the kind of power that could be used to press the regime into taking actions. Rather, my argument is that in both 1980 and 1990 Saddam Husayn identified foreign sources of, or foreign support for, the threats to his domestic hold on power. The objects of his attacks were seen, directly or indirectly, as the causes of his domestic problems: Iran's encouragement of Iraqi Shi'i opposition and Kuwait's oil policies, which were aimed at reducing the resources he needed to maintain his regime. While the capture of Kuwait was undoubtedly popular in Iraq, the aim of Saddam's decision to invade was not to increase his popularity domestically, but rather to break up what he saw as a regional/international effort to weaken and destabilize him at home. For Saddam, these wars were not "diversionary." They were aimed directly at the causes of his domestic problems.

This case will be made through two comparisons. The first is between Iraqi perceptions of the present and immediate future in the periods before the 1980 and 1990 decisions. I contend that similar perceptions of threat posed by the continuation of the regional/domestic status-quo for regime stability existed, and these similar fears led Saddam Husayn to make the same, risky decision — to go to war to change both his regional and domestic situations. This comparison is more suggestive than decisive, because it cannot decisively refute the major alternative explanation for these decisions, offensive realism, that will be discussed in the next section. In both 1980 and 1990 Iraq's enemy appeared weak and unable to resist a military assault, and thus an "easy" target. However, the similarity between Saddam's perceptions of his situation in 1980 and 1990 lends credence to a regime security explanation, and the offensive realist explanation has a number of other problems that will be discussed below.

The second comparison is between the Iraqi failure to withdraw from Kuwait in the face of near-certain defeat in 1991 and the Iraqi acceptance of the Iranian *diktat* embodied in the 1975 Algiers Accord. This comparison provides more decisive support for a regime security explanation of the 1990 war decision. The key difference was that in 1975 Iran promised to stop supporting Baghdad's Kurdish opponents as the *quid pro quo* for Iraqi acceptance of Iranian territorial demands along the Shatt al-Arab. Bowing before superior military power was seen as strengthening the Ba'th's grip on power at home. In 1991 Iraqi leaders believed that a withdrawal from Kuwait would not end the international pressure on the regime, and might weaken the regime domestically.

ALTERNATIVE EXPLANATIONS FOR THE 1980 AND 1990 DECISIONS

The most popular explanation for Saddam Husayn's decision to invade Kuwait in 1990 is some combination of offensive realism[7] and the innate aggressiveness of dictators in general and Saddam in particular. Similar motives are also attributed to his decision to invade Iran in 1980. From an offensive realist perspective, the 1990 invasion seemed perfectly sensible. Kuwait was both rich and weak, a valuable strategic and financial prize and no match on the battlefield. No other regional power could block the move: Iran was immobilized after its own war with Iraq, other Arab countries were aligned with Iraq and thankful for its efforts against Iran; and strong currents of Arab public opinion would support such a move, characterized as both a step toward Arab unity and a challenge to Israeli and American dominance of the area. The United States exhibited little interest in the brewing crisis; the famous comments by US Ambassador to Iraq April Glaspie to Saddam in their meeting of July 25, 1990 (discussed further below) are usually cited as the proof that Saddam thought the US would not actively resist the move.[8]

While less studied than the 1990 case, the dominant explanation of the 1980 Iraqi invasion of Iran also is strongly flavored with (often unstated) offensive realist assumptions.[9] With Iran bogged down in its own internal problems and Egypt isolated in the Arab world because of its peace treaty with Israel, it seemed a perfect time for the ambitious new president of Iraq to reassert Iraqi border claims, teach the new Iranian regime a lesson, and establish Iraq as the dominant regional power in the Gulf.[10] In both cases, according to the offensive realist position, Saddam saw few risks in, and enormous gains to be made by, going to war.

The offensive realist explanation of these two war decisions is usually accompanied by a subsidiary argument about Saddam Husayn's own personal predilections toward violence. Saddam's personality is seen as the motive force behind Iraqi willingness to leap through these windows of opportunity. The strongest version of this

7 "Offensive," or "aggressive," realism posits that the anarchical nature of the international system pushes states to expand their power when the possibility of doing so presents itself. It is distinguished from defensive realism, which posits that states seek security, not power *per se*. For the best bibliographical source on this distinction, see Gideon Rose, "Neoclassical Realism and Theories of Foreign Policy," *World Politics*, Vol. 51, No. 1 (1998), p. 146.

8 For offensive realist explanations for the 1990 decision, see Richard Hermann, "Coercive Diplomacy and the Crisis over Kuwait, 1990-91," in Alexander L. George and William E. Simons (eds.), *The Limits of Coercive Diplomacy*, 2nd ed. (Boulder: Westview, 1994), p. 235; Amatzia Baram, "The Iraqi Invasion of Kuwait: Decision-Making in Baghdad," in Amatzia Baram and Barry Rubin (eds.), *Iraq's Road to War*, (New York: St. Martin's Press, 1993), pp. 16-26. On the centrality of the Glaspie meeting, see Mohamed Heikal, *Illusions of Triumph: An Arab View of the Gulf War*, (London: Harper Collins, 1993), pp. 232-242; Karsh and Rautsi, "Why Saddam Hussein Invaded Kuwait."

9 An interesting exception is Stephen Walt's account of the 1980 war as an example of the general phenomenon of post-revolutionary international conflict, spurred on by dual misperceptions on the part of the attacker: misperceived weakness in the opponent and misperceived threat of the spread of revolution to the attacker's domestic society. *Revolution and War*, (Ithaca: Cornell University Press, 1996), Chapter 6. I disagree with his contention that the Iranian revolutionary threat to Saddam's regime was a misperception.

10 For examples, see Efraim Karsh, "Geopolitical Determinism: The Origins of the Iran-Iraq War," *Middle East Journal*, Vol. 44, No. 2 (1990); Daniel Pipes, "A Border Adrift: Origins of the Conflict," in Shihrin Tahir-Kheli and Shaheen Ayubi (eds.), *The Iran-Iraq War: Old Conflicts, New Weapons*, (Westport: Greenwood, 1983); and Chubin and Tripp, *Iran and Iraq at War*, Chapter 2.

argument sees Saddam's willingness to use force as part of a long tradition of dictators going to war to increase their own glory.[11] The more prevalent version is the less ambitious, more specific contention that Saddam Husayn is a violent megalomaniac.[12]

Some elements of this argument are beyond dispute. Iraq seemed to have the military advantage over Iran in 1980; it certainly had such an advantage over Kuwait in 1990. However, those facts in and of themselves are not helpful in explaining the *timing* of the two war decisions, and thus in helping us understand the motives which drove Saddam Husayn to take those decisions. Iran began to experience internal problems as early as 1977. While the Shah was in power the Iraqi regime not only did not exploit his weakness, but sought to support his rule. Baghdad expelled Ayatollah Ruhollah Khomeini from Iraq in 1978 and held talks on internal security matters with both Saudi Arabia and the Shah as revolutionary pressures mounted in Iran.[13] Even with the fall of the Shah in early 1979, and the decimation of the Iranian officer corps that followed the establishment of the Islamic Republic, Iraq did not exploit its neighbor's apparent weakness. Only in September 1980, after numerous statements by the new Iranian leaders encouraging revolt in Iraq, tangible efforts by Iran to encourage such revolt, and serious evidence of domestic discontent by Iraqi Shi'is did Iraq go to war. If offensive realist motivations were the driving force in Iraq's decision to invade Iran, it would have occurred earlier.

Likewise, an offensive realist rationale for an Iraqi attack on Kuwait has existed for decades. Iraqi governments since independence have claimed Kuwait as part of Iraq. The strategic value of Kuwait for Iraq, both in terms of extending Iraq's Gulf coastline and Kuwaiti oil, was a constant. Iraq's decisive military superiority over Kuwait long predates 1990. Even more telling, if offensive realist rationales were central in animating the Iraqi decision to invade Kuwait, it is puzzling that Baghdad did not postpone the invasion for a year or two until its nuclear program had generated a small arsenal of weapons. Its ability to confront international opposition to its annexation of Kuwait would have been vastly enhanced had it done so.

It is unlikely Iraq would have taken the war decisions of 1980 and 1990 without the belief that there was a good chance to win the wars. However, to accept the offensive realist explanation, one must also accept the fact that this explanation *failed* to explain Iraqi restraint at other times, when the potential gains of attacking Iran or Kuwait were equally high and the risks less obvious. Therefore, offensive realism cannot provide an adequate explanation for why these war decisions happened *when they did.* The belief that war was winnable was a necessary condition for the decisions of 1980 and 1990; it was not a sufficient condition.

Adding the personality factor to the offensive realism case does not solve this

11. Laurie Mylroie, "Why Saddam Hussein Invaded Kuwait," *Orbis*, Vol. 36, No. 1 (1993).

12. Jerrold Post, "Saddam Hussein of Iraq: A Political Psychology Profile," *Political Psychology*, Vol. 12, No. 2 (1991); Samir al-Khalil [pseudonym for Kanaan Makiyya], *Republic of Fear: The Politics of Modern Iraq*, (Berkeley: University of California Press, 1989), pp. 270-76; Karsh and Rautsi, *Saddam Hussein: A Political Biography.*

13. *New York Times*, June 18, 1978, p. 1.

problem. Saddam Husayn has been the dominant force in Iraqi decision-making since at least the mid-1970s, and President since mid-1979.[14] The war with Iran began a full year after his assumption of the presidency, and from the outset of revolutionary turmoil in Iran he had been in effective control of Iraqi foreign and military policy. It was under Saddam's leadership that Iraq repaired its historically tense relations with Kuwait in the late 1970s, and developed very close relations with that country during the Iran-Iraq War. He did not always take advantage of favorable conditions to attack neighbors. He also accepted a cease-fire in the Iran-Iraq War in 1988, when Iraqi troops had the battlefield advantage. It is likewise incorrect to see Saddam as someone unable to accept defeat, or at least tactical retreat in the face of superior power. He personally signed the 1975 Algiers Accord with Iran, and repeatedly called after 1982 for an end to the Iran-Iraq War based on a return to the *status quo ante*. It is undoubtedly true that he is brutal, ambitious and more than willing to solve his problems through violence. But Saddam's personality is a constant in the Iraqi equation from at least the late 1970s. It cannot explain why he chose to go to war specifically in 1980 and 1990.

The other major alternative explanation for the Iraqi war decision of 1990, which does not apply to the 1980 case, is economic crisis.[15] There is no disputing the fact that Iraq faced serious economic problems in the wake of the Iran-Iraq War. That war left Iraq with an enormous debt burden, usually estimated at $80 billion (roughly $40 billion to the Gulf monarchies, which would not be paid back but remained on the books, and $40 billion to other governments and creditors).[16] Moreover, during the war Iraq had engaged in a large-scale economic liberalization and privatization program. The unintended consequences of this program included "high levels of inflation, unemployment, shortages in basic goods, growing and highly visible economic inequality, and the emergence of a brisk black market in foreign currencies."[17] According to Chaudhry, the regime by the beginning of 1990 faced "such chaos that not even the experienced repressive apparatus of the Ba`th Party could guarantee domesic political stability."[18] The quick way out, this argument goes, of this economic crisis was for Saddam to get his hands on Kuwaiti oil and Kuwaiti financial assets.

Once again, the facts here are not in dispute. There were serious economic prob-

14. Marion Farouk-Sluglett and Peter Sluglett, *Iraq Since 1958*, (London: I.B. Tauris, 1990) , pp. 205-09; Phebe Marr, *The Modern History of Iraq*, (Boulder: Westview, 1985), pp. 217-20, 228-29; Charles Tripp, *A History of Iraq*, (Cambridge: Cambridge University Press, 2000), p. 215.

15. Kiren Aziz Chaudhry, "On the Way to Market: Economic Liberalization and Iraq's Invasion of Kuwait," *Middle East Report*, No. 170 (1991); and Fred Lawson, *The Social Origins of Egyptian Expansionism during the Muhammad 'Ali Period*, (New York: Columbia University Press, 1992), pp. 156-171.

16. Lawrence Freedman and Efraim Karsh, *The Gulf Conflict 1990-91*, (Princeton: Princeton University Press, 1993), p. 39; Patrick Clawson, "Iraq's Economy and International Sanctions," in Baram and Rubin, *Iraq's Road to War*, p. 81, note 12; Baram, "The Iraqi Invasion of Kuwait," p. 7; Heikal, *Illusions of Triumph*, p. 178.

17. Chaudhry, "On the Way to Market," p. 17.

18. Chaudhry, "On the Way to Market," p. 14.

lems in Iraq on the eve of the invasion of Kuwait. Iraqi leaders, including Saddam, referred to them regularly.[19] Iraq's demand in the lead-up to the invasion for $10 billion from Kuwait as a "reconstruction loan" and complaints about Kuwaiti and UAE oil production policies keeping the price of oil "artificially" low are clear indications that economic issues weighed heavily on Saddam Husayn and other Iraqi leaders. But the assertion that it was the economic crisis itself that drove Iraq to invade Kuwait is open to two major criticisms.

First, there is an unstated assumption in this argument that economic problems lead naturally to aggressive foreign policies. If that were the case, the world would have seen many more wars than it has, and Iraq certainly would have gone to war again in the decade after 1991. If the economic issue were central, it would have been just as logical for Saddam to have accepted the Kuwaiti offer of $500 million and the Kuwaiti agreement to return to its OPEC oil production quota, pocketed his gains and then come back to the table later with other demands.[20] Second, financial crisis clearly played no role in the other Iraqi war decision. In 1980 Iraq was flush with money, after the oil price increases of the 1970s. Accepting the economic crisis argument would, in effect, be asserting that financial problems lead Iraq to war but financial comfort does not lead it to peace. While it is possible that the two war decisions were made for completely different reasons, an explanation that encompasses both, demonstrating that similar circumstances led to similar outcomes, has more inherent plausibility.

The economic crisis did play a major role in pushing Saddam Husayn toward war, but not in the way this argument proposes. Saddam and his inner circle saw these problems not as an outcome of mistaken government economic policy and the operation of supply and demand in the world oil market, but rather as part of a more general plan aimed at weakening Iraq and destabilizing the Ba'th regime, orchestrated from the outside and encompassing a number of domestic, regional, and international actors. It was that encompassing sense of threat, including but not limited to the economic crisis, and similar to the sense of threat to regime stability perceived in 1980, that best explains Iraqi war decisions.

The remainder of this article, then, addresses two tasks: 1) to show that the timing of Iraq's war decisions in 1980 and 1990 was coterminous with a perception of increasing threats to regime stability, and that the Iraqi regime saw war as a way to strike at the foreign sources of domestic problems; and 2) to demonstrate that Iraq's probability of victory and benefits from victory in these wars did not noticeably increase in the periods leading up to the decisions to go to war, in order to show that the offensive realist explanation cannot account for the 1980 and 1990 war decisions.

THE IRAQI WAR DECISION OF 1990

Establishing with certainty when Saddam Husayn decided to attack Kuwait is a difficult task. There are indications from Iraqi sources themselves that the decision

19. Freedman and Karsh, *The Gulf Conflict*, pp. 45-50

20. The Kuwaiti financial offer is mentioned in Heikal, *Illusions of Triumph*, p. 209, and confirmed by author's interviews in Kuwait..

was made only a few months before the actual invasion. Sa'd al-Bazzaz, editor of one of the major Iraqi papers in the period before and during the Gulf War, reports on a series of meetings beginning in mid-June 1990 that formulated the plan for the invasion. Two plans were set out, one calling for the occupation of the border area and the Kuwaiti islands Warba and Bubiyan, and the other calling for a complete occupation of Kuwait. It was only on July 29, 1990, according to al-Bazzaz, that Saddam Husayn decided to implement the second plan.[21] Deputy Prime Minister Tariq 'Aziz, shortly after the Gulf War, also spoke of a June war decision and a last minute choice by Saddam to occupy all of Kuwait.[22]

Other sources place the decision to invade slightly earlier in 1990. Iraqi opposition circles claim that they have evidence that the decision was made in March 1990.[23] Wafiq al-Samarra'i, deputy director of Iraqi military intelligence during the war, told the producers of the BBC/PBS documentary *The Gulf War* that he thought the decision was made in April 1990, though he did not offer any direct evidence supporting this claim.[24] Al-Samarra'i, who saw Saddam with some frequency, wrote that in the first quarter of 1990 Saddam realized that his policies were failing: "This was reflected in the psychological state of the president, and led him to make statements and follow policies that were more and more spasmodic."[25] Hasan Ruhani, then head of the Iranian National Security Council, reportedly told Arab officials that Saddam Husayn sent a message to Iran in May 1990 that "certain events" would soon take place in the Gulf that Iran should not interpret as being directed against it.[26] While many people in the countries aligned against Iraq in the Gulf War believe that Saddam had a long-term plan to invade Kuwait, no source that refers specifically to the timing of the decision places it earlier than the spring of 1990.

The haste with which the decision was made was indirectly reflected in some of the (very mild) self-criticism exercised by Iraqi leaders after the invasion. At a meeting of the Iraqi Revolutionary Command Council and Ba'th party leadership on January 24, 1991 Taha Yasin Ramadan told his colleagues "I am not saying that August 2, 1990 was the best day for the mother of battles. We had not studied the situation for a year, or even for months, preparing for the mother of battles. But it was the will of

21. Sa'd al-Bazzaz, *Harb Tulid Ukhra* [One War Gives Birth to Another], (Amman, Jordan: Al-Ahliyya lil-Nashr wa al-Tawzi', 1993), pp. 27-36; Sa'd al-Bazzaz, *Al-Janaralat Akhir man Ya'lam* [The Generals Are the Last to Know], (Amman, Jordan: Al-Ahliyya lil-Nashr wa al-Tawzi', 1996), pp. 47-55.

22. Milton Viorst, *Sandcastles: The Arabs in Search of the Modern World*, (New York: Alfred A. Knopf, 1994), pp. 340-41, 344.

23. Falih 'Abd al-Jabbar, *"Jaysh al-Miliyun Yuriq Iqtisad al-Dawla al-Mudammar,"* ("The Million-Man Army Lies Heavily on the Destroyed Economy of the State"), *al-Hayat*, May 30, 1991, p. 5; Falih Abd al-Jabbar, "Roots of an Adventure: The Invasion of Kuwait – Iraqi Political Dynamics," in Victoria Brittain (ed.), *The Gulf between Us: The Gulf War and Beyond*, (London: Virago Press, 1991).

24. http://www.pbs.org/wgbh/pages/frontline/gulf/oral/iraqis.html.

25. Wafiq al-Samarra'i, *Hatam al-Bawaba al-Sharqiyya* [The Destruction of the Eastern Gate], (Kuwait: Dar al-Qabas, 1997), p. 214.

26. *New York Times*, March 20, 1991, pp. A1, A12.

God that decided the date."[27] One would not dare to criticize a decision of Saddam Husayn in his presence, the implication being that Saddam himself recognized that the decision was rushed. The Ba'th party also circulated an analysis to ranking party members after the war ended in February 1991 that admitted that Iraq was forced to take a quick decision to invade Kuwait, because of the pressures it was under, even though all necessary preparations for the confrontation had not been made.[28] There is every indication that the decision to invade Kuwait was made relatively shortly before the invasion, under feelings of time pressure. What had happened in the period leading up to the decision to trigger it?

The Iraqi belief that, as Foreign Minister Tariq 'Aziz said after his meeting with Secretary of State James Baker in Geneva on January 9, 1991, "this crisis did not begin on August 2," is now well documented. The Iraqi leadership has steadfastly maintained that there was an international conspiracy against it, meant to weaken Iraq internationally and destabilize it domestically. The first sentence of al-Bazzaz's first book on the Gulf War reads, "There is no doubt that Iraq was feeling danger and the existence of a conspiracy against it."[29]

The outlines of the Iraqi argument are well known. Its economic problems were blamed on lower oil prices, which were in turn blamed on the "overproduction" of Kuwait and the UAE, clients of the United States. Small shifts in American policy (like limits on US credits for Iraqi purchases of US rice exports and Congressional resolutions condemning Iraq for human rights violations) and damaging revelations (such as Iraq's use of the Atlanta branch of an Italian bank to launder arms purchase money) after the end of the Iran-Iraq war were read as evidence that the United States had adopted a hostile attitude toward Iraq. Media attention to the Iraqi nuclear program, and subsequent British and American efforts to block the export of dual-use technology to Iraq, were seen as part of a concerted effort to build a case against Iraq as a prelude to more severe measures. Lurking behind many of these efforts, in the Iraqi view, was Israel, seen as preparing for a strike on the Iraqi nuclear establishment similar to the one it conducted in 1981.[30] Al-Samarra'i says that at the beginning of 1990 military intelligence began receiving a wave of warnings from Saddam's office about Israeli plans to strike at Iraqi nuclear, chemical, and biological weapons facilities.[31] Al-Bazzaz reports that the Iraqi leadership fully expected an Israeli military attack sometime in August 1990.[32]

27. Al-Bazzaz, *Al-Janaralat*, p. 200, quoting from minutes of the meeting.

28. Al-Bazzaz, *Harb*, p. 34.

29. Al-Bazzaz, *Harb*, p. 17.

30. Freedman and Karsh, *The Gulf Conflict*, Chapters 2-3; Heikal, *Illusions of Triumph*, pp. 158-231; Baram, "The Iraqi Invasion of Kuwait."

31. Al-Samarra'i, *Hatam*, p. 365.

32. Al-Bazzaz, *Harb*, p. 345. One does not have to accept the fact that there was a conspiracy against Iraq to accept that this belief could have profoundly affected Iraqi decision-making. It is clear that there was no concerted American policy aimed against Iraq. American policy in this period might better be described as working at cross-purposes. See Bruce W. Jentleson, *With Friends Like These: Reagan, Bush and Saddam, 1982-1990*, (New York: W. W. Norton, 1994); Zachary Karabell, "Backfire: US Policy Toward Iraq, 1988-2 August 1990, *Middle East Journal*, Vol. 49, No. 1 (1995).

Saddam himself bluntly described this "conspiracy" to al-Samarra'i in March 1990:[33] "America is coordinating with Saudi Arabia and the UAE and Kuwait in a conspiracy against us. They are trying to reduce the price of oil to affect our military industries and our scientific research, to force us to reduce the size of our armed forces...You must expect from another direction an Israeli military airstrike, or more than one, to destroy some of our important targets as part of this conspiracy."

What is less well known is the internal aspect of the Iraqi regime's fears. In either late 1988 or early 1989 scores of officers, many decorated for heroism in the war with Iran, were arrested and executed on the charge of membership in a secret organization working to bring down the government. Hundreds of high-ranking officers indirectly connected to the accused were forced to retire.[34] Iraqi ruling circles came to believe during 1989 that they had evidence that a number of foreign powers, including Iran, Saudi Arabia, and the United States, were attempting to infiltrate Iraqi society to collect intelligence and pressure the government.[35] In May 1989, Iraqi Defense Minister 'Adnan Khayrallah, who was also Saddam's cousin and brother-in-law, died in what official accounts called a helicopter crash (after a number of prominent Iraqi generals had died in similar helicopter accidents since the end of the Iran-Iraq war). Rumors circulated in Iraqi opposition circles, however, that Khayrallah was killed on Saddam's orders, either because of his involvement in these foreign machinations or because of his inability to insulate the army from them.[36] Other sources report a failed coup attempt in September 1989 and the exposure of a coup attempt, coupled with a plan to assassinate Saddam, in January 1990.[37]

While Saddam Husayn increasingly saw his domestic political and economic situation in 1989 deteriorate, events in the larger world during that year reinforced his growing sense of crisis. The fall of the Soviet client states in Eastern Europe increased his fears about the future of his own regime. Al-Bazzaz reports that in the summer of 1989 Saddam and a number of Iraqi experts debated the effect on Iraq of the end of the bi-polar world, and concluded that Iraq's international enemies would use the events in Europe to create an impression among Iraqis that it would be easy for them to change their government.[38] This belief helps to explain the intensely negative reaction of the Iraqi government to a Voice of America (VOA) Arabic service broadcast in January 1990 which implicitly compared Saddam to the fallen Eastern European lead-

33. Al-Samarra'i, *Hatam*, pp. 222-223.34. Baram, "The Iraqi Invasion of Kuwait," p. 8; al-Bazzaz, *al-janaralat* pp. 36-37, 89-90; al-Samarra'i, *Hatam*, pp. 184-85; Tripp, *A History of Iraq*, pp. 249-50.

35. Al-Bazzaz, *Harb*, pp. 159-160, 210-213.

36. Tripp, *A History of Iraq*, p. 250; Said K. Aburish, *Saddam Hussein: The Politics of Revenge*, (New York: Bloomsbury Publishing, 2000), p. 263.

37. Freedman and Karsh, *The Gulf Conflict*, pp. 29-30; Amatzia Baram, "Neo-tribalism in Iraq: Saddam Hussein's Tribal Policies, 1991-1996," *International Journal of Middle East Studies*, Vol. 29, No. 1 (1997), pp. 5-6; Amatzia Baram, *Building Toward Crisis: Saddam Hussein's Strategy for Survival*, Policy Paper No. 47, The Washington Institute for Near East Policy, Washington, D.C., 1998, p. 27; al-Samarra'i, *Hatam*, p. 185.

38. Al-Bazzaz, *Harb*, p. 392.

ers, and to the February 1990 State Department human rights report, and to Congressional resolutions critical of his human rights record.[39] They were seen as proof of American involvement in the plots against Saddam's regime. Saddam himself referred to the VOA broadcast in his July 25, 1990 meeting with US Ambassador April Glaspie.[40]

Saddam Husayn's sense that international and regional forces were conspiring with his domestic opponents against him had reached the point that, in October 1989, Tariq 'Aziz raised this issue in his meeting with Secretary of State Baker in Washington.[41] Baker reiterated the US desire to work with Saddam's government. President Bush passed the same message to Saddam, through Saudi Arabia's ambassador in Washington Prince Bandar bin Sultan Al Sa'ud, in April 1990, when Saddam, through the Saudis, raised similar suspicions of American intentions.[42] These reassurances had little effect on Saddam, who since the revelation of the Iran-contra affair in 1986 had come to view official American pronouncements with suspicion.[43]

These domestic and international events had by early 1990 convinced Saddam Husayn that his regime was being targeted. This belief was reflected in the changes in his rhetoric and the tone of Iraqi foreign policy. Before 1990, Iraqi foreign policy was aimed at reaping the economic and strategic benefits of Iraq's "victory" in the Iran-Iraq war, but through regional alliances and normal diplomatic efforts. Iraq's initiative in forming the Arab Cooperation Council (a loose grouping of Iraq, Jordan, Egypt and North Yemen) in February 1989 and its April 1989 non-aggression pact with Saudi Arabia (aimed at neutralizing Saudi fears of the new Arab grouping) can be seen in that light. Iraqi diplomacy toward Kuwait, while clearly aiming at securing territorial concessions and economic benefits from its neighbor, was not overtly threatening. The border issue was raised in a number of meetings in 1989 — sometimes by the Iraqi side, sometimes by the Kuwaiti side, according to a variety of sources — but the Kuwaitis did not seem to think that it was particularly pressing and the Iraqis did nothing to indicate otherwise.[44] As late as September 1989, when Amir Jabir al-Ahmad Al-Sabah, the ruler of Kuwait, visited Baghdad he was awarded Iraq's highest honor amid lavish praise for Kuwait by Saddam himself.[45] Likewise, Iraqi rhetoric on the United States and on Israel in 1989 was in the moderate vein adopted by Baghdad during the Iran-Iraq War.

The change in Saddam's rhetoric and the tone of Iraqi foreign policy comes in

39. Freedman and Karsh, *The Gulf Conflict*, p. 31.

40. See text of meeting released by Iraq, reproduced in Micah L. Sifry and Christopher Cerf (eds.), *The Gulf War Reader*, (New York: Random House, 1991), pp. 123-24.

41. James A. Baker, III, *The Politics of Diplomacy*, (New York: G.P. Putnam's, 1995), p. 265; Heikal, *Illusions of Triumph*, p. 159; al-Bazzaz, *Harb*, p. 154-160.

42. Bob Woodward, *The Commanders*, (New York: Simon and Schuster, 1991), pp. 203-04.

43. Saddam himself referred to this in his interview with Ambassador Glaspie. Sifry and Cerf, *The Gulf War Reader*, p. 123. See also al-Samarra'i, *Hatam*, p. 128.

44. al-Bazzaz, *Harb*, pp. 41-45.

45. See coverage in Kuwaiti newspapers *al-'Anba'*, September 24, 1989, p. 1; and *Al-Ra'i al-'Am*, September 24, 1989, p. 1; September 25, 1989, p. 1.

early 1990. In February 1990 Saddam launched an attack on the United States military presence in the Gulf at the founding summit of the Arab Cooperation Council and devoted much of the speech to criticism of Israel.[46] The famous "burn half of Israel" comment followed in early April 1990. The rhetorical temperature escalated from there. At the same time, Iraqi rhetoric toward Kuwait and the other Gulf states hardened, and in January 1990 Iraq first proposed that Kuwait "loan" it $10 billion, as well as write off Iraqi debts incurred during the war with Iran.[47] This culminated with the harsh Iraqi language at the Arab summit of May 1990, where Saddam likened overproduction of OPEC quotas to an act of war against Iraq, and in Iraq's July 16 memorandum to the Arab League.[48]

This shift in Iraqi foreign policy, the beginning of the process that led to the invasion of Kuwait, came when Saddam concluded that there were international efforts afoot to destabilize him domestically. As he began to perceive that the future could hold serious challenges for his rule, his foreign policy became more aggressive. Further support for this reading of Saddam's motivations in attacking Kuwait comes from statements by Saddam himself and by one of his closest aides at the meeting of the Revolutionary Command Council/Ba'th Party leadership held on January 24, 1991. It should be noted that, while these statements came after the invasion of Kuwait, and thus might be read as *ex post facto* justifications, they came before Iraq had clearly been defeated in the war, just one week into the air campaign against it and a full month before the allied ground assault. Saddam told those present that perhaps they had heard some people suggesting that Kuwait should be returned as a way to avoid war. He went on,[49]

> You must say to those people, 'What were things like before August 2 when Iraq was without Kuwait? They were conspiring against you to starve you, after they had deprived us of our economic capacity...We did not want, before August 2, anything but to preserve our status as it was and develop it gradually, perhaps 1% per year or 2% per year, but they wanted to force our status backwards, something we could not accept for Iraqis for whom we want better. Even our standard of living at the time, they were planning to push it backwards in their despicable conspiracies, to crush us spiritually and force us to abandon our role.

Taha Yasin Ramadan, Iraq's First Deputy Prime Minister, said at the same meeting (after he had voiced the belief that August 2, 1990 was not the ideal time for the attack on Kuwait, mentioned above):[50]

> Yes, the battle is inevitable. If it happened later our revolutionary base would not have been consolidated and would have been weaker in confronting the size of the hostile forces as they are now. Imagine if we had waited two years, and the

46. Ofra Bengio (ed.), *Saddam Speaks on the Gulf Crisis: A Collection of Documents*, (Tel Aviv: The Moshe Dayan Center for Middle Eastern and African Studies, 1992), pp. 37-49.
47. Heikal, *Illusions of Triumph*, p. 209.
48. Freedman and Karsh, *The Gulf Conflict*, pp. 46-48.
49. Al-Bazzaz, *Al-Janaralat*, pp. 227-28.
50. Al-Bazzaz, *Al-Janaralat*, pp.198-99.

> Gulf oil policy had continued as it is. Iraq is $50 billion in debt and the price of oil does not meet 50% of our even minimal needs, with our indebtedness eating up 50% of our earnings at this price and with the interest on the loans. The Western states and America decided to stop exporting technology to us after April 1990, and America stopped agricultural facilities [subsidized exports to Iraq] in March 1990 by Congressional decision. It is possible that this boycott might have spread to other European countries as the technology boycott spread.
>
> How were we going to maintain the loyalty of the people and their support for the leader if they saw the inability of the leadership to provide a minimal standard of living in this rich country? In this situation, could you lead the army and the people in any battle, no matter what its level and under any banner? I think not. I am not deviating from my deep faith in victory in this battle, but whatever the outcome, if death is definitely coming to this people and this revolution, let it come while we are standing.

WHY NOT WITHDRAW? 1991 AND 1975 COMPARED

The centrality of the perceived threat to regime security in the Iraqi decision to go to war in August 1990 can be highlighted by a comparison of the Iraqi refusal to withdraw from Kuwait in 1991 with the 1975 decision by Iraq to accept a change in its borders with Iran to avoid a military confrontation. In each case, a militarily stronger power demanded a territorial concession from Iraq to end hostilities. But acquiescence would be costly, politically, even while saving Iraqi military and economic assets.

The difference in the two episodes, which explains the different Iraqi decisions in each, is the perceptions of how these decisions would affect domestic regime security. The key element for Baghdad in the Algiers Accord of 1975 was an Iranian commitment to end its support (and that of Israel and the United States, which had used Iranian territory as a conduit for their support) for the Iraqi Kurdish groups opposing Ba'thist rule. The Ba'th regime had, since coming to power in 1968, tried both negotiations and military force to bring the Kurdish areas under its control, but to no avail. With the end of external support to the Iraqi Kurds, Baghdad was able to bring its Kurdish opponents to heel.[51] Regime security was more important than minor border modifications. Saddam Husayn personally signed the Algiers Agreement on behalf of Iraq, and defended it repeatedly as necessary to consolidate the regime's hold on power domestically.[52]

From the August 2, 1990 occupation of Kuwait through the end of 1990, despite the build-up of American and coalition forces and the series of UN Security Council votes that went against Iraq, Saddam believed that he could avoid military defeat. Al-

51. F. Gregory Gause, III, "Gulf Regional Politics: Revolution, War and Rivalry," in W. Howard Wriggins (ed.), *Dynamics of Regional Politics*, (New York: Columbia University Press, 1992), pp. 44-50.

52. See for example his speech to the Iraqi National Assembly in November 1980, shortly after he had initiated the war with Iran, FBIS-MEA-80-216, November 5, 1980, p. E2.

Samarra'i reports that the political team around Saddam was convinced that the United States would hesitate to use force against Iraq, and the crisis would turn into a long political contest of wills. He says that Saddam described the strategy on a number of occasions as "holding on to the elephant's trunk," patiently waiting and drawing things out.[53] Saddam believed that even in the event of an attack, his forces could hold out long enough for the coalition to fragment, as Arab public opinion rose up against Arab members of the coalition and as France and the Soviet Union split from the United States. As American forces took casualties, Saddam believed, public opinion in the United States would also turn against the war. As late as January 10, 1991 he told officials from military intelligence that "America is a weak country...We will not retreat from Kuwait and we will prevail."[54]

The beginning of the air war on January 16, 1991, was the turning point in Saddam's perceptions. Al-Samarra'i reports that Iraqi military intelligence sent numerous reports to Saddam as the January 15 UN deadline approached that war was inevitable and that Iraq was not well positioned to resist. As the air war began, these reports warned of the very negative consequences for Iraq of a ground war. Iraqi military intelligence also informed Saddam, during the air war, of large-scale Iraqi army defections within Kuwait.[55] While Saddam was receiving these reports and the air war began, he also saw that the coalition remained intact and that Arab coalition members were able to manage whatever public opinion problems they had. According to al-Samarra'i, shortly after the air war began Saddam ordered 'Ali Hasan al-Majid, his cousin , whom he appointed military governor of Kuwait, to prepare for the destruction of the Kuwaiti oil wells and other Kuwaiti installations, an indication that he was at least entertaining the prospect that he could not hold on to Kuwait.[56]

With his original strategy revealed as badly flawed, one might expect that Saddam would reassess the costs and benefits of remaining in Kuwait, and seek out a diplomatic solution, much as Iraq had done in 1975. The reason he did not was that the Iraqi leadership did not believe that withdrawal from Kuwait would end what it saw as the international conspiracy against it. On January 14, 1991, Saddam told a visiting Yemeni delegation, "We have no guarantees if we withdraw...Why should we surrender at the last moment?"[57] On the eve of the ground war, after enduring a month of air attacks, Saddam told Soviet envoy Yevgeny Primakov, "If America decided on war it will go to war whether I withdraw from Kuwait or not. They were conspiring against us. They are targeting the leadership for assassination. What have the Iraqis lost? They might yet gain!"[58] After the war, Tariq 'Aziz was asked on the PBS documentary *The Gulf War* why Iraq did not withdraw when defeat seemed inevitable. He replied:

53. Al-Samarra'i, *Hatam*, pp. 321-22.
54. Al-Samarra'i, *Hatam* pp. 237-38.
55. Al-Samara'i, *Hatam*, pp. 237, 246-47; al-Samara'i interview with PBS, http://www.pbs.org/wgbh/pages/frontline/gulf/oral/samarrai/1.html
56. Al-Samarra'i, *Hatam*, pp. 284-85.
57. Al-Bazzaz, *Harb*, p. 265.
58. Al-Bazzaz, *Harb*, p. 399.

"Iraq was designated by George Bush for destruction, with or without Kuwait. Inside Kuwait or outside Kuwait. Before the 2nd of August or after the 2nd of August."[59] The belief that simply withdrawing from Kuwait would not end the pressure on the Iraqi regime was reflected in Iraq's February 15, 1991 proposal for ending the crisis (a month into the air war, with Iraqi military prospects growing increasingly bleak). Not only did Iraq demand a cease-fire, Western military withdrawal from the Gulf, Israeli withdrawal from occupied territories, and the abrogation of all UN Security Council resolutions adopted against Iraq, but also the end of "all resolutions and measures of boycott and embargo, as well as the other negative resolutions and measures that were adopted by certain countries against Iraq unilaterally and collectively before August 2, 1990, which were the real reasons for the Gulf crisis."[60]

Iraq's belief that retreat in 1975 would strengthen the regime's domestic position helps explain why it would accept a defeat at the international level. Saddam Husayn's belief that withdrawal from Kuwait would not end the pressures on his domestic position emanating from abroad explains the different outcome in 1991.

Offensive Realism and the Gulf War Case

To complete the argument, we need to return to the major alternative explanation for the 1990 decision, an offensive realist reading. It needs to be shown that the invasion decision was not the product of an Iraqi calculation that August 1990 was the height of Iraqi military power and thus the period of highest probability of victory. Rather, there are good reasons to believe that the Iraqi leadership did not see August 1990 as the ideal time to launch a war in terms of the regional and international situation, but was driven by fears of the domestic consequences of the continuation of the status-quo.

A good case can be made that Saddam Husayn himself realized that Iraq in 1990 was not as strongly positioned for war as it might be later. Consider the Iraqi nuclear program. UN investigators have established that the Iraqis were closer to developing a rudimentary nuclear arsenal than was thought in 1990. If Saddam had waited until he had achieved nuclear capacity, he would have had a higher probability of deterring an attack by opponents of the invasion of Kuwait. This is the logic underlying the comments by Iraqi leaders cited earlier that August 2, 1990 was not the ideal time for the "mother of battles." Consider also Saddam's own publicly stated assessment of the international balance of forces. In his February 1990 speech to the Arab Cooperation Council summit he talked about the erosion of Soviet power and support for the Arabs, and about the eventual recalibration of international politics with the rise of Europe and Japan to challenge American dominance. However, he cautioned his listeners, "we believe that the US will continue to depart from the restrictions that govern the

59. PBS interview with 'Aziz, http://www.pbs.org/wgbh/pages/frontline/gulf/oral/aziz/2.html.

rest of the world throughout the next five years until new forces of balance are formed."[61]

In Saddam's view, Iraq would have been stronger and the United States would have been weaker after 1990, but clearly he did not feel that he had the time to wait for his position to improve. The case that Saddam would have been better off invading Kuwait in 1989 is more speculative, but still compelling. Iraq's international position was much better in 1989. The threat from Iran during the previous war was more immediate, both to the Arab world and the United States. In early 1989 Baghdad had organized the Arab Cooperation Council and reaffirmed its relations with Saudi Arabia through the non-aggression pact. Its relations with the United States were better, and the Soviet Union might have been expected to be more supportive, as the Cold War had not yet clearly ended. Iraq's economic position was stronger, as oil prices were higher in 1989 than in 1990. If Saddam was just looking for the best time to attack Kuwait, there were numerous reasons why 1989 looked like a good time. Yet we know that his aggressiveness toward Kuwait increased as his own perception of his regional and domestic strength declined and as the risks Iraq faced grew.

The meeting between Saddam Husayn and US Ambassador April Glaspie is frequently cited in offensive realist accounts as playing a decisive role in Iraq's decision-making. Glaspie's supposed failure to warn Saddam of the American reaction to an Iraqi military move on Kuwait is presumed to have given Saddam the confidence that the US would not resist his plans. This argument is suspect on two counts. First, Tariq 'Aziz subsequently reported that Iraq did expect an American reaction to the invasion, and that Iraq did not read the Glaspie interview as a "green light" from Washington.[62] Second, Saddam told the Iraqi leadership on August 2 that he had avoided the mistake made by `Abd al-Karim Qasim, the Iraqi leader who laid claim to Kuwait in 1961. He said that Qasim had been too clear about his intentions, allowing the British and the Egyptians to send troops to Kuwait before Iraq could move. This time, Saddam said, Iraq was doing things differently.[63] The meeting with Glaspie, conciliatory telephone calls with Egyptian President Husni Mubarak and Saudi King Fahd, Iraqi agreement to the July 31 meeting with Kuwaiti representatives in Jidda, the lowering of the rhetorical heat in Iraq's official media in the week before the attack, can all be seen as part of a deliberate Iraqi policy of tactical deception in order to retain the element of surprise. In this reading of the evidence, Saddam had already made his decision to attack Kuwait before he had met with Ambassador Glaspie.

60. *New York Times*, February 16, 1991, p. 5.

61. Bengio, *Saddam Speaks*, p. 41.

62. See account of Tariq 'Aziz's interview with Turkish newspaper *Milliyet* in *Al-Hayat*, May 31, 1991, p. 4. Al-Bazzaz, *Al-Janaralat*, p. 76, while saying that Iraqi leaders were taken by Glaspie's soft tone, says that in the end the meeting was not a major turning point in the Iraqi decision process.

63. Al-Bazzaz, *Al-Janaralat*, pp. 73-74.

THE IRAQI WAR DECISION OF 1980

That the Iranian Revolution presented a serious threat to the stability of neighboring Iraq, and was a major cause of Saddam Husayn's decision to invade Iran in September 1980, is generally, though not unanimously, agreed upon.[64] At the same time, there were also substantial benefits, for Iraq strategically and for Saddam personally, to be had in a short and victorious war against Iran. A quick victory would have reversed the set-back of the Algiers Accord of 1975, consolidated Iraq's leadership in the Arab world (after Egypt's peace treaty with Israel in 1979) and the Gulf, and possibly led to the fall of the revolutionary regime in Tehran. Iran was convulsed by revolution, had decimated its officer corps and cut off its relations with its superpower ally. The probability of victory seemed high.

As in the case of the Gulf War, the evidence indicates that it was the sense of threat, more than the opportunity of gain, that drove Saddam's decision calculus. The benefits from and enhanced possibility of victory had been present for some time before Saddam launched Iraq's war against Iran in September 1980. However, the salience of the threat of revolutionary contagion from Iran had increased markedly in the period just before the decision to go to war was made. Less has come out of Iraq itself about the regime's decision-making on this question than is the case with the 1990 decision. Of necessity this analysis will rely more on the chronological record than direct attributions of Saddam's changing view. Still, the case for a threat-based reading is strong, and the similarities to the 1990 decision striking.

The weakening of Iran in conventional power terms, which began in late 1977 as the revolutionary movement gathered steam, did not immediately excite Iraqi ambitions. On the contrary, Baghdad expelled Ayatollah Khomeini from Iraq in October 1978[65] and engaged in security consultations with the Shah's government.[66] When the monarchical regime fell in February 1979, Iraq's first reactions were mildly welcoming to the new regime. Iraq's president, Ahmad Hasan al-Bakr, sent a congratulatory

64. Walt, *Revolution and War*, Chapter 5; R. K. Ramazani, *Revolutionary Iran: Challenge and Response in the Middle East*, (Baltimore: Johns Hopkins University Press, 1986), Chapter 4; W. Thom Workman, *Social Origins of the Iran-Iraq War*, Chapter 4; Majid Khadduri, *The Gulf War*, (New York: Oxford University Press, 1988); Chubin and Tripp, *Iran and Iraq*, Chapter 3; Stephen Pelletiere, *The Iran-Iraq War: Chaos in a Vacuum*, (New York: Praeger, 1992), Chapter 2; Karsh and Rautsi, *Saddam Hussein*, pp. 136, 142-148.

65. Official Iraqi sources insist that their decision to expel Khomeini in 1978 was not made at the behest of the Shah's regime, which asked Iraq to allow Khomeini to stay. Rather, the decision stemmed from an Iraqi fear that, if Khomeini continued his anti-Shah activities, it would give Iran a pretext to break the Algiers Agreement, to obstruct the transfer of territory still held by Iran to Iraq under the terms of that Agreement and to renew efforts to meddle in Iraqi domestic politics. This logic was set out in a series of articles on Iran-Iraq relations published in the *al-Thawra* newspaper in June 1979 (see specifically FBIS-MEA-79-116, June 14, 1979, p. E4-E5) and reiterated in Arab Ba'th Socialist Party-Iraq, *The Central Report of the Ninth Regional Congress, June 1982*, (Baghdad: nd, 1983), pp.172-173.

66. *New York Times*, June 18, 1978, p. 1.71.

message to Khomeini upon the establishment of the Islamic Republic. Iraq publicly supported Iran's withdrawal from CENTO and offered to support Iranian membership in the Non-Aligned Movement. As late as July 1979 Iraq extended an invitation to the Iranian prime minister, Mehdi Bazargan, to visit Baghdad.[67]

Relations soon deteriorated, however. In June 1979 Ayatollah Muhammad Baqir al-Sadr, the most politicized of the major Iraqi Shi'i religious leaders, was arrested on the eve of a scheduled trip to Tehran. Violent demonstrations ensued in the cities of Najaf and Karbala, and in Shi'i areas of Baghdad itself, which were suppressed with force. Al-Sadr was released shortly thereafter to house arrest in Najaf.[68] Several prominent Iranian ayatollahs, including Khomeini, condemned the Iraqi regime as "despotic" and "criminal," warning Iraq's rulers of "the wrath of God and the anger of the Muslim people."[69] Iranian media charged Iraq with instigating disturbances among Iranian Arabs in Khuzestan province.[70] The Iraqi media answered in kind, accusing the new Iranian government of following in the footsteps of the Shah in trying to exploit domestic divisions within Iraq.[71] Border clashes in the Kurdish areas ensued, with Iraq using its airforce against Iranian villages where Iraqi Kurdish insurgents were based.[72] In July 1979 Mas'ud and 'Idris Barazani, the sons of Iraqi Kurdish leader Mustafa Barazani, crossed the border into Iran and received support from the revolutionary government.[73]

In the midst of these events, Saddam Husayn became President of Iraq on July 16, 1979. An explanation that focused purely upon Saddam's ambitions would expect a militant change in Iraqi policy toward Iran from that time. The internal chaos in Iran at that time would also have encouraged Saddam, were he determined from the outset to go to war, to take advantage of those circumstances. However, that did not happen. On the contrary, the two governments sought in the short term to deescalate the tensions. Border skirmishes subsided. In early September 1979 Iranian Foreign Minister Ibrahim Yazdi even invited Saddam to visit Iran.[74] To some extent, this deescalation might be attributable to Saddam's need to consolidate his own power. He found it necessary, upon taking office, to execute 22 leading party and government figures for plotting, ostensibly with the Syrian branch of the Ba'th party, against him.[75] Tensions

67. Ramazani, *Revolutionary* Iran, pp. 57-59.

68. Tripp, *History of Iraq*, pp. 220-21; Chubin and Tripp, *Iran and Iraq at War*, p. 25.

69. David Menashri, *Iran: A Decade of War and Revolution*, (New York: Holmes and Meier, 1990), p. 101.

70. FBIS-MEA-79-107, June 1, 1979, p. E1.72.

71. See the series of articles in *al-Thawra*, FBIS-MEA-79-115, June 13, 1979, pp. E1-E3; FBIS-MEA-79-116, June 14, 1979, pp. E1-E6; FBIS-MEA-79-117, June 15, 1979, pp. E1-E5.

72. Anthony H. Cordesman and Abraham R. Wagner, *The Lessons of Modern War – Volume II: The Iran-Iraq War*, (Boulder: Westview, 1990), p. 27.

73. Dilip Hiro, *The Longest War: The Iran-Iraq Military Conflict*, (New York: Routledge, 1991), p. 35.

74. Mehashri, Iran, p. 102.

75. Khadduri, *The Gulf* War, pp. 72-78.

in the Iraqi Shi'i community remained high, with Ayatollah Muhammad Baqir al-Sadr still under house arrest.

But the deescalation can also be attributed to Saddam's view of the political situation in Iran. He clearly differentiated between the government of Iran, headed by prime minister Mehdi Bazargan, who tried to maintain correct relations with Iran's neighbors , and the revolutionary circles around Ayatollah Khomeini. In the long analysis of Iran-Iraq relations published in *al-Thawra* in June 1979, it was acknowledged that "there were many trends, wings and groupings in the Iranian Revolution...What caught our attention was the fact that a large percentage of the direct and indirect offenses were made by the group that is considered to belong to Khomeini."[76] The series of articles ended with the admonition that "there is still sufficient patience to give certain Iranian circles the opportunity to arrange matters and to favor the ideas of good and reason over forces, intrigues and malice."[77] Saddam himself averred to this analysis in a speech in early January 1980, when he criticized the attitude of "some of the Iranian leaders" toward Iraq.[78] He later reiterated the point to the Iraqi National Assembly in November 1980: "During the first few months after Khomeini came to power — during the first four months to be precise — I used to make all kinds of excuses to my colleagues to give the Persians more time. I used to remind my colleagues in the leadership of the state of confusion and chaos we were in in 1963."[79] A correct relationship with the Bazargan government was seen in Baghdad as a way to weaken the revolution's more radical tendencies.

This is not to say that Saddam was averse to exploiting the chaotic situation in Iran for Iraq's geopolitical benefit. In the fall of 1979 Iraq began to raise the issue of the three islands in the Gulf taken by the Shah from the UAE in 1971, to rally Gulf Arab support for Iraq by offering Iraqi protection to Gulf Arab regimes, and to highlight the situation of Arab residents of Iran.[80] In February 1980 Saddam issued his "Arab National Charter," seeking acknowledgement from other Arabs of Iraq's leadership in the Arab world. Iran continued to accuse Iraq of supporting Arab Iranian opposition groups in Khuzestan province. But there were as yet no concrete indications of Iraqi military ambitions against Iran.

The brief deescalation in tensions with Iran after Saddam's accession to the presidency did not lead to any lessening of political ferment among Iraq's Shi'i majority. Just the opposite. In July 1979, while under house arrest, Ayatollah Muhammad Baqir al-Sadr called for violent opposition to the regime. Shortly thereafter the major Iraqi Shi'i political groups announced the formation of the "Islamic Liberation Movement," ready to "resort to all means" to bring down the Ba'thist regime. In October 1979 the organization of the Iraqi 'ulama, which previously had been leery of overt political

76. FBIS-MEA-79-116, June 14, 1979, p. E2.

77. FBIS-MEA-79-117, June 15, 1979, p. E5.

78. FBIS-MEA-80-005, January 8, 1980, p. E4.79. FBIS-MEA-80-216, November 5, 1980, p. E3.

79. FBIS-MEA-80-216, November 5, 1980, p. E3.

80. Chubin and Tripp, *Iran and Iraq at War*, p. 28; *Washington Post*, November 1, 1979: A29; Saddam Husayn interview with *Al-Mustaqbal* magazine, FBIS-MEA-79-200, October 15, 1979, pp. E1, E2.

opposition, declared its support for the use of violence against the government. Al-Da'wa, the major Iraqi Shi'i party, formed a military wing by the end of the year.[81]

The regime, usually loath to admit to any domestic opposition (at least within the Arab population of Iraq), was by their standards quite open about the threat that Shi'i militancy posed. In May 1980, the Iraqi Interior Minister told an interviewer that, while there were fewer than 1,000 members of al-Da'wa, "the number of misguided supporters and religious sympathizers is considerable."[82] The report of the 1982 Ba'th Party regional congress was even more open. It admitted that the Iranian revolution had led to an upsurge of support for religious-political movements in Iraq, and that the intense social transformations that Iraq had undergone presented a large number of potential recruits to those movements.[83] It even acknowledged that the religious-political movement had infiltrated the junior ranks of the party itself: "The spread of fake religious practices among some Party members down to the level of Junior Supporters had created a state of relative entanglement between the base of the Party and those of religious-political parties...This had contributed to weakening the Party's alertness towards the growth of the hostile religious-political phenomenon."s[84]

In the midst of this rising tide of Shi'i opposition in late 1979, Iranian politics took a militant turn. Prime Minister Bazargan resigned in November 1979, in the wake of the take-over of the American embassy in Tehran. With the victory of the more radical forces around Khomeini, there was a marked turn in Iranian government rhetoric against the Ba'thist regime.[85] Statements about the need to export the Iranian revolutionary model around the region became more frequent, and by 1980 there were explicit calls by Iranian government officials for the Iraqi people to overthrow the Ba'th regime.[86] In March 1980, the Iranian ambassador left Baghdad and a number of Iranian consulates in the country were closed. Iran claimed the ambassador was recalled; Iraq contended that he was asked to leave "due to his blatant interference in Iraq's internal affairs."[87]

In late March 1980, the Iraqi government retroactively made membership in al-Da'wa a capital offense, after it had executed earlier that month 96 members of the organization. On April 1, a member of one of the Shi'i opposition groups attempted to kill Deputy Prime Minister Tariq 'Aziz while he was speaking at al-Mustansiriyya University in Baghdad.[88] A number of students attending the talk were killed, though 'Aziz himself escaped injury. During the funeral procession for some of those killed

81. Joyce R. Wiley, *The Islamic Movement of Iraqi Shi'as*, (Boulder: Lynne Rienner, 1992), pp. 54-55; Tripp, *History of Iraq*, p. 229.

82. FBIS-MEA-80-097, May 16, 1980, p. E2.

83. Arab Ba'th Socialist Party – Iraq, *Central Report*, pp. 272-77.

84. *Central Report,* p. 280.

85. Ramazani, *Revolutionary Iran*, p. 59, identifies the fall of Bazargan as the turning point in Iranian-Iraqi relations.

86. Khadduri, *The Gulf* War, p. 82; Walt, *Revolution and War*, pp. 238-39; Chubin and Tripp, *Iran and Iraq at War*, p. 34; Menashri, *Iran*, p. 157-58.

87. FBIS-MEA-80-048, March 10, 1980, p. E1.

88. Wiley, *Islamic Movement*, p. 55.

on April 6, according to the Iraqi media, a bomb was thrown from a window of an "Iranian school" in Baghdad as the procession went past.[89] In retaliation the Iraqi government executed Ayatollah Muhammad Baqir al-Sadr and his sister, who was also a Shi'i activist, publicized what it alleged were the organizational links between al-Da'wa and the Iranian regime, and began to expel tens of thousands of Iraqi Shi'a of Iranian origins from the country.

These events of March-April 1980 were the final straw for Saddam Husayn, and the spur for the Iraqi war decision. His rhetoric underwent an immediate change. Previously, he warned the Iranians not to interfere in Iraq, but he had never attacked the Iranian leadership personally nor had he overtly threatened war. From this point, he began to threaten Iran in the most obvious way. He went to al-Mustansiriyya University after the attack on Tariq 'Aziz and told the students: "It is not our tradition to make threats, but we tell you, by God, by God, by God and in the name of every particle of earth in Iraq that the pure blood that was shed in al-Mustansiriyah University will not go in vain...[W]e tell those cowards and dwarfs who try to avenge al-Qadisiyah [the battle in which the Muslim Arab army defeated the Persian Empire in 637 CE] that the spirit of al-Qadisiyah as well as the blood and honor of the people of al-Qadisiyah who carried the message on their spearheads are greater than their attempts."[90] This was the first use by Saddam of the imagery of the battle of al-Qadisiyya to clothe his conflict with the Iranians in the banners of Islam and Arab history.

Later in April, Saddam referred to the Iranian ruler as "that mummy Khomeini" and told his listeners "we hope that the Iranian people will find someone other than Bani-Sadr [then president of Iran] who will have better relations with the Arabs and the Muslims, and find someone else than this rotten man Khomeini." He said that "it is not a bad thing to avoid a clash with Iran, unless a clash is a patriotic and national duty. When a clash is a patriotic and national duty, we shall wage it in all its forms." He concluded this speech: "We are sure that you are prepared to shed your blessed blood in defense of the nation."[91] By late July 1980, Saddam was all but promising a war: "We are not the kind of people to bow to Khomeini . He wagered to bend us and we wagered to bend him. We will see who will bend the other . . . We are not a trigger happy people, but whoever attacks our outside fence from afar, we will knock at his front door a thousand times. Whoever violates our border, we will answer him in the manner he understands. And they know how we retaliate."[92]

When news of Ayatollah Baqir al-Sadr's execution reached Iran, in mid-April 1980, the Iranian reaction matched the hostility now being exhibited by Saddam. Ayatollah Khomeini reiterated his previous calls to the Iraqi people and the Iraqi army to overthrow the regime, accusing the Ba'th of launching a "war against Islam."[93] A clash at the Iranian town of Qasr-e Shirin, in the central border area, was reported in

89. FBIS-MEA-80-068, April 7, 1980, pp. E5-E7.
90. FBIS-MEA-80-066, April 3, 1980, p. E3.
91. FBIS-MEA-80-076, April 17, 1980, pp. E2-E5.
92. FBIS-MEA-80-144, July 24, 1980, pp. E4-E5.
93. Quoted in Hiro, *The Longest War*, p. 35.

mid-April,[94] and the Iraqi media began reporting regular border clashes in June 1980.[95] Iraq also increased more unconventional attacks on Iran. On April 29, gunmen attempted to assassinate Iranian Foreign Minister Sadeq Qotbzadeh, who was on a visit to Kuwait; on April 30, Iranian Arabs backed by Iraq seized the Iranian embassy in London, holding it for five days; in early June a bomb exploded at the Iranian embassy in Kuwait. Iraq had been in contact with Iranian political exiles since 1979, but in March 1980 two Iranian generals were received in Baghdad and allowed to use Iraq as a base to plan a military coup. Former Iranian Prime Minister Shahpur Bakhtiyar was permitted to set up a radio station in Iraq, which began broadcasting in May 1980.[96]

Sources that have reported on the timing of the Iraqi decision to go to war almost unanimously place the decision in the spring of 1980, after the events of April.[97] Fuad Matar, Saddam's official biographer, said that the Iraqis "had prepared for the eventuality of war some months before it happened" and identified early April as the turning point in Saddam's approach toward Iran.[98] According to R. K. Ramazani, Arab diplomats pointed to April 1980 as the time that Iraq began its preparations to go to war with Iran.[99] Edgar O'Ballance reports that Saddam ordered his Chief of Staff to start preparations for war in May 1980, though he provides no source for that information.[100] Anthony Cordesman and Abraham Wagner, based on interviews with those involved, state that Iraqi officials began to "consult" with colleagues in the Arab Gulf states in May 1980 about Iraq's intentions to attack Iran, and that Saddam informed several Arab Gulf state leaders about his plans in July and August.[101]

The gap between the war decision and the actual initiation of conflict in the beginning of September 1980, with the full-scale invasion on September 22, 1980, is attributable to two factors. The first is planning and organization, which would take some months to achieve. The second is the effort by Iranian exiles in Iraq to organize a military coup to overthrow the Islamic regime in Tehran. That effort, termed the

94. Cordesman and Wagner, *Lessons*, p. 29.

95. FBIS-MEA-80-109, June 4, 1980, p. E1; FBIS-MEA-80-118, June 17, 1980, p. E4; FBIS-MEA-80-123, June 24, 1980, p. E1.

96. Mark Gasiorowski, "The Nozeh Coup Attempt and Iranian Politics," unpublished paper, Louisiana State University, 1999, pp. 3-4. This is a section of Gasiorowski's forthcoming book on Iranian politics.

97. The one source that places the Iraqi war decision substantially earlier than the spring of 1980 is Mark Gasiorowski, Chapter 3 of his forthcoming book on Iranian politics. He reports that in October 1979 "a CIA officer gave a briefing in Tehran to members of the Bazargan government in which he presented satellite photographs and other evidence of Iraqi invasion preparations..." No Carter Administration source in their various books and memoirs reports on this early American intelligence conclusion. If Iraq were prepared as early as October 1979 to go to war, it is difficult to explain why Saddam waited a full year finally to initiate the invasion.

98. Fuad Matar, *Saddam Hussein: The Man, the Cause and the Future*, (London: Third World Centre for Research and Publlishing, 1981), pp. 13, 135.

99. Ramazani, *Revolutionary Iran*, p. 60.

100. Edgar O'Ballance, *The Gulf War*, (London: Brassey's Defence Publishers, 1988), p. 48.

101. Cordesman and Wagner, *Lessons*, pp. 38-39, notes 25 and 26.

"Nozeh coup" for the air force base from which it was launched, was fully supported by Iraq and planned on Iraqi territory. Begun on July 9, 1980, it was a spectacular and immediate failure, leading to a number of arrests and an extensive purge of the Iranian armed forces.[102] Coming after the fall of the Bazargan government in November 1979, the collapse of the Nozeh coup was the last chance that the Iraqis saw for eliminating the Khomeini regime from within. Had the coup succeeded, Saddam might still have seen some geopolitical benefit in a short and victorious war with a weak and chaotic Iran. We will never know. However, the failure of the coup reinforced to the Baghdad leadership the durability of the Islamic revolutionary regime on its border, a regime that was actively working to bring about the demise of Ba'thist rule in Iraq.

The Iraqi war decision of 1980 is best explained by the change in Saddam Husayn's framing of the issue of how to deal with Iran that came to a head in March-April 1980. Up to March-April 1980, in Saddam's brief period in office and before, when he was the dominant actor in the al-Bakr government, domestic unrest was not directly attributed to Iranian efforts to destabilize the regime. Thus the June 1979 riots, while souring the bilateral atmosphere, did not lead to armed conflict between Tehran and Baghdad. In fact, Saddam attempted to lessen tensions with Iran immediately after coming to power. However, with the changes in Iran after November 1979 and the more open calls for the export of the Islamic Revolution, domestic unrest in Iraq came to be seen as orchestrated by Tehran. Saddam's regime could only look forward to further efforts by Tehran to foment revolution against it, if nothing changed in Tehran. Facing that prospect, Saddam chose the risky path of war. He certainly thought he and Iraq would gain by victory, but the elements that made victory likely had been in place for some time. What had changed was his belief that a continuation of the status quo would only bring him more domestic problems.

OFFENSIVE REALISM AND THE 1980 CASE

There is no denying that Saddam Husayn saw both great benefits and comparatively high probabilities of victory in a 1980 attack on Iran. One event, relatively close to his decision to go to war (which I take to be in the spring of 1980), that increased the chances of victory in attacking Iran was the American hostage crisis that began in November 1979. With that event, any possibility that the United States would aid Iran in the event of an Iraqi attack disappeared, greatly reducing the risks of defeat for Iraq. The fact that Saddam assumed complete control of the Iraqi government so shortly before the decision to go to war further complicates the case against the offensive realist argument, as the personal ambition of Saddam is usually seen as a corollary to such an explanation of Iraqi war decisions.

However, there is nothing in the record to indicate that Iraqi leaders, including Saddam, seriously contemplated military action against Iran in the first two years of

102. For a complete account of the Nozeh coup attempt, based on interviews with those involved, see Gasiorowski, "The Nozeh Coup Attempt."

steady Iranian weakening — from the beginning of the revolutionary upheaval in the summer of 1977 through the November 1979 fall of the Bazargan government. During this period the Iranian government collapsed, the army fell apart through desertions in the ranks and purges of the officer corps, and the attentions of Iranian governments (both imperial and Islamic) were focused on serious internal security threats. If Iranian weakness were the spur to Iraq's decision to go to war, the war would have occurred sooner than September 1980.

Likewise, there is no reasonable reading of the international indicators that would lead one to the conclusion that Iraqi power *vis a vis* Iran had peaked in the spring of 1980. On the contrary, trends at the international level seemed to be moving Iraq's way. Iran had cut its final link to the United States, while Iraq still had excellent relations with both the Soviet Union and France. Oil prices were at all-time highs, bringing Iraq vast riches while the Iranian oil sector struggled to recover from the chaos of the revolution. Iraq's standing in the Arab world was on the rise. The Gulf states looked to it for protection against revolutionary Iran; with Egypt out of the Arab equation because of peace with Israel, Iraq's major rival for Arab leadership was neutralized. Time seemed to be on Iraq's side. It was Saddam's perception that, because of his *domestic* problems, he could not wait to deal with the Iran issue, that best explains why he chose to go to war when he did.

CONCLUSIONS

Regime security considerations provide a better explanation for the Iraqi decisions to go to war in 1990 and 1980 than any suggested alternative. Perceptions of threats to domestic regime security, directly linked to outside actors, pushed Saddam Husayn to take the risky course of war in both cases. He certainly saw potential gains in victory in each war, or he would not have undertaken them. Belief that victory was both highly probable and beneficial was a necessary condition of the war decisions, but not a sufficient condition. What drove him to take advantage of seemingly propitious international circumstances for war in each case was the belief that striking at foreign parties would end their efforts to destabilize him domestically. Saddam Husayn is certainly an ambitious, violent and aggressive man. But to understand his war decisions, we must focus first on his perceptions of regime insecurity.

[12]

Ballistic Missiles and Chemical Weapons

The Legacy of the Iran-Iraq War

Thomas L. McNaugher

Three phrases—chemical weapons, ballistic missiles, and the Iran-Iraq war—became inextricably linked over the course of that war, and especially in the wake of its last harrowing months. Between late February and mid-April, 1988, Iraq launched some 160 extended-range SCUD missiles (called the "Al-Husayn" by the Iraqis) toward Tehran alone, and scores of missiles toward cities like Isfahan and Qom. Notwithstanding the relatively small conventional warheads atop these missiles, Iraq's missile strikes caused panic and contributed to a striking decline in civilian morale.[1] Beginning in April 1988, in counterattacks against Iranian forces occupying Iraqi territory, Iraq's forces reportedly used chemical weapons successfully to create panic among Iranian soldiers. Chemical weapons, Robin Wright concluded, were "clearly . . . a major factor in Baghdad's stunning victories."[2] Both technologies have been credited with bringing Iran to the negotiating table after eight years in which Iranian intransigence, rather than confusion and panic, had been the norm.[3]

Blatant and apparently effective use of these technologies produced a sharp increase in concern for their proliferation and the effects their presence might have on future battlefields and strategic balances, especially those in the volatile Middle East.[4] Impressed by mass evacuations from Iranian cities

The author thanks Richard Betts, Jae Engelbrecht, Elisa Harris, Andrew Mack, Andrew Portocarrero, Uzi Rubin, Roger Smith, Jessica Stern, and Michael Sterner for their helpful comments on earlier drafts.

Thomas L. McNaugher is a Senior Fellow at the Brookings Institution. He is the author of Arms and Oil: U.S. Military Strategy and the Persian Gulf *(Brookings, 1985), and* New Weapons, Old Politics: America's Military Procurement Muddle *(Brookings, 1989).*

1. Then–Majlis Speaker (now President) Hashemi Rafsanjani himself referred to the mass departure of citizens from Tehran, Isfahan, and Qom due to Iraq's missile threat in one of his Friday sermons. *Foreign Broadcast Information Service*, FBIS-NES-88-074, April 18, 1988, p. 68. Elaine Sciolino referred to the "psychological terror caused by Iraqi bombings of Tehran"; Sciolino, "On the Iranian Home Front, a Time of Turmoil," *International Herald Tribune*, April 22, 1988, p. 2.
2. Robin Wright, *In the Name of God: The Khomeini Decade* (New York: Simon and Schuster, 1989), p. 184.
3. I use the term "technologies" to refer to ballistic missiles and chemical weapons collectively, because missiles are not weapons but delivery vehicles.
4. On ballistic missile proliferation see, for example, Aaron Karp, "Ballistic Missiles in the Third

during Iraq's "war of the cities," one military analyst has suggested that missiles are "highly effective when attacking large urban centers."[5] Another worries that leaders in the so-called "Third World" will be encouraged by Iraq's example to acquire chemical weapons as "the poor man's nuclear bomb."[6] And still another fears that putting the two technologies together to produce chemically-armed missile arsenals in the Middle East will be highly destabilizing.[7]

The last argument, of course, strays well beyond the realities of the Iran-Iraq war, in which chemical warheads were never mounted on ballistic missiles. But even those who focus more closely on the war itself seem to be overreacting to a surprising series of events that brought about Iran's virtual collapse in the first six months of 1988. There is little doubt that chemical weapons and especially ballistic missiles played a role in sharply lowering the morale of Iranian citizens and soldiers. Yet up until 1988 these technologies had been considerably less effective for either belligerent; indeed, in some cases their use was counterproductive. This suggests that the effectiveness of these technologies in 1988 rested largely on factors lying beyond their technical capabilities. Observers are misreading a series of dramatic events whose cause lay much more with Iran's internal problems and international pressure than with Iraq's weapons.

The Iranians themselves seem to be making the same mistake. President Hashemi Rafsanjani, for example, has called missiles "the most important and the most essential weapons of the world."[8] And Robin Wright's comment, noted above, about the decisiveness of chemical use in Iran's final battles was based partly on interviews with Iranian officials. Iranians may have been as surprised as everyone else at what happened to their country in 1988, and in response may be simplifying complex events that they themselves do not yet fully understand. They may also be dissembling. In signing

---

World," *International Security*, Vol. 9, No. 3 (Winter 1984/85); W. Seth Carus, "Chemical Weapons in the Middle East," *Policy Focus*, Washington Institute for Near East Policy, Research Memorandum No. 9, December 1988; Joseph S. Nye, Jr., "Arms Control After the Cold War," *Foreign Affairs*, Vol. 68, No. 5 (Winter 1989/90), pp. 54–62; and Martin S. Navias, "Ballistic Missile Proliferation in the Middle East," *Survival*, Vol. 31, No. 3 (May/June 1989), pp. 225–239.

5. W. Seth Carus, "Missiles in the Middle East: A New Threat to Stability," *Policy Focus*, Washington Institute for Near East Policy, Research Memorandum No. 6 (June 1988), p. 9.

6. Brad Roberts, "Chemical Weapons: A Policy Overview," *Issues in Science and Technology*, Vol. 2, No. 3 (Spring 1986), p. 109.

7. See, for example, Gerald Steinberg, "The Middle East in the Missile Age," *IDF Journal*, No. 19 (Winter 1990), p. 33.

8. *Foreign Broadcast Information Service*, FBIS-NES-88-186, September 26, 1988, p. 56.

UN Cease-fire Resolution 598 in August 1988, Iran's leaders took grave political risks, since in doing so they tacitly admitted that the decision to continue the war, taken over the summer of 1982, was a costly strategic error. What better way to avert domestic political turmoil than to blame defeat on Iraq's inhuman use of special weapons and on the world's willingness to condone it? The same logic could be applied at the tactical level. In the last months of the war, Iraqi ground forces violated Iranian territory virtually at will. Calling attention to Iraq's special weapons may have been a way to cover up the incompetence or lack of morale among Iran's soldiers. Finally, those among Iran's leadership who were anxious to bring the war to a close may have exaggerated the devastating effects of missiles and chemicals to strengthen their arguments.

Western analysts also seem to overlook such history as has been amassed on past use of ballistic missiles and chemical weapons in war. Using V-2 rockets, Germany ran its own "war of the cities" against London in 1944, but with much less dramatic results than Iraq achieved with its Al-Husayn missiles in 1988. Chemical weapons were used extensively during World War I, and in lesser quantities by the Italians in Ethiopia in the 1930s and by the Japanese in Manchuria during World War II. They have been used infrequently since, most notably by Egypt in North Yemen in the mid-1960s. They have been tactically useful, but they have produced the kind of panic attributed to Iranian troops in 1988 only against ill-equipped and untrained soldiers.

Although we have no empirical grounds for assessing the stability or instability of a regional balance involving chemical warheads atop ballistic missiles, it is not clear that a world in which adversaries deploy such weapons will suffer an increase in instability. Part of the problem in arguments that stress growing instability stems from the tendency to talk of chemical weapons as if they were little different from nuclear weapons, the "poor man's nuclear bomb." In fact, chemical weapons differ fundamentally from nuclear weapons, in ways that favor stability in regional balances. Another part of the problem stems from the application of a nuclear strategic logic, developed by the superpowers to describe their own situation, to regional balances where distances between adversaries are small and populations are both small and concentrated. The differences markedly alter incentives to preempt or strike first, which suggests that regional balances involving chemical weapons on missiles may be stable even in the absence of the kind of "secure second strike capabilities" the superpowers developed.

These arguments are not meant to dismiss concern for the proliferation of chemical and missile technologies. No one should welcome the introduction of these technologies on a wider scale. Still, overstating their effectiveness risks encouraging states to acquire them. It also risks diluting the focus of arms control policy by placing a host of new technologies on the list of those that should be subject to export controls. The analysis that follows leads me to conclude that missiles are far less important than the warheads they carry, and that nuclear warheads remain more worrisome than chemical warheads.

Missiles, Chemicals and the End of the Iran-Iraq War

In July 1988 the Ayatollah Khomeini, a man who for six years had sought nothing more ardently than total victory over Iraq, agreed to sign United Nations Cease-fire Resolution 598. The announcement surprised nearly everyone, although it had been a rough spring for Iran. Iraq's missiles had rained down on Tehran, Iraqi soldiers had recaptured strategic territory, and U.S. naval vessels had sniped at Iranian speedboats and frigates. In a major tragedy early in July, the USS *Vincennes* accidentally downed an Iranian commercial airliner. But Iran had been down before; its resilience went almost unquestioned.[9] Because Iraq's missiles and chemical weapons inspired panic in Tehran and on battlefields during this final spasm of the war, these technologies have figured prominently in explanations of Iran's sudden change of heart. But a broader view of the role these technologies played in the war suggests that future historians will be less impressed with their effectiveness, and more impressed with the social and international context within which they were used.

THE WAR OF THE CITIES AS STRATEGIC BOMBING

Given fairly clear evidence that Iraq's 1988 missile strikes caused panic in Tehran, the "war of the cities" would seem to support arguments offered by proponents of strategic bombing that strikes on civilians will help to break the enemy's will and induce an end to the war. One informed observer of

9. Press reports from Baghdad after Iraq had recaptured al-Fao, for example, suggested that "few observers in Baghdad believe a peaceful solution is any closer than before." Quoted in Jim Muir, "Catching Them Napping," *Middle East International*, April 30, 1988, p. 5. After the Airbus tragedy, U.S. officials, while recognizing that Iran faced "the absence of any good options," nonetheless expected an increase in Iranian terrorist attacks, perhaps on U.S. military personnel stationed in the Gulf region. *Middle East Policy Survey*, No. 204 (July 8, 1988), pp. 1–2.

the Iran-Iraq war has thus concluded that ballistic missiles are "cost-effective . . . (*vis-à-vis* the manned bomber) if the relative cost per missile was measured against its political/psychological impact."[10] Yet the leap from the specific to the general case ignores contrary evidence from earlier years in the Iran-Iraq war, as well as from the broader history of strategic bombing. And it downplays the importance of context in establishing the effectiveness of strategic bombing.

The war of the cities had started long before 1988, yet until 1988 it had remarkably little effect on the war goals of either belligerent. Earlier strikes on cities often involved bombing raids and artillery barrages, but Iraq also used ballistic missiles as early as 1982. Press reports suggest, for example, that Iraq's SCUD and FROG missiles did at least as much damage to Dezful between 1982 and 1983 as they did to Tehran in 1988, yet few besides the Iranians noticed.[11] Starting in spring of 1985, 39 Iraqi SCUDs fell on Isfahan, 14 of them direct hits to the center of the city.[12] By Spring 1985 Iran had acquired its own SCUDs from Syria and Libya, and in 1985 and 1986 Iran reportedly fired 22 of these at Iraq, three at Baghdad.[13]

Whether the work of missiles or bombers, these strikes do seem to have affected civilian morale initially; Iraqi bomber strikes on Tehran in 1985, for example, reportedly produced a certain amount of war-weariness.[14] And Iranian missile strikes on Iraqi cities cut to the heart of Baghdad's strategy of insulating its people from the battlefield, suggesting that before 1988, Iraq may have been more disturbed by missile strikes than was Iran.[15] In the end, however, neither these missile strikes, nor Iraq's frequent bombing raids, nor Iran's continuous shelling of Basra, brought an end to the war.

---

10. Shahram Chubin, "Iran and the Lessons of the War with Iraq: Implications for Future Defense Policies," to appear in Geoffrey Kemp and Shelley Stahl, eds., *The Arms Race in the Middle East and South Asia* (New York: St. Martin's, forthcoming).
11. Vahe Petrosian, "Iran claims new missile in its armoury," *Middle East Economic Digest*, April 2, 1988, p. 7.
12. James A. Bill, "Why Tehran Finally Wants a Gulf Peace," *Washington Post*, August 28, 1988, p. B1.
13. "Iran's growing missile forces," *Jane's Defence Weekly*, July 23, 1988, p. 130.
14. See "The bombs knock holes in morale," *The Economist*, June 22, 1985, p. 35. Replete with references to "a curious mixture of despondency and panic," Iranians fleeing the city at night, and demonstrations against the war, this article could have been written in 1988. That bombers rather than missiles produced such consequences raises questions about the special characteristics of missiles used against civilians.
15. This point was alluded to by Efraim Karsh when he noted that Iranian counterstrikes on Baghdad "did expose the fragility of Iraqi national morale by frustrating the regime's attempts to insulate its rear from the war." See Karsh, "Military Lessons of the Iran-Iraq War," *Orbis*, Vol. 33, No. 2 (Spring 1989), p. 218.

Indeed, military analysts writing before 1988 tended to dismiss the ongoing war of the cities rather quickly. Writing before the 1988 counter-city exchanges had begun, David Segal concluded that "from a military point of view, the value of the attacks on Iranian population centers, the 'war of the cities,' has been virtually nil."[16] Anthony Cordesman agreed: "Neither side's strikes at population centers," he wrote in 1986, "had much political or economic impact once their initial shock wore off."[17] One observer, writing in 1985, even suggested that "far from undermining Iranian resolve, the attacks on civilians actually boosted morale."[18]

All of this clearly changed in 1988. But the change was notable only in Iran. While the world's attention focused on Iraq's missile strikes on Tehran, Iran fired some 70 SCUDs at Iraq, most at Baghdad.[19] Yet as one reporter noticed at the time, "whereas civilian life in Tehran is reported to have been disrupted by the [Iraqi missile] attack, life in the centre of Baghdad appears to be proceeding much as normal."[20] Somehow Iran's SCUDs lacked the psychological effectiveness of Iraq's missiles. The difference may have been due partly to the smaller number of real hits in Baghdad; about half of Iran's SCUDs missed the city, the result either of aiming errors or of an Iranian decision to avoid civilian areas, as "counterproductive."[21] But against this must be placed the fact that Iran's SCUDs carried the full 800-kilogram warhead, while Iraq's missiles carried a substantially lighter warhead in order to increase the missile's range.

If the effectiveness of Iraq's spring 1988 missile campaign seems out of proportion to the effects of earlier Iraqi missile and bomber campaigns, it

16. David Segal, "The Iran-Iraq War: A Military Assessment," *Foreign Affairs*, Vol. 66, No. 5 (Summer 1988), p. 658.
17. Anthony H. Cordesman, *The Iran-Iraq War and Western Security, 1984–1987: Strategic Implications and Policy Options* (London: Jane's, 1987), p. 75.
18. Edmund O'Sullivan, "The Gulf war: no end in sight," *Middle East Economic Digest*, September 14, 1985, p. 15.
19. Although Iran's supply of SCUD missiles was initially quite limited, Iran was able to purchase roughly 100 SCUDs from North Korea in 1987. These were delivered early in 1988. See "Iran's growing missile forces," *Jane's Defence Weekly*, July 23, 1988, p. 130.
20. Alan Rusbridger, "Iraq walkover may be due to chemical war," *Guardian* (London), April 23, 1988, p. 20. Rusbridger, reporting from Baghdad, admitted that part of the explanation may have been that "the first 20 or so Iranian missiles landed near power installations and refineries on the southern outskirts of Baghdad. . . . More recently, though, they have hit an elementary school and, in the past fortnight, a crowded funeral party near the centre of the city."
21. Petrossian, "Iran claims new missile in its armoury," p. 7. See also "Iran's growing missile forces," *Jane's Defence Weekly*, July 23, 1988, pp. 130–131, which holds that 61 of a total of 77 Iranian SCUDs were fired at Baghdad, while the remainder were fired at Mosul, Kirkuk, Takrit, and Kuwait.

was wildly disproportionate to the effects of earlier strategic bombing campaigns. The 1988 war of the cities ranks as one of the smallest strategic bombing campaigns in history. Assuming that each of the 160 modified SCUDs that landed in Tehran carried a warhead of roughly 190 kilograms (418 pounds, or roughly a quarter-ton), then over seven weeks Tehran was treated to about forty tons of high explosives.[22] By comparison, during the Christmas bombing of North Vietnam in December 1972 (Linebacker II), the U.S. Air Force dropped 20,000 tons of bombs on the Hanoi-Haiphong area in just eleven days. Between January and August 1945, the U.S. Twenty-first Air Force dropped 153,000 tons of ordnance on Japan in a systematic effort to destroy Japan's cities. In 1944 alone, British and American aircraft dropped nearly 1.2 million tons of bombs on Germany, albeit not all of it on cities. And in a last desperate gambit near the end of the war, Germany fired over one thousand V-2 missiles, each armed with a one-ton warhead, at London.[23] Not surprisingly, damage to Tehran in 1988 was minimal in comparison to earlier cases in which entire cities were razed. And the civilian death toll of 2000, while deplorable, hardly compares to the 185,000 casualties produced by the U.S. firebombing of Tokyo in March 1945, to name only the worst of many possible examples.[24]

These comparisons make the psychological effects of Iraq's 1988 campaign all the more puzzling, since earlier bombing campaigns have tended to highlight the surprising extent to which civilians can absorb pain and suffering *without* breaking. The U.S. Air Force virtually flattened Pyongyang during the Korean War, with no discernible effect on the willingness of North Korea

---

22. The most authoritative source on Iraq's modification of the SCUD to produce the Al-Husayn missile is W. Seth Carus and Joseph S. Bermudez, Jr., "Iraq's *Al-Husayn* Missile Programme," *Jane's Soviet Intelligence Review*, Vol. 2, No. 5 (May 1990), p. 205. See also Robert D. Shuey, et al., *Missile Proliferation: Survey of Emerging Missile Forces*, Congressional Research Service (CRS), Report for Congress No. 88-642-F, October 3, 1988, p. CRS-7. Although it is now generally agreed that Iraq reduced the weight of the missile's warhead to help increase its range (as opposed to using strap-on boosters, for example), there remains some conjecture on the warhead's final weight. The comparisons in this paragraph hold, however, even if the standard SCUD warhead weight of 800 kg (1,760 pounds) is used.

23. Tonnages in this paragraph are taken from: Mark Clodfelter, *The Limits of Airpower: The American Bombing of North Vietnam* (New York: Free Press, 1989), p. 194; R.J. Overy, *The Air War, 1939–1945* (New York: Stein and Day, 1980), p. 100 (Japan), p. 120 (Germany); and Winston S. Churchill, *Triumph and Tragedy*, Vol. 6 of *The Second World War* (Boston: Houghton Mifflin, 1953), p. 52.

24. On casualties in Tehran, see *Middle East Economic Digest*, May 6, 1988, p. 1. On casualties in Tokyo, see United States Strategic Bombing Survey, *Summary Report (Pacific War)* (Washington, D.C.: U.S. Government Office [U.S. GPO], 1946), p. 20.

or China to negotiate an end to the war.[25] The strategic bombing of Germany during World War II began to work only when the allies stopped bombing cities and began to target Germany's oil infrastructure, after which the German war machine literally ran out of gas. An exception might be Japan, in which U.S. strategic bombing helped bring the Japanese to the brink of surrender even before atomic bombs were dropped on Hiroshima and Nagasaki. Even in this case, however, the evidence is mixed, since bombing only compounded the decline in morale resulting from the long-standing U.S. blockade.[26]

To be sure, the 1988 war of the cities was conducted with missiles rather than bombers, and the difference is not trivial. In a purely technical sense, missiles armed with conventional warheads should be less effective than bombers, since missiles can only be used once, and generally carry less explosive tonnage than even a small bomber. From a psychological perspective, however, the record of Germany's V-2 strikes on London in 1944 suggests that the edge goes to missiles. Winston Churchill wrote that Germany's missile war:

> imposed upon the people of London a burden perhaps even heavier than the air-raids of 1940 and 1941. Suspense and strain were more prolonged. Dawn brought no relief, and cloud no comfort. The man going home in the evening never knew what he would find. . . . The blind impersonal nature of the missile made the individual on the ground feel helpless. There was little that he could do, no human enemy that he could see shot down.[27]

In short, missiles had a more powerful psychological effect than bombers because they struck without warning at any time of day or night, making it impossible for London's citizens to scramble for shelter as they had learned to do during Germany's earlier bombing campaign against them.[28] These effects were exacerbated by the fact that no terminal defense was possible

25. Max Hastings, *The Korean War* (London: Pan Books, 1988), pp. 326–327.
26. See Overy, *The Air War, 1939–1945*, pp. 98–101.
27. Churchill, *Triumph and Tragedy*, p. 39. Churchill was referring to attacks by the air-breathing, slow-flying V-1 as well as by the ballistic V-2; by changing its air defenses Britain ultimately was able to shoot down 6 out of 7 V-1s as they approached London. See ibid., p. 48.
28. Because they travel faster than sound, ballistic missiles cannot be heard until after they have struck. Not giving any audible warning, ballistic V-2s produced 4.8 casualties per round landing in Greater London, against 2.2 for the relatively noisy air-breathing V-1s that survived Britain's defenses. See Frederick I. Ordway III and Mitchell R. Sharpe, *The Rocket Team* (New York: Thomas Y. Crowell, 1979), p. 244.

against the ballistic V-2; Britain's effort to harass German launch sites proved to be both expensive and unsuccessful until allied armies captured the sites.[29]

Although these data support the argument that missiles have a special "terrorizing" quality in comparison to strategic bombers, it is worth remembering that Germany's missiles failed to produce the decisive end to the war that Hitler expected of them. The same holds for the missiles of Iran and Iraq—until 1988, that is, when Iraqi missiles suddenly helped bring Iran to the edge of collapse. This suggests that other factors, unique to the circumstances in Iran in 1988, must have been at work.

Churchill may have put his finger on the central issue when he noted that the V-2s produced few complaints in London: "Everyone saw we just had to lump it, an ordeal made easier by our hopes of a successful advance in Normandy."[30] Iran's SCUDs had virtually no effect on life in Baghdad because Iraq in 1988, like Great Britain in 1944, smelled victory. Iraq had defeated the most serious of Iran's offensives early in 1987. Iraqi officials were keenly aware of growing morale problems in Iran's armed forces as well as among its citizenry. And they knew themselves to be the beneficiaries of growing international support—so much so that the UN was willing to overlook excesses like Iraqi use of chemical weapons and attacks on population centers.

Conversely, Iraq's 1988 missile campaign came at a time when few in Iran were willing to "lump it" in hopes that battlefield advances would soon bring an end to the war, or at least to Iraq's missile strikes. Iran's military fortunes had peaked with the capture of al-Fao in February 1986. The Karbala series of offensives that began late in 1986 and continued through March 1987 had produced nothing but Iranian casualties, and in their wake even high-level officials had begun to question Iran's war goals. War weariness was evident even among Iran's Revolutionary Guards, several hundred of whom joined Iranian civilians in demonstrating against the war in the streets of Tehran just after the Karbala defeat.[31] Meanwhile, Iran was finding it increasingly difficult to deal with the U.S. naval vessels that had begun protecting reflagged Kuwaiti tankers in July 1987. For Iranians, the U.S. Navy was making

29. Postwar British studies suggested that it cost Britain more in terms of lost war production and attacking missile launch sites than it cost Germany to develop and operate its missiles. Ibid., pp. 245–247.
30. Churchill, *Triumph and Tragedy*, p. 43.
31. Wright, *In the Name of God*, p. 157.

the Gulf safe for Iraqi strikes on Iranian tankers, and thus had clearly entered the war on Iraq's side.

Iran also was gripped by a growing economic and political crisis. Iran's economy was probably not much worse in 1988 than it was the year before, while the worst year, in terms of oil income, was 1986, when the price of oil fell briefly below $10 a barrel. But since Iraq's strikes on Iranian tankers and oil infrastructure placed an artificially low cap on Iran's oil exports, the Iranian economy was not getting any better, nor was it likely to. Iraq's economy, on the other hand, was very definitely getting better. The steady growth in Iraq's oil pipeline network slowly circumvented Iran's blockade of Iraq's coast. Early in 1988, Iraqi oil export volume passed that of Iran; the bromide that "time favored Iran" in its war with Iraq rang less and less true, a point not lost on citizens in Tehran.[32]

It was also becoming increasingly clear to Iranians that Tehran's ability to deal with its economy was hamstrung by political debate between those seeking state control of the economy (notably Prime Minister Mir Hussein Musavi) and conservative clerics on the Council of Guardians opposed to such moves. Early in 1988 Khomeini moved to resolve this dispute by announcing, in effect, that the state's rights superseded those of the church, but the decision was so controversial that little was done to implement it.[33] For the public, this embarrassing series of events may have increased a sense that incompetence was at the helm. For the government, it may have signaled that the country's economic problems were beginning to threaten the revolution.

But for many Iranians, the most telling evidence of incompetence in Tehran was the government's failure to accept a cease-fire resolution that was both near at hand and unobjectionable. UN Resolution 598, which had been on the table since July 1987, posed no requirement for Iran to cede territory of

32. Former Prime Minister Mehdi Bazarghan summarized these points in an open letter to the Ayatollah Khomeini distributed early in June 1988: "You have denounced the policy of the United States, and they are now installed solidly at our gates in the Persian Gulf. You have spoken of the failure of Iraq and the crumbling of its regime, but thanks to your misguided policies, Iraq has fortified itself, its economy has not collapsed, and it is we who are on the verge of bankruptcy." The letter went on to accuse Khomeini of "a despotism worthy of the Pharaohs." Bazarghan was a frequent critic of Iran's regime, but the wide distribution given this letter and its apparent reception among citizens in Iran suggests that in this case his message coincided with growing public sentiment. See Youssef M. Ibrahim, "Ex-Premier Attacks Khomeini as Despot," *New York Times*, June 3, 1988, p. A3.
33. See Vahe Petrossian, "Khomeini opens the way for economic reform," *Middle East Economic Digest*, January 16, 1988, p. 10.

its own to Iraq, to pay reparations, or to suffer occupation. Quite the opposite: it mentioned the need for aid to Iran, and called for the formation of a commission to investigate the origins of the war, items high on Iran's agenda. One of the reasons often given to explain German and Japanese resistance to strategic bombing campaigns in World War II was the fear in both countries that pillage and plunder were almost certain to follow allied victory. In sharp contrast, international diplomacy in the Iran-Iraq case relieved Iranians of any good reason to endure the hardships of war, brought home to Tehran by Iraq's missiles, however small their numbers.[34]

Given these circumstances, it is arguable that Iraqi bombers would have produced much the same result as Iraqi missiles in the spring of 1988, and conversely that neither bombers nor missiles aimed at Tehran would have produced the same result much before 1988. It is impossible to assert that missiles in the Iran-Iraq war before 1988 had special psychological or terrorizing effects, let alone to use the experience of that war to bolster a general argument for the special effectiveness of missiles. The Iran-Iraq war simply does not produce a valid measure of missile effectiveness. But the strong inference is that far more important than the missiles in producing psychological effects were social and political circumstances in Iran, which changed dramatically between 1986 and 1988. Iraq's missiles did not so much destroy Iranian morale as hasten a deterioration in morale that was the work of other factors and that was already well underway when the 1988 missile campaign began. If there is a general case to be made, it is that missiles are an effective "coup de grace" weapon, decisive only when target states are already near collapse.

THE EFFECTIVENESS OF CHEMICAL WEAPONS

Chemical weapons (CW) can have two effects: physical (they can kill and injure), and psychological (they can scare and produce panic). They can have these effects at two levels. CW used at the tactical level can influence the shape and outcome of battle. Used strategically against the enemy's cities, CW, like ballistic missiles, can produce physical and psychological effects that may undermine morale and produce a decision to end the conflict.

34. Advocates of strategic bombing rarely notice the usefulness of "escape routes" in helping to produce the collapse in popular will that bombing is presumed to cause. From this perspective, the war of the cities in 1988 has much in common with the Christmas bombing of Hanoi and Haiphong, which helped persuade the North Vietnamese to sign a peace agreement they had almost signed two months before. See Clodfelter, *The Limits of Airpower*, pp. 177–202.

Although the evidence on CW use in the Iran-Iraq war is a good deal murkier than that available on the use of ballistic missiles, the war experience with chemical weapons sheds some light on all aspects of their possible effects, but especially on their effects at the tactical level.

What follows concentrates on Iraqi use of CW because Iraq was far and away the major CW user during the war. There is general agreement that Iran also used CW, but when, where, and how much remains uncertain.[35] There is no question, however, that Iran introduced CW much later than Iraq, used it far more modestly, and (despite news reports to the contrary) had only a fledgling CW production capability even at the time the cease-fire was signed.[36] What follows relies on news articles and other secondary sources, UN reports, and interviews with U.S. government officials. These leave many questions unanswered, but overall tend to confirm a pattern similar to that found in the case of ballistic missiles: chemical weapons were helpful though not decisive through most of the war, but seem to have acquired startling effectiveness in Iraq's 1988 campaigns, due less to their particular characteristics than to changes in the quality and morale of Iranian forces.

35. David Segal reports that Iran started using mustard gas and phosgene in 1987, for example, while Seth Carus asserts that the first documented use of CW by Iran did not occur until 1988. Segal, "The Iran-Iraq War," p. 956; Carus, "Chemical Weapons in the Middle East," p. 4. A United Nations investigation team that visited both belligerents in mid-1988 remained skeptical about the evidence Iraq presented to confirm its claim about Iran's use of CW. See Patrick E. Tyler, "U.N. Cites Iraqi Use of Toxic Gas," *Washington Post*, August 2, 1988, pp. A1, A17.

36. Iranian use of CW presents a real puzzle as to motivations and decisionmaking. Iranians claimed to have a CW capability late in the war. Why wasn't it used as their front collapsed in 1988? There were suitable military targets for Iranian CW, especially late in the war—for example, in June 1988, when Iraq massed its tanks for the final assault on Majnoun Island (occupied by Iran in 1983). But if, as some Iranians suggested, Khomeini himself had serious moral qualms about using chemical weapons and preferred to retain the "moral high ground" on this issue, why did Iran use CW at all? In the end Iran seems to have chosen the worst of both worlds: its desultory and indecisive use of chemicals late in the war harmed its own moral position more than it harmed Iraq's military forces. Explanations for this puzzle may lie in factional infighting in Tehran and among Iran's military commanders, driven by differing assessments of how CW would affect the battlefield and Iran's deteriorating international position. But the available data on Iranian use may simply be incorrect or so incomplete as to be misleading. Especially problematic in this regard is the claim that Iran, as well as Iraq, used CW against Kurdish civilians at Halabjah in April 1988 (see note 52 below). On Iran's CW capability, see the widely noted piece by Jack Anderson and Dale Van Atta, "Iran May Turn Chemical Tables on Iraq," *Washington Post*, October 2, 1985, p. F11, which claimed that Iran's CW capability was ready in 1985. Iranians themselves did not make such a claim until late in 1987: see "Iran Declares Chemical Weapons in Production, Missiles Deployed," *Washington Times*, December 28, 1987, p. 10. For reports on Khomeini's moral qualms, see "Chemical war fear hangs over the Middle East," *Independent*, March 23, 1988, p. 12; and Simon O'Dwyer-Russell, "Iran ready to unleash gas revenge on Iraqis," *Sunday Telegraph* (London), March 27, 1988, p. 7.

IRAQ'S TACTICAL USE OF CHEMICAL WEAPONS. Iraq's initial tactical experiments with chemical weapons in the Iran-Iraq war were with nonlethal riot control agents like tear gas. These were introduced into battle in northern Iraq in 1982, with disastrous results; a change in wind direction blew the gas cloud back toward Iraq's own troops.[37] In 1983 Iraq tried again, avoiding disaster but again relying primarily on riot control agents rather than lethal agents. The first major use of lethal agents confirmed by UN investigators was in 1984 on the southern front, in which Iraq used both mustard gas and a nerve agent against Iran's human-wave assaults. Given that the 1984 "offensive to end all offensives" was an especially effective Iranian attack, chemical weapons probably helped shore up Iraq's as yet shaky defensive lines. This should not be surprising. Iran's troops were unequipped with masks or the protective clothing necessary against mustard gas, which burns exposed skin. Morale was high; at this point in the war religious fervor gave Iranian youths the courage to walk through minefields, and no doubt made them fearless of gas clouds as well. Zealotry could not keep them alive, however, so Iraqi chemical weapons no doubt took their toll in Iranian casualties.[38]

David Segal is probably correct in attributing Iraq's initial decision to use CW to a ruthless logic of cost-effectiveness: "[The Iraqis] had to choose between the possible adverse effects of offending world opinion and the certain effects of being overrun by Iranian soldiers. Under the circumstances, the choice was easy."[39] He is on less firm ground when he asserts that "since 1984, the Iraqis have been able to stop Iran's offensives without blatant violations of the 1925 Geneva Protocols."[40] It is true that, with the exception of the surprise capture of the al-Fao peninsula in February 1986, Iran's post-1984 offensives tended to be unimaginative, and that Iraq's defensive capabilities, its barrier system, and it ability to move forces laterally behind its lines all improved measurably between 1984 and 1988. Yet while these factors reduced Iraq's need to use chemical weapons, there is little doubt that Iraq did, in fact, continue to use CW after 1984.[41]

37. See David B. Ottaway, "In Mideast, Warfare With A New Nature," *Washington Post*, April 5, 1988, p. A14.
38. There is considerable evidence, in the form of unexploded chemical shells on the Iranian side of the battlefield, that Iraq was far from mastering the technology. By presenting the Iraqis with easily spotted and massed troop formations, Iran's human wave tactics no doubt helped compensate for Iraq's lack of technical mastery.
39. Segal, "The Iran-Iraq War," p. 956.
40. Ibid.
41. See, for example, Wright, *In the Name of God*, pp. 132–134. UN investigation teams visiting the front during this period uniformly confirmed the use of chemical weapons.

The same ambiguity applies to Iraq's use of CW in its spring 1988 offensives. There is general agreement that Iraq used chemical weapons in these offensives, which succeeded in recapturing Iranian-held territory at al-Fao and Majnoun in the south, plus border mountaintops in the north. A UN team visiting Majnoun and Kurdistan in June of that year concluded that "heavy use of chemical weapons by Iraq played a role in two stunning military successes over Iran earlier this year." The June 25 attack on Majnoun, they concluded, was "preceded by a two-hour artillery bombardment with shells filled with cyanide and nerve gases," while in a three-day assault in spring 1988 on Iranian positions around Halabjah, Iraqi aircraft and artillery both used chemical munitions.[42] Robin Wright suggests that Iranians panicked at the sight of gas clouds, citing a "popular, but unconfirmed, story told by foreign military attaches in Tehran [which] claimed that Iraq used only smoke bombs in one attack; the psychological impact of chemical weapons was such that the Iranians had fled at the mere sight of the clouds."[43] Interviews with U.S. government officials suggest, on the other hand, that these reports were overstated; Iraq possessed the capability to use CW in these battles, but found extensive use unnecessary.

In fact, a strong argument can be made that chemical weapons, even if used, were not essential to Iraq's success in these final battles. Military analysts surveying Iraq's performance in recapturing al-Fao in April reportedly found "no reason why [chemical weapons] should have been needed in what they described as a very well-planned and well-executed operation."[44] Iraq in this case achieved surprise against a depleted Iranian defending force of perhaps less than 10,000 in strength. Rather than cut these defenders off from the single bridge still standing over the Sha'at al-Arab, the Iraqis sensibly drove them toward it.[45] In the Majnoun campaign, chemical weapons probably added marginally to the impact of Iraq's massive tank force in the attack. In this case Iraq gave up the chance for surprise in favor of sheer mass.

Evidence from the middle years of the Iran-Iraq war accords with the experience of belligerents in World War I. In both cases, CW was tricky to

42. Cited in Tyler, "U.N. Cites Iraqi Use of Toxic Gas," pp. A1, A17.
43. *In the Name of God*, p. 184.
44. Quoted in Jim Muir, "Catching them napping," *Middle East International*, April 30, 1988, p. 4.
45. See ibid., and Andrew Gowers, "Iraqi victory at Fao may have been won by stealth," *Financial Times*, April 23, 1988, p. 2.

use and somewhat unpredictable. Yet in both cases chemical weapons were sometimes effective, especially (and not surprisingly) against unprotected troops. Indeed, on average chemical weapons used in World War I produced from 1.6 to 5 times as many immediate casualties as a similar amount of high explosive (HE), prompting some analysts in the postwar era to extol CW's usefulness. Averages were misleading, however, in that they tended to be driven up by high casualty rates incurred when gas was used against unprotected troops. This was especially the case in 1915, when gas was first introduced, and again in 1917, both because mustard gas, introduced that year, required protective clothing as well as masks, and because newly-arrived U.S. forces lacked the equipment and training for gas warfare, despite the fact that CW had been in use for three years.[46] Properly trained and protected troops took much lower casualties, save in the case of mustard gas, which lingered often imperceptibly in the air longer than gas masks of the era could be comfortably worn.[47] In both wars CW tended to produce non-lethal casualties rather than deaths, at least in comparison to other forms of firepower.[48]

---

46. Although extensive German use of mustard gas after July 1917 produced high casualties among British troops, the British took their heaviest gas casualties in the battles of 1915, when protection was not available. The most costly gas attack in the war was Germany's early use of phosgene mixed with chlorine, and reflected the inadequacy of early British helmet-masks. See Dorothy Kneeland Clark, *Effectiveness of Chemical Weapons in WWI*, ORO-SP-88 (Bethesda, Md.: Johns Hopkins University, Operations Research Office, November 1959), pp. 99–100, 109, 140. As for the experience of American troops, Charles E. Heller has argued that "the real inhumanity of chemical warfare in World War I lay in the blindness of U.S. civilian and military leaders who . . . ignored the real and present threat posed by gas," allowed extraction of "a high toll at the front." Major Charles E. Heller, *Chemical Warfare in World War I: The American Experience, 1917–1918*, Leavenworth Paper No. 10 (Fort Leavenworth, Kan.: U.S. Army Command and General Staff College, Combat Studies Institute, September 1984), p. 94. For statistics on U.S. casualties see Clark, *Effectiveness of Chemical Weapons*, pp. 1, 109, 137–139.

47. Britain's widely-used "SBS" mask, for example, resisted breathing so much that it could be worn for no more than about eight hours, at which point soldiers had to be evacuated. Clark, *Effectiveness of Chemical Weapons*, p. 123. But General Robert Bullard, commander of the U.S. First Infantry Division, said of the British mask, initially issued to U.S. troops, that he could not wear the mask longer than 3 minutes without feeling smothered. See Heller, *Chemical Warfare in World War I*, p. 65.

48. Clark notes that the rate of death per casualty was 2 to 4 times higher for fire wounds than for CW wounds. *Effectiveness of Chemical Weapons*, p. 1. Although no data are available on relative casualties in the Iran-Iraq war, Iran reported only 262 deaths through April 8, 1987, out of a total of 27,000 Iranians affected by Iraqi chemical weapons of all types. See Carus, "Chemical Weapons in the Middle East," p. 7. Noting the U.S. Army Surgeon General's conclusion that in World War I "the man wounded by gas had about twelve times the chance to live, in comparisons with his fellow soldier suffering from the effects of the traditional weapon," one writer asserted that those who "have opposed gas weapons on the basis of inhumanity have long since been halted by the facts." Colonel Alden H. Waitt, *Gas Warfare: The Chemical Weapon,*

In its initial and final use, however, Iraq's experience with chemical weapons in battle diverges from studies of chemical warfare in World War I. When Germany first introduced gas in World War I, it caused great panic among British and French forces, but thereafter World War I combatants tended to become inured to the CW threat and also acquired protective equipment. Chemical weapons normally caused panic only among troops that were untrained, ill-equipped, or low in morale for other reasons.[49] The introduction of mustard gas in July 1917 disturbed this pattern, but not for long. In the Iran-Iraq war, the reverse seems to have been the case; unprotected Revolutionary Guard troops apparently rushed heedlessly into gas clouds in 1983 and 1984, but in 1988 Iranian troops equipped with masks and presumably familiar with Iraq's use of CW are said to have panicked and run.

To the extent that panic occurred—and there remains considerable debate on this issue—it can be explained by remembering that in 1984 Khomeini's vision of heaven provided Revolutionary Guards with "protection," which made them fear gas clouds as little as they feared machinegun nests and minefields. By 1988, many of the more zealous Revolutionary Guards had been killed in earlier battles, leaving a force composed largely of conscripts.[50] These soldiers were equipped with gas masks, but many reportedly failed to shave their beards and thus were unable to achieve a proper seal with their masks, suggesting lack of training, and perhaps also low morale. Indeed, by 1988 Iran's soldiers probably realized that the war could not be won, that missile attacks had produced panic among civilians and government officials in Tehran, and that even Revolutionary Guard units were protesting against the war's continuation. In short, the cause might no longer have seemed worth dying for, especially not in the grisly way associated with chemicals.

Its Use, and Protection Against It (New York: Duell, Sloan and Pearce, 1942), p. 5. Because gas casualties tied up enemy evacuation, hospitalization, and logistics facilities, it was argued that "gas responds to an outstanding degree to every requirement of an effective military agent." Lt. Colonel Augustin M. Prentiss, *Chemicals in War: A Treatise on Chemical Warfare*, 1st ed. (New York: McGraw-Hill, 1937), p. 670. On the other hand, gas casualties also tended to return to combat faster than soldiers wounded by rifles and shrapnel. Thus Dorothy Kneeland Clark concluded that "if long-lasting reduction in the enemy's forces by death and incapacitation is desirable, as it must be in a war of more than a few weeks' duration," then "gas in WWI was decidedly less effective in casualty production than the other casualty-causing agents." Clark, *Effectiveness of Chemical Weapons*, p. 109.

49. Clark, *Effectiveness of Chemical Weapons*, p. 86.

50. Wright, *In the Name of God*, p. 27. On Iran's recruiting problems, see Alan Cowell, "Gulf Lull: An Iran Offensive on Hold," *New York Times*, January 26, 1988, p. A10, which notes that the forces Iran had massed for its 1988 offensive were substantially smaller than those of previous years.

This suggests that, as in the case of ballistic missiles, chemical weapons acquired their 1988 reputation for dramatic effectiveness from special circumstances that took several years to emerge. That those circumstances can be reproduced only with great difficulty, and certainly not quickly, undermines conclusions about the stunning usefulness of chemical weapons that have been drawn from their use in 1988. Chemical weapons have serious tactical uses; the official British history's conclusion that gas "made war uncomfortable . . . to no purpose," is overly glib.[51] But it is closer to the truth than the contention that chemical weapons are nearly magical devices that invariably cause large casualties and inspire panic.

THREATS OF STRATEGIC CHEMICAL USE. Neither Iraq nor Iran used chemical weapons directly against each other's major population centers. To be sure, Iraq, and possibly Iran, did use chemicals against civilians, notably Kurds residing in Halabjah, a small town in northeastern Iraq that had the misfortune to become the center of a major battle in April 1988.[52] And Iraq alone used such weapons again in September, after UN Cease-fire Resolution 598 had been signed, to drive Iraqi Kurds away from their villages near Iraq's common border with Turkey.[53] Although these cases say something about

51. This is the only mention of CW in the official British history of World War I, and is confined to a footnote. Noted in Clark, *Effectiveness of Chemical Weapons*, p. 139.
52. U.S. officials insist that Iran as well as Iraq used CW at Halabjah, on the basis principally of photographic evidence suggesting the presence of cyanide, a chemical believed *not* to be in Iraq's inventory. See Patrick E. Tyler, "Both Iraq and Iran Gassed Kurds in War, U.S. Analysis Finds," *Washington Post*, May 3, 1990, p. A37. Why Iran would have done so, however, remains unclear, and this raises questions about the assertion. There is little doubt that at this point Iran possessed CW, yet Iran used chemicals sparingly, for tactical purposes. It may be that Iran saw Halabjah as a strategic objective so important as to be worth the risk of "collateral damage" to civilians. Or perhaps Iran mistakenly fired on civilians while trying to hit Iraq's forces. Mistakes like this are always possible in combat, especially in a battle that, government officials suggest, was among the more confused of the war. With missiles hitting Tehran and morale sagging there as well as at the front, an Iranian victory would have been worth a great deal. On the other hand, at this point Iran was seeking to hold the "moral high ground" while generating international condemnation of Iraq. Thus questions of Iranian use and the motivation for it remain open.
53. Conclusive evidence of Iraqi CW use is lacking in this case, since both Iraq and Turkey, to which Kurdish refugees were forced to migrate, denied UN investigating teams access to areas in which Kurds claimed CW was used. Yet the circumstantial evidence derived from photographs of wounds and interviews with refugees was damning. Late in 1988 a soil sample from the area where Iraq reportedly used CW was obtained by a British film crew that secretly entered Iraq in the fall of 1988 to make a film entitled "The Winds of Death." Lab reports showed that the sample contained three chemicals found in mustard gas and little else. See "Poison Gas Traces Are Found in Iraq," *New York Times*, December 4, 1988, p. 2. Meanwhile, in a fiery session on September 8, 1988, with Iraqi Deputy Premier Saadoun Hammadi, Secretary of State George Shultz claimed to have "incontrovertible evidence" (developed, it was later made clear, by

Iraqi-Kurd relations, they prove nothing but the obvious point that chemical weapons are grimly effective when used against unprotected civilians.

But the effectiveness of chemical weapons resides in their psychological as well as their physical effects, and threats to use CW against enemy civilians during the Iran-Iraq war did, apparently, produce a drop in morale. In mid-March, 1988, the director of the Cairo office of the Iraqi News Agency stated publicly that "an authorized Iraqi spokesman" had told him that "Iraq might choose a number of large Iranian cities to be the targets of chemical weapons as a deterrent and punitive measure."[54] It is not clear that at this point Iraq had the technical capability to mount chemical warheads on missiles; the threat probably referred to bomber attack, which, given the state of Iran's air defenses, would no doubt have succeeded in delivering chemicals to Tehran. Whatever the case, U.S. government officials assert that this announcement produced a marked increase in the sense of panic that already gripped Tehran's residents, the result of causes described previously. For a city already in panic, the threat of CW seems to have exacerbated the problem, perhaps substantially.[55]

Other evidence from the Iran-Iraq war tends to support the argument that chemical weapons have special emotional and psychological effects on observers. Journalists interviewing Kurdish rebel leaders after the Halabjah massacre found them "fatalistic" in thinking that Iraq was willing to "leave Kurdistan a desert."[56] Meanwhile, the response of U.S. politicians to Iraq's

---

eavesdropping on Iraqi pilots delivering the weapons) that Iraq was using chemical weapons against its Kurds. See Don Oberdorfer, "U.S. Says Iraq Used Poison Gas on Kurds," *Washington Post*, September 9, 1988, p. 1.

54. As reported in "Iraq, Saying It Hit Tankers, Threatens Chemical Strikes," *International Herald Tribune*, March 30, 1988.

55. Significantly, Iran's counterthreat to mount CW on its own missiles seems to have had little effect in Baghdad. In an interview on March 23, 1988, Iran's war information minister, Mr. Kamal Khorazi, said that his country "may be forced" to use chemical weapons against Iraq in retaliation for Iraq's use against Iran. See Patrick E. Tyler, "Iran Charges Iraq With a Gas Attack and Its Grisly Toll," *International Herald Tribune*, March 24, 1988. That this threat produced no visible effect in Baghdad may be seen as support for the argument advanced earlier that these weapons acquired their effectiveness from the context in which they were used. But it is not clear that Iraq took the threat seriously; Iran's CW capabilities at this point were fairly rudimentary. Moreover, the Iranians also stated that the decision to fire chemical warheads at Baghdad would be based partly on whether or not international organizations continued their "deathly silence" in response to Iraq's use of CW. This suggests that Iran may have been trying to pressure the UN rather than to mount a credible threat. Without more evidence, there is not much to be made from this example.

56. Quoted in Jonathan C. Randal, "Kurds, Fearing Iraqis, Ponder Cutting Ties to Iran," *International Herald Tribune*, May 5, 1988, p. 4.

use of CW against its own civilians outweighed their response to Iraq's use of chemicals against Iranian troops or its use of missiles against Tehran. In August 1984, for example, the Senate Foreign Relations Committee issued a report on the Iran-Iraq war that barely mentioned Iraq's use of gas against Iran, despite the fact that such use had recently been verified by a UN team; it focused instead on the recently-begun tanker war, which threatened a possible halt to the flow of oil through the Strait of Hormuz. By contrast, the Halabjah massacre prompted a sharp rebuke from this and other congressional committees. In April 1988—after the Iraqi CW attack on its Kurdish civilians at Halabjah, but before its attacks in September—Senate Resolution 408 was introduced, explicitly condemning Iraq for its use of CW.[57] In reaction to the September attacks, the Senate introduced the "Prevention of Genocide Act of 1988," while both houses of Congress sought to impose sanctions on Iraq.

These were legitimate expressions of concern and fear, to be sure. Yet it would be difficult to argue that Iraq's use of gas against its Kurdish population was especially cruel in the historical context of the Iraqi-Kurd relationship. For decades the Iraqis have tried to subdue dissident Kurds. Viewing these Kurds as traitors, the Iraqis have used brutal means that undoubtedly have produced more misery and desolation among Kurds than the Halabjah massacre and the September attacks. If concern for the Kurds were the only motivation for Kurdish fears or congressional concern, then it could have been expressed in the early 1970s, when Iraq conducted major conventional military operations against the Kurds, or even earlier in the Iran-Iraq war. It seems clear that killing civilians with chemicals is more emotionally revolting than killing them with high explosives, or for that matter with any number of other techniques.

Martin van Creveld is correct in asking, from an "objective" point of view, "why the use of high explosives for tearing men apart should be regarded as more humane than burning or asphyxiating them to death." In his words, this is "one more example of an irrational attitude to technology."[58] Still, the attitude is no less real for being irrational, and it is at least worth speculating whether Iraq tried to manipulate the revulsion that chemical weapons provoke, well before it made the threat of strategic retaliation explicit in 1988.

57. *Congressional Record*, Senate, April 12, 1988, p. S3827.
58. Martin L. van Creveld, *Technology and War: From 2000 B.C. To the Present* (New York: Free Press, 1989), pp. 72, 177.

Arguably by 1986, as Iran realized the size and sophistication of Iraq's CW arsenal, it became increasingly aware that an Iranian breakthough in the war would likely bring, in retaliation, a massive Iraqi CW attack on cities as well as soldiers. The mere existence of Iraq's growing chemical arsenal thus may have helped convince Iran—or at least the more pragmatic among the country's revolutionary leadership—that the war could not really be won or, more to the point, that the war would not be worth winning.

Into the Future with Chemical Weapons and Ballistic Missiles

Even if we accept the argument that ballistic missiles and chemical weapons were not especially effective in the Iran-Iraq war, there remain good grounds for concern about the proliferation of these technologies. In part, such concern springs from the classical concerns of arms control, notably, to reduce tensions and to limit war's destructiveness should it occur. But the driving interest in control of these technologies is the vision of a strategic marriage in which chemical warheads atop ballistic missiles place leaders of the less-developed countries in command of the "poor state's nuclear arsenal."[59] This has evoked fear of a "delicate balance of terror," akin to that said to exist between the superpowers in the first decade of the Cold War. U.S. strategists then argued that unprotected nuclear delivery vehicles invited preemption, raising the possibility of instability, especially in crises, as well as the risk of nuclear war. Israeli strategist Gerald Steinberg has recently made much the same argument with regard to antagonists in the Middle East: "The introduction of missiles reduces critical decision making time to a few minutes. In a crisis, states that face the possibility of a missile attack involving chemical or biological, not to mention nuclear warheads will be strongly induced to launch a preemptive strike."[60] Until secure second strike capabilities are developed, the proliferation of chemically-armed ballistic missiles will add a new and dangerous source of instability in a region already well-known for its volatility.

59. See Roberts, "Chemical Weapons: A Policy Overview," p. 109. The discussion of strategic stability that follows is based on the assumption that, insofar as chemical weapons are seen as weapons of mass destruction, instability will arise from their potential for use against strategic targets like cities. The proliferation of extremely accurate ballistic missiles poses a separate and potentially serious problem, but not one that stems from the presence of chemical warheads. To the contrary, improvements in missile accuracy increase the tactical utility of conventional warheads (against targets like command and control centers, for example), while reducing the need to escalate to the chemical or nuclear levels.
60. Steinberg, "The Middle East in the Missile Age," pp. 32–33.

Although the strategic marriage of chemical weapons and ballistic missiles was not consummated during the Iran-Iraq war, use of the technologies separately clearly piqued concern for proliferation, prompting the creation of new institutions aimed at curbing the export of key constituent elements of these technologies. In 1984, as it became clear that chemicals were being used in the war, the United States and nine of its allies adopted measures banning the export of certain chemicals.[61] Within a year Australia took the lead in focusing this broad agreement on specific chemicals, and since then the "Australia Group" (which now includes twenty states) has identified nine chemicals on its "core list" and forty-one chemicals on its "warning list." These lists are circulated to industry in an effort to control sales.[62] In 1987 the United States and six of its allies (Canada, the Federal Republic of Germany, France, Italy, Japan, and the United Kingdom) launched the Missile Technology Control Regime (MTCR) in an effort to limit exports of missile technologies to countries with strictly peaceful plans for them.[63]

Without judging the potential effectiveness of export controls to curb the spread of chemical and missile technologies, to a great extent these technologies have already spread, especially in the Middle East. Iran and Iraq both now possess, in some degree, the domestic wherewithal to manufacture chemical weapons and ballistic missiles. Israel, Syria, Egypt, and Saudi Arabia possess missiles, while by almost every report, Israel, Syria, and Egypt also possess chemical facilities. Export controls may slow or even halt the development of more accurate missiles and more deadly chemicals, but the basic capabilities are already in place. Thus there is every reason to consider a future in which ballistic missiles and chemical weapons are available to major regional powers in greater quantities than is currently the case.

What are the dangers of proliferation of these two technologies? At the tactical level, ballistic missiles are relatively unimportant, since artillery and aircraft remain far more efficient ways to deliver conventional as well as

61. In addition to the United States, the original group included Australia, Canada, Japan, the Netherlands, Denmark, the Republic of Ireland, the United Kingdom, and the Federal Republic of Germany. The Soviet Union agreed in January 1986 to control its own chemical exports, but is not a member of the Australia Group.

62. See Elisa D. Harris, "Chemical Weapons Proliferation: Current Capabilities and Prospects for Control," in Aspen Strategy Group, *New Threats: Responding to the Proliferation of Nuclear, Chemical, and Delivery Capabilities in the Third World* (Lanham, Md.: University Press of America, 1990), p. 78; and Herschel Hurst, "Chemical weapons: a declaration of 'war'," *Pacific Defence Reporter*, Vol. 16, No. 4 (October 1989) pp. 16–17.

63. The agreement focuses on missiles capable of carrying at least 500 kg (1100 lbs) to a range of at least 300 km (186 miles), and seeks voluntary curbs on the export of production facilities as well as missile components. See Shuey, *Missile Proliferation*, pp. CRS-96–97.

chemical warheads. Thus the focus below is, first, on the chemical battlefield, examining particularly the often-heard argument that chemical weapons favor defense over offense and thus help to stabilize the battlefront.

At the strategic level, ballistic missiles are not important in and of themselves—they remain an inefficient means for delivering high explosives—but acquire importance with chemical (or nuclear) warheads, for which they are a quick and sure means of delivery. In contrast to the conclusions of the section below on chemical weapons (and also to the fears of some strategists), I conclude below while hardly a desirable prospect, that the proliferation of chemically-armed ballistic missiles, may not be as destabilizing and dangerous as many observers expect.

CHEMICAL WEAPONS AND THE TACTICAL DEFENSE

In World War I, the use of chemical weapons forced troops to don protective clothing that severely degraded combat effectiveness.[64] New technologies may change this, as masks become less resistant to breathing while protective suits become lighter and cooler, but at the moment U.S. Army protective equipment makes breathing and movement exceedingly difficult. Current U.S. Army manuals note that "wearing protective gear causes performance degradation because of heat buildup, fatigue, and stress."[65] Even with newer equipment, the need for relatively prompt decontamination or disposal of protective outfits will impose some cost on effectiveness, especially for troops on the offensive, moving forward and away from established positions. This encourages some observers to conclude that, by reducing tactical mobility, the threat of CW use will favor the defense over the offense, helping to stabilize battlefronts.[66]

While the evidence from World War I does not refute this conclusion, neither does it strongly support it. At the tactical level, defense dominated offense in World War I because of increases in firepower rather than the threat of chemical use, and also because, as Dorothy Kneeland Clark ob-

64. Clark, *Effectiveness of Chemical Weapons*, p. 3.
65. Headquarters, Department of the Army, FM 3-100, *NBC Operations*, September 17, 1985, p. 1-12. "NBC" stands for nuclear, biological, and chemical.
66. This argument may apply especially to U.S. troops, whose relatively obsolete chemical protection equipment poses special impediments to tactical mobility. See Evan E. Koslow, "Would *You* Go to War Wearing *This* Equipment?" *Armed Forces Journal International*, May 1990, pp. 55–58.

served, "World War I armies were not organized and equipped for rapid exploitation in any offensive.[67]

Logically gas would seem to favor the tactical defense even more strongly in the Middle East, where intense heat and humidity can be counted on to make protective gear all the more uncomfortable. Still, evidence from the Iran-Iraq war is not conclusive on this point. Iraq's use of gas in 1984 seems to have benefited the defense, but it did so by killing the Iranian attackers rather than by forcing them to don bulky protective outfits. In the next few years Iran acquired protective equipment, but Iraq acquired more conventional firepower, while it also constructed elaborate barriers along likely invasion routes. Iraq's defenses generally held, but it is impossible to distinguish the role of gas from the role of these other defensive measures. Iraq went on the offensive in 1988, but the chemical threat from Iran at that point was minimal.

There are two reasons to question the notion that the proliferation of chemical weapons will help to stabilize battlefronts. One is the historical tendency to overcome increases in firepower, which favor the tactical defense, with surprise strategic maneuver that forces one's enemies onto the tactical offense. Advances in firepower helped to produce the stalemate notable in World War I. The weapons that produced that stalemate had begun to appear in the mid-nineteenth century, however, and toward the end of the American Civil War some battles foreshadowed the trenches and carnage of World War I. In both the Franco-Prussian and Austro-Prussian wars, soon after the American Civil War, Germany compensated for the increased power of the tactical defense with strategic maneuver that forced its enemies onto the tactical offense.[68] Germany's bid to exploit surprise and strategic maneuver in 1914 was stymied by the particular failures of the Schlieffen plan. However, neither Iran nor Iraq is bothered by Germany's acute two-front problem, and thus both may still seek to achieve a surprise penetration, seize positions in the enemy's rear, and defend them with chemical weapons.

Surprise might also favor the offense even without strategic maneuver. While U.S. Army manuals recognize the performance degradation caused by

67. Clark, *Effectiveness of Chemical Weapons*, p. 134.
68. See Martin L. van Creveld, *Military Lessons of the Yom Kippur War: Historical Perspectives, Washington Paper*, Vol. 3, No. 24 (Washington, D.C.: Center for Strategic and International Studies, 1975), pp. 41–43.

wearing chemical protective gear, these are seen to apply more to infantry than to mechanized forces; tanks and personnel carriers can be equipped with filters to protect their occupants, for example. Few military officials felt that the presence of chemical weapons in Europe dampened Soviet plans for surprise offensive operations deep into NATO territory. To the contrary, Soviet chemical weapons were seen to be "directly related to their doctrine for offensive action," and would be used to open holes in NATO's lines, screen flanks, create barriers along likely routes for NATO's counterattacks, and put NATO's airfields out of action.[69] The same logic could be applied to Iran and Iraq as they acquire more elaborate chemical arsenals.

Given the absence in Iran and Iraq of the kind of political obstacles that made barrier construction difficult on the inter-German border, if Iran and Iraq do not reach a political accommodation, we should not be surprised to see a continuation of the barrier construction begun by Iraq around Basra during the war. Trenches, water barriers, minefields, and walls are likely to multiply as both countries seek to eliminate the possibility for offensive surprise and strategic maneuver. Barriers and obstacles, more than the threat of CW, will strengthen defenses and limit offensive action. To the extent that both sides will continue to look for ways to run surprise offensive maneuvers, however, the presence of chemical weapons will increase the premium on surprise and extremely rapid maneuver.

HOW DELICATE THE BALANCE?

Iraqi President Saddam Hussein's recent threat to use chemical weapons to "eat up half of Israel" if Israel were to strike Iraq's chemical or fledgling nuclear facilities has piqued concern about the emergence of a chemical balance of terror in the Middle East.[70] But Iraq's chemical weapon program has a great deal less to do with Israel than with Iran, and springs largely from Iraq's relative lack of strategic depth. Iraq's two major cities, Baghdad and Basra, are perilously close to the Iran-Iraq border; Tehran, by comparison, is some 300 miles from the Iraqi border, so far that it took Iraq several years to develop missiles capable of hitting it at all. For Iraq, virtually any Iranian conventional attack is likely to threaten national extinction. For Iran,

69. U.S. Army Command and General Staff College, Student Test 3-1, *Fundamentals of NBC Operations*, Fort Leavenworth, Kan., July 1986, p. 1-4.
70. For the full speech, see *Foreign Broadcast Information Service*, FBIS-NES-90-064, April 3, 1990, pp. 32–35.

by contrast, Iraqi attacks like the one that opened the Iran-Iraq war in 1980 are likely to threaten no more than a chunk of Iranian territory far from most of Iran's major cities. Iraq's acquisition of chemical weapons and ballistic missiles represents an effort to counter this fundamental imbalance by posing the threat of "extinction" to Iran with advanced weapons. In this sense, Iraq and Israel, both lacking strategic depth, seek "weapons of mass destruction" for the same reason.

The problem is that Iraq cannot seek a retaliatory capability against Iran without also threatening Israel. Therein lie the wellsprings of Israel's June 1981 preemptive strike on Iraq's Osiraq nuclear reactor, and probably also the source of Saddam's recent threat. Although Saddam's motivations are unclear, it is likely that he is less concerned with Israel as an object of attack than with the possibility that the Israelis will eliminate Iraq's margin of superiority in these special weapons over Iran, Iraq's real enemy. With Iraq's acquisition of long-range ballistic missiles and chemical warheads, the Iran-Iraq balance has thus become intimately tied to the broader Middle East missile balance.

Does this promise instability until secure second strike capabilities are in place? Gerald Steinberg argues that it does, and that the geographic proximity of antagonists in the region makes the problem of crisis instability even more dangerous in the Middle East than it was between the superpowers.[71] It seems more logical, however, to argue that the physical proximity of Middle Eastern powers makes it inappropriate to transfer superpower logic to local confrontations. Precisely because Iran and Iraq share a common border, the variety of delivery vehicles for weapons of mass destruction available to these antagonists goes well beyond ballistic missiles and strategic bombers, to include trucks, small aircraft, small tactical missiles, and even artillery. Although Israel and Iraq are separated by Jordan, weapons of mass destruction could still be delivered by trucks, by aircraft, and by small missiles trucked into Jordan. Proximity multiplies the range of available delivery means, and thus reduces the viability of preemption as a defense. Put another way, under these conditions "secure second strike" capabilities of some sort are much more easily available. The acquisition of ballistic missiles does not fundamentally change this equation.

Population distributions in many of these countries reinforce this logic. Populations in Israel, Iraq, and Syria are highly concentrated; Israel and Iraq,

71. Steinberg, "The Middle East in the Missile Age," p. 32.

in particular, are little more than "two-target" states for real "city-busting" weapons. U.S. strategists who contemplated the usefulness of preemption never saw it as 100 percent effective; at best it would have held casualties to a minimum. For two-target states, however, *any* remaining weapons of mass destruction promise disaster. Thus preemptive attacks on an adversary's delivery means are unlikely to save any Middle Eastern power from nuclear destruction, assuming the adversary possesses a nuclear arsenal of reasonable size. Nor should regional leaders feel compelled to fire nuclear weapons first to eliminate opponents' missile delivery systems. Assuming that those leaders wish to fire nuclear weapons in the first place, their decision to do so will depend mainly on their assessment of their adversary's retaliatory capabilities, which will be impossible to eliminate.

The argument thus far assumes that chemical weapons fall with nuclear weapons into the category of "weapons of mass destruction." Yet this is not a valid assumption. In terms of relative destructive power, chemical weapons used tactically fall somewhere between high explosives and battlefield nuclear weapons, but closer to conventional ordnance. Indeed, CW is not very effective at all against troops equipped with protective gear and trained in its use.

We have, fortunately, no empirical data on the effects of chemical weapons used massively against cities. Presumably some of the special qualities of chemical weapons suit them well for such uses. Chemical agents would seep into basements and bunkers that conventional bombs would leave untouched. (But conventional bombs knock things over and crush nearby buildings, creating kinds of damage chemicals do not.) Chemicals would also persist, especially if caught in closed rooms, extending their lethality (depending on the agent) over hours, days, or perhaps weeks. Finally, whereas chemical weapons used in open terrain tend to be dissipated by wind, dispersion in cities might be constrained by walls and buildings. Given the correct weather conditions, a deadly cloud of gas might form, the urban equivalent of what was called "pancaking" in World War I. These arguments suggest that chemical weapons may be more destructive, and thus more like nuclear weapons, when used against cities than when used tactically.

Even so, the difference between chemical and nuclear destructiveness probably remains substantial. One analyst has suggested very tentatively that "500 CW-loaded Iraqi SCUD missiles exploded over Iranian cities might have . . . direct effects comparable to 5 or 6 Hiroshima-type nuclear attacks."[72]

72. Andrew Mack, "Missile Proliferation in the Asia-Pacific Region," paper presented to the

Obviously such calculations are uncertain; to some degree they will always be uncertain, since chemical dispersion is more dependent than nuclear destruction on weather conditions. At the moment it is hard to envisage casualty levels from a sizable chemical attack that would compare to those likely to be suffered during even a small atomic attack.

The amount of human misery and death that gas can bring to a city will also depend, as it does on the battlefield, on the degree of protection afforded the target population. Here is a fundamental difference between nuclear and chemical weapons: there is no defense against nuclear *blast*, hence the defensive emphasis has always been on preemption or missile defense. In sharp contrast, a mask and protective suit constitute a reasonable defense against chemical weapons. Citizens must have the gear and enough time to don it, of course, but to the extent that chemicals spread their destruction relatively slowly rather than instantly, warning is feasible even after a surprise chemical detonation. Indeed, to the extent that mounting a full-scale chemical attack on a city might require repeated missile launchings over, say, many hours, evacuation alone would constitute a defensive measure for at least part of the target population.[73]

The difference between chemical and nuclear weapons affects the stability of the strategic chemical balance. With a variety of defenses available and chemical destruction unfolding over a relatively long period of time, pressures to preempt are reduced. The ability of a state to "ride out" or otherwise absorb an incoming chemical attack on its cities is thus substantially higher with chemical than with nuclear warheads. The balance of terror is thus less delicate.

This assumes, of course, that states purchase protective gear for their citizens. Although Israel has equipped its citizens with masks (but not suits),

---

"Workshop on Chemical and Missile Proliferation: Implications for Asia-Pacific," Australian National University, Canberra, March 8–9, 1990, p. 17. Anthony Cordesman also asserts that "the persistent nerve agents that Iraq now seems ready to deploy would have to be delivered in such large amounts that they are not suitable for use against cities." He goes on to assert, however, that "biological agents like botulism and anthrax are true weapons of mass destruction." See Anthony H. Cordesman, "Creating Weapons of Mass Destruction," *Armed Forces Journal International*, February 1989, p. 56.

73. Uzi Rubin has argued, for example, that "75 simultaneous [missile warhead] impacts require 100 launches from approximately 120 launchers. . . . To launch this barrage from just 20 launchers would take about 30 hours, which is longer than can be expected for weather patterns to hold still (in fact, [the attack] would take even longer, considering that the launchers need some maintenance and refurbishing after a certain number of firings). The same [calculation] for 500 warheads is 6.7 times more difficult or more protracted and thus less effective." Personal correspondence with the author, May 27, 1990.

similarly equipping the population of the Middle East as a whole may sound preposterous. Technically, however, it is far less ambitious and expensive than pursuing elaborate anti-missile systems, as in the Strategic Defense Initiative (SDI). Moreover, research and development into chemical protective gear funded with, say, some of the billions of dollars that have gone into SDI might produce gear far more comfortable and less obtrusive than suits and masks currently available. The problem is social more than technical. But if CW is truly seen as a serious threat, social resistance to the distribution of protective gear might diminish substantially.

Conclusions

If the arguments presented here are even close to correct, then there has been a tendency to overstate the usefulness of ballistic missiles and chemical weapons in the Iran-Iraq war. Perhaps this should not be surprising. Iran's near-collapse in 1988 surprised nearly all observers, prompting a search for special explanations. Ballistic missiles and chemical weapons are new and special weapons, in some senses. And there was certainly a substantial amount of evidence to confirm that missiles produced visible panic in Tehran, and some evidence suggesting that chemical weapons had a similar effect at the battlefront. Meanwhile, it seems clear that chemical weapons in particular provoke from observers a visceral reaction out of proportion to their lethality.

A less exaggerated assessment of the role these technologies played in the Iran-Iraq war by no means implies that they are useless. Chemical weapons are generally effective when first introduced into battle against unsuspecting and unprepared adversaries. That chemical weapons are lethal to the unprotected is hardly surprising, but neither is it trivial; this capability proved invaluable to Iraq in 1984, when its defenses verged on collapse. Chemicals are less effective against prepared forces. Iraqi uses of gas against protected but demoralized Iranian troops provoked panic, but did not contribute much to Iraq's stunning 1988 victories. Still, even against protected troops the use of gas, or the threat to use gas, fundamentally alters the battlefield. In short, gas seems to be at least a marginally cost-effective weapon.

The evidence of the Iran-Iraq war on ballistic missiles is less compelling. Despite the plausibility of Churchill's argument about the demoralizing quality of Germany's V-2 strikes on London, it finds scant support in the various "wars of the cities" in the Iran-Iraq conflict. Ballistic missiles would appear to be no substitute for a good air force. Indeed, if conventional weapons were the only ones available, it might be useful to encourage states to buy missiles rather than bombers, since doing so would be a relative waste of money, leaving fewer resources available for war.

Chemical weapons may soon supplant conventional warheads on many of the Middle East's ballistic missiles. This prospect may seem disturbing, but fears of instability associated with an emerging chemical/missile balance—a regional balance of terror—may be exaggerated. Such fears overstate the special qualities of ballistic missiles under regional circumstances in which there are few targets and several means for delivering weapons to them. Moreover, the crucial differences between chemical and nuclear weapons make it less likely that such a balance will be highly unstable. Chemical weapons are less destructive and more easily defended against than nuclear weapons; the two technologies should not be lumped together as essentially similar weapons of mass destruction.

If accepted, my arguments may dampen incentives for Third World states to purchase or develop chemical and missile technologies. But there will still be proliferation incentives. Limited usefulness is still usefulness. Gas used against an unprepared foe can yield a substantial tactical payoff. And gas remains one of several ways to deal with internal dissidence, like that which the Iraqi government associates with its Kurds. These arguments apply more to gas than to missile technology, but this may change as much more accurate ballistic missiles become available, since these may be useful in specific military scenarios not seen in the Iran-Iraq war—conventional strikes against command and control facilities, for example.

Should proliferation worry us? Should not arguments that belie the effectiveness of new technologies and the instability their proliferation may cause also leave us less worried when these technologies find their way into military arsenals? Even if chemical weapons and ballistic missiles are less effective than is widely believed, there is still cause for concern. In a region as potentially volatile as the Middle East, no one should welcome the introduction of more and different means of destruction. The classic goals of arms control—to reduce the probability of war, and to reduce war's destructiveness should it occur—remain relevant.

It remains to be seen whether the export control mechanisms set in place during the Iran-Iraq war will be able to halt the spread of chemical weapon and missile technologies. All else being equal, however, the effectiveness of such controls will improve if policymakers are clear about priorities and focus the control effort accordingly. They should focus relatively less on ballistic missile proliferation than on the proliferation of chemical weapons. Missiles do not constitute a very cost-effective way to deliver conventional ordnance, despite the success of Iraq's 1988 missile strikes on Tehran, since conventional warheads do relatively little damage except when used in large numbers. Ballistic missiles with nuclear warheads appear to be especially worrisome, since they ensure delivery of what are truly weapons of mass destruction. But even less certain delivery means are likely to put a few of the weapons on target, and for states with small and concentrated populations this would be disastrous. This suggests that it would be wiser to worry about the warhead than the delivery vehicle.

Which warheads should worry us most? Nuclear weapons remain far more worrisome than gas. Even if logic suggests that regional nuclear and chemical balances are likely to be fairly stable, the widespread possession of devices capable of truly massive destruction in a volatile region like the Middle East is grounds for concern. Chemical weapons simply are not as destructive, and means exist for defense against them. It might be disturbing to contemplate the distribution of protective gear, but not as disturbing as thinking about the effects of nuclear weapons. Possessor states like the United States would do well to focus their attention above all on nuclear technologies, which are, fortunately, fairly specialized and thus easier to spot than chemical exports.

The Gulf War, 1990–1991

[13]

Saddam's Security Apparatus During the Invasion of Kuwait and the Kuwaiti Resistance

Ibrahim Al-Marashi

"New Terrorism" or Modern-day Fanaticism?

In Iraq's modern history, Saddam Hussein had presided as the head of state longer than all his predecessors, despite three costly wars, numerous internal insurrections, coup attempts and crippling international sanctions. Iraq's security apparatus was one of the largest instruments of state control in Saddam's regime, essential to the preservation of his rule. This vast, complex and wide-ranging labyrinth of organizations, with mutually independent intelligence and military units, ensured the protection of the president and his regime, by pervading all layers of Iraqi society.

Insight into the type, nature and breadth of the Iraqi regime's security organizations is essential to understanding the efforts Saddam made to transform Kuwait into an Iraqi province. The three primary objectives of the security apparatus in serving Saddam's strategy of the incorporation of Kuwait as an Iraqi province, were maintaining internal security in the 19th province, erasing the existence of Kuwait as an independent nation, and preparing it for an impending military conflict with coalition forces. This strategy was carried out by Saddam through the coordination of four Iraqi intelligence units, *Jihaz al-Amn al-Khas* (The Special Security Apparatus), *Mudiriyyat al-Amn al-'Amm* (The General Security Service), *Dairat al-Mukhabarat al-'Amma* (The General Intelligence Directorate), *Mudiriyyat al-Istikhbarat al-'Askariyya* (The Military Intelligence Directorate). In addition to these four agencies, there was a myriad of Ba'ath Party security agencies, civil police forces, paramilitary militias and special military units that interacted together in a complex network that formed Iraq's security apparatus, with the ultimate goal of preserving and defending its Kuwait Province.

It was the aforementioned instruments of Iraqi state security that transformed Kuwait into an integral part of Iraq's *mukhabarat*[1] state. The concept of the *mukhabarat* state in terms of Iraq referred to the overall police state that was in the service of the Saddam regime through the use of domestic coercion.

1 *Al-Mukhabarat* is often used as a catch-all phrase for intelligence units in Arabic.

While some police states protect the interests of the existing government, in the case of Iraq, the security, intelligence, militia, and military organizations that formed the Iraqi security apparatus were dedicated to serving the individual, Saddam, who portrayed himself as the personification of the nation.

The jurisdictions of the agencies in Kuwait were designed to be duplicative, to maintain competition and to ensure that no one security service would emerge too powerful. A clear division of labor among the Iraqi security agencies in Kuwait did not exist, and in fact some agencies specifically monitored the activities of the others. The collective responsibilities of the agencies were to maintain internal security by countering domestic dissention in Kuwait, preventing external threats to the regime's interests in Kuwait, as well as monitoring the Iraqi military stationed in the 19th province. Generally, Military Intelligence dealt exclusively with military matters, General Intelligence dealt with foreign intelligence gathering, while General Security focused on civilian affairs, and Special Security had the authority to deal with all of the aforementioned domains. The four intelligence agencies were headquartered in Baghdad, while General Intelligence, Military Intelligence and General Security maintained field offices in Kuwait.

The coordination and command of these security agencies was ostensibly a family affair, as one account wrote, "So important was this responsibility that the small number of men selected for the task were chosen only after the most careful vetting of their families, their tribal ties, their absolute loyalty to Saddam."[2] As a general rule, relatives of Saddam, members of his al-Bu Nasser tribe, as well as those from Tikrit (Saddam's home town) formed the majority of these security forces. These security and military forces were also drawn from the towns of Dur, Sharqat, Huwayja, Bayji, Samarra and Ramadi, located in what is known as "the Sunni Arab Triangle." Sunni tribes and families that played a powerful role in the security apparatus during this period included the Dulaym, the Jubur (mixed Shi'a/Sunni) and the 'Ubayd tribe, as well as members of the Duri, and Samarrai families.[3]

2 Andrew Cockburn and Patrick Cockburn, *Out of the Ashes: The Resurrection of Saddam Hussein* (New York: Harper Collins, 1999), 107.

3 Amazia Baram, "Between Impediment and Advantage: Saddam's Iraq," United States Institute of Peace Special Report, http://www.usip.org; and "Saddam Husayn Between His Power Base and the International Community," *MERIA Journal*, 4.4 (December 2000): 11-12; http://meria.idc.ac.il/journal/2000/issue4/jv4n4a2.html.

The Intelligence Network

Special Security

Saddam created *Al-Amn al-Khas* (Special Security)[4] during the Iran-Iraq War to serve as a super-secret organization and it evolved into the most powerful agency in the security apparatus. It emerged from within General Security in 1982 to provide bodyguards to the President after a failed assassination attempt on Saddam.[5] Hussein Kamil, who was Saddam's cousin, son-in-law, as well as Minister for Military Industrialization[6] was instrumental in the creation of this agency and only selected loyal and devoted agents from the General Security, Military Intelligence and General Intelligence to serve in this ultra-elite intelligence unit. After graduating from his studies in 1988, Qusay Hussein, the son of Saddam was placed as deputy director. There were an estimated 5,000 members[7] in this organization mostly from the towns of Tikrit, Huwayja and Samarra.[8]

Special Security guarded the President at all times, especially during travel and public meetings, and secured all presidential facilities, such as palaces and offices. Since Saddam confined himself to the environs of Baghdad, Special Security played a limited role from Baghdad in supervising the security services in Kuwait.

Special Security conducted surveillance of General Intelligence, Military Intelligence, and General Security, essentially spying on Iraqi "spies." As opposed to the other three major security organizations, it served as the inner most intelligence agency of the regime, functioning as the nerve center of the Saddam's security agencies. By creating this agency, Saddam was able to maintain greater control over the various units in the security network. According to one source, "It is the eyes and ears of the President, as well as the hand to implement, directly or indirectly, the President's security directives.

4 *Al-Amn al-Khas* (Special Security) is also known as *Mudiriyyat al-Amn al-Khas* (The Special Security Directorate) or *Jihaz al-Amn al-Khas* (The Special Security Apparatus, The Special Security Organization or The Special Security Service). It is also referred to as *Jihaz Mukhabarat al-Ra'isa* (The Presidential Intelligence Apparatus, The Presidential Affairs Department or The Presidential Intelligence Bureau). In some publications it is abbreviated by the acronym, SS, SSS or SSO.

5 Dhilip Hiro, *Neighbors, Not Friends: Iraq and Iran After the Gulf Wars* (New York: Routledge, 2001), 55.

6 Charles Tripp, *A History of Iraq* (Cambridge: Cambridge Univ. Press, 2000), 254.

7 Hiro, *Neighbors, Not Friends*, 56. This figure is also claimed by Federation of American Scientists, see "Iraq's Intelligence Agencies", http://www.fas.org.

8 Anthony Cordesman, *Iraq and the War on Sanctions* (Westport, CT: Praeger, 1999), 152.

This body was in charge of collecting information about the activities of all high ranking officials and even information about members of the President's immediate family."[9]

During the 1991 Gulf War, this agency controlled units of Iraq's Chemical Corps, which was responsible for deploying Iraq's chemical weapons arsenal.[10] Special Security was put in charge of controlling SCUD missiles that could deliver chemical munitions, as well as concealing them from coalition air strikes.[11] Saddam deliberately chose to entrust Special Security, a security/intelligence agency, with the responsibility of deploying these weapons rather than a regular military unit to prevent the possibility of a coup being launched from within the military.[12] Post and Baram, two experts on the Saddam's personality, writes of this agency that, "In other words, these people, who will be the ones to initiate nuclear, biological, chemical and/or radiological weapons use, are the closest to what would see as an extension of Saddam's self."[13]

During the 1991 Gulf War, Fanar Zibin Hassan al-Tikriti served as the official director of *al-Amn al-Khas*,[14] but was replaced only a year later by Qusay. Fanar, a Major General in the Iraqi military played a minimal role in the events that transpired in this six-month period in Kuwait, as Qusay who was serving as its Deputy Director at the time, signed most of the correspondence from the Special Security to the Kuwait province.

9 Mustafa Alani, "Saddam's Support Structure" in Sean McKnight, Neil Patrick and Francis Toase, eds., *Gulf Security: Opportunities and Challenges for the New Generation* (London: The Royal United Services Institute for Defence Studies, 2000), 43.

10 Ibid.

11 Scott Ritter, *Endgame: Solving the Iraq Problem Once and For All* (New York: Simon and Schuster, 1999), 102.

12 Timothy V. McCarthy and Jonathan B. Tucker, "Saddam's Toxic Arsenal", in Peter R. Lavoy, Scott D. Sagan and James J. Wirtz, *Planning the Unthinkable: How New Powers Will Use Nuclear, Biological, and Chemical Weapons* (Ithaca, NY: Cornell Univ. Press, 2000), 48.

13 Jerrold Post and Amazia Baram, "Saddam is Iraq: Iraq is Saddam", *Countreproliferation Papers No. 17* (Maxwell Air Force Base, Alabama: USAF Counterproliferation Center, 2002), 55.

14 Michael Eisenstadt, *Like a Phoenix From the Ashes: The Future of Iraqi Mlilitary Power* (Washington, D.C.: Washington Institute for Near East Policy, 1993), 11.

The General Security Service

Al-Amn Al-'Amm (General Security),[15] the oldest security agency in the country dates back to 1921, when it was created during the British Mandate era.[16] In 1973, Nadhim Kazzar, head of General Security attempted a coup against both the President Hassan al-Bakr and then vice-President Saddam. In response to this unexpected attack, Saddam transferred many of General Security's responsibilities to his newly formed General Intelligence Directorate[17] and arranged for the KGB to aid in a reorganization and modernization of General Security.[18]

General Security remained under the Ministry of the Interior, as a civilian police force until the late 1970s, when it was established as an independent agency reporting directly to the Presidential Palace.[19] In the late 1980s, a number of detectives were transferred to General Security from the investigative section of the civilian police.[20] In 1980, Saddam appointed 'Ali Hassan al-Majid as its director to instill the ideology of the Ba'ath Party into the agency,[21] where upon it was then delegated more political-intelligence responsibilities during the Iran-Iraq War.[22]

The headquarters of General Security was located in Baghdad, from which it guided the work of its branches in each Iraqi governorate, including the branch in Kuwait, referred to as *al-Amn al-Kuwait*. It was further divided into sub-branches for every neighborhood in Kuwait, such as *al-Amn al-Damshiyya*, referring to the Damshiyya district in Kuwait City. In the 19th province, it functioned as a political security police force, responsible for detecting dissent among the Kuwaiti general public and eliminating political criminal activity, such as economic subversion, illegal smuggling and manifestations of Kuwaiti opposition to the regime. General Security monitored the day-to-day lives of the

15 It is also known as *Mudiriyyat General Security* (General Security Directorate or General Security Service) and also referred to as "The Secret Police," and is sometimes written with the acronym GS or GSS.

16 Hiro, *Neighbors, Not Friends*, 54.

17 Ritter, *Endgame*, 62.

18 Kanan Makiya, *Republic of Fear: The Politics of Modern Iraq* (Berkeley, CA: University of California Press, 1998), 12.

19 Human Rights Watch/Middle East Watch, *Genocide in Iraq: The Anfal Campaign Against the Kurds*, (New York: Human Rights Watch, 1993), 3.

20 Human Rights Watch/Middle East Watch, *Bureaucracy of Repression: The Iraqi Government in its Own Words* (New York: Human Rights Watch, 1994), 3.

21 Hiro, *Neighbors, Not Friends*, 55.

22 In 1987, Saddam appointed 'Ali Hassan al-Majid as the head of the Ba'ath Party's Northern Bureau to suppress the Kurdish insurrection (and thereafter earning his odious title, "Chemical Ali").

Kuwaiti population creating a pervasive local presence,[23] through an extensive network of informers in Kuwait, under the 1970 Law no. 83 "The Law of Securing the Trustworthy in Defending the Revolution." General Security coordinated its operations with the civilian police force and maintained a unit in every police station of Kuwait City.[24]

General 'Abd al-Rahman al-Duri replaced 'Ali Hassan al-Majid as the head of General Security after 1987 and continued holding this post through the 1991 Gulf War. The director of its departments in the Kuwait province was Sa'dun Sabri,[25] who was required to collect intelligence for Al-Majid, who then responded with specific directives to be implemented. When General Security was given such commands, it often used the Special Forces[26] to conduct arrests or executions of the Kuwaiti dissidents.[27]

The Iraqi Intelligence Service

While General Security and Military Intelligence were created during the period of Iraq's monarchy, General Intelligence (*al-Mukhabarat al-'Amma*)[28] emerged from within the Iraqi Ba'ath Party. After escaping arrest from an unsuccessful attempt to assassinate Iraq's ruler 'Abd al-Karim Qasim in October 1959, Saddam fled to Syria and then Egypt, where he assumed a position in the Party leadership. Under his exclusive control, he created a small internal, security intelligence organization consisting of some of the party's younger members. This unit was known by the codename, *Jihaz al-Khas* ("The Special Apparatus") or *Jihaz al-Hanin* ("The Yearning Apparatus") was known for assassinat-

23 Makiya, *Republic of Fear*, 12.
24 Sean Boyne, "Inside Iraq's Security Network, Part One," *Jane's Intelligence Review* 9 (July 1997): 312.
25 Kuwait Data Set (KDS) Folder 90809 File 908-1-38 p.139. The Kuwait Data Set consists of 300,000 classified Iraqi state documents left by retreating forces in Kuwait during the 1991 Gulf War. They are being transcribed by the Iraq Research and Documentation Project (IRDP) at Harvard University. They can be viewed on the IRDP website: www.fas.harvard.edu/~irdp
26 After the Gulf War, Al-Amn was assigned its own paramilitary force known as the Emergency Forces, *al-Quwwat al-Tawari'*. However, during this period, the Special Forces served *al-Amn*, even though the Special Forces were under the jurisdiction of the Ministry of Defence.
27 KDS Folder 90809 File 908-3-10 p.170
28 It is also known as *al-Mukhabarat al-Amma* (General Intelligence), and is also referred to as *Da'irat al-Mukhbarat al-'Amma* (The General Intelligence Directorate, The General Intelligence Department, The General Intelligence Service or The Iraqi Intelligence Service). It is sometimes written with the acronym, IIS, GID or GIS.

ing members of other political groups as well as suspect fellow Ba'ath Party members.[29] Saddam's experience in controlling this small intelligence unit allowed him gradual control over the party organization and taught him valuable lessons when he would assume the presidency.[30]

Saddam's unit aided the Ba'ath Party coup on July 17, 1968, where upon he Saddam expanded it into an official organization, known as the Public Relations Office, to rival the already existing General Security. In 1973, Saddam transformed it officially into the *Da'irat al-Mukhabarat al-'Amma* (The General Intelligence Department) in response to the failed coup attempt by General Security director Nadhim Kazzar.[31]

After the coup, Saddam appointed his half-brother Barzan Ibrahim Hassan al-Tikriti to a prominent role in General Intelligence, while Sa'dun Shakir, Saddam's cousin served as its head. In 1982 Barzan replaced Sa'dun as the director due to the latter's failure to pre-empt an assassination attempt on Saddam's life in the town of Dujayl north of Baghdad.[32] The 1982 assassination attempt would also serve as one of the primary forces behind the creation of The Special Security Organization.

Barzan's appointment did not last long when in 1983 Saddam demoted him to Iraq's ambassador to the UN in Geneva. Barzan was succeeded by an academic, Fadil Barak al-Tikriti who held this position until 1989, when he was arrested on espionage charges on behalf of the Soviets and later executed. Afterwards, Barzan's brother, Saba'wi Ibrahim Hassan al-Tikriti served as the director of General Intelligence during the invasion of Kuwait and the 1991 Gulf War.[33]

General Intelligence was roughly divided into two departments, responsible for internal and international operations. Its internal activities in Kuwait were coordinated through its district headquarters, while its international operations in influencing world public opinion during the invasion of Kuwait were conducted from various Iraqi embassies. Its internal activities in Kuwait included suppressing the Kuwaiti opposition, counter-espionage, and targeting hostile foreign individuals and groups inside of Kuwait. It monitored foreign

29 Makiya, *Republic of Fear*, 15. Jihaz al-Khas should not be confused with Jihaza al-Amn al-Khas, a security organization that emerged during the Iran-Iraq War.

30 Alani, "Saddam's Support Structure," 42.

31 Eberhard Kienle, *Ba'ath v Ba'ath: The Conflict Between Syria and Iraq, 1968-1989* (New York: St. Martin's Press, 1990), 85.

32 Helen Chaplin Metz, "Iraq: A Country Study" in Leon M. Jeffries ed. *Iraq: Issues, Historical Background, Bibliography* (New York:Nova Science Publishers, 2003), 245.

33 Eisenstadt, *Like a Phoenix From the Ashes*, 11.

embassies and foreigners in Kuwait as well as maintained an internal network of informants, that rivaled General Security's network.

Its external activities during the invasion of Kuwait included providing disinformation and attempts to exploit or use Arab and other media; and maintaining an international network of informants, by using popular organizations abroad such as the Union of Iraqi Students.[34]

There were five regional directorates during the Gulf War. Directorates 21 through 26 were responsible for monitoring various regional districts in Iraq. Directorate 21, the residency located in Baghdad, was in charge of security issues in the capital as well as issuing residence permits for foreigners and Arabs in Iraq. Directorate 23, the Southern District based in Basra, conducted operations in the south of Iraq, while Directorate 24, the Northern District, did the same in northern Iraq and Iraqi Kurdistan. Based in Mosul, with an office in Kirkuk, it was responsible for infiltrating the opposition Kurdish parties in Iraqi Kurdistan. Directorate 25, the Western District, was located in Ramadi and maintained a network of informants in Syria and Jordan. Directorate 26, the Eastern District operated in the Karbala Governerate and monitored the Shi'a community there.[35] A separate regional directorate was established in Kuwait, referred to in the documents as *al-Mukhabarat al-Khalij* (Gulf Intelligence). The General Intelligence director in Baghdad would often given commands to the Gulf (ie Kuwait) General Intelligence branch, headed by 'Isam Khadir.[36]

Military Intelligence

Mudiriyyat al-Istikhbarat al-'Askariyya al-'Amma (or The General Military Intelligence Directorate) was created in 1932,[37] to report to the Ministry of Defense, although in the early 1980s it was subordinated to report directly to the Presidential Palace. Its responsibilities included tactical and strategic reconnaissance of regimes hostile to Iraq, assessing threats of a military nature to Iraq, monitoring the Iraqi military and ensuring the loyalty of the officer corps, and protecting military and military-industrial facilities.

Military Intelligence maintained regional headquarters throughout the country, in five administrative areas known as *manthumat*. These areas of jurisdiction for collecting intelligence included the Kirkuk office responsible for the northern Iran border region and the Kurdish region of northern Iraq, the Mosul office responsible for Turkey and Syria, the Basra office, responsible for the Gulf states (excluding Kuwait) and the southern Iranian border, and the Baghdad office (separate from the main headquarters), which monitored Jordan,

34 Hiro, *Neighbors, Not Friends*, 56.

35 See Federation of American Scientists, "Iraq's Intelligence Agencies" http://www.fas.org.

36 KDS Folder 90809 File 908-1-38 p.139.

37 Hiro, *Neighbors, Not Friends*, 56.

Syria, and the central Iranian border region.[38] The Baghdad headquarters delegated all Military Intelligence responsibilities to its Kuwait director, Colonel Hatim Haylan, responsible for *manthumat Kuwait* ("Kuwait military district").

Saddam appointed Sabir 'Abd al-'Aziz al-Duri[39] as head of *Military Intelligence* during the 1991 Gulf War,[40] a non-Tikriti, from the tribal alliance that traditionally supported the regime.[41]

Maintaining Internal Security in Kuwait

The Role of Ali Hassan al-Majid

Saddam placed the responsibility of eliminating the threats of the Kuwaiti resistance as well as erasing the very identity of Kuwait on 'Ali Hassan al-Majid, his paternal cousin. The role of al-Majid, ("Chemical Ali") was instrumental in the process of the "Ba'athification" of Kuwait. As Secretary of the Ba'ath Party Northern Bureau, al-Majid was given sweeping government sanction to suppress rebellious Kurdish activity from 1987 to 1988, known as the *Anfal* campaign, where the regime embarked upon a policy of eliminating the Kurdish opposition through sustained, institutionalised repression against both combatants and civilians. "Under RCC decree no.160 of March 29, 1987, the Northern Bureau's Secretary General, Ali Hassan al-Majid, was given extraordinary powers over all other state, party, military and security agencies."[42] In the 19th province, al-Majid had the same powers over the organizations he managed in the north of Iraq.

In Iraqi Kurdistan, he employed tactics that included the use of informants, forced "disappearances", arrests, mass detentions, executions, deportations, and a systematic campaign to eliminate villages that were located in politically sensitive areas. In other words, it created a pervasive sense of internal fear by penetrating society in Iraqi Kurdistan at every level, from the government agencies to menial places of work. As the governor of Kuwait, al-Majid was responsible for crushing the Kuwaiti resistance, as he had the Kurdish resistance, and for erasing Kuwait's identity, as had erased that of the Kurds. He administered Kuwait as his own fiefdom, using many of the tactics he employed as the architect of the *Anfal* campaign in Iraqi Kurdistan.

38 Human Rights Watch, *Bureaucracy of Repression*, 4.

39 Sabir Abd al-Aziz al-Duri was then placed as head of *al-Mukhabarat* after the 1991 Gulf War.

40 Eisenstadt, *Like a Phoenix From the Ashes*, 11.

41 Amazia Baram, "Saddam Husayn Between His Power Base and the International Community," 4.4 *MERIA Journal* (December 2000): 12, http://meria.idc.ac.il/ journal/ 2000/issue4/jv4n4a2.html.

42 Human Rights Watch, *Genocide in Iraq*, 355.

While Al-Majid was appointed as governor of the Kuwait province, his appointment was not made public.[43] In fact, he never signed any of the documents as governor of Kuwait, but as a Member of the Regional Leadership of the Arab Socialist Ba'th Party (*Adu al Qiyada al Qatriyya li Hizb al-Ba'th al'Arabi al-Ishtiraki*).[44] Despite the differences in titles, al-Majid was granted all encompassing powers by Saddam, chairman of the RCC, as he had when he was charged with suppressing the Kurdish insurrection in the north of Iraq.[45]

The documents demonstrate that the ultimate intelligence and security agency was not Special Security but al-Majid's office, as the vast majority of all communication between the intelligence and security agencies was directed to al-Majid. Any intelligence obtained by these networks, had to be immediately reported to al-Majid.[46] The Popular Army and Special Forces sent daily reports on their progress to him.[47] Usually al-Majid would respond to such requests with hand written notes on the document that were later translated into official policy papers. Thus, al-Majid emerged as the mastermind behind the incorporation of Kuwait into Iraq's *mukhabarat* state.

The Kuwaiti Resistance

The Beginnings of The Kuwait Resistance

One of the greatest challenges to Saddam's strategy of maintaining internal security in the 19th province was the Kuwaiti resistance, which according to the Gulf war literature only played a minor role during this conflict. However, a large number of the documents were devoted solely to the repression of their activities, which speaks to the fact that the Kuwaiti resistance posed a significant challenge. The elite Special Forces, as well as many Popular Army units were tasked with eliminating the Kuwaiti resistance rather than preparing for the defence of the province from an impending war. The commander of the Iraqi Special Forces, Bariq 'Abdallah al-Haj Huntah, was charged with suppressing the Kuwaiti resistance. In all his communications, the Special Forces essentially functioned as a security intelligence unit, rather than a purely military unit, reporting to both Ali Hassan al-Majid and Saba'wi Ibrahim Hassan, director of the General Intelligence Service, rather than the Ministry of Defence. Smaller units of Huntah's 65th, 66th , and 68th Brigades conducted patrols in the urban areas, monitoring the Kuwaitis. During these patrols, each of the Special Forces units was augmented by 100 troops from the Popular

43 Ofra Bengio, "Iraq" in Ami Ayalon, ed., *Middle East Contemporary Survey, 1990* (Boulder, San Francisco and Oxford: Westview, 1992), 404.

44 KDS Folder 90809 File 908-2-26 p.11.

45 Saddam granted him all encompassing powers in the north of Iraq, through a RCC decree on 29 March 1987.

46 KDS Folder 90809 File 908-3-19 p. 30-33.

47 KDS Folder 90809 File 908-2-17 p. 2.

Army for the simple purpose of providing brute force. However, these auxiliary Popular Army troops were not required to report to the Special Forces but directly to al-Majid and Saba'wi. Most correspondence of both the Special Forces and Popular Army usually began with "We received information" and follows with specific intelligence, usually military, and how they should act upon the given information, usually provided by Military intelligence. The vast majority of all their correspondence issued a command that would end with, "Please look into the matter and do what is necessary."[48] This wording however, was more than a polite request.

One tool of the Kuwaiti resistance was to wage war against the rhetoric of their invaders with their own propaganda campaign. A significant document that appeared in one of the offices of the Iraqi security apparatus was a leaflet, entitled, "A Call to Arms from Kuwaiti Resistance to Everybody in Kuwait."[49] It begins, with the heading, "God (*Allah*), the Nation (*al-watan*), and the Prince (*al-amir*)." In this instance, the Kuwaitis demonstrated how they invoked the term *al-watan* opposed to the Iraqi term *al-qawm*, as a their response in a war of rhetoric with the Iraqis. The Kuwaiti decision to use the term al-watan is significant as it is a direct challenge to the Iraqi appeal to al-qawm. With this simple choice of words, the Kuwaiti resistance reasserts its independence and nationhood. They are a sovereign state under God and their Prince.

The leaflet opened with a verse from the Quran, "They took over the land, and increased the corruption, and your God made them suffer, your God is waiting for them."[50] The verse essentially casts the Iraqis as conquerors and aggressors, placing God on the Kuwaiti side. The leaflet continues is the same theme, "People of the resilient Kuwait, do not give up on the dear motherland, do not cross your brother because the motherland is reliant on all, whether Kuwaiti or non-Kuwaiti Arab." Perhaps, the author or authors of the leaflet intended to reach non-Kuwaitis, realizing that most of Kuwait's population were expatriate Arabs. This was an appeal to them not be taken by Iraqi suggestions of greater "Arabhood" but to be loyal to the country that has been a home to them. "For that, at this time we ask all young men Kuwaiti or any other Arab who would like stand side by side and defend this dear motherland, to return it to his Excellency, the prince Jabir Ahmad al-Sabah, and his noble government. We ask God to make this land free and secure and to damn those who have occupied this land and stole its blessings." The leaflet was signed in the name of the "State of Kuwait" and the Kuwaiti resistance, who used the title, *al-Shabab al- Murabitun.*[51]

48 KDS Folder CD 17 File 132-1-003 p. 11.

49 KDS Folder 90810 File 908-3-116 p. 390.

50 Sura, verse. ???? COULDN'T FIND THE VERSE !!!

51 KDS Folder 90810 File 908-3-116 p. 390.

The Kuwaitis also issued another flyer to appeal to Iraqi soldiers asking for their insubordination to the orders of the Iraqi government. The flyer began, "To the Iraqi soldier, o Iraqi hero who defended the Arab nations from the greedy and sacrificed your bodies. We the people of Kuwait did not forget that, for that we ask for you to consult your conscience and you will know that the honourable heroic soldier will not be content with the hostile actions taken against your brothers and neighbours."[52] This statement was an indirect reference to the sacrifices Iraqi soldiers made during the Iran-Iraq War and a reminder of Kuwaiti financial support during this conflict. The flyer continued in this vane, "In the name of God we ask you to consult your conscience and mind and ask is it just for an Arab to attack his Arab neighbour and this land that had supported you in the past."[53]

The flyer then made several implorations to the Iraqi forces. "O Iraqi hero, do not kill yourself and do not kill your Arab brothers and do not point your weapon to their chests. O Iraqi, do not steal the security from the eyes of mothers, and do not steal the opportunity from the eyes of children, O honourable hero."[54] In this instance, the Kuwaitis employed the same theme of the common ethnic bond between the two Arab antagonists that Iraq used to make a case for absorbing Kuwait to argue for its rights to independence. It continued with this theme, "We the Kuwaitis are your family, and if you throw your weapons you can share our food and live among us you and your family. And God will grant victory."[55] The flyer ended with a rejection of Iraq's attempt at forming a new Kuwaiti government, stating, "Dear Iraqi: we the Kuwaitis will not be content except living under any government except for our legitimate government, under the leadership of our prince, Jabir, may God bless him."[56]

The first documented instances of coordinated Kuwaiti demonstrations and armed resistance began in late August. On 27 August 1990, the Popular Army command in Kuwait issued directives to all its units about a group of Kuwaitis that were carrying out assassination and kidnapping operations against the soldiers and members of the Popular Army. It stated that the resistance was using the mosques during prayer times to organize their activities, and furthermore, that they also used women to assault their fighters. These challenges would not go on unchecked, and a decision was made to "target such incidents to teach them manners and prevent this phenomena."[57]

52 KDS Folder 90810 908-3-122 p. 167.
53 Ibid.
54 Ibid.
55 Ibid.
56 Ibid.
57 KDS Folder CD 17 File 132-1-003 p. 11

Criticisms of the Iraqi president were taken seriously and dealt with brutally. A Kuwaiti lady and her 15-year old son were arrested for cursing the President while waiting in line to get her pension check.[58] These forces also arrested two sisters caught writing hostile propaganda (i.e. graffiti) a boy's elementary school with 'hostile propaganda'[59] such as "Allah, the Nation, Jabir", and *'ash Kuwait,* "Kuwait lives", and *la lil-Ba'th wa la li-Saddam,* "No to the Ba'th, no to Saddam."[60] A 50 year old Kuwaiti woman was arrested for cursing the president leader, saying, *allah ya'khud 'umr Saddam*, "May God take the life of Saddam."[61] One Kuwaiti was apprehended for having "inimical flyers in his desk", which included a letter in his daughter's writing cursing the "heroic" *al-batil* Iraqi army and was sent to the General Security Service for questioning.[62] While an Iraqi would never consider cursing the President knowing full well that such reckless behaviour would inevitably lead to incarceration and interrogation, it seems the Kuwaitis failed to realize the severity of such actions in what they still perceived as their own country under occupation. Indeed, by this point the "Republic of Fear" and all aspects of Saddam's *tarhib* had emerged in Kuwait.

Saudi and Syrian Support for the Resistance

The Iraqi security services attributed most of the acts of Kuwaiti resistance to support from Saudi Arabia. A memorandum on 29 August from the Popular Army Command in Kuwait states, "a group of Saudi Arabian elements" were financing the "saboteur groups" in Kuwait by providing them with weapons, medium machine guns, hand guns with silencers, and wireless "sticky" American mines that could be exploded by using a wireless device, which could be attached to the military cars and internal security vehicles.[63]

The Ba'ath Party informed 'Ali Hassan Al-Majid that the Saudi government was sending 500 dinars monthly to every Kuwaiti family that resided in Kuwait. Members of the Bedouin tribes were smuggling Kuwaitis to Saudi Arabia and vice versa for 5000 dinars a person. The Party reported that large numbers of Kuwaitis did not respond to the orders to exchange their Kuwaiti ID's to Iraqi ID's because they feared physical harm from other Kuwaitis or assassination by Saudi agents.[64] The letter warned that these specific armed groups would attack the armed forces located in the Kuwaiti neighbourhoods of Al-Nada' and Al-Salimiyya and that, "This sabotage action will happen in

58 KDS Folder 90809 File 908-1-19 p.1-4.
59 Ibid.
60 KDS Folder 90809 File 908-3-31 p. 4.
61 KDS Folder 90809 File 908-s-61 p.7.
62 KDS Folder 90809 File 908-2-77 p. 2.
63 KDS Folder CD 17 File 132-1-003 p. 35.
64 KDS Folder CD001A File 104-1-17 p. 184.

coordination with the expected deployment of the American troops, for the purpose of confusing the domestic battle."[65]

Just a few days after the invasion, a Military Intelligence report warned that the Syrians might also play a role in aiding these Kuwaiti elements.[66] On 29 August 1990, the Ba'ath Party wrote to the Presidential Secretariat to inform them that the Syrian Embassy in Kuwait had distributed 200 Russian made rifles to the Syrians residing there.[67] An Iraqi informant stated that the stores used to repair flat tires were in fact fronts for Syrian intelligence officers.[68] The Iraqi police in Kuwait sent an alert to all the security agencies in Kuwait, including the Popular Army, Special Forces, General Intelligence, and General Security that a former Kuwaiti officer was leading a group of resistance fighters from the Syrian neighbourhoods in Kuwait in an effort to avenge the death of his father. This same officer had brought in doctors to open small clinics in these Syrian residences to treat resistance fighters and their sympathizers.[69] According to another informant, a Syrian with Kuwait nationality worked for the Syrian intelligence based in Saudi Arabia.[70] Such reports form a larger picture of the Iraqi siege mentality, where foreign forces, both Arab and Western, were conspiring to undermine the Iraqi mission in Kuwait.

The Iranian Connection of the Kuwait Resistance

One author wrote that the first armed resistance to emerge in Kuwait City among civilians was started by Kuwaiti Shi'a of Iranian origin.[71] "Individuals, mostly Shia, sniped at Iraqi troops but this was soon countered by tanks."[72] The Kuwaiti Shi'a, many of whom were loathe to seek refuge in ultra-Sunni Saudi Arabia stayed in Kuwait to defend their nation, and perhaps take revenge against the country that waged war against their native homeland, Iran.

The Iraqi intelligence officers made an effort to differentiate the Kuwaitis who were Sunni Arabs, and Kuwaitis of Iranian origin. They were particularly aware of those Kuwaitis originally form the province of Khuzestan, bordering southern Iraq, since they formed the bulk of the Kuwaiti resistance. For example, a Popular Army unit reported on a "destruction net", another euphemistic term for the Kuwaiti resistance, whose ten-member cell was almost entirely Kuwait Shi'a of Iranian origin. Some of the members were Ghulum Ali

65 KDS Folder CD 17 File 132-1-003 p. 35.
66 KDS Folder C 09 File 104-15-001b p. 38.
67 KDS Folder 1215-1429 File 908-3-112 p. 104.
68 KDS Folder 90809 File 908-3-19 p. 30-33.
69 KDS Folder 90809 File 146-1-10 p. 2.
70 KDS Folder 90809 File 908-3-19 p. 37- 41.
71 Adel Darwish and Gregory Alexander, *Unholy Babylon: The Secret History of Saddam's War* (New York: St. Martin's Press, 1991), 279.
72 Ibid., 281.

Hussain and his son-in-law, Jabir Ghulum who worked with the Kuwaiti Police, and Anwar 'Abd al-Zahra Al-Hahad, whose names would only be used by Shi'as.[73] The other members included six members with the last name Dashti, another prominent Kuwaiti family of Persian descent.[74] According to one source, the Dashti family ran their own resistance group, comprised primarily of their family members.[75] General Security had also tried to find a Kuwaiti of Iranian origin who belonged to a movement called the "panthers" which consisted of 11 members.[76]

A letter from General Security to 'Ali Hassan al-Majid detailed the extent of the Iranian connection to the resistance. It provided an account of a kidnapped Iraqi, who had been a long time resident of Kuwait. The resident reported being kidnapped by four Kuwaitis who interrogated him, insisting that he confess that he was an agent of the (Iraqi) *Istikhbarat*. In an exchange between one of the Kuwaitis and the Iraqi, the Kuwaiti jokingly threatened to keep him hostage, until the latter's son would conduct a "terrorist operation" (*'amaliyya takhribiyya)* against the Presidential Palace in Baghdad. The Iraqi noticed in this basement another Iraqi soldier of the Popular Army was also being held hostage. He then overheard how the resistance fighters intended to kidnap an officer in the Republican Guard. Somehow he was able to distinguish that only one of the men was Kuwaiti and that three men were "Iranians", and he counted 40 rifles in their possession. They released him on the condition that he leave Kuwait immediately, but instead he reported to General Security and offered to lead them to the house of his captors.[77]

The Persian background of many of the resistance fighters certainly compounded Saddam's "siege mentality" of Iranian collaboration to undermine his position in Kuwait. The Popular Army had reported that the Iranian embassy in Kuwait had armed all those Kuwaitis with Iranian background with arms.[78] Such fears were not unfounded, as the Kuwait resistance claimed that Iran aided them and even trained them in making explosives.[79]

73 The name Ghulum can be translated as "slave", in this case the "slave of Ali", (ie Imam Ali). The other member was Anwar 'Abd al-Zahra Al-Hahad, whose middle name, 'Abd al-Zahra, is usually used only by the Shi'a ("the servant of Zahra").

74 KDS Folder 90809 File 161-1-22 p. 2

75 Andrew Whitley, "Kuwait: The Last Forty-Eight Hours", in Robert B Silvers and Barbara Epstein, eds., *A Middle East Reader* (New York: New York Review of Books, 1991), 174.

76 KDS Folder 90809 File 908-3-10 p. 170.

77 KDS Folder 90809 File 908-3-19 p. 44.

78 KDS Folder 90809 File 908-2-71 p. 2.

79 Miron Rezun, *Saddam Hussein's Gulf Wars: Ambivalent Stakes in the Middle East* (Westport, CN, and London: Praeger, 1992), 74.

Non-Kuwaitis in the Resistance

Besides Syrians, other Arabs took part in the Kuwaiti resistance. For example, a former Egyptian soldier and resident of Kuwait led a unit that "facilitated some of the negative and terrorist elements in this area."[80] Another member of the Kuwaiti resistance and his sister were arrested, and later it emerged that in fact they were originally Iraqi.[81] According to a Popular Army report, the Kuwaiti resistance was offering persons of all nationalities first grade Kuwaiti citizenship[82] and 30,000 Kuwaiti dinars for joining the their ranks.[83] The same unit issued a report that the "removed Kuwaiti Qarun government" had contacted a network of suicide persons in Japan and allegedly had paid very large sums of money to their families for carrying out suicide missions against the Iraqi forces in Kuwait City to demonstrate to the world that there "is a national resistance movement in Kuwait."[84] Non-Kuwaiti organizations also offered some resistance to the Iraqi forces. General Intelligence began to pursue members of the Iraqi Da'wa Party and Communist Party in the first hours of the invasion.[85] On the 26 August 1990, regular army units received directives from Military Intelligence that the Iraqi Communist party intended to blow up various vital targets in Iraq.[86] In one instance, the Ba'ath Party forwarded to 'Ali Hassan Al-Majid information that an escaped communist was currently in Kuwait and had joined an Iranian group of "bad elements."[87]

The other target of the security apparatus was the Iraqi Islamist organization, al-Hizb al-Da'wa ("The Call Party") and their network in Kuwait. Iraqi informants had offered lists of their names and their residences to General Security.[88] One informant also named another employee of a Kuwaiti oil company who was a member of the Lebanese Hizbullah, and that two employees in the same oil company that had collected donations that were sent

80 KDS Folder 1215-1429 File 908-3-112 p. 104
81 KDS Folder 90809 File 908-3-10 p. 170
82 First Grade citizenship is reserved for those Kuwaitis, who can trace their lineage back to three generations of inhabitants of the emirate. Such a rank is usually reserved for Kuwaiti Arab Sunnis, who ironically for the most part originated in the Najd area of the Arabian Penninsula.
83 KDS Folder CD001A File 148-1-1 p. 106 and KDS Folder CD 17 File 132-1-003 p. 2
84 KDS Folder CD001A File 148-1-1 p. 107 and KDS Folder CD 17 File 132-1-003 p. 29
85 Darwish and Alexander, *Unholy Babylon*, 281.
86 KDS Folder 90810 File 525-3-40 p. 181
87 KDS Folder 1215-1429 File 908-3-112 p. 105
88 KDS Folder 90809 File 908-3-19 p. 18.

to Iran.[89] Eleven Iraqi informants, who had been residing in Kuwait, provided all the information on the Da'wa cells.[90]

The multi-faceted nature of the resistance in Kuwait, most likely gave the impression to Saddam and his security apparatus that numerous agents, including the US, Israel, the Gulf states, as well as Iraqis themselves had formed an axis to undermine the rightful "re-unification" of Iraq and Kuwait. From August to the beginning of the Gulf War in January, Saddam worked on developing a strategy to eliminate this threat emerging from within the 19th province, as he had done to eliminate the internal threats to his state that emerged from within Iraqi Kurdistan during the Iran-Iraq War.

The Iraqi Informant Network

The General Security Service recruited most of the informants working on behalf of the Iraqi security apparatus to report on subversive Kuwaiti activities. Correspondence on August 16th 1990, from Ali Hassan al-Majid (at that time the head of the Ba'th party in Kuwait) orders Husayn Kamil, head of MIMI, to forward the intelligence gathered from their eavesdropping equipment and stations, which are under control of the Centre of Technical Research.[91]

On 29 August 1990, General Security informed Ali Hassan al-Majid, that an Iraqi living in Kuwait for six years had provided them with the information that a Kuwaiti company was hiring American intelligence operatives who were posing as engineers. The informant also provided the location of escape routes used by the Kuwaitis, as well as the locations of the Amir's financial assets. He confirmed that religious cooperatives were attached to the former regime and that they were the "right hand to Qarun al-Kuwait," since they smuggled food to Kuwaiti families and served as conduits of information from Saudi Arabia to Kuwait.[92]

General Security Branches in Iraq usually had records of Iraqis living in Kuwait and recruited them due to their value of having familiarity with the 19th province. The director of General Security sent a 12-page report to Ali Hassan Al-Majid, on 29 August, which included secret information gathered by General Security branches in the Iraqi provinces of Dhi Qar and al-Muthanna. The report gave the names of the people that could serve as informants and detailed the information that they had already provided to General Security. Among the informants was an Iraqi, who claimed to know the location of the houses of two Kuwaiti officials, the homes of the owners of ten car dealerships, as well as owners of eight bank branches, all of whom were connected to the

89 Ibid.
90 Ibid.
91 KDS Folder 90809 File 908-1-38 p. 46.
92 KDS Folder 90809 File 908-3-19 p. 30-33.

"old regime."[93] A second Iraqi, who was had worked twenty years with a construction company that belonged to Muhammad Khalifa al-Sabah, brother of the former Kuwaiti oil minister, offered his help in locating the homes of the bodyguard of Ahmad Jabir al-Sabah.[94] A third Iraqi, who had lived in Kuwait for 17 years, offered information on Kuwaiti officers in his area, as well as the locations of satellite dishes and transmission facilities and also reported how officers were escaping to Saudi Arabia through unguarded desert passages.[95] Another Iraqi businessman, who had lived in Kuwait for 20 years offered information on hidden Kuwaiti financial assets and the whereabouts of a former Kuwaiti National Guard officer operating a smuggling ring for Kuwaitis and foreigners to Saudi Arabia.[96] A striking feature shared by all these Iraqi informants is that they had been living in Kuwait for more than a decade, a span of time seemingly long enough for these informants to have made formed loyalties to their host nation. Under these circumstances, it raises the question of whether these Iraqis unwillingly provided information by force or under the threat of torture or if they volunteered information in exchange for offers of security and financial incentives.

While it is difficult to make generalizations about why Iraqi expatriates in Kuwait chose to act as informants, the incidences of Iraqis coming forth with information are numerous. Another woman informed General Security that she was approached by the Resistance to work with them. She told General Security that she was willing to be a double agent and report on the activities of the resistance fighters and locations of their weapons[97] One Iraqi informant promised General Security that he knew the whereabouts of the former Kuwaiti director of state security as well as Kuwaiti intelligence officers working in the Directorate.[98] The Division of Arab Volunteers in the Kuwait Province (*Qati'ah al-Mutatawa'in al-Arab min Muhafazat al-Kuwait fil Wasit*) (where is al-Wasit), wrote to the Popular Army that a volunteer Arab fighter among their forces who used to work in a Kuwaiti army base of the "previous regime" had reported that high ranking military officers were still hiding in Kuwait and provided the addresses of eight of them.[99]

A Ba'ath party official wrote to al-Majid of a friend he had known since his childhood, who was a Lebanese businessman with a wide network of "Kuwaiti, Arabs and foreign" acquaintances. The official admitted, "he has a lot of secret information and some are rumours, some stories", but among them that the

93 Ibid., 9-15.
94 Ibid.
95 Ibid.
96 Ibid.
97 Ibid., 15.
98 Ibid., 17.
99 KDS Folder 90809 File 908-3-18 p. 2

Kuwaitis who have remained still used Kuwaiti money, and were not being charged in cooperatives, but rather kept tabs.[100]

Informants usually had to swear a signed affidavit to the Ba'ath party vowing he or she had no ill intentions in providing information. In one of these letters, the informant who was apparently a Kuwaiti exile, "hopes that the leader (Saddam) will let him return to his brothers in Kuwait, after he has been deprived of living in Kuwait for the last two and half years, from the tyrant of Kuwait." He stated that he had connections with very important people still in Kuwait, almost twenty people, including the head of the Kuwaiti Fadhul tribe.[101]

General Security approached the *khatib*, "the prayer leader" of Riqa mosque and drew him to their side. "He will have an active role in the general education of the population especially in the Friday sermons."[102] One of the most prominent informants was the head of a Kuwaiti tribal federation, who gave information on two houses in the neighbourhood, one of which was the home of the head of a Kuwait government official and whose sons had changed their identity and were responsible for anti-Iraq graffiti. He also reported that Kuwaitis living in Saudi Arabia had to obey the Saudis otherwise they would be exiled to the "empty quarter", *ruba' khali*. He then confirmed that all the tribes that left to Saudi Arabia were against the Iraqi government but that the ones that stayed are loyal to the President.[103]

These documents demonstrate how the Iraqis formed an informant network in Kuwait similar to that of Iraq and perhaps one of the most notorious features of Iraq's "Republic of Fear." The facets and pillars of Iraq's mukhabarat state had manifested itself in Kuwait, ultimately serving Saddam's goal of returning Kuwait to its "Iraqi roots."

September 1990

By September 1990, the tactics of the resistance shifted from civil disobedience to launching a series of assassinations and bomb attacks, and the Iraqi severity in responding to such incidents also increased.

The resistance had various methods of targeting Iraqis soldiers, including sabotage. The Popular Army command in Kuwait forwarded intelligence they received that the resistance was booby-trapping civilian cars that were left in the streets with explosives to target the soldiers that are trying to "benefit" from the deserted vehicles.[104] They also warned that the resistance would booby-trap cigarette boxes, or poison them and throw them onto the street for the un-

100 KDS Folder 90809 File 908-3-19 p. 5.
101 Ibid., 6.
102 KDS Folder 90809 File 908-2-61 p. 6.
103 KDS Folder 90809 File 908-3-19 p. 37- 41.
104 KDS Folder CD 17 File 132-1-003 p. 8.

expecting soldier to smoke.[105] A circular was passed from Military Intelligence to all Iraqi military units to warn them that the Kuwaitis had been poisoning food, such as drinks and watermelons and then would distribute it to the Iraqi fighters. If an Iraqi soldier were to accept such items in the future, the punishment was severe, assuming that in fact the food was not poisoned, and he could live long enough to meet his punishment.[106] As a response all soldiers were forbidden from purchasing food supplies from the local markets, and had to rely on their military rations. Some samples of poisoned food were sent to Military Intelligence for testing.[107]

Other acts of sabotage included arson. The Popular Army command in Kuwait reported that the resistance had burned centres that were of "economic value to the state,"[108] such as warehouses of food, military preparations and weapons.[109]

The nature of armed attacks launched by the resistance indicated that they were receiving lethal military assistance from outside sources. A Military Intelligence report indicated the level of armament available to the resistance, when security units discovered a GMC van, that had forty TNT explosives, five gas masks, seven 60mm rifles, one anti-tank mine, one American canon, and one anti-tank missile.[110] Some of the armed resistance attacks were intense. An army lieutenant had run out of ammunition, after expending 150 rounds repulsing one attack.[111] The resistance had managed to even obtain missiles, which were shot at an Iraqi civilian airliner.[112] When such weapons were not available, the resistance improvised. For example, a person was arrested for attempting to drive a Mercedes into an airplane as it tried to take off from Kuwait airport. In fact, an Iraqi residing in Kuwait was suspected of taking part in this plot and was sent to General Security for his involvement. [113]

A letter dated on the 19th of September from Ba'ath party security office, *Maktab Aman Sir al-Qatar* demonstrated that the security service could obtain detailed intelligence about resistance fighters. The Ba'ath office informed al-Majid of a member of the Kuwaiti resistance, who was carrying a fake ID card from the Ninawa Province and drove a red Cadillac. He would invite Iraqi

105 KDS Folder 90809 File 908-3-56 p. 3
106 KDS Folder Iraq-Docs File 73-003 p. 2.
107 KDS Folder 90810 File 525-3-40 p. 165.
108 KDS Folder CD 17 File 132-1-003 p. 3.1
109 Ibid., 9.
110 KDS Folder 90810 File 908-3-122 p. 120.
111 KDS Folder 90809 File 646-1-8 p. 14.
112 KDS Folder 90810 File 908-3-122 p. 11-13.
113 KDS Folder 90809 File 908-2-18 p. 5.

soldiers to his house for food, kill them and take their weapons. According to Ba'ath, he had killed 11 soldiers.[114]

The Iraqi Response to Resistance Tactics

Several meetings were held in the beginning of September to coordinate the tactics for maintaining internal security and suppressing the resistance in Kuwait. On the 2 September 1990, a three-day meeting of the Special Forces of the Gulf Area (i.e. Kuwait) was convened along with representatives from the Police, General Intelligence and General Security.[115] The meeting agreed that clear lines of responsibilities needed to be established between the Special Forces and other security agencies. Any instances of the Special Forces dealing with foreigners had to be directed to General Intelligence, the agency tasked with dealing with foreign matters. If the Special Forces captured any soldiers in the Kuwaiti Army, they were to be surrendered to the Popular Army, or if they discovered a case of civilian theft or anyone not using Iraqi dinars, the apprehended were sent to the police.[116] The nature of the meeting indicated that perhaps clear lines of division and responsibility had not existed, or that al-Hintah was exceeding his mandate in Kuwait.

On the following day, another meeting took place between the Kuwait-based commanders of the Special Forces, The Popular Army, General Security, the Police, General Intelligence and Military Intelligence. The gathering took place to discuss preventive steps against the Kuwaiti residents on the occasion of the Prophet Muhammad's Birthday, which would take place on 30 September 1990.[117] It was decided that all security services would send personnel to every mosque in Kuwait to monitor the sermons. The Special Forces, in coordination with General Security, were to be on full alert to deal with any potential instances of demonstrations, public gatherings, and "fiery" or "hostile" prayer speeches. Afterwards, the committee discussed the issue of building a new garage.[118]

On the third and final day of the conference, it was decided that General Security, General Intelligence and the police should only be carrying out their duties and refrain from taking part in military operations in the cities. Their duties were limited only to providing information and observation and they could only respond militarily to armed attacks.[119]

114 KDS Folder 90809 File 908-3-19 p. 49.

115 KDS Folder 90810 File 148-1-3 p. 255.

116 Ibid., 221.

117 Abdulla Hammadi, *The Big Terrorism of Nations Destroyers: Documented Stories and Atrocities during the Savage Iraqi Invasion of the State of Kuwait* (Kuwait: n.p., 1992), 350.

118 KDS Folder 1215-1429 File 565-2-3(2) p. 197-98.

119 KDS Folder 90809 File 908-1-38 p. 139.

The Special Forces and the Popular Army emerged as the key armed, security units charged with crushing the resistance. During this period, they produced vast volumes of documents, detailing various *tarhib* tactics. The documents are characterized by a bureaucratic rigidity, where acts such as executions were reported in a sanitized manner. For example, a Special Forces report states as, "A lady driving the car who was out of control was shot in the head."[120] The Kuwaiti resistance threatened Kuwait shopkeepers to close their stores at certain times and if they did not they would confiscate materials from their merchandise.[121] In response to such tactics, the Special Forces wrote, "The resistance ordered businesses to be closed, we will force them open."[122] Special Forces Brigade 66 in a report stated, "Gunfire was reported in an area. A unit was dispatched, found the weapons, and demolished four houses."[123] In a report of the Special Duties Forces 894, "A unit went to the Arabian Gulf to search some of the suspected house and to execute some of the criminals of the previous regime, who had been responsible for the killings of Iraqi soldiers. Some of the houses were burnt and then we returned to HQ without incident." One must wonder if executions do not constitute an incident for these soldiers, then what would. The officer of this unit then requested to be provided with bulldozers and flamethrowers to burn homes in the future.[124] Another document indicating the stark and callous discourse of the Special Forces' tactics said, "Hostile elements fired on us from deserted buildings. Our forces returned fire and called for help. Within minutes more of our forces arrived at the location and opened fire at the building with RPGs[125] and 100 bullets, bringing the situation under control."[126]

Demonstrations of Disobedience

The Special Forces were used as al-Majid's long arm for internal repression in Kuwait. Hintah employed harsh collective punishments to quell the slightest incidents domestic disobedience,

The Kuwaitis would take part in a coordinated act of civil disobedience, by screaming "God is Great" from the top of their roofs at the same time. Hintah writes, "We have received information that on the first of September from 1200 noon to 1230 that the Kuwaitis will go to their roofs to pray. This can be remedied (*tu'allij*) by opening fire on the roofs that has people praying, with all

120 KDS Folder 90809 File 908-3-25 p. 3-5.
121 KDS Folder 90809 File 107-3-026 p. 48.
122 KDS Folder 90809 File 590-1-9 p. 3.
123 KDS Folder 90809 File 908-3-10 p.111.
124 KDS Folder 90809 File 908-2-49 p. 2 and p. 6.
125 Rocket Propelled Grenades.
126 KDS Folder 90809 File 908-3-48 p. 6.

weapons."[127] Iin another report it mentioned how five Kuwaitis had been arrested for "hostile prayer" *(aswat takbir mu'adi)* from their rooftops.[128]

On 8 September, al-Hintah sent orders to the 65th Brigade on how to handle demonstrations that were to be held the on the day of the the Gorbachev-Bush summit, which resulted in a joint US-Soviet condemnation the invasion of Kuwait. He ordered the brigade to find the location and size of demonstration. He then command them to formulate a plan:

> You need to walk to the demonstration area, without vehicles and quietly, quietly, quietly get close to the demonstrators from behind and close their alternative routes of escape. Then open fire at the same time with everything you have including rifles, automatic weapons, light cannons, and flame-throwers for the objective of killing all the demonstrators so to serve as an example to all others. [129]

He then ordered them to prohibit any civilians from approaching the demonstration site, and to be stationed there for 20 hours on the day of the 8th.[130] While the Popular Army was charged with surrounding the area in the case of this "hostile demonstration," the Special Forces were charged with "pacifying" the gathering in this manner.[131]

Home Demolitions

By September, the Special Forces began a policy of burning down homes in response to various crimes. By 10 September 1990, more than a month after the invasion, the Special Forces intensified their operations stating, "Due to the circumstances and the need for harsh measures with the saboteurs the soldiers are allowed to burn and destroy any house that contains weapons. Wherever Iraqi soldiers casualties are found, the entire block is to be burned and destroyed using flamethrowers. The women and children are to be allowed to leave the homes."[132]

Homes could be destroyed if the walls had "writings or propaganda (ie graffiti) from the buried (*maqbura*) government."[133] The Special Forces would destroy the home whether or not the homeowner was the author of the graffiti.

127 KDS Folder 90809 File 590-1-9 p. 3.
128 KDS Folder 90809 File 908-2-61 p. 4.
129 KDS Folder 90809 File 590-1-8 p. 2.
130 Ibid.
131 Ibid.
132 KDS Folder CD001A File 148-1-1 p. 198.
133 KDS Folder 90809 File 908-2-49 p. 2 and p. 6.

In one instance a Kuwaiti's house was burnt for displaying a picture of the Amir.[134]

In one circular, Al-Hintah gave guidelines on how to search homes. "Knock first. If no answers, break in. If family is inside take them outside – burn the house in front of them, interrogate them, then take them to Basra prison."[135] Al-Hintah is required to place a reminder in the text of these guidelines. "We ask our brothers and comrades that when burning or demolishing any house to abstain from taking any property, regardless of price or need."[136]

Executions

Executions became more common by the first week of September. It was decided at the three day conference held from the 2 to 4 September that, "At any time the units are fired upon, the area will be searched and the people in the area will be arrested, and hanged publicly in front of the people, immediately. Their families will be deported out of Kuwait and their belongings will be confiscated."[137] On the 26th another meeting was held, ordering, "When searching a home, and there are explosives or hand grenades, the person should be executed immediately in front of his neighbours and to inform the neighbours of these crimes."[138] On the same day, such commands were implanted. IN this case the Special Forces reported on "terrorists" who opened fire on them from a house. As a result, "The force went into the house and arrested six people which they sent to Amn al-Kadhima and burned the house immediately." In a handwritten note by al-Majid, he wrote of their fate. "After swift investigations they were involved in the shooting and executed in the same area."[139]

While numerous Kuwaitis were executed, three documented cases indicated the Iraqi justifications for such a sentence. During one Special Forces patrol, they arrested the manager of a cooperative society, since he closed the cooperative on the one-month commemoration of "the liberation of Kuwait", for displaying pictures of the former government in his office, and for "his hate and his bad relations" with the Iraqi soldiers and forbidding them from entering the cooperative. He continued to meet with the members of the resistance, refused to hang a picture of Saddam on the store and refused to use the Iraqi currency. Additionally they found a letter belonging to his daughter where she criticized the Iraqi army. In this instance, a handwritten note in Ali Hassan al-

134 KDS Folder 90809 File 908-2-35 p. 2-3.
135 KDS Folder 90809 File 605-1-014 p. 2.
136 KDS folder 1215-1429 file 146-1-8 p. 4-5.
137 KDS Folder 90809 File 908-1-38 p.139.
138 KDS Folder 90810 File 908-3-116 p. 5-9.
139 KDS Folder 90809 File 908-3-25 p. 3-5.

Majid's writing said, "If these accusations are proven, he should be hanged in front of the cooperative for six hours and to confiscate his assets."[140]

On the 14 September 1990 an informant reported to a police station that five Kuwaitis fired upon Iraqi troops. A joint detachment of Iraqi Special Forces and the police found three bodies, some belonging to the Republican Guard. The informant provided the name of the Kuwaitis and the location of the home involved in the attack. "The house was vacant. We searched meticulously and then burned the house down." A joint team of The Special Forces, the Police, and Popular Army later ambushed a suspect and questioned him. In this instance he was most likely tortured but such terms were always absent in official correspondence, but the document describes the graphic nature of his execution, by first burning him with gasoline and then shooting him through the head. His body was thrown in front of the building where the attack took place to serve as an example to others. Later they captured the second suspect who was also executed him by shooting him in front of the building where the crime was committed.[141]

On 29 September 1990, the Popular Army Shatt al-Arab Unit saw a group of children holding Kuwaiti flags and cheering against the Iraqis. They chased the kids, who then entered a house. The owner of the house and his wife came outside and spoke in "hostile language" to the unit. They were arrested and sent to General Security office in al-Jahra. In handwriting, al-Majid ordered the execution of the owner of the house and his wife.[142] The words in the handwritten comment were transferred into an official letter dated 30 September 1990, ordering the execution of the couple in public and to burn their belongings.[143] In only of case of the aforementioned executions, was the accused a member of the Kuwaiti resistance. The other two incidents involved simply cursing the Iraqis.

The fate of those captured by the Iraqi security apparatus was rarely mentioned in the documents. The word "torture" would never be used, but rather the term "interrogation." For example, General Security received information of a member of the Kuwaiti resistance involved in the killing of a soldier and officer of the Special Forces, and the burning of a tank. "He has been captured and is under interrogation, where it was revealed that he was a former captain of the Kuwait police."[144] Given the list of his "crimes" he was most likely tortured while under interrogation, but this was assumed rather than mentioned directly.

140 KDS Folder 90809 File 908-1-38 p. 145-48.
141 KDS Folder 90809 File 533-1-002 p. 18.
142 KDS Folder 90809 File 908-3-98 p. 5.
143 Ibid., 3.
144 KDS Folder 90809 File 908-3-21 p. 3-4.

Conclusion

The collective punishments employed by the Iraqi security apparatus had the desired effects by the end of September 1990. The exiled Kuwaiti monarch called on the Kuwaiti resistance to halt such attacks by October due to the suffering of the Kuwait civilian population. While acts of resistance continued, the intensity of such acts declined after this plea, as well as the result of the brutal suppression used by Saddam.

Saddam had a tendency to assign similar duties to each of the agencies of this apparatus following the policy that competition would produce better results, thus decentralizing the network ostensibly centred around the Special Security Organization. While agencies competed with each other for intelligence in the field of foreign and domestic operations, the documents of the Kuwait Data Set demonstrated that Saddam still managed to develop a highly centralized and personalized security network in Kuwait. The nature of the Iraqi security apparatus was a reflection of the personality and mindset of the individual of Saddam Hussein and his personal desire to impose an even firmer grip on the vast network that maintained his regime. Saddam's tendency to form separate security bodies from within existing security organizations, revealed his perceived ineffectiveness of existing intelligence agencies or from a potential internal threat from within his apparatus of state control. It was in this atmosphere of mistrust and insecurity that Saddam went on to create various other intelligence agencies. Saddam managed to survive for decades by using such cleavages to his advantage, preventing one agency, or one of his own relatives for that matter from attaining the necessary strength to challenge his rule. In his view, competition between rival agencies, rather than cooperation and information sharing worked to guarantee his survival.

[14]

The Myth of Air Power in the Persian Gulf War and the Future of Warfare

Daryl G. Press

On February 24, 1991, U.S. ground troops, supported by British, French, and Arab forces, moved north from Saudi Arabia to liberate Kuwait and destroy the Iraqi military.[1] Four days later the job was apparently done: The Kuwaiti flag once again flew over Kuwait City, and what remained of the Iraqi military was in full retreat. Astonishingly, only sixty-three Americans were killed in a ground operation that most analysts expected would cause thousands of U.S. casualties.[2] How did the coalition's ground forces destroy the Iraqi army so quickly and with so few coalition casualties? Does the Persian Gulf War herald a future of U.S. military dominance and low-cost U.S. military operations? What are the lessons from the fighting for U.S. foreign policy?

The conventional wisdom among historians, military analysts, and foreign policy decisionmakers is that air power neutralized the Iraqi military before the ground war began.[3] This interpretation of the Gulf War has important im-

Daryl G. Press is Assistant Professor in the Government Department and a Research Fellow at the Rockefeller Center at Dartmouth College. He is also an Associate of the John M. Olin Institute for Strategic Studies at the Weatherhead Center for International Affairs at Harvard University.

The author would like to thank Stephen Biddle, Stephen Brooks, Daniel Byman, Eliot Cohen, Eugene Gholz, David Kang, Thomas Keaney, Jennifer Lind, Thomas Mahnken, Michael Mastanduno, Karl Mueller, Kenneth Pollack, Barry Posen, Joshua Spero, Allan Stam, and William Wohlforth for helpful comments on earlier drafts of this article. Previous versions of this article were presented at the Dartmouth Seminar on International Politics, the John M. Olin Institute for Strategic Studies, the Stanford University Center for International Security and Cooperation, the Columbia University Institute of War and Peace Studies, and the New York Military Affairs Symposium.

1. Thirty-six countries joined the coalition against Iraq, but the ground offensive can be fairly characterized as a U.S. operation with British support. The coalition's ground force comprised nine and two-thirds U.S. divisions, two-thirds of a British division (two brigades), and two-thirds of a French division (two brigades), along with several less-capable Arab divisions. Only the U.S. and British forces were given important roles in the offensive.
2. The number of U.S. soldiers and marines killed during the ground war is from Stephen T. Hosmer, *Psychological Effects of U.S. Air Operations in Four Wars, 1941–1991* (Santa Monica, Calif.: RAND, 1996), p. 155. This number excludes the twenty-eight Americans killed by a Scud missile that hit a barracks in Dhahran, Saudi Arabia, during the ground war. For an account of prewar estimates of U.S. casualties, see Benjamin S. Lambeth, *The Transformation of American Air Power* (Ithaca, N.Y.: Cornell University Press, 2000), p. 2.
3. For the most prominent analyses that support this view, see Eliot A. Cohen, Director, *Gulf War Air Power Survey* (Washington, D.C.: U.S. Government Printing Office [GPO], 1993), 5 vols. (hereinafter cited as *GWAPS*), plus *Summary Report*. See specifically *GWAPS Summary Report*, pp. 116–117;

plications for U.S. foreign policy and for U.S. military procurement strategies. For U.S. foreign policy, the Gulf War seems to show—and the 1999 Kosovo conflict appears to confirm—that air power is now so lethal, and American air power so dominant, that the United States can win nearly cost-free military victories against its foes. For U.S. military procurement debates, the lessons are equally clear: The United States should significantly alter its military procurement priorities to favor air power over ground forces; in the future, decisive battles will be won from the air.[4]

U.S. confidence in air power is soaring, but the conventional wisdom about its decisiveness in the Gulf War has never been rigorously tested. It is undeniable that for six weeks—during the period now known as the air campaign—coalition aircraft dropped tons of bombs and missiles on Iraqi targets. It is also undeniable that Iraq's ground forces were totally ineffective against the coalition's ground forces. But those facts do not prove that the bombing caused Iraqi ineffectiveness. The questions that must be answered are: Why were the Iraqi ground forces so incapable during the ground war? Were they neutral-

---

GWAPS, Vol. 2, Pt. 2, p. 107; James A. Winnefeld, Preston Niblack, and Dana J. Johnson, *A League of Airmen: U.S. Air Power in the Gulf War* (Santa Monica, Calif.: RAND, 1994), pp. 5, 285, 287; Thomas A. Keaney, "The Linkage of Air and Ground Power in the Future of Conflict," *International Security*, Vol. 22, No. 2 (Fall 1997), p. 147; Michael R. Gordon and Bernard E. Trainor, *The Generals' War: The Inside Story of the Conflict in the Gulf* (Boston: Little, Brown, 1995), p. 474; Rick Atkinson, *Crusade: The Untold Story of the Persian Gulf War* (Boston: Houghton Mifflin, 1993), p. 495; Hosmer, *Psychological Effects of U.S. Air Operations;* Robert A. Pape, "The Limits of Precision-Guided Air Power," *Security Studies*, Vol. 7, No. 2 (Winter 1997/1998), pp. 93–114; Robert A. Pape, "The Air Force Strikes Back: A Reply to Barry Watts and John Warden," *Security Studies*, Vol. 7, No. 2 (Winter 1997/1998), pp. 191–214; Aharon Levran, *Israeli Strategy after Desert Storm: Lessons of the Second Gulf War* (London: Frank Cass, 1997), p. 29. The only academic studies that place substantial emphasis on the importance of the ground campaign are Stephen Biddle, "Victory Misunderstood: What the Gulf War Tells Us about the Future of Conflict," *International Security*, Vol. 21, No. 2 (Fall 1996), pp. 139–179; Stephen Biddle, "The Gulf War Debate Redux," *International Security*, Vol. 22, No. 2 (Fall 1997), pp. 163–174; and my response to Biddle's article in Daryl G. Press, "Lessons from Ground Combat in the Gulf: The Impact of Training and Technology," ibid., pp. 137–146. Several U.S. Army-sponsored accounts place emphasis on the ground campaign. See, for example, Robert H. Scales, *Certain Victory: The U.S. Army in the Gulf War* (Washington, D.C.: Brassey's, 1997); Frank N. Schubert and Theresa L. Kraus, eds., *The Whirlwind War: The United States Army in Operations Desert Shield and Desert Storm* (Washington, D.C.: Center for Military History, 1995); and Richard M. Swain, *"Lucky War": Third Army in Desert Storm* (Fort Leavenworth, Kans.: U.S. Army Command and General Staff College Press, 1997).

4. This position appears to be guiding the current review of U.S. defense policy mandated by U.S. Secretary of Defense Donald Rumsfeld. See Thomas E. Ricks, "Rumsfeld Outlines Pentagon Overhaul," *Washington Post*, March 23, 2001, p. 1; Tom Bowman, "Pentagon Faces Transformation," *Baltimore Sun*, March 13, 2001, p. 1A; Greg Jaffe, "Pentagon Lists Potential Cuts in About 30 Weapons Programs," *Wall Street Journal*, March 26, 2001, p. A3; and Mortimer B. Zuckerman, "Rethinking the Next War," *U.S. News and World Report*, March 5, 2001, p. 64.

ized by the air campaign? Or were they simply outmatched by vastly superior U.S. and British ground forces?[5]

To test the proposition that air power neutralized the Iraqi ground forces, I divide the argument into five different versions:[6] (1) air power prevented the Iraqis from maneuvering, which is critical in desert combat; (2) air power disrupted Iraqi command, control, communications, and intelligence (C3I); (3) air power severed Iraqi supply lines; (4) air power attrited the Iraqi forces too heavily for them to fight effectively; and (5) air power broke Iraqi morale. Each version makes predictions about the conduct of the Iraqi ground forces during the ground operation. In this article I use detailed evidence from the ground campaign to test these five explanations.

The evidence from the ground campaign shows that the conventional wisdom about the Gulf War is wrong. Although air power played an important role in the coalition's victory, its role has been exaggerated and misunderstood. I make two primary arguments about air power during the Gulf War. First, air power was not decisive; it did not neutralize the Iraqi ground forces. At the end of the air campaign, Iraqi ground forces could still maneuver, and they still had the C3I, supplies, numbers, and morale to fight.

Second, the six-week air campaign did not play a necessary, enabling role that made the ground war "easy" for U.S. forces. Had there been no air campaign, U.S. and British fatalities in the ground war would probably have been slightly higher. But evidence strongly suggests that with or without the air campaign, the coalition's ground attack would have led to a rout of historic proportions. In sum, air power contributed to the coalition's effort, but the air campaign was neither sufficient nor necessary for the very one-sided victory.

If air power did not neutralize the Iraqi army, why were the Iraqi ground forces totally ineffective during the ground campaign? The answer is that they were simply overmatched by the U.S. and British ground forces. In 1991 the Iraqis fielded a military that was mediocre even by third world standards, and during the Gulf War they were facing the most powerful military forces in the world on terrain ideally suited to U.S. and British military strengths. Furthermore, Iraq's timing was terrible; Iraq invaded Kuwait just after the Reagan mil-

5. An obvious third possibility is that a combination of air bombardment and the coalition's superiority on the ground neutralized the Iraqi ground forces. I test this third possibility along with the other two in this article.
6. By "air power" I mean fixed-wing aircraft, not helicopters; helicopters are fully integrated into ground forces.

itary buildup was completed and just before America's post–Cold War downsizing had begun. In sum, the United States and Great Britain took the forces that they built to fight while heavily outnumbered against the best divisions in the Soviet military, and turned them loose against Iraq. The Iraqi ground forces never had a chance.[7]

To be clear, this article does not argue that air power was irrelevant in the Gulf War. First, air power was critical for getting combat power to the Gulf quickly after the Iraqi invasion of Kuwait. When it briefly looked like Iraq might continue south from Kuwait and invade Saudi Arabia, the United States rushed combat aircraft to the Arabian Peninsula to prevent further Iraqi attacks. Transport aircraft also flew a brigade of the 82d Airborne Division to Saudi Arabia to demonstrate U.S. resolve. Because of air power, the United States was able to quickly "draw a line in the sand" and establish an early capability to defend Saudi Arabia.[8] Second, during the coalition's offensive to liberate Kuwait, U.S. aircraft successfully carried out all of their important traditional roles: They gained air superiority, flew successful reconnaissance missions, were significantly involved in coordinating the actions of coalition air and ground forces, conducted strategic attacks on targets deep in Iraq, and reduced and disrupted Iraq's military. Air power fulfilled its traditional missions; it just did not neutralize the Iraqi forces.

What does the Gulf War tell us about U.S. foreign policy and procurement priorities? The conventional interpretation of the war has been used to support the view that there has been a fundamental shift in the relative decisiveness of air and ground forces in modern combat. Modern air power, it is now widely believed, has rendered heavy ground forces obsolete. The United States—with its large, advanced air force—should have little difficulty in defeating potential opponents, as long as America commits the full weight of its air power to the fight. This interpretation also suggests that the United States should cut heavy divisions in favor of lighter ground forces, tactical aircraft, long-range bombers, and cruise missiles.

7. This article is part one of a larger project to determine why the Iraqi ground forces were defeated so easily. In the second part, I use detailed studies of the major battles of the ground war to determine what about the ground campaign was so decisive. For example, I assess the relative importance of U.S. training, ground war technology, tactical surprise, operational surprise, and Western tactics.

8. On the pace of the deployment to the Gulf and the role of air power in Desert Shield, see Winnefeld, Niblack, and Johnson, *A League of Airmen*, pp. 25–54; and *GWAPS*, Vol. 2, Pt. 1, pp. 11–21.

My interpretation of the Gulf War, in contrast, paints a more complicated picture of the emerging relationship between air and ground forces. Air power failed to neutralize the Iraqi ground forces because destroying a largely static, defensive force from the air is inherently difficult, even in the era of information-age intelligence and precision-strike weapons. The lesson of the Gulf War is not that air power is a weak instrument of national military power, but that the capabilities of air power against mechanized ground forces on the offensive are substantially greater than air power's capabilities against defensive forces.

The implication of my analysis for U.S. foreign policy is that air power may play a decisive role in future U.S. operations to halt an enemy's mechanized assault on a U.S. ally. It will not likely be decisive, however, if the United States or its allies need to conduct an offensive to take enemy-controlled territory. For example, if North Korea attacks South Korea, or if Iraq invades Kuwait and Saudi Arabia, U.S. air assets may play a leading role in the destruction of the invading forces. But if the U.S. objective in these contingencies is to launch a counteroffensive into North Korea, or to once again evict Iraq from Kuwait, air power will be far less effective against defensively oriented North Korean, or Iraqi, forces. The force structure implications of this analysis are straightforward: If the United States envisions launching offensive operations to defeat its enemies, it will still require a balanced military that includes substantial heavy ground forces. Overemphasizing air assets may prove very costly.[9]

This article is divided into six main sections. First, I lay out five mechanisms through which air power may have neutralized the Iraqi military. In the second section, I describe what happened during the four-day ground offensive. In the third section, I use the history of the ground war to test each of the five mechanisms. In the fourth section, I test the argument that the air campaign was a necessary "enabler" for U.S. ground forces and made the ground war far

9. Maintaining an exaggerated view of the lethality of air power may impose two types of cost on the United States and its allies. First, undue hopes for a cheap victory from the air may encourage the United States to become involved in conflicts, only to be surprised when air power is not decisive. At this point the United States may be forced to choose between accepting defeat or escalating to include ground operations, both of which may be less attractive than staying out in the first place. Second, if the United States and its allies believe that air power is more lethal than it really is, they may unbalance their military forces and eliminate (or substantially reduce) heavy ground forces in favor of air assets and light ground forces. If a war occurs, they may pay a cost in higher casualties if they have too few heavy ground forces.

less costly than it would have been otherwise. In the fifth section, I address the counterargument that air power has grown much more lethal since the Gulf War, and that this war therefore tells us little about the future of warfare. In this section I describe the reasons that air power was unable to neutralize the Iraqi ground forces in 1991, and I argue that the problems that coalition air power faced in the Gulf War have not yet and will not soon go away. I conclude this argument with a brief review of the Kosovo air campaign. Finally, in the sixth section, I describe the implications of my findings for the future of U.S. force structure and American foreign policy.

Myths of Air Power in the Gulf War

Military analysts are nearly unanimous in their view that air power was decisive in the Gulf War. One of the best histories of this war explains the rout of the Iraqi ground forces by noting that the air campaign "had clearly battered the enemy to near senselessness."[10] Another excellent historical account concludes that "air attacks made it impossible for the Iraqis to mount an effective defense."[11] According to air power advocate Richard Hallion, "Simply (if boldly) stated, air power won the Gulf War."[12] Robert Pape, author of the seminal work on modern air power, wrote that "Iraq's ground forces were decimated" by the air campaign.[13] The two most detailed studies of the air war, the *Gulf War Air Power Survey* (*GWAPS*) and the RAND book *A League of Airmen*, came to the same conclusion. The authors of *GWAPS* concluded that "air power was the decisive factor" in the "crushing military defeat" of the Iraqis, and that "air power made [Iraqi] resistance disorganized and totally ineffective."[14] The authors of the RAND study largely agreed, arguing that "air power was indeed decisive in neutralizing Iraqi forces."[15]

Despite the agreement that air power was decisive in the Gulf War, there is no consensus on how air power accomplished this objective. One view is that

10. Atkinson, *Crusade*, p. 495.
11. Gordon and Trainor, *The Generals' War*, p. 474.
12. Richard P. Hallion, *Storm over Iraq: Air Power and the Gulf War* (Washington, D.C.: Smithsonian Institution Press, 1992), p. 2.
13. Pape, "The Limits of Precision-Guided Air Power," p. 102, n. 13; and Pape, "The Air Force Strikes Back," p. 201, n. 24.
14. For the first quote, see *GWAPS*, Vol. 2, Pt. 2, p. 104. For the second quote, see *GWAPS Summary Report*, p. 117.
15. Winnefeld, Niblack, and Johnson, *A League of Airmen*, p. 285.

air power pinned down the Iraqi ground forces and prevented them from maneuvering. As a result, the Iraqis were unable to defend themselves from the coalition's flanking attack—the "left hook."[16] A second possibility is that Iraqi forces were rendered ineffective because air power disrupted their C3I. Air strikes against Iraqi command-and-control bunkers, power grids, telephone exchanges, and radio transmitters may have "blinded" the Iraqi ground forces. By the time of the ground war, even those Iraqi units still able to communicate were afraid to do so, because any transmission invited coalition attention in the form of air strikes. With their communications severely disrupted, the Iraqi ground forces were unable to coordinate their response to the coalition attack, allowing themselves to be destroyed piecemeal.[17]

A third version of the air power theory is that air strikes cut Iraqi supply lines by targeting Iraqi supply dumps and, more important, destroying thousands of Iraqi trucks meant to carry supplies to the forward Iraqi divisions. By the time the ground war began, the Iraqis were not looking for a fight; they were looking for food.[18] A fourth explanation focuses on the attrition of the Iraqi forces in the Kuwaiti theater of operations (KTO). Stated simply, air power may have destroyed enough Iraqi equipment and killed enough troops

16. See *GWAPS Summary Report*. The authors of the report, Thomas A. Keaney and Eliot A. Cohen, wrote that "the most important contribution of air power . . . and a prime reason why the ground campaign was so short and so overwhelming, was the success of air interdiction in preventing the heavy divisions from moving or fighting effectively" (p. 116), and that air attacks "immobilized" the Republican Guard (p. 119). Winnefeld, Niblack, and Johnson, *A League of Airmen*, p. 153, write that "the Iraqi troops were so pinned down that it was virtually impossible to move, either day or night," and that "this air-supported maneuver dominance was an important contributor to the coalition success during the ground war." Thomas G. Mahnken and Barry D. Watts, "What the Gulf War Can (and Cannot) Tell Us about the Future of Warfare," *International Security*, Vol. 22, No. 2 (Fall 1997), pp. 158–159, also claim that the Iraqis were "pinned down" by air power. See also Keaney, "The Linkage of Air and Ground Power in the Future of Conflict," pp. 148–149; and Gordon and Trainor, *The Generals' War*, pp. 287–288, 431.

17. Many analysts claim that air attacks on Iraqi C_3I had a major effect on the outcome of the fighting, even though they do not generally argue that these effects in themselves were decisive. See, for example, Mahnken and Watts, "What the Gulf War Can (and Cannot) Tell Us about the Future of Warfare," pp. 158–159, who write that the air campaign left the Iraqis "blinded, bloodied, and pinned down"; Winnefeld, Niblack, and Johnson, *A League of Airmen*, pp. 130–131, argue that attacks on Iraqi C_3I "probably significantly affected the course of the war." See also ibid., pp. 152, 156–157; U.S. News and World Report, *Triumph without Victory: The Unreported History of the Persian Gulf War* (New York: Random House, 1993), p. 409; and Tom Clancy with Gen. Fred Franks, Jr. (Ret.), *Into the Storm: A Study in Command* (New York: Berkley, 1998), p. 242.

18. For arguments that place substantial emphasis on the effects of air power on the Iraqi supply situation, see Winnefeld, Niblack, and Johnson, *A League of Airmen*, p. 156; Pape, "The Limits of Precision-Guided Air Power," p. 113; *GWAPS*, Vol. 2, Pt. 1, p. 323; *GWAPS Summary Report*, p. 116; and Keaney, "The Linkage of Air and Ground Power in the Future of Conflict," pp. 148–149.

that when the ground war was launched, there were not enough Iraqi forces left in the theater to resist effectively.[19]

A fifth explanation is that six weeks of air attacks broke Iraqi morale. This explanation includes elements of the other explanations: Air attacks caused serious attrition and supply shortages, breaking the will of the Iraqi army.[20] The Iraqi army may still have had the capability to resist the coalition after the air campaign, but they lacked the will.[21] To determine if any of these explanations is correct, we need to look at what happened during the ground war.

The Ground War

On January 17, 1991, the United States and its coalition partners launched Operation Desert Storm to liberate Kuwait. For the next thirty-eight days, thousands of coalition air strikes hit targets throughout Iraq, including military command bunkers, bridges, telephone exchanges, electrical power plants, and government buildings. In the KTO, coalition aircraft pummeled Iraqi forces and interdicted their supply lines. During the air campaign, the only ground fighting was a skirmish near the Saudi city of al-Khafji.[22] On February 24 the

19. Ironically, despite the significant attrition that coalition aircraft inflicted on Iraqi ground forces, this explanation gets less attention than the others. For an argument that emphasizes the effects of attrition, see Pape, "The Limits of Precision-Guided Air Power," p. 113. Keaney, "The Linkage of Air and Ground Power in the Future of Conflict," pp. 148–149, notes the effects of attrition, though he says this was less important than preventing the Iraqi forces from maneuvering.

20. For the best evidence supporting this argument, see Hosmer, *Psychological Effects of U.S. Air Operations*, pp. 141–173. See also *GWAPS Summary Report*, pp. 107–108; Pape, "The Limits of Precision-Guided Air Power," p. 113; Keaney, "The Linkage of Air and Ground Power in the Future of Conflict," pp. 148–149; and Levran, *Israeli Strategy after Desert Storm*, p. 29. John Mueller, "The Perfect Enemy: Assessing the Gulf War," *Security Studies*, Vol. 5, No. 1 (Autumn 1995), pp. 108–111, argues that low Iraqi morale was decisive, but that Iraqi morale had been broken even before the air campaign began.

21. Note that the explanations focusing on attrition, C_3I, and supply all overlap with the morale explanation because severe attrition, lack of adequate command and control, or lack of adequate supply could all contribute to a collapse of Iraqi morale. I distinguish between these explanations in the following manner: If air power caused a collapse of Iraqi morale through any mechanism—incessant bombardment, attrition, supply shortages, or disruption of C_3I—I credit the collapse to the morale theory because air power had shattered the Iraqis' will, not their capability. I credit the other theories only if air power destroyed Iraqi capability. For example, if air strikes damaged Iraqi C_3I to the point that Iraqi units could not effectively coordinate their actions, this would be evidence for the C_3I explanation. But if attacks on Iraqi C_3I made Iraqi soldiers feel isolated and abandoned in the theater, and made them unwilling to fight, I code this as success for the morale explanation.

22. The al-Khafji battle was significant because three Iraqi divisions attempted to advance into Saudi Arabia but were repelled by coalition air power and severely attrited. This battle demonstrates the potential lethality of U.S. air power against forces advancing in the open, but it stands in stark contrast to the inability of U.S. air power to destroy static units, and to the failure to inter-

coalition launched a ground offensive to liberate Kuwait and destroy as many of the remaining Iraqi ground forces as possible. Only four days later, a cease-fire was initiated and the war was over.

IRAQI GROUND FORCES AND DEPLOYMENTS

To understand the ground war, it is necessary to understand the ground forces, deployments, and strategy of the Iraqi military. On the eve of the ground war, Iraq had fifty-one divisions in the KTO. Its best forces in the theater were eight Republican Guard divisions manned by professional soldiers and handpicked for their loyalty to Saddam Hussein's Ba'th Party. They received the best training and equipment of all the Iraqi armed forces. The heavy divisions[23] of the Republican Guard were armed with the best vehicles in the Iraqi army, principally T-72 main battle tanks and BMP infantry fighting vehicles. Nevertheless, these forces were substantially inferior to U.S. and British forces in both equipment and training.[24]

The second-best Iraqi forces in the KTO were the nine heavy divisions of the regular army. Like the Republican Guard, these divisions were composed prin-

---

dict the maneuvers of the other Iraqi divisions during the ground war (see below). For a good account of the al-Khafji battle, see Gordon and Trainor, *The Generals' War*, pp. 267–288; and Lambeth, *The Transformation of American Air Power*, pp. 121–124.

23. "Heavy division" refers to any division with several hundred armored vehicles. Typically, there are two types of heavy division: armored divisions and mechanized infantry divisions. Each generally has 100 or more tanks and other armored vehicles. "Light divisions," by contrast, are typically infantry divisions without substantial numbers of armored vehicles

24. The enormous technological advantages enjoyed by U.S. forces over the Republican Guard can be glimpsed by comparing the U.S. M1 tank with the Iraqi T-72. M1s were able to detect Iraqi vehicles out to 4,000 meters and hit them—usually with their first shot—beyond 3,700 meters. T-72s, by contrast, had difficulty detecting U.S. vehicles beyond 1,000 meters. Furthermore, U.S. tank rounds had no problem penetrating Iraqi tanks from any side out to their maximum range; even the machine guns on U.S. armored vehicles could destroy Iraqi T-72s from 700 meters. At that range, T-72s could not penetrate the frontal armor of M1s with their main guns, let alone with their machine guns. For data on the ability of U.S. weapons to penetrate the armor of Iraqi vehicles, see Battlefield Assessment Team, "Armor/Antiarmor Operations in Southwest Asia," Research Paper No. 92–0002 (Quantico, Va.: Marine Corps Research Center, July 1991), p. 15; U.S. Army Armor Center, "Desert Shield and Desert Storm Emerging Observations" (Fort Knox, Ky.: U.S. Army Military History Institute, October 7, 1991), pp. 1–12. For evidence on the inability of Iraqi rounds to penetrate the front of M1 tanks, see U.S. General Accounting Office (GAO), "Operation Desert Storm: Early Performance Assessment of Bradley and Abrams," January 1992, GAO/NSIAD-92-94, p. 24; and Ezio Bonsignore, "Gulf Experience Raises Tank Survivability Issues," *Military Technology*, Vol. 16, No. 2 (February 1992), p. 67. For data on the range disparities between U.S. and Iraqi tanks, see U.S. GAO, "Operation Desert Storm," p. 33; and U.S. Army Armor Center, "Desert Shield and Desert Storm Emerging Observations," pp. 1–10. A great overview of the breadth of the technological advantages enjoyed by the West over Soviet-designed tanks was written before the Gulf War. See Malcolm Chalmers and Lutz Unterseher, "Is There a Tank Gap? Comparing NATO and Warsaw Pact Tank Fleets, *International Security*, Vol. 13, No. 1 (Summer 1988), pp. 23–45.

cipally of professional soldiers, but their equipment and training were not as good. Heavy divisions of the regular army were typically armed with T-55 main battle tanks and older armored personnel carriers (APCs). Although adequate for a third world military, these forces were inferior to the Republican Guard and substantially inferior to their U.S. and British counterparts.

The third class of Iraqi soldiers in the KTO—the frontline infantry—were conscripts deployed in thirty-four divisions near the Saudi border. They were given poor equipment and perfunctory training. Furthermore, because they were drawn predominantly from Iraq's rebellious Kurdish and Shi'a populations, who generally opposed Saddam's Ba'th Party, their enthusiasm for Saddam's war over Kuwait was suspect.[25] Even by third world standards, the frontline infantry divisions were extremely poor military units.

Figure 1 illustrates the approximate positions of the Iraqi forces at the beginning of the coalition's ground offensive. Iraqi frontline infantry divisions were deployed along the border between Kuwait and Saudi Arabia, along the Iraqi-Saudi border, and up the Kuwaiti coastline. Thick obstacle belts, including barbed wire and minefields, were built to buttress their positions. The frontline infantry were supported by the heavy divisions of the regular army, 25–50 kilometers to their rear. Behind these were three heavy divisions from the Republican Guard—the Tawakalna, Madinah, and Hammurabi—considered the most capable Iraqi units in the KTO.

Iraq had a reasonable strategy for defending Kuwait. The frontline infantry would slow the coalition's attack and channel it into two or three breakthrough sectors. Heavy divisions of the regular army would then counterattack into the coalition's breakthrough sectors. Finally, the heavy divisions of the Republican Guard would reinforce any parts of the Iraqi line being penetrated and counterattack the coalition forces.[26] Although the Iraqi deployment left their western flank exposed, the Iraqis understood how difficult it would be for the coalition forces to navigate and maintain supply lines across the nearly featureless desert. The Iraqis apparently did not anticipate the effectiveness of the U.S. global positioning system (GPS) and other navigation equipment, which

25. "According to an MIT dissertation on the Iraqi Army, by Ahmed Hashim, the front line divisions of the Iraqi Army were 70 percent Shiite and 20 percent Kurd." Cited in U.S. News and World Report, *Triumph without Victory*, p. 404.
26. Gordon and Trainor, *The Generals' War*, p. 287; and Kenneth M. Pollack, "The Influence of Arab Culture on Arab Military Effectiveness," Ph.D. dissertation, Massachusetts Institute of Technology, 1996, pp. 333–334.

Figure 1. Iraqi Ground Force Deployments.

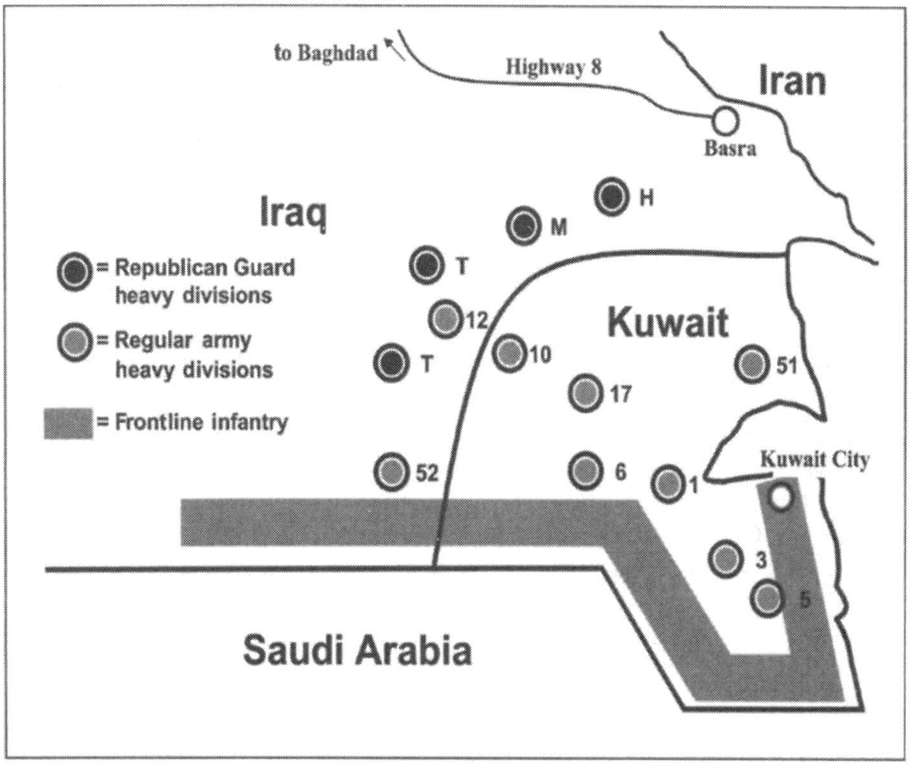

NOTE: The Tawakalna, Madinah, and Hammurabi Divisions are indicated by T, M, and H, respectively. Note that the Tawakalna Division was divided and deployed in two locations separated by a few kilometers; it is therefore indicated in two places. Because the focus in this article is on the Iraqi heavy divisions, Iraq's light Republican Guard divisions are not indicated. See Central Intelligence Agency, *Operation Desert Storm: A Snapshot of the Battlefield,* Report IA 93-10022 (Washington, D.C.: Government Printing Office, September 1993); and Michael R. Gordon and Bernard E. Trainor, *The Generals' War: The Inside Story of the Conflict in the Gulf* (Boston: Little, Brown, 1995), p. 341.

allowed U.S. forces to move off-road across the desert without getting lost. But even U.S. commanders, who had a better sense of American capabilities, wondered whether the left hook was logistically possible.

The Iraqi leadership did not expect to "beat" the coalition ground forces. Rather, they hoped that by maneuvering their best units into the path of the coalition's offensive, they might be able to inflict several thousand U.S. casualties. At that point, the Iraqis believed, the United States might be willing to negotiate a settlement to the war.[27]

COALITION GROUND FORCES AND DEPLOYMENTS

The coalition plan for the ground war had two parts. First, two U.S. Marine divisions would attack up the middle into the teeth of the Iraqi defenses (see Figure 2). The primary goals of this attack were (1) to convince the Iraqis that this was the main coalition effort; (2) to draw the Iraqi reserves—especially the Republican Guard—south to counterattack the marines; and (3) to advance north and help the coalition's Arab partners liberate Kuwait City.

The second part of the coalition's plan was the U.S. Army operation. During the air campaign, the army secretly shifted its ground forces west so that they were no longer deployed south of Iraq's main defenses in Kuwait. From this position, the army prepared to launch its left hook. The army's offensive was supposed to start one day after the marine operation. If everything went according to plan, the Iraqis would respond to the marine attack by moving their reserves south. As this happened, the army would race around the Iraqi flank and envelop the Republican Guard.[28]

Two U.S. Army Corps were tasked with executing the left hook. The VII Corps, the main effort, was ordered to find and destroy the Republican Guard. A lighter force, the XVIII Corps, was deployed farther west. Its mission was to advance rapidly across the mostly empty desert, get behind the Iraqi forces in the KTO, and prevent an Iraqi retreat.[29]

DAY 1—FEBRUARY 24

The marine ground attack started early on the morning of February 24 and began better than U.S. planners had expected. The marines met little resistance as

27. Gordon and Trainor, *The Generals' War*, p. 180; and *GWAPS*, Vol. 2, Pt. 1, pp. 111–113, 125–126. For a statement that supports this view of Iraqi strategy, see Saddam Hussein, "Speech to the Islamic Conference," *Baghdad Voice of the Masses*, January 11, 1991, in Foreign Broadcast Information Service (FBIS)–Near East 91-009; and "Interview with Yevgenni Primakov," *Paris Europe Number One*, April 28, 1991, in FBIS–Soviet Union 91-083, p. 11. Both are cited in Pollack, "The Influence of Arab Culture on Military Effectiveness," p. 97.

28. Some army units, such as the 82d Airborne and the 101st Air Assault Division, were supposed to start the same day as did the marines, but the main army effort was to begin the day after the marines attacked.

29. The VII Corps began the ground campaign with four heavy divisions—three and one-third U.S heavy divisions and two-thirds of a British armored division. On the third day of the ground war, the corps was given another heavy U.S. division. The XVIII Corps, by contrast, was much lighter. Designed for quicker movement, it had one and one-third heavy divisions plus an airborne division, a helicopter–mobile air assault division, and a medium-weight French division.

Figure 2. Coalition Forces and Deployment Plans.

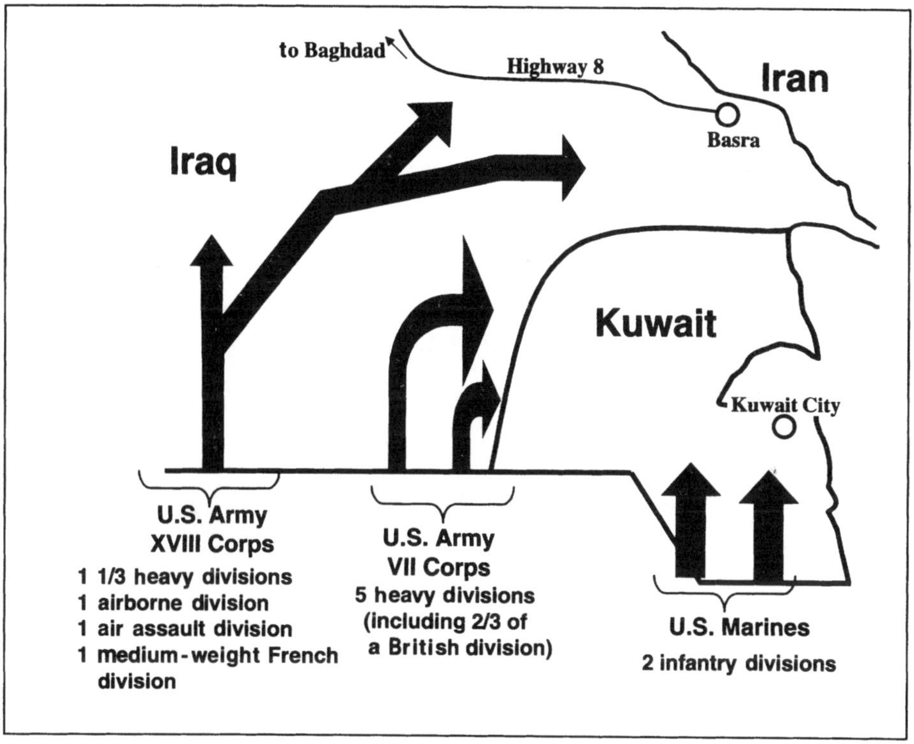

they advanced north into Kuwait. The Iraqi frontline infantry showed no interest in fighting, and thousands surrendered without firing a shot. The biggest problem for the marines on the first day was finding a way to keep track of, and care for, the Iraqi prisoners.[30] In fact, the marine offensive was going so well, and the Iraqis were showing so few signs of resistance, that U.S. commanders began to worry. If the Iraqis did not resist, the marine attack might not draw the Iraqi reserves south; instead the Iraqis might flee north before the U.S. Army had a chance to unleash the left hook.[31] By midday, U.S. command-

30. See Charles J. Quilter II, *U.S. Marines in the Persian Gulf, 1990–1991: With the I Marine Expeditionary Force in Desert Shield and Desert Storm* (Washington, D.C.: Headquarters, United States Marine Corps [USMC], 1993), pp. 75–91; Gordon and Trainor, *The Generals' War*, pp. 355–363; and Scales, *Certain Victory*, pp. 221–222.
31. Gordon and Trainor, *The Generals' War*, p. 362.

Figure 3. Day 1—February 24.

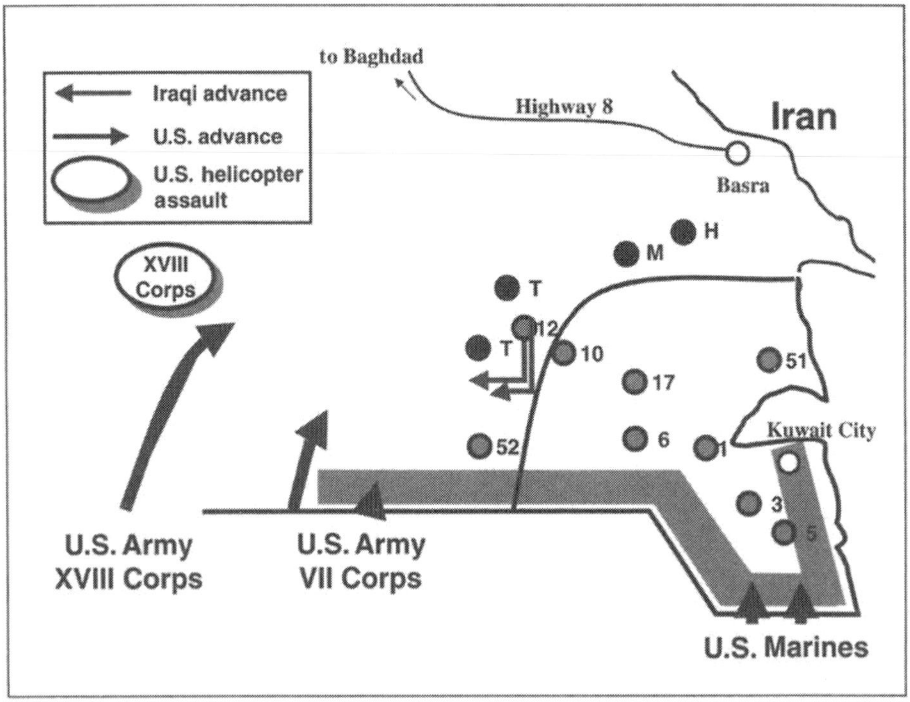

ers decided to move up the army's attack by one day, beginning its advance at dusk on February 24 (see Figure 3).

The XVIII Corps made great progress on February 24. They were deployed so far west that there were few Iraqis in front of them, so the ground element of the XVIII Corps charged quickly ahead into the Iraqi desert. A helicopter assault moved other units from the XVIII Corps 100 kilometers into Iraq toward Highway 8, the main road connecting Basra with Baghdad.[32]

The VII Corps, the U.S. Army's main force, proceeded more slowly than either the marines or the XVIII Corps. Facing nothing but open desert, the western half of the VII Corps (like the XVIII Corps to the west) charged ahead for the first few hours of the attack. But the eastern side of the VII Corps had to breach minefields and other obstacles before they could advance into Iraq.

32. Scales, *Certain Victory*, pp. 217–220, 254–256.

Rather than allowing his forces to conduct the dangerous breaching operation at night, the VII Corps commander halted both wings of the corps until dawn.[33]

At the end of the first day of ground combat something unexpected happened. Late on the evening of February 24, only six hours after the army had launched its surprise attack in the west, the Iraqi military began to react to the left hook. Senior Iraqi military leaders ordered two brigades of the 12th Armored Division to move west to defend against a possible coalition attack coming from that direction (see Figure 3).[34] The movement of the Iraqi brigades was tracked by U.S. JSTARS surveillance aircraft.[35]

DAY 2—FEBRUARY 25

The second day of the coalition attack involved serious fighting for the marines, but no significant engagement for the U.S. Army or British forces in the left hook. The day began with a nasty surprise for the marines. Early on the morning of February 25, as the marines prepared to advance north, forces from two Iraqi divisions counterattacked. One Iraqi brigade (from the 5th Mechanized Division), which had been deployed in the thick smoke around a burning Kuwaiti oil field, attacked the flank of a marine division. Although initially caught off guard, the marines quickly recovered and routed the Iraqis, suffering no fatalities in the process.[36] In the second battle, a brigade from

33. Gordon and Trainor, *The Generals' War*, pp. 375–380.
34. A brigade is approximately one-third the size of a division. Some histories have argued that a brigade of the Tawakalna Division also moved on February 24. I do not believe that this is correct, but if it is true, it strengthens my argument. I benefited from discussions with Kenneth Pollack on this point. The movement of the Iraqi reserves is described in Central Intelligence Agency (CIA), *Operation Desert Storm: A Snapshot of the Battlefield*, Report IA 93–10022 (Washington, D.C.: GPO, September 1993); Scales, *Certain Victory*, p. 238; and Swain, *"Lucky War,"* pp. 230, 233.
35. The Iraqi order to maneuver west was intercepted by U.S. intelligence. See Atkinson, *Crusade*, pp. 438–439. JSTARS, which stands for joint surveillance target attack radar system, is an aircraft with sophisticated radar equipment that allows it to track the movement of vehicles on the ground over several hundred square kilometers. JSTARS tracked the movement of the 12th Armored Division. A reproduction of a JSTARS radar picture of this Iraqi maneuver can be found in Scales, *Certain Victory*, p. 248.
36. For details on this battle, see Press, "Lessons from Ground Combat in the Gulf," pp. 139–143, which is based on the two best high-resolution descriptions of the battle: an unpublished history by John H. Turner, USMC, "Counterattack: The Battle at Al Burqan," March 1993; and Charles H. Cureton, USMC Reserve, *U.S. Marines in the Persian Gulf, 1990–1991: With the 1st Marine Division in Desert Shield and Desert Storm* (Washington, D.C.: Headquarters, U.S. Marine Corps, 1993), pp. 91–96. Turner was the operations officer in the regiment of the 1st Marine Division that bore the brunt of the main Iraqi attack.

another Iraqi division attacked the 2d Marine Division. The Iraqis were easily defeated.[37]

Out west, the XVIII Corps continued its rapid advance through Iraq. U.S. light infantry units conducted a helicopter assault into positions astride Highway 8, and the ground force continued to race ahead, essentially unopposed, through the desert (see Figure 4).

The VII Corps advanced into Iraq and fought two minor battles. The British 1st Armored Division (under command of the VII Corps) overran the Iraqi 52d Armored Division, which had been targeted during the air war for especially heavy air attack, and defeated it without much of a fight. The U.S. 2d Armored Cavalry Regiment, in the western half of the VII Corps advance, encountered the lead brigade of the Iraqi 12th Armored Division, the unit that had been ordered to move west the day before. Already significantly attrited by air power as it moved west[38]—30 of the 90 Iraqi tanks had been destroyed[39]—the Iraqi brigade was easily defeated.

At the end of the second day of the ground war, Iraq made two surprising responses to the coalition attack. First, Baghdad radio announced that the Iraqi army, having been victorious, would pull out of Kuwait; the Iraqi army immediately began its retreat.[40] Second, Iraq shifted more of its forces west to block the left hook. Two brigades of the Tawakalna Republican Guard Division, a brigade of the 12th Armored Division, the 17th Armored Division, and two other heavy divisions of the Republican Guard—the Madinah and the Hammurabi Divisions—all began moving west.[41]

As the Iraqis maneuvered, U.S. surveillance aircraft detected a column of Iraqi vehicles moving north out of Kuwait City; hundreds of coalition aircraft attacked the Iraqi column along what would be known as the Highway of Death. The air attack claimed 1,400 vehicles, mostly military trucks but also some civilian vehicles. Only about 30 Iraqi armored vehicles were among those destroyed in the attack.[42]

---

37. For details on the 2d Marine Division's battle, see Quilter, *With the I Marine Expeditionary Force*, pp. 91–93; and Dennis P. Mroczkowski, *U.S. Marines in the Persian Gulf, 1990–1991: With the 2nd Marine Division* (Washington, D.C.: Headquarters, U.S. Marine Corps, 1993), p. 53.

38. Air power's interdiction of this Iraqi brigade is discussed later.

39. CIA, *Operation Desert Storm;* Scales, *Certain Victory*, pp. 233–235; and Clancy and Franks, *Into the Storm*, pp. 369–370.

40. Swain, *"Lucky War,"* p. 250; and *GWAPS*, Vol. 2, Pt. 1, p. 307.

41. CIA, *Operation Desert Storm;* and Clancy and Franks, *Into the Storm*, pp. 369–370.

42. Despite the graphic pictures of the carnage, few Iraqis died along the Highway of Death. Mueller, "The Perfect Enemy," pp. 91–92. Nevertheless, the gruesome pictures of Iraqis killed in

Figure 4. Day 2—February 25.

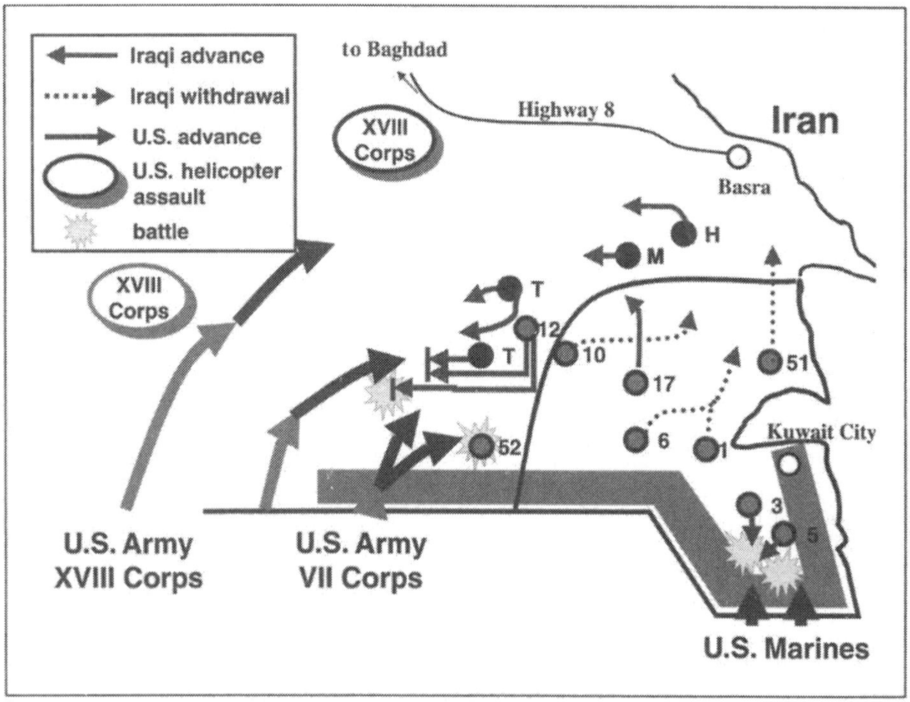

At the end of the second day of the ground war, the battlefield was in great flux. Virtually all of Iraq's mobile forces in the Kuwaiti theater were on the move: Some were heading west to block the left hook; others were pulling out of the theater altogether. Meanwhile, the coalition forces were racing to close in on the Iraqis from the south and the west.

DAY 3—FEBRUARY 26

The third day saw the most serious ground combat of the war. U.S. Marines fought a series of small battles outside Kuwait City and then escorted Arab coalition forces to the outskirts of the city, so that the Arabs could officially liberate Kuwait. For the two marine divisions, the offensive was over.

---

these air attacks apparently had a powerful effect on the image of the battlefield that crystallized in the minds of war planners in Riyadh, leaders in Washington, and Americans in their living rooms. On the type of vehicles destroyed along the Highway of Death, see CIA, *Operation Desert Storm.*

For the U.S. Army, the war was just beginning. Two days after the Iraqi frontline infantry surrendered en masse, and a day after the Iraqis announced their withdrawal from Kuwait, the army finally found and engaged the Iraqi Republican Guard at their new, western-oriented lines of defense (see Figure 5).

On the western flank, the XVIII Corps continued advancing. Its heavy ground forces crossed Highway 8, fought a series of small engagements near the highway, and seized two Iraqi airfields. The 24th Infantry Division discovered supply depots filled with all kinds of Iraqi military equipment undamaged by the air campaign.[43]

For the VII Corps, February 26 was the climactic day of the war. Throughout the day, five major battles were waged between elements of the VII Corps and two Iraqi divisions—the Tawakalna and the remnants of two brigades of the 12th Armored Division.[44] In every one of these battles, the Iraqis were on the defensive. The quality of the Iraqi defensive positions varied. Some were quite good, involving reverse-slope defenses,[45] minefields, dismounted infantry in trenches, and in some cases deep holes from which Iraqi tanks could briefly emerge, fire, and return to safety.[46] Other Iraqi positions were poor and suggest that some Iraqi units had moved into these positions only minutes before being attacked by the coalition.

The Iraqis encountered by the VII Corps on February 26 did not surrender, and in many cases they fought with courage; all their efforts, however, were futile. U.S. helicopters and tanks smashed the Iraqi armored forces they encountered; missiles fired from American helicopters, and the main guns on U.S. tanks, routinely destroyed Iraqi armored vehicles from thousands of meters away. In the dark of night, the Republican Guard infantry made easy tar-

43. These supply dumps are described in more detail below. See Scales, *Certain Victory*, pp. 259, 306; and Atkinson, *Crusade*, p. 455.
44. Each of the five battles on February 26 that I call "major engagements" involved battalion-sized units or greater on both sides. An armored or mechanized battalion comprises approximately 50 armored vehicles (tanks, APCs, and infantry fighting vehicles). Of these battles, good data on both U.S. and Iraqi forces are available for three of them. These data are summarized in Table 1.
45. A reverse-slope defensive position is one established on the rear side of a hill near its crest. One purpose of this type of deployment is to neutralize an enemy's superior weapon range. As the enemy crests the hill, defensive forces can engage them at short range.
46. Scales, *Certain Victory*, pp. 261, 270, 272–273, 284–285; U.S. News and World Report, *Triumph without Victory*, pp. 338–339; Clancy and Franks, *Into the Storm*, pp. 398–400; U.S. Army Armor Center, "Desert Shield and Desert Storm Emerging Observations," pp. 2–33; and Pollack, "The Influence of Arab Culture," pp. 346, 348, 352.

Figure 5. Day 3—February 26.

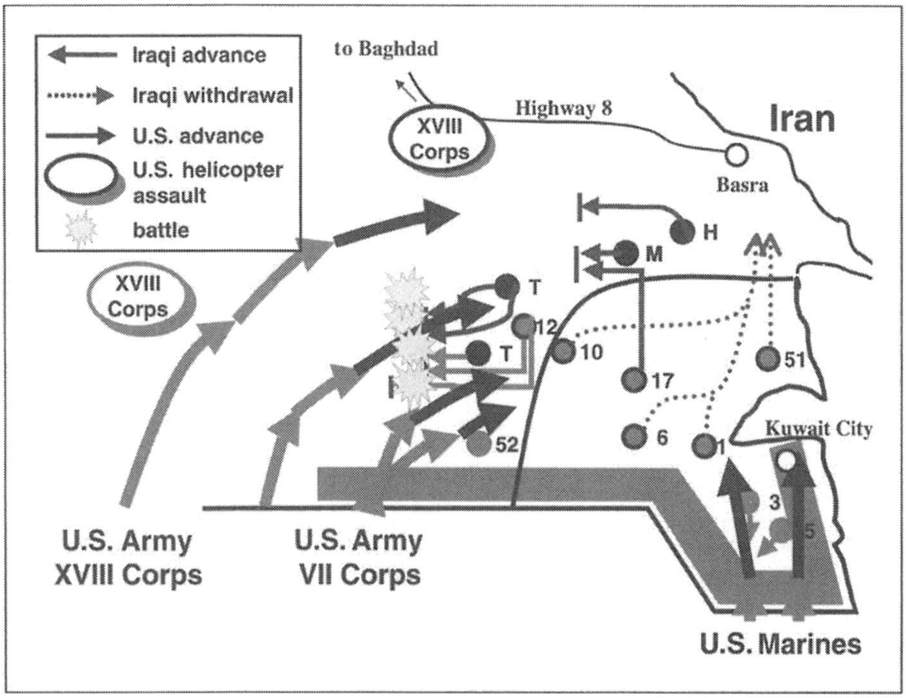

gets for U.S. and British tank gunners with thermal sights. And repeatedly, advanced armor protection on U.S. tanks saved American lives. Iraqi rounds ricocheted off U.S. vehicles or disabled the vehicles without killing the soldiers inside.[47]

Despite the Iraqis' willingness to fight, these battles were astonishingly one-sided. In the five major battles on this day, approximately 350 Iraqi armored vehicles were destroyed. Hundreds of Iraqi soldiers in the vehicles and fighting on foot nearby were undoubtedly killed. Astonishingly, U.S. losses in these battles numbered one killed by enemy fire and twelve killed by friendly fire (see Table 1).[48]

47. For a description of the technological advantages enjoyed by U.S. ground forces, see n. 24.
48. See Biddle, "Victory Misunderstood," p. 147; Scales, *Certain Victory*, pp. 267–270, 272–281, 285; Atkinson, *Crusade*, p. 462; and U.S. News and World Report, *Triumph without Victory*, pp. 368–370.

Table 1. Major Ground Engagements in Operation Desert Storm.

Battle Number and Date	U.S. Unit (engaged elements)	Iraqi Unit	Iraq on: (offense/ defense meeting)	Force Ratio (United States: Iraq)	U.S. Fatalities
#1* February 25	1st Marine Division (2 full-strength battalions)	5th ID (2 brigades at half strength)	meeting	1:1	0
#2 February 25	2d Marine Division (elements)	3d AD (elements)	meeting	NA	NA
#3 February 26	2d Marine Division	3d AD (elements)	defense	NA	NA
#4* February 26	2d ACR (3 troops, 65 armored vehicles)	Tawakalna Division (96 armored vehicles)	defense	1:1.5	1
#5 February 26	3d AD (elements of 1st and 2d Brigades)	Tawakalna Division (number of armored vehicles unknown)	defense	NA	5 (friendly fire)
#6* February 26	1st AD (1 battalion of 3d Brigade: 41 armored vehicles)	Tawakalna Division (45 armored vehicles)	defense	1:1	0
#7* February 26	1st ID (2 battalions of 1st Brigade; 120 armored vehicles)	Tawakalna Division and 12th AD (120 armored vehicles)	defense	1:1	1 (friendly fire)
#8 February 26	1st ID (3d Brigade)	12th AD	defense	NA	6 (friendly fire)
#9* February 27	1st AD (elements of 1st and 2d Brigades; 165 armored vehicles)	Madinah Division and 17th AD (110 armored vehicles)	defense	1.5:1	1

As the third day of the ground war came to a close, and the Tawakalna and the Iraqi 12th Armored Divisions were being decimated, the Madinah and Hammurabi Republican Guard Divisions continued to move west into blocking positions. Elements of the 17th Armored Division deployed alongside the Madinah. Other Iraqi units in the KTO continued to flee (see Figure 5).

DAY 4—FEBRUARY 27

The fourth day of the ground war was anticlimactic. The XVIII Corps continued its advance east, fighting several minor battles and preparing to surround the remaining Iraqi forces from the north (see Figure 6). The VII Corps had one more battle to fight. One of its divisions—the U.S. 1st Armored Division—crested a hill and discovered a brigade of the Madinah Division 3 kilometers in the distance. At that range, U.S. armored vehicles could hit the Iraqis, but Iraqi fire could not reach the Americans. The U.S. tanks, supported overhead by attack helicopters, slowly advanced while shooting at every Iraqi vehicle they could see. The 1st Armored Division destroyed the brigade of the Madinah Division; only one U.S. soldier was killed by enemy fire in this battle.[49] A cease-fire took effect early on the morning of February 28.

NOTE: The five battles with an asterisk next to the date (1, 4, 6, 7, and 9) are those for which there are good data about the forces on both sides. All force numbers are approximate and are based on author's analysis. In the column listing U.S. fatalities, the notation "friendly fire" indicates that all U.S. fatalities in that battle were from friendly fire.

ABBREVIATIONS: NA means the data are not available. ID is infantry division, AD is armored division, ACR is armored cavalry regiment, which equals approximately 1/3 of an armored division.

SOURCES: The data for this table were compiled from Charles H. Cureton, USMC Reserve, *U.S. Marines in the Persian Gulf, 1990–1991: With the 1st Marine Division in Desert Shield and Desert Storm* (Washington, D.C.: Headquarters, U.S. Marine Corps, 1993), pp. 91–96; Charles J. Quilter II, *U.S. Marines in the Persian Gulf, 1990–1991: With the I Marine Expeditionary Force in Desert Shield and Desert Storm* (Washington, D.C.: Headquarters, United States Marine Corps [USMC], 1993), pp. 91–93; Dennis P. Mroczkowski, *U.S. Marines in the Persian Gulf, 1990–1991: With the 2nd Marine Division* (Washington, D.C.: U.S. Marines Headquarters, 1993), p. 53; John H. Turner, "Counterattack: The Battle at Al-Burqan," unpublished manuscript, March 1993; Robert H. Scales, *Certain Victory: The U.S. Army in the Gulf War* (Washington, D.C.: Brassey's, 1997), pp. 221–222, 270, 281–285, 292–300; Rick Atkinson, *Crusade: The Untold Story of the Persian Gulf War* (Boston: Houghton Mif[?]in, 1993), pp. 466–467; Tom Donnelly and Sean Naylor, *Clash of Chariots: The Great Tank Battles* (New York: Berkley, 1996), pp. 259, 266–272; U.S. News and World Report, *Triumph without Victory: The Unreported History of the Persian Gulf War* (New York: Random House, 1993), pp. 377–391; Central Intelligence Agency, *Operation Desert Storm: A Snapshot of the Battlefield,* Report IA 93-10022 (Washington, D.C. Government Printing Office, September 1993); Richard M. Bohannon, "Dragon's Roar: 1-37 Armor in the Battle of 73 Easting," *Armor,* Vol. 101, No. 3 (May–June 1992), pp. 11–17; Gregory Fontenot, "Fright Night: Task Force 2/34 Armor," *Military Review,* Vol. 7, No. 1 (January 1993), pp. 38–52; and W.M. Christenson and Robert A. Zirkle, "73 Easting Battle Replication—A Janus Combat Simulation," P-2770 (Alexandria, Va.: Institute for Defense Analyses, September 1992), pp. A-8, 11.

---

49. Scales, *Certain Victory,* pp. 292–300.

Figure 6. Day 4—February 27.

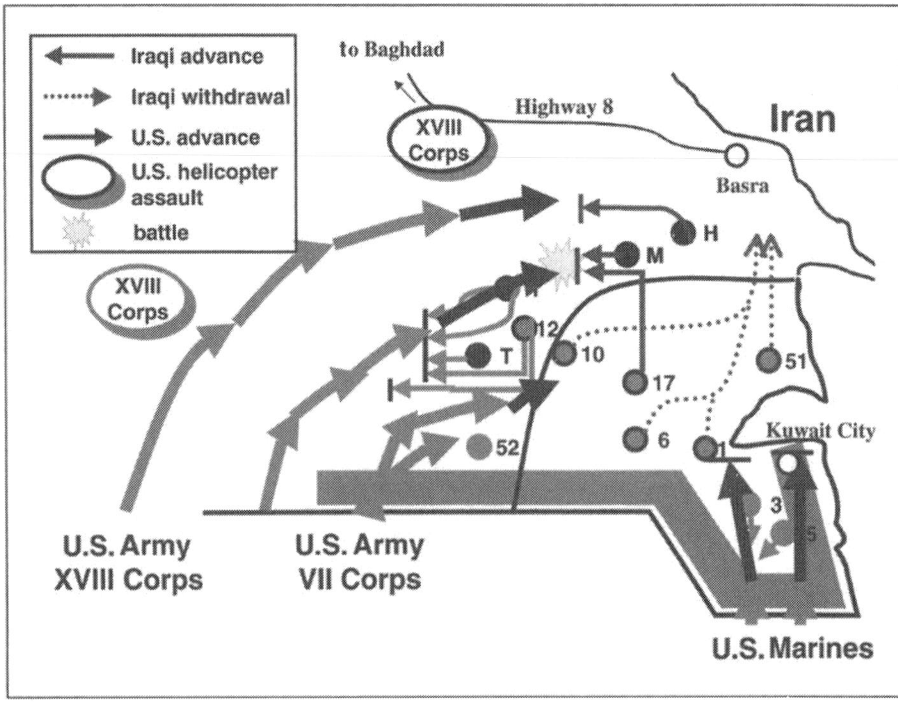

Testing the Air Power Explanations

In this section I test five explanations for how air power might have neutralized the Iraqi ground forces. These explanations are that air power (1) prevented the Iraqi forces from maneuvering, (2) severed their C3I, (3) cut Iraqi supply lines, (4) attrited the Iraqi forces, and (5) broke Iraqi morale. At the end of this section, I consider the possibility that air power neutralized the Iraqi ground forces through the cumulative effects on Iraqi maneuver, C3I, supply, force numbers, and morale.

DID AIR POWER PREVENT THE IRAQIS FROM MANEUVERING?

The predictions from this explanation are straightforward: If this explanation is correct, the Iraqi forces, especially the mobile reserves, should have been largely stationary during the ground war. If they tried to move, this explanation predicts that they should have been attacked savagely by coalition aircraft, either destroying them or forcing them to hunker down and abandon their attempts to maneuver.

The claim that coalition air power neutralized the Iraqi ground forces by preventing them from maneuvering is clearly wrong. During the ground war, virtually all of Iraq's mobile divisions—which comprised nearly all of Iraq's combat power—were on the move. Of the nine heavy divisions of the regular army in the KTO, two counterattacked south into the marines, two moved west into blocking positions to oppose the left hook, and four fled north after the general retreat was ordered. Only one (the 52d Armored Division) did not move; it was deployed very close to the Iraqi-Saudi border and was overrun within hours of the beginning of the ground war. All three heavy divisions of the Republican Guard moved west to block the path of the VII Corps. Coalition air power did not pin down the heavy Iraqi divisions in the KTO.[50]

Furthermore, the Iraqis were not savaged by air power during their maneuvers. More than 3,000 Iraqi armored vehicles were on the move during the ground war, and only about 150 of these were destroyed in concentrations along the roads by coalition aircraft. More important, almost all of the vehicles destroyed along the roads were moving north to withdraw from the theater rather than west to oppose the left hook.[51] For example, on the night of February 25, the U.S. Air Force decimated a column of retreating Iraqis along the Highway of Death; but while this was happening, all three Republican Guard heavy divisions and two heavy divisions of the Iraqi army were moving west to block the left hook. The heavy divisions moving west to fight were hardly hit as they maneuvered.[52]

Clearly, coalition air power was capable of destroying Iraqi vehicles moving along the roads; air power did this very effectively during the battle of al-Khafji and on the Highway of Death. But coalition air power failed to pin down the good units in the Iraqi military (the Republican Guard or heavy divisions of the army) as they moved west to fight the VII Corps.[53]

---

50. Air power did not pin down the light divisions of the Republican Guard either. The Adnan Division moved west to block the left hook; the Republican Guard Special Forces Division and the Baghdad Division were shifted north toward Baghdad to protect the capital, and the al-Faw Division was sent toward Basra. See CIA, *Operation Desert Storm;* and Pollack, "The Influence of Arab Culture," pp. 344–345.
51. CIA, *Operation Desert Storm.*
52. Only one concentration of 30 Iraqi vehicles was destroyed along the road by coalition air power as it headed west. Those were from the Iraqi 12th Armored Division. They were attacked as they maneuvered west during the first day of the ground war. Ibid.
53. Why did coalition aircraft fail to destroy the Republican Guard as it maneuvered west to fight the VII Corps? This is still a mystery, and an important one, but there are at least four plausible ex-

DID AIR POWER SEVER IRAQI C3I?

According to this explanation, air power neutralized the Iraqi army by cutting their C3I, effectively blinding them to the coalition's moves and making it impossible for the Iraqis to coordinate their defense. If this explanation is correct, we would expect to see an Iraqi army that either did not react to coalition moves, did not react quickly, or did not react well. The Iraqi ground forces should have been confused and disorganized.

There are signs that Iraqi C3I suffered from coalition attacks. When Saddam Hussein ordered the attack on al-Khafji, he called the corps commander who would execute the attack back from the KTO to Baghdad to give him the orders in person. This may or may not have caused problems with the al-Khafji attack, but it is a sign that Saddam Hussein no longer trusted his C3I capability. Another sign of Iraq's degraded C3I capability appeared during the ground war: The Iraqi corps commander communicated emergency orders to redeploy several Iraqi divisions without encryption, allowing U.S. intelligence to intercept the conversation.

Despite the attacks on Iraq's C3I system, the Iraqi senior military commanders retained surprisingly good command and control throughout the war.[54] For example, they identified the left hook with remarkable speed. Only six hours after the U.S. Army launched this operation, the Iraqi corps commander had a strong enough sense that something important was happening out west to maneuver two heavy brigades to new west-facing blocking positions. A day later, apparently convinced (correctly) that the main U.S. effort was coming in the left hook, he deployed his best divisions west. That he correctly identified the location of the main coalition attack so quickly is remarkable because Iraqi defenses were collapsing on all fronts. Given the speed with which Iraq's southern Kuwaiti defenses were crumbling, it is surprising that Iraqi leaders could tell that the marines' attack was not the main effort.

Not only did Iraq's senior military commanders identify the left hook quickly, but they successfully ordered a reasonable response. The Iraqi reserves

---

planations: (1) U.S. intelligence assets did not detect the movement of the Republican Guard or did not process the data quickly enough; (2) poor weather conditions prevented coalition aircraft from attacking the Republican Guard; (3) coordination problems existed between the coalition's air planners and the ground commanders, and the fear of fratricide prevented air attacks; and (4) old-fashioned "fog of war" prevented senior U.S. commanders in Riyadh from developing a clear picture of the chaotic battlefield, so they never recognized that the Iraqis were establishing a coherent blocking position to oppose the left hook.

54. Hosmer, *Psychological Effects of U.S. Air Operations*, p. 48, notes that Iraqi C_3I survived the air campaign in reasonably good shape. See also Scales, *Certain Victory*, pp. 332–333.

obeyed the order to maneuver into the oncoming coalition forces despite the risks involved. In sum, despite the weeks of air bombardment, Iraqi commanders still had enough capability to identify the coalition maneuver, formulate a good response, communicate it to their forces, and get the desired reaction. Iraqi commanders got their best divisions into the right place to defend against the left hook; the Iraqi military was not neutralized by air attacks on the C_3I network.

DID AIR POWER CUT IRAQI SUPPLY LINES?

There are two versions of the argument that air power neutralized the Iraqi ground forces by severing Iraqi supply lines. One version claims that air power degraded the Iraqi supply situation so significantly that it broke Iraqi morale. The second posits that air power denied Iraqi forces the supplies that they needed to fight effectively. Version one of the "supply" argument is a variant of the "morale" argument, which I address later. Here I test the second argument.

When the ground war began, Iraqi supplies in the KTO were plentiful. Many Iraqi supply dumps were enormous, with thousands of individual bunkers scattered over tens of acres.[55] Air strikes could whittle away at the enormous stockpiles of supplies, but even after five weeks of bombing, the stockpiles were only slightly down.[56] This should not be surprising; the Iraqis had five months to amass supplies in the theater before the air campaign began. Furthermore, the Iraqi ground forces were in stationary defensive positions from August 1990 through February 24, 1991, so they did not consume many of their supplies.[57] As a result, when U.S. forces advanced through Kuwait, and especially through southeast Iraq, they overran supply dumps totaling thousands of bunkers filled with food, fuel, and ammunition.[58]

---

55. For example, one U.S. division found "an immense dump, seventy-odd square miles covered with more than a thousand storage bunkers—enough ammunition to supply an army for many months." Atkinson, *Crusade*, p. 455. At another site, the Iraqi air base at al-Tallil, there "was the center of a ten-mile-square network of deep, well-camouflaged bunkers full of weapons, ammunition, and other supplies that had been reserved specifically to provision and maintain Iraq's army in Kuwait." U.S. News and World Report, *Triumph without Victory*, p. 376.
56. Winnefeld, Niblack, and Johnson, *A League of Airmen*, p. 152, write: "Supply depots were so numerous and large that they could not be eliminated; however, they were methodically attacked throughout the war, resulting in moderate reduction in stored materials.
57. *GWAPS Summary Report*, p. 99.
58. For example, on the third day of the ground war, the U.S. 1st Armored Division captured "more than 100 tons" of munitions at an Iraqi supply depot near the town of al-Busayya. Schubert and Kraus, *The Whirlwind War*, p. 188. This logistics dump "contained enough fuel, ammunition, and food to support an entire armored corps." U.S. News and World Report, *Triumph without Victory*, p. 325. On the first day of the ground war, the British 1st Armored Division "overran several

Not only did the supply dumps survive the air war, but the supplies got to the Republican Guard and the heavy divisions of the regular army. Iraq deployed the Republican Guard and several of the army heavy divisions near major supply dumps; in many cases these divisions were located on top of huge complexes of supply bunkers.[59] The proof that the Iraqi armored and mechanized divisions had adequate supplies to maneuver and fight is that when the war started, they did maneuver and they did fight. There is no sign that supply shortages prevented the mobile Iraqi divisions from moving where they wanted to go. And rather than surrender when they made contact with coalition forces, the Republican Guard and the other heavy divisions fought.

The supply situation was much bleaker for the Iraqi frontline infantry. These divisions suffered from severe shortages of food, water, and other necessities. It is not possible to know whether the frontline infantry still had the capability to fight when the ground war began because they lacked the will to resist. Whether it was air power that broke the morale of the frontline infantry—by denying them supplies, by attriting them, or by some other effect—is discussed below in the section on morale.

Air power did not neutralize the Iraqi army by severing its supply lines. Air attacks greatly reduced the flow of critical supplies to the frontline infantry, but the heavy divisions of the Iraqi army—and all the forces in the Republican Guard—still had the food, fuel, and ammunition they needed to fight when the ground war began.

DID AIR ATTACKS ATTRIT THE IRAQIS TOO HEAVILY FOR THEM TO RESIST?

There are two versions of the argument that air power neutralized Iraq's ground forces through attrition: (1) attrition during the air campaign destroyed

---

huge logistics sites" in Iraq within a few kilometers of the Saudi border. Scales, *Certain Victory*, p. 245. See also Swayne, "*Lucky War*," p. 244; Scales, *Certain Victory*, p. 259; and Clancy and Franks, *Into the Storm*, pp. 357–358.

59. When the Tawakalna Division moved west to block the left hook, it deployed into an area that had been prepared for a defense. The new area had bunkers full of supplies. When the U.S. Army's 2d Armored Cavalry Regiment hit the Tawakalna, U.S. artillery knocked out 27 ammunition bunkers (Scales, *Certain Victory*, p. 262) containing "large stockpiles of fuel, ammunition, and other supplies," including about 65 trucks (Clancy and Franks, *Into the Storm*, pp. 357–358). A few hours later, when the U.S. 1st Infantry Division attacked another element of the Tawakalna and the Iraqi 12th Armored Division, it reported that the Iraqis were deployed next to ammunition bunkers. Secondary explosions from these bunkers proved that they were full of supplies. See U.S. News and World Report, *Triumph without Victory*, pp. 349–350; and Scales, *Certain Victory*, p. 285. When the Madinah Division deployed west to block the left hook along with the Iraqi 17th Armored Division, they both deployed next to a large Iraqi supply dump. See Scales, *Certain Victory*, 293; and Clancy and Franks, *Into the Storm*, pp. 369–370.

Iraqi morale, leading them to surrender; and (2) Iraqi ground forces were attrited so heavily during the air campaign that they were hopelessly outnumbered in the major battles of the ground war. The first version is addressed in the next section. Here I consider the second explanation.[60]

Iraq's ground forces were heavily attrited during the air campaign. The best analyses of Iraqi attrition from air attacks conclude that 40 percent of Iraq's armored vehicles in the KTO were neutralized during the air war.[61] This figure probably overstates the percentage destroyed by air power,[62] but the attrition suffered by the Iraqi ground forces from air attacks was clearly very heavy.

Despite the heavy losses, the Iraqis had enough armored vehicles in the KTO to defend themselves from the coalition ground attacks. There were nine major ground engagements during the Gulf War. High-resolution data are available for five of these, making it possible to estimate the force ratios involved in these engagements (see Table 1).[63] One of these battles was a meeting engagement: That is, the two armies ran into each other as they moved across the desert, so neither side was on the defensive. In the other four, the Iraqis were on the tactical defense.

---

60. A related argument concedes that air power did not neutralize Iraq's ground forces through attrition, but argues that this attrition made the ground battles much more one-sided than they would have been otherwise. This argument is addressed later in this article; here I focus on the question: Did air attacks reduce the size of the Iraqi ground forces to the point that they were hopelessly outnumbered?

61. These estimates were made using the following method: Any Iraqi vehicle that neither moved nor fought during the ground war was coded as a casualty of the air war. Even if there was no sign that the vehicle was hit by an air-delivered weapon, it was credited as an air-war kill. This counting rule correctly accounts for both the direct and indirect effects of the air attacks. Vehicles directly destroyed by an air-dropped bomb would be included in this total, as would any vehicle indirectly neutralized by air power for one of the following reasons: its crew was killed in an air attack, the crew was scared away by an air attack, or the vehicle could not be maintained properly because its crew feared an air attack. See CIA, *Operation Desert Storm.*

62. Although the counting rule for assessing Iraqi vehicles neutralized during the air war appropriately counts both air power's direct and indirect effects on the Iraqi ground forces, it should be noted that this method probably overstates the effects of air power, because it also credits air attacks for vehicles disabled during the air war by things that had nothing to do with air power. For example, Iraqi crews who abandoned their vehicles on February 24 because they feared the ground attack would be scored as an air-war kill using this metric. So a large fraction of the 40 percent of abandoned Iraqi vehicles may have been from non-air power effects. Forty percent then should be taken as a maximum for the attrition of Iraqi ground forces during the air war.

63. In this section I examine only the numerical balance of Iraqi and coalition ground forces in the major battles; I exclude the qualitative differences between these two forces. I do this to separate two different causal explanations for the one-sided rout of the Iraqis. One explanation is that the Iraqis lacked sufficient numbers to fight, largely because of attrition during the air campaign. A competing argument is that the Iraqis were hopelessly outmatched by U.S. ground force capabilities (e.g., training and technology). By focusing here on raw numbers—specifically, the quantity of armored vehicles in these battles—I distinguish between the effects of air power (which reduced the number of Iraqi vehicles) and the effects of superior ground forces.

Military professionals and analysts have long recognized that in tactical engagements, defenders can often hold out against substantially larger attacking forces. Some analysts argue that defenders can usually stave off an enemy that is up to three times bigger; others argue that three to one is not a meaningful threshold. But most analysts agree that defenders enjoy some tactical advantages that allow them to defend against larger attacking forces.[64]

In all four of the major battles in which the Iraqis were on the defensive, Iraq had sufficient numbers to fight effectively. In one of these engagements (battle 9 in Table 1), the Iraqis were slightly outnumbered (by 1.5:1)—a ratio that should have been satisfactory for Iraq given that their forces were dug into defensive positions. But in this battle, approximately 165 American armored vehicles destroyed 110 Iraqi tanks and infantry fighting vehicles, and the Iraqis killed only one U.S. soldier.[65] In two of these (battles 6 and 7), the U.S. and Iraqi forces had essentially equal numbers. In the other engagement (battle 4), the Iraqi forces had a 1.5:1 numerical advantage over the United States.[66] De-

64. In this article I do not rely on the 3:1 rule; I simply argue that given defensive advantages, Iraq had sufficient numbers in all five well-documented ground battles to fight effectively. On defender advantages, see John J. Mearsheimer, "Assessing the Conventional Balance: The 3:1 Rule and Its Critics," *International Security*, Vol. 13, No. 4 (Spring 1989), pp. 54–89; John J. Mearsheimer, "Why the Soviets Can't Win Quickly in Central Europe," *International Security*, Vol. 7, No. 1 (Summer 1982), pp. 3–39; and Barry R. Posen, *Inadvertent Escalation: Conventional War and Nuclear Risks* (Ithaca, N.Y.: Cornell University Press, 1991), pp. 72–73 (especially n. 3), pp. 119–120 (especially Table 3.7).

65. The U.S. forces were from the 1st Armored Division; the Iraqis included elements of the Madinah Republican Guard Division and the 17th Armored Division. The main engagement in this battle was between the 2d Brigade of the U.S. 1st Armored Division and the 2d Brigade of the Madinah. The U.S. forces attacked with nine tank companies in a line, totaling approximately 165 M1 tanks; they destroyed about 110 Iraqi armored vehicles (about 65 tanks and 45 armored personnel carriers). See Atkinson, *Crusade*, pp. 466–467; CIA, *Operation Desert Storm;* U.S. News and World Report, *Triumph without Victory*, pp. 377–391; Tom Donnelly and Sean Naylor, *Clash of Chariots: The Great Tank Battles* (New York: Berkley, 1996), pp. 266–272; and Scales, *Certain Victory*, pp. 292–300.

66. In battle 6, the U.S. 1st Armored Division attacked elements of the Tawakalna Division. One battalion-sized armored task force from the 3d Brigade of the U.S. 1st Armored Division attacked elements of the 29th Brigade of the Tawakalna Division. The U.S. battalion attacked with 41 M1 tanks. Almost all the sources on the battle, including firsthand reports, claim that the Iraqis lost between 40 and 45 armored vehicles. See CIA, *Operation Desert Storm;* Donnelly and Naylor, *Clash of Chariots*, p. 259; and Richard M. Bohannon, "Dragon's Roar: 1–37 Armor in the Battle of 73 Easting," *Armor*, Vol. 101, No. 3 (May–June 1992), pp. 11–17. The only account that lists substantially higher numbers for the Iraqis is Scales, *Certain Victory*, p. 270, who reports that 150 Iraqi armored vehicles were destroyed.

In battle 7, the U.S. 1st Infantry Division fought elements of two Iraqi divisions: the Tawakalna and 12th Armored Division. There is more uncertainty about the numbers in this battle, but the biggest engagement between these divisions involved two U.S. battalion-sized task forces from the 1st Brigade of the 1st Infantry Division (which totaled about 120 armored vehicles) against 100–140 Iraqi armored vehicles. The number of Iraqi vehicles can be estimated from Scales, *Certain Victory*,

spite Iraq's defensive advantages and these favorable force ratios, the Iraqis managed to kill only one U.S. soldier in these four battles, while they lost more than 250 armored vehicles.[67]

In the meeting engagement (battle 1), the Iraqis and the marines had roughly equal numbers of forces. Nevertheless, the Iraqis were defeated in another extraordinarily one-sided fight. More than 100 Iraqi vehicles were destroyed, and no marines were killed. The Iraqis had sufficient numbers in this battle, but they were nevertheless totally ineffective.[68]

Taken together, the one-sidedness of all nine battles is shocking. Judging by force size alone, the Iraqis should have won most of the major engagements. Instead they were defeated in all nine, lost more than 600 armored vehicles, and killed only two U.S. soldiers.[69] The Iraqis were entirely ineffective against U.S. ground forces, but air power had not neutralized them by reducing their numbers too far; Iraqi ground forces were simply unable to compete with the better-equipped and better-trained U.S. and British divisions.[70]

DID AIR POWER BREAK IRAQI MORALE?

It is impossible to measure the morale of the Iraqi ground forces directly because morale exists in the minds of people. Morale must therefore be measured indirectly by examining behavior. More specifically, to assess the morale of any unit, one must evaluate the actions of the unit along two key dimensions: Did the unit follow orders, and did the unit fight? Based on these criteria, the morale of the heavy divisions of the Iraqi regular army, and especially the Republican Guard, appears to have been more than adequate when the ground war began.

---

pp. 281–285; and a firsthand account by Gregory Fontenot, "Fright Night: Task Force 2/34 Armor," *Military Review*, Vol. 73, No. 1 (January 1993), pp. 38–52.

Battle 4, between the U.S. 2d Armored Cavalry Regiment and elements of the Tawakalna Division and the 12th Armored Division, is one of the most carefully studied ground battles of the war. U.S. forces comprised 29 tanks and 37 Bradley armored vehicles. The Iraqis had approximately 96 armored vehicles, including 58 tanks and 38 APCs. See W.M. Christenson and Robert A. Zirkle, "73 Easting Battle Replication—A Janus Combat Simulation," P-2770 (Alexandria, Va.: Institute for Defense Analyses, September 1992), pp. A-8, 11.

67. Ibid.

68. See Press, "Lessons from Ground Combat in the Gulf," pp. 139–143; Turner, "Counterattack"; and Cureton, *U.S. Marines in the Persian Gulf War*, pp. 91–96.

69. The data on the number of Iraqi vehicles destroyed in each battle are undoubtedly imperfect. But because these engagements were *so* one-sided, errors in the size of the Iraqi forces are unlikely to change the conclusions.

70. For a preliminary exploration of some of the aspects of coalition ground forces that gave them such huge advantages over Iraqi ground forces, see Press, "Lessons from Ground Combat in the Gulf." For a conflicting view, see Biddle, "Victory Misunderstood."

Obeying some orders requires more esprit de corps than obeying others. For example, an order to "hold position and fight" requires greater morale from the soldiers executing it than an order to withdraw. An order to maneuver into the path of an enemy force requires an even higher level of morale. This is the order that the heavy divisions of the Republican Guard, and half the heavy divisions of the Iraqi army, received and obeyed on February 25 and 26.

All three heavy divisions of the Republican Guard received orders to move west to block the left hook, which they obeyed immediately.[71] In fact, the morale of the Republican Guard was surprisingly high, given that they had just endured five weeks of aerial bombardment that had significantly reduced their forces. It would not have been surprising if the Republican Guard units had disobeyed orders and held their positions or fled north out of Kuwait. But two of these heavy divisions (the Tawakalna and the Madinah) maneuvered and then fought against the coalition; the other (the Hammurabi) moved toward the coalition and then withdrew after the cease-fire. Perhaps even more surprising, four of the heavy divisions in the regular army maneuvered to fight the coalition ground forces. Two of them maneuvered west, as ordered, and fought alongside the Republican Guard against the left hook; the other two counterattacked south.[72] The four heavy divisions of the regular army that withdrew north were ordered to do so; there is no evidence that any of the heavy divisions—from the Republican Guard or the regular Iraqi army—refused to fight.[73]

Furthermore, in the battles between the U.S. Army and the Iraqi heavy ground forces, there are numerous reports of Iraqi soldiers fighting with notable courage, especially among the Republican Guard. Dismounted Iraqi Republican Guard infantry attacked U.S. tanks from bunkers and trenches,

---

71. Two light Republican Guard divisions were also ordered to block the left hook, which they did. CIA, *Operation Desert Storm*.

72. The Iraqi 12th and 17th Divisions moved west to help block the left hook. The 3d and 5th Divisions attacked south into the marines.

73. In fact, the heavy divisions of the Iraqi army that stayed and fought were, on average, attrited more heavily during the air war than those that pulled out of the theater. This confirms that the heavy divisions that pulled out did so because of the orders we know they received, not because their morale was broken. This is based on data in ibid. and my analysis. The only heavy division of the Iraqi army that remained stationary during the ground war was the 52d Armored Division. It was deployed so close to the Saudi border that it was overrun almost immediately by the British 1st Armored Division without a fight. The British attack occurred after an intense aerial bombing and fierce artillery raid against the 52d Armored Division.

repeatedly hitting their targets with their ineffective handheld antitank weapons.[74] On several occasions, Iraqi infantry feigned death and then, once bypassed, jumped up and shot at the rear of U.S. vehicles.[75] Some Iraqi infantry jumped onto passing U.S. tanks, only to be shot off by trailing U.S. armored vehicles.[76] Several U.S. divisions reported encountering Iraqi tanks waiting with their engines off to avoid detection by the thermal sights on American armored vehicles. When the U.S. tanks passed the Iraqi positions, the Iraqi tanks would fire at the more vulnerable side or rear of the American vehicles.[77] And in several instances, Iraqi infantry and armor repeatedly counterattacked U.S. armored formations after American tanks had destroyed entire battalions of Iraqi armor in a matter of minutes.[78] Iraqi soldiers—even the professionals—were not well trained. Their shooting was extremely inaccurate, and they were hopelessly outgunned by U.S. technology. But the evidence does not suggest that the Iraqis were cowards. Many of them—particularly those in the Republican Guard—fought and died bravely.

Unlike the Republican Guard and the heavy divisions of the Iraqi army, the Iraqi frontline infantry melted away at first contact with coalition ground forces. There are no reports of the frontline infantry fighting with even moder-

74. U.S. tankers reported that when Iraqi antitank weapons (RPGs) hit their tanks, the weapons merely damaged the personal effects that the tank crews frequently strapped to the tops of their vehicles. Scales, *Certain Victory*, pp. 284–285.

75. The 2d Armored Cavalry Regiment reported Iraqis feigning death and then shooting antitank weapons at U.S. tanks from the rear. Atkinson, *Crusade*, pp. 443–445; and U.S. News and World Report, *Triumph without Victory*, p. 338. The 1st Infantry Division reported Iraqi infantry allowing themselves to be bypassed and then taking shots with their RPGs at the rear of U.S. tanks. Scales, *Certain Victory*, pp. 284–285; and Clancy and Franks, *Into the Storm*, p. 396.

76. See Clancy and Franks, *Into the Storm*, pp. 394–396; and Atkinson, *Crusade*, pp. 443–445.

77. The U.S. 1st Infantry Division experienced this with the Tawakalna. See Scales, *Certain Victory*, pp. 284–285. The 2d Armored Cavalry Regiment found Iraqis sitting with their vehicle engines off. U.S. News and World Report, *Triumph without Victory*, pp. 445–447.

78. The U.S. 3d Armored Division reported that infantry from the Tawakalna kept fighting and counterattacking its positions even after their armor was destroyed. Scales, *Certain Victory*, p. 279. The U.S. 1st Infantry Division reported that Iraqi antitank infantry teams from the Tawakalna kept hitting U.S. tanks, but to no effect. Then, "throughout the night, stubborn Iraqi antitank teams repeatedly emerged from previously bypassed positions to stalk the tanks and Bradleys. . . . American tankers had little trouble detecting the approaching Iraqis through thermal sights, and they cut them down like wheat with long bursts of coaxial machine-gun fire." For a description of Iraqi resistance and counterattacks, see ibid., pp. 284–285; and U.S. News and World Report, *Triumph without Victory*, pp. 368–370. The 2d Armored Cavalry Regiment reports that there were Iraqi company-sized counterattacks after the Battle of 73 Easting. Atkinson, *Crusade*, pp. 445–447. The U.S. 3d Armored Division reported a battalion-sized Iraqi armored counterattack early on the morning of February 27. Clancy and Franks, *Into the Storm*, pp. 398–400.

ate conviction, let alone demonstrating heroism.[79] What is clear about the frontline infantry is that they had very low morale when the ground war began; what is less clear is the reason for this low morale.

The best evidence in support of the argument that air attacks shattered the will of the Iraqi frontline infantry comes from Stephen Hosmer of the RAND Corporation. Hosmer reviewed transcripts of thousands of interviews with Iraqi prisoners of war conducted by the United States and Saudi Arabia in 1991. Most of the prisoners were frontline infantry, and almost all of them had surrendered without putting up a fight. The majority reported that the air bombardment had played a large role in their decision to surrender.[80]

Hosmer's research is excellent, but other evidence suggests that air power may not have been the sole cause of the frontline infantry's collapse. The frontline infantry divisions suffered very different levels of bombing. Some along the Kuwaiti-Saudi border were attacked by thousands of coalition sorties, including large numbers of devastating B-52 strikes, and they surrendered. Others were hit with much less intensity, and few B-52 strikes, yet they also surrendered.[81] In other words, there was considerable variation across the theater in how hard the Kurdish and Shi'a conscripts were bombed. Some of the divisions were not hit hard, but the result was the same everywhere: The frontline infantry did not fight.

More generally, the intensity with which a division was hit during the air war does not correspond to that division's willingness to fight. The heavy divisions of the Republican Guard were hit the hardest, but they were willing to fight. The Iraqi frontline infantry ahead of the marines, and many of the heavy divisions of the Iraqi regular army, were hit almost as hard as the Republican Guard; the heavy divisions fought, but the frontline infantry surrendered.[82]

Skeptical readers might argue that the fact that there were so few U.S. casualties in these battles between U.S. forces and heavy Iraqi divisions proves that

79. See Quilter, *With the I Marine Expeditionary Force*, pp. 75–91; Gordon and Trainor, *The Generals' War*, pp. 355–363; and Scales, *Certain Victory*, pp. 221–222.
80. Hosmer, *Psychological Effects of U.S. Air Operations*, pp. 152–176.
81. The frontline infantry divisions far out in the west along the Iraqi-Saudi border were not hit hard during the air war because the coalition did not want to warn Iraq about the left hook, and because it wanted to maximize its effort to weaken the divisions in front of the marines. Yet even these western Iraqi divisions also surrendered immediately. For maps showing the location of each Iraqi division in the KTO and the distribution of coalition attacks against these forces, see *GWAPS*, Vol. 2, Pt. 1, pp. 211, 287.
82. Ibid.

the Iraqis did not fight.[83] But history shows that very one-sided outcomes such those in the Gulf War have happened before, when forces from the world's leading military power fights troops from the less developed world. When tactics and terrain allow the leading military power to bring its superior technology and training to bear, the results can be catastrophic for the locals. For example, in 1879 at the Battle of Ulundi in southern Africa, a British force using early machine guns achieved an exchange ratio in excess of 100:1 against the Zulu army, even though the Zulus reportedly fought bravely. A few years later, at the Battle of Omdurman in northern Africa, the British achieved an exchange ratio greater than 200:1.[84] The Gulf War was not a colonial war to acquire empire for the United States, but from the standpoint of military technology and training, the comparison is useful. Like the colonial troops, U.S. soldiers used superior technology and training to decimate their enemy in engagements in which the enemy had little chance of reciprocating. Even when the locals fought bravely, they never had a chance.[85]

In sum, did the air campaign affect Iraqi morale? The answer is yes. Air power certainly contributed to the utter collapse of the frontline infantry, and it might be true that it single-handedly caused the infantry to disintegrate. But even after five weeks of bombing, the best Iraqi units—the Republican Guard and the heavy units of the regular army—were willing to maneuver into the path of U.S. forces and fight. Air attacks did not neutralize the Iraqi force by crippling their morale.

THE CUMULATIVE EFFECTS OF AIR POWER ON THE IRAQI GROUND FORCES

Air power did not neutralize the Iraqi army by destroying its ability to maneuver or its C3I, supply, force numbers, or morale. But perhaps air attacks weak-

83. "You can hardly march two armored divisions past each other during peacetime without producing more casualties than those so-called battles," quipped a friend.

84. Bryan Perrett, *The Battle Book: Crucial Conflicts in History from 1449 to the Present* (New York: Sterling, 1992), pp. 225, 292; and Thomas Harbottle, *Dictionary of Battles* (New York: Stein and Day, 1971), pp. 208–209, 290.

85. The list of the U.S. technological advantages over Iraqi ground forces—including advanced sights that allowed U.S. tanks and APCs to see Iraqi vehicles four times as far as the Iraqis could see U.S. forces; fire-control computers that allowed U.S. tanks to kill Iraqi vehicles on the first shot four times further than the Iraqis could hit U.S. vehicles; thermal sights that allowed U.S. forces to see the enemy equally well at night and during the day, while the Iraqis were nearly blind in the dark; ammunition that allowed U.S. tanks to destroy Iraqi tanks from the front, side, or rear, out to 4 kilometers, compared to Iraqi tank ammunition that could not penetrate the frontal armor of U.S. tanks at any range; and the navigation equipment that allowed the U.S. forces to advance off-road through the barren desert, compared with Iraqi forces that were anchored to the roads—suggests

ened each of these, and the net effect led to the collapse of the Iraqi forces. If this hypothesis is correct, Iraqi ground forces should have been in disarray prior to their major engagements with the coalition ground forces.

This explanation fails, too. Although the Iraqi frontline infantry did collapse prior to the ground war, Iraq's heavy divisions—those in the Republican Guard and in the regular army—were functioning, organized military units until they were engaged and destroyed in ground combat.[86] As proof, when the ground war began, Iraqi commanders successfully maneuvered their best divisions into defensive positions in the path of the coalition's left hook. The Iraqi soldiers in these divisions followed orders and fought against the coalition ground forces. Despite their defensive positions and favorable force ratios, the Iraqis had no chance in the open desert against the much better trained and well-equipped U.S. and British divisions. The problem that the Iraqi divisions inevitably faced was not the effects of sustained bombardment on their ability to maneuver, C3I, supply, force numbers, and morale. Their problem was that they were simply out of their league.

Did Air Power Make the One-Sided Victory Possible?

The conventional wisdom about the Gulf War is that air power neutralized the Iraqi ground forces. Above I refuted this argument and showed that, at the end of the air war, the best Iraqi divisions were still ready and willing to maneuver and fight. But there is a weaker version of the air power argument that must be considered: Although air attacks did not neutralize the Iraqis, perhaps they were necessary for the one-sided coalition victory. In other words, it could be argued that without the air campaign, many more coalition soldiers and marines would have been killed.

The weaker version of the air power argument also appears to be untrue. Evidence from the Gulf War suggests that even if there had been no significant air campaign, the ground war would not have been substantially different. There may have been a few more coalition casualties (maybe an additional 20–200 dead), but the result would still have been a rout of historic proportions. U.S. commanders were wise to conduct the lengthy air campaign because they

that the analogy between Gulf War armored battles and British battles against the Zulus may be fully justified.

86. The light divisions of the Republican Guard, not the focus of this article, retained their cohesion too, despite heavy bombardment. CIA, *Operation Desert Storm*.

could not be sure that coalition casualties would be so low. But if the left hook had been launched without a long air campaign to soften up the Iraqi forces, the coalition ground forces would still have greatly overmatched their Iraqi opponents.

The most significant effect that the air campaign had on the ground war is that it substantially attrited the Iraqi ground forces before the ground war began. At most, 40 percent of Iraqi armored vehicles were neutralized by the air campaign. But a close look at the ground battles strongly suggests that these vehicles would not have caused many additional coalition casualties had there been no air war.

One of the most surprising aspects of the major ground battles in the Gulf War is that U.S. losses correlate with neither the total number of Iraqi forces in the battle nor with the force ratios between the Iraqis and the coalition. In the five battles for which there is good information, only two U.S. soldiers were killed by enemy fire. In the battle in which the United States enjoyed the biggest numerical advantage, one American died. And in the one battle in which the Iraqis had numerical superiority, the United States lost one soldier.

The consistent pattern across all of the battles is that greater numbers of Iraqi forces simply produced greater numbers of Iraqi casualties. This suggests that had there been 40 percent more Iraqi tanks in these battles (i.e., had the air war not happened), there would have been 40 percent more Iraqi casualties during the ground campaign. The evidence from the major ground battles does not support the argument that many more coalition soldiers would have been killed.

But suppose, despite the evidence to the contrary, that if the Iraqis had 40 percent more tanks and other armored vehicles in each of the major ground battles, they were able to inflict 40 percent more U.S. fatalities. If we increase the number of U.S. fatalities during the ground war (the total was sixty-three) by 40 percent, the total number of U.S. soldiers and marines killed in the ground war rises to eighty-eight.[87]

Predicting exact numbers of losses in a war is impossible, but given what we now know about the one-sided outcome of every ground engagement during the Gulf War, it is difficult to support the claim that the air campaign was necessary to soften up the Iraqi forces for a successful ground operation. The

87. Many battlefield relationships are nonlinear; so to be even more conservative, if 40 percent more Iraqi vehicles could have increased American losses by 300 percent, total U.S. fatalities in the operation from Iraqi attacks would have risen to only 189.

ground attack would likely have gone very well for the coalition even if there had been no extended air war.

Air Power in the Gulf and over Kosovo

The lessons from the Gulf War about military technology and the relationship between air and ground forces are still relevant and will likely remain so. In this section I briefly discuss why coalition air power was unable to neutralize the Iraqi ground forces and argue that these factors have not changed. I then support this argument with a brief description of NATO's air campaign against the Serbian army in Kosovo. I show that the same problems that the coalition experienced as it tried to target the Iraqi army in 1991 greatly hampered NATO's efforts against Serbia in 1999.

Coalition air power was unable to do more damage to Iraqi ground forces during the air campaign because of one key fact: The Iraqis were deployed in static defensive positions. Static forces, especially those not engaged in combat, require fewer supplies than either units on the attack or units conducting a mobile defense. They burn far less fuel, require fewer spare parts, and consume less food and water. They also require minimal communication bandwidth. A division sitting still in the desert, or even conducting a single countermaneuver (e.g., the Iraqi response to the coalition's left hook), requires relatively little communication among its officers or between the division's leaders and their superiors. Local commanders do not need to communicate nearly as much among themselves, or with senior commanders out of the theater, as they would in a more mobile defense or an offensive operation. Thus, even if air attacks can choke off most of an enemy's supply throughput around the battlefield, and can significantly reduce its C3I bandwidth, they may have little effect on an enemy in static positions.

This is precisely what happened in the Gulf War. Coalition air power successfully reduced road traffic into Kuwait by 90 percent,[88] and it hindered Iraqi C3I. But because the Iraqis were not conducting long, complicated maneuvers, and because the ground battles were so short, their stockpiles were sufficient to sustain their forces.[89] Similarly, the mission given to the Iraqi ground forces was relatively simple: Defend current positions and then maneuver west to

88. *GWAPS*, Vol. 2, Pt. 1, p. 281; and *GWAPS Summary Report*, pp. 95–96.
89. *GWAPS Summary Report*, p. 99.

block a coalition flanking attack. Given the simplicity of these operations, Iraq's degraded C3I system could still be effective. The implication is that whenever the United States faces static, defensively oriented foes, air power will have great difficulty hampering—let alone neutralizing—them by attacking supply and C3I targets.

A second fundamental problem resulted from Iraq's defensive posture: Static forces are very difficult to detect from the air, even when they are in the desert and even with information-age sensors. Static forces give few signs of their location; they emit little heat, noise, or radio traffic. In addition, it is easy to create simple decoys that are nearly indistinguishable from stationary military forces. From an aircraft flying at 10,000 feet, it is very difficult to discern a tank half-buried in the sand from a sheet of corrugated metal half-buried in the sand. Stationary forces are both inherently difficult to observe and even more difficult to distinguish from simple decoys.

Identifying stationary vehicles and discerning real vehicles from decoys were big problems for coalition aircraft during Desert Storm. From high altitude, distinguishing between "living" and "dead" Iraqi vehicles was also difficult—so much so that it led to multiple hits on the same target.[90] Even worse, the Iraqis learned that U.S. aircraft could easily be tricked and constructed decoys to mimic live targets, often putting wreckage near "living" vehicles to make them appear "dead."[91]

The problems that coalition air power faced in its attacks on stationary Iraqi ground forces were complex, and it appears that the United States has made little progress in solving them over the last decade. The same problems hamstrung NATO efforts to destroy Serbian ground forces in Kosovo in 1999.[92] For

90. Much has been made of the "tank plinking" in the Gulf War, in which U.S. aircraft used their infrared sensors to detect the warmth radiated by Iraqi tanks partially buried in the sand. What is not usually mentioned is that "living" Iraqi tanks create the same heat signal when warmed by the desert sun as vehicles already destroyed. Furthermore, both "living" and "dead" Iraqi vehicles were very hard to distinguish from simple metal decoys, such as sheets of corrugated metal attached to short sections of thick pipe, buried in the sand.

91. The best account of Iraqi efforts to use decoys to protect their ground forces from coalition air attacks is William F. Andrews, *Airpower against an Army: Challenge and Response in CENTAF's Duel with the Republican Guard* (Maxwell Air Force Base, Ala.: Air University Press, 1998), pp. 42–46.

92. One implication of this article is that initial analyses of wars—frequently conducted before adequate information is available for careful analysis—should be treated with substantial skepticism. This applies to the analyses of the Kosovo war on which I rely here. The problems that NATO experienced during the Kosovo war seem similar to those that the coalition experienced in the Gulf War, but the conclusions about the war in Kosovo are preliminary.

eleven weeks NATO aircraft flew thousands of sorties over Kosovo to disrupt the Serbs' C3I and to attrit their forces through precision attacks, leading NATO spokespeople to confidently assess that NATO air strikes had reduced Serbian ground forces by one-third. But after the war, it became apparent that NATO had greatly overestimated the effectiveness of its attacks.[93] In fact, NATO analysts who inspected the hulks of destroyed Serbian weapons reportedly found that "fewer than 20 Serbian tanks, a similar number of artillery pieces, and fewer than ten armored personnel carriers" had been knocked out.[94]

NATO's bombing raids against Serbian ground forces ran into many of the same problems that coalition forces faced in the Gulf War. Serbian forces were easy to hide from patrolling aircraft, and the Serbs made extensive use of decoys that look like real targets from 10,000 feet.[95] A decade has passed since the Gulf War, but the problems inherent in finding a static, defensively oriented army from the air, and distinguishing real targets from decoys, remain.

Implications for U.S. Foreign Policy and Force Structure

Historians, military analysts, and policymakers have drawn the wrong lessons from the Gulf War, and this may have serious implications for U.S. foreign policy and U.S. force structure. Air power is now significantly more lethal against ground targets than it was before. Almost any target that is detected and identified can be destroyed from the air with precision munitions.[96] But air

93. For a detailed discussion of the various NATO estimates over time, and for alternative estimates of the damage that NATO did to the Serb army, see Barry R. Posen, "The War for Kosovo: Serbia's Political-Military Strategy," *International Security*, Vol. 24, No. 4 (Spring 2000), pp. 64–66.
94. Richard J. Newman, "A Kosovo Numbers Game: So Where Are All the Tanks NATO Killed?" *U.S. News and World Report*, July 12, 1999, p. 36. See also Steven Lee Myers, "Damage to Serb Military Less Than Expected," *New York Times*, June 28, 1999, p. 1; "NATO's Game of Chicken," *Newsweek*, July 26, 1999, p. 59; and Tim Butcher and Patrick Bishop, "NATO Admits Air Campaign Failed," *London Daily Telegraph*, July 22, 1999, p. 1.
95. For brief descriptions of Serb decoys, see Myers, "Damage to Serb Military Less Than Expected"; and "NATO's Game of Chicken," *Newsweek*, July 26, 1999, p. 59.
96. There are several important caveats to this generalization. First, bad weather continues to inhibit air attacks because most sensors still cannot "see" through clouds and because laser-designators, a key component of many precision weapons, cannot penetrate clouds. GPS-guided weapons can accurately hit targets in bad weather but depend on weather-sensitive reconnaissance to determine the precise location of the targets. Second, some targets can be protected from air attack by locating them near civilians sites, raising the risks of collateral casualties to an unacceptable level. Third, targets can be protected from air strikes by burying them in deep bunkers. Despite these ca-

power still has limited effectiveness against defensively oriented enemy ground forces. Locating enemy ground forces in the desert is maddeningly difficult; in forest or mountainous terrain, or in urban areas, it is even harder. And distinguishing real targets from decoys frustrated the United States in both the Gulf War and Kosovo.

There are many contingencies in which the United States or its allies may need to destroy a defensively oriented enemy ground force. A U.S.–South Korean counterattack into North Korea; a U.S.-led counteroffensive into Iraq; a U.S. invasion of Kosovo, Montenegro, or Serbia; or an Israeli offensive to retake the Golan Heights or southern Lebanon would probably all require overcoming defensively oriented enemy ground forces. In all of these cases, air power would probably have only limited effects on enemy defenses; success would hinge on the effectiveness of ground forces.

At least two policy implications follow from this analysis. First, the United States and its allies should not overestimate the effectiveness of air attacks and undertake offensive military operations with the expectation that air power will provide a cheap victory. When enemy ground forces must be ejected from the territory they occupy, success or failure will be determined on the ground, and the price will depend on the ability of U.S. and allied ground forces to overmatch the enemy.

Second, the geography and foreign policy of the United States require that it maintain a balanced military force structure. Because the United States has global military commitments, it must have a military that can deploy rapidly to defend faraway allies. Air power is ideal for this mission: It can get to distant battlefields quickly and—as al-Khafji and the Highway of Death show—it can be lethal against enemy ground forces on the move. However, because its allies are far away, the United States often joins wars late. Allied territory often must be recaptured, and sometimes enemy territory must be taken. For these missions, the United States needs ground forces that can dominate the battlefield. Unless the United States military maintains large, well-trained, and well-armed ground forces, it will not be prepared to achieve more one-sided victories like the Gulf War.

A decade has passed since the Gulf War, but the six weeks of fighting—the long air campaign and the short ground war—still offer the clearest view we have of the capabilities of modern, high-technology militaries, especially when

---

veats, the generalization is still valid—if an enemy target is identified and is within range of precision-strike munitions, it can usually be destroyed.

they are pitted against third world opponents. Air power is a more powerful tool of national policy than it once was, but it still does not dominate the battlefield. Understanding the relationship between air and ground forces is necessary to make sound foreign policy decisions, to create a force structure that can meet the challenges that the United States will face in the coming decades, and in the end, to save American lives.

The 2003 War in Iraq

[15]

INTERNATIONAL LAW AND THE WAR IN IRAQ

John Yoo

In his speech before the United Nations (UN) in September 2002, President George W. Bush characterized the possible use of force against Iraq as necessary to enforce existing Security Council resolutions and to eliminate a dangerous threat to international peace and security.[1] The Security Council responded by adopting Resolution 1441, which found Iraq to be in material breach of previous Security Council resolutions and threatened serious consequences for further intransigence.[2] When Iraq refused to fully comply with these resolutions, the United States led an ad hoc "coalition of the willing" that invaded Iraq on March 19, 2003, quickly defeated Iraq's armed forces, and ended the regime of Saddam Hussein and the Ba'ath party. On May 1, 2003, President Bush announced that major combat operations in Iraq had ended. At the time of this writing, the United States has assumed the position of an occupying power that is responsible for rebuilding Iraq, as recognized by the Security Council in Resolution 1483.

Despite these actions, other leading nations (primarily France, Germany, and Russia) and many international scholars have argued that international law did not justify the war in Iraq. The first part of this paper will explain why their view failed to properly read existing Security Council resolutions to authorize the use of force. Even putting the United Nations to one side, the war was further justified as an exercise of self-defense. Under basic notions of customary international law, properly understood, the United States could have attacked Iraq

[23] For its part, the United States made clear that it would have preferred the Council to adopt a second resolution. Its view was that such a resolution would have operated as an important demonstration of resolve by the international community to increase the pressure on Iraq and, ultimately, to reduce the likelihood that resort to force would be necessary. While the Council's inability to come together for such a resolution may have misled Iraq into believing it could wait out the process in New York, the absence of Council action in no way diminished the effect of what the Council had previously authorized.

* Mr. Taft is the Legal Adviser of the United States Department of State; Mr. Buchwald is the Assistant Legal Adviser for Political-Military Affairs of the United States Department of State.

[1] George W. Bush, UN General Assembly in New York City Address (Sept. 12, 2002), 38 WEEKLY COMP. PRES. DOC. 1529 (Sept. 16, 2002), *available at* <http://www.whitehouse.gov/news/releases/2002/09/20020912-1.html> [hereinafter Bush UN Address].

[2] SC Res. 1441 (Nov. 8, 2002).

even without Security Council authorization. The second part of this paper will discuss why the toppling of Saddam Hussein's regime was justified as an exercise of anticipatory self-defense.

This paper is divided into three sections. Part I lays out the background to the conflict with Iraq. Part II discusses the UN Security Council's authorization for the use of military force against Iraq. Part III explains why the United States could use force in anticipatory self-defense. I conclude by sketching out the implications of these arguments for a future threatened by weapons of mass destruction (WMD), rogue nations, and terrorism.

I. BACKGROUND

Examination of the factual and legal setting leading up to the Iraq war will show that sufficient legal authority existed for the 2003 conflict with Iraq. On August 2, 1990, Iraq invaded Kuwait. The next day, the Security Council adopted Resolution 660, the first of many resolutions to condemn Iraq's actions and demand withdrawal from Kuwait.[3] After diplomacy failed, the Security Council adopted Resolution 678, which gave Iraq until January 15, 1991, to implement Resolution 660 fully.[4] Paragraph 2 of Resolution 678 authorizes member states "to use all necessary means to uphold and implement resolution 660 (1990) and all subsequent relevant resolutions and to restore international peace and security in the area."[5] Iraq refused to withdraw from Kuwait before the January 15th deadline, and Operation Desert Storm began the next day. Iraq was expelled from Kuwait on February 27, 1991.

On April 3, 1991, the Security Council adopted Resolution 687, which established the conditions for a formal cease-fire suspending hostilities in the Persian Gulf. Resolution 687 required Iraq to: (1) destroy its chemical and biological weapons and ballistic missiles and agree to on-site inspections; (2) not use, develop, construct, or acquire such WMD and their delivery systems; (3) not acquire or develop nuclear weapons or nuclear-weapons-usable material or components; and (4) accept on-site inspection and destroy nuclear-related weapons or materials.[6] To carry out the inspections, the resolution established a United Nations Special Commission (UNSCOM) to cooperate with the International Atomic Energy Agency (IAEA), which was to take custody of all of Iraq's nuclear-weapons materials. On April 6, 1991, Iraq officially accepted the terms, and a formal cease-fire went into effect between Iraq, Kuwait, and the nations that had cooperated with Kuwait, including the United States.[7]

[3] *See* SC Res. 660 (Aug. 2, 1990). *See also* SC Res. 661 (Aug. 6, 1990) (establishing oil embargo and sanctions regime against Iraq and Kuwait); SC Res. 662 (Aug. 9, 1990) (deciding that Iraq's annexation of Kuwait has no legal validity); SC Res. 664 (Aug., 18, 1990) (demanding that Iraq permit and facilitate the immediate departure from Kuwait of third-party nationals); SC Res. 665 (Aug. 25, 1990) (calling upon allied nations to use necessary measures to enforce embargo established in SC Res.661); SC Res. 666 (Sept. 13, 1990) (implementing sanctions regime and the humanitarian needs exception); SC Res. 667 (Sept. 16, 1990) (demanding that Iraq immediately protect safety of diplomatic and consular personnel and premises in Kuwait); SC Res. 669 (Sept. 24, 1990) (relating to Jordan's request for relief from effects of implementing oil embargo and sanctions regime); SC Res. 670 (Sept. 25, 1990) (regulating aircraft transporting cargo to Iraq or Kuwait and reaffirming Iraq's liability for grave breaches of the Geneva Convention Relative to the Protection of Civilian Persons in Time of War); SC Res. 674 (Oct. 29, 1990) (condemning Iraqi mistreatment of Kuwaiti and third-party nationals and inviting collection of materials to bring charges against Iraq for its violations of international law); SC Res. 677 (Nov. 28, 1990) (condemning Iraqi attempts to destroy demographic composition of Kuwait and Kuwaiti civil records).

[4] SC Res. 678 (Nov. 29, 1990).

[5] *Id.*

[6] SC Res. 687 (Apr. 3, 1991).

[7] Toward the end of the Gulf war, Iraq brutally suppressed Kurdish insurrections in Iraqi Kurdistan, causing the flight of hundreds of thousands of civilians to Iran and Turkey. In response, the Security Council adopted SC Res. 688 (Apr. 5, 1991), which condemned "the repression of the Iraqi civilian population" and found that the consequences of such repression caused "a massive flow of refugees towards and across international frontiers and . . . cross-border incursions" that threatened international peace and security in the region. To aid the refugees who had fled northern Iraq, the coalition established a no-fly zone prohibiting Iraqi military aircraft from flying north of the 36th parallel. Due to further repression of the civilian population in 1992, British, French and U.S. coalition forces established an additional no-fly zone south of the 32nd parallel.

From the beginning, Iraq resisted UNSCOM's efforts to perform its mandate. On August 15, 1991, little more than four months after the adoption of Resolution 687, the Security Council "condemn[ed]" Iraq's "serious violation" of a number of its obligations regarding the destruction and dismantling of its WMD program and of its agreement to cooperate with UNSCOM and the IAEA and stated that the violation "constitutes a material breach of the relevant provisions of [Resolution 687] which established a cease-fire and provided the conditions essential to the restoration of peace and security in the region."[8] Iraq's intransigence continued for years. In June of 1996, the Security Council "deplored" the refusal of Iraq to allow access to UNSCOM inspectors and Iraq's "clear violations" of its resolutions.[9] In June 1997, the Council again "condemn[ed] repeated refusal of Iraqi authorities to provide access" to UNSCOM inspectors as a "clear and flagrant" violation of its resolutions and demanded full, immediate, and unconditional compliance.[10] In the fall of 1997, Resolution 1137 "condemn[ed] . . . the continued violations by Iraq of its obligations under the relevant resolutions to cooperate fully and unconditionally with [UNSCOM]," found that the situation continued to constitute a threat to international peace and security, and warned that "serious consequences" would result if Iraq failed to comply with its international obligations.[11]

In February 1998, UN Secretary-General Kofi Annan secured a memorandum of understanding confirming Iraq's acceptance of all relevant Security Council resolutions and its reaffirmation to cooperate fully with UNSCOM and the IAEA. Nonetheless, Iraq formally halted all cooperation with UNSCOM at the end of October. The Security Council responded by condemning Iraq's decision as a "flagrant violation of resolution 687 . . . and other relevant resolutions."[12] On December 15, UNSCOM reported that it could not complete its mandate due to Iraq's obstructionism. The next day, the United States and Britain launched a seventy-hour missile and aircraft bombing campaign against approximately one hundred targets in Iraq. For the next four years, Iraq refused to permit UN inspections. In December 1999, the Security Council decided to disband UNSCOM and replace it with the United Nations Monitoring, Verification and Inspection Commission (UNMOVIC).[13]

The terrorist attacks on September 11, 2001, and the United States' military response in Afghanistan against the Qaeda terrorist organization and the Taliban militia that harbored it, led to a sharper focus on the Iraq problem.[14] The attacks on the World Trade Center and the Pentagon, carried out by Al Qaeda operatives trained and led from their bases in Afghanistan, demonstrated the threat posed by terrorists who could seek safe haven in rogue nations with potential access to WMD. As President Bush said in his January 2002 State of the Union address:

> States like these, and their terrorist allies, constitute an axis of evil, arming to threaten the peace of the world. By seeking weapons of mass destruction, these regimes pose a grave and growing danger. They could provide these arms to terrorists, giving them the means to match their hatred. They could attack our allies or attempt to blackmail the United States. In any of these cases, the price of indifference would be catastrophic.[15]

[8] SC Res. 707 (Aug. 15, 1991).

[9] SC Res. 1060 (June 12, 1996).

[10] SC Res. 1115 (June 21, 1997).

[11] SC Res. 1137 (Nov. 12, 1997).

[12] SC Res. 1205 (Nov. 5, 1998).

[13] SC Res. 1284 (Dec. 17, 1999).

[14] I have discussed the domestic legal authority for the war in Afghanistan elsewhere. *See* Robert J. Delahunty & John C. Yoo, *The President's Constitutional Authority to Conduct Military Operations Against Terrorist Organizations and the Nations that Harbor or Support Them*, 25 HARV. J.L. & PUB. POL'Y 487 (2002).

[15] George W. Bush, State of the Union Address (Jan. 29, 2002), 38 WEEKLY COMP. PRES. DOC. 125, 135 (Feb. 4, 2002), *available at* <http://www.whitehouse.gov/news/releases/2002/01/iraq/20020129-11.html>.

On September 12, 2002, President Bush challenged the United Nations to address the threat posed by Iraq as highlighted by its continuing defiance of the Security Council.[16] On November 8, the Security Council unanimously approved Resolution 1441 to address "the threat Iraq's non-compliance with Council resolutions and proliferation of weapons of mass destruction and long-range missiles poses to international peace and security."[17] The resolution "deplor[es]" the absence of international inspections in Iraq since December 1998 and Iraq's continued failure to renounce international terrorism and cease the repression of its civilian population, and gives Iraq "a final opportunity to comply with its disarmament obligations under relevant resolutions of the Council."[18] It reminded Iraq that the Security Council has repeatedly warned that "serious consequences" would result from the continued violation of its obligations.[19]

Although Iraq responded to the resolution by permitting the resumption of inspections, it never took advantage of its final opportunity to comply with its international obligations. Iraq submitted a declaration on December 7, 2002, but the declaration was incomplete, inaccurate, and composed mostly of recycled information. Iraq's declaration clearly failed to address any of the outstanding disarmament questions that previous disarmament inspectors had publicly documented. The reports submitted by UNMOVIC to the Council confirmed these shortcomings.[20] Iraq's submission of a declaration that did not comply with Resolution 1441 was a further material breach of its obligations. As President Bush stated on March 6, 2003, Iraq continued to produce missiles that violate the restrictions in Resolution 687 and to hide biological and chemical agents to avoid detection by international inspectors.[21] No permanent member of the UN Security Council claimed that Iraq had lived up to its obligations in Resolution 1441. On February 5, 2003, Secretary of State Colin Powell delivered a comprehensive presentation to the Security Council demonstrating Iraq's ongoing WMD efforts and their concealment from UN inspectors.[22] As Dr. Blix indicated to the Security Council on January 27, 2003, Iraq had not come "to a genuine acceptance, not even today, of the disarmament which was demanded of it."[23]

The conclusion is inescapable that at the time of the outbreak of the 2003 conflict, Iraq had decided to refuse to comply with its disarmament obligations. This placed Iraq in material breach of Resolution 1441 as well as Resolution 687. At the time of this writing, coalition forces in Iraq continue to search for WMD sites; while no weapons have yet been discovered,

[16] Bush UN Address, *supra* note 1.

[17] *See* SC Res. 1441, *supra* note 2, at 1.

[18] *Id.* at 3.

[19] *Id.* at 5.

[20] On February 28, 2003, UNMOVIC Executive Chairman Dr. Hans Blix stated that the declaration contained "little new significant information . . . relating to proscribed weapons systems." Annex of the Twelfth Quarterly Report of the Executive Chairman of the United Nations Monitoring, Verification and Inspection Commission in Accordance with Paragraph 12 of Security Council Resolution 1284, para. 7, *available at* <http://www.un.org/depts/unmovic/documents/2003-232.pdf>. On January 27, 2003, Dr. Blix stated that the Iraqi declaration did not contain "any new evidence that would eliminate" the "many open disarmament issues." Update on Inspection to the UN Security Council (Jan. 27, 2003), at 3. On March 7, 2003, Dr. Blix repeated his conclusion that, "It was a disappointment that Iraq's declaration of 7 December did not bring new documentary evidence." Oral Introduction of the Twelfth Quarterly Report of UNMOVIC, at 2.

[21] Iraq failed to provide access to all officials for interview by inspectors; intimidated witnesses with threats; undertook massive efforts to deceive and defeat inspectors, including cleanup and transshipment activities at nearly thirty sites; failed to provide numerous documents requested by UNMOVIC; failed to list any WMD personnel beyond a period ending in 1991 or to include many of those identified previously by UNSCOM; and failed to account for seventeen empty chemical warheads subsequently discovered by weapons inspectors, as well as over 2000 pages of Iraqi documents regarding past uranium enrichment programs. The Bush administration's more detailed description of Iraq's refusal to comply with relevant Security Council resolutions can be found in 149 CONG. REC. H1957-60 (daily ed. Mar. 19, 2003) (Report in Connection with Presidential Determination Under Public Law 107–243).

[22] *U.S. Secretary of State Colin Powell Addresses U.N. Security Council* (Feb. 5, 2003), *available at* <http://www.whitehouse.gov/news/releases/2003/02/20030205-1.htm>.

[23] The Security Council: 27 January 2003: an Update on Inspection (as delivered) by Executive Chairman of UNMOVIC Dr. Hans Blix, *available at* <http://www.unmovic.org>.

it may take months if not years to learn the fate of Iraq's WMD stockpile. It has become clearer since the war that Saddam Hussein had allowed Al Qaeda-linked terrorists to operate from Iraq. Nonetheless, these ex post developments do not directly bear on the justification for the use of force ex ante. What is important for *jus ad bellum* purposes is what the United States and its allies reasonably understood the facts to be at the start of hostilities, not what turned up afterwards.

II. UN SECURITY COUNCIL AUTHORIZATION FOR THE USE OF FORCE AGAINST IRAQ

In light of this background, two independent sources of law provided the United States and its allies with authority to use force in Iraq: UN Security Council resolutions and the right to self-defense. Resolution 678 authorized member states "to use all necessary means to uphold and implement resolution 660 (1990) and all subsequent relevant resolutions and to restore international peace and security in the area."[24] One of the most significant "subsequent relevant resolutions" was Resolution 687. Pursuant to Resolution 678, the United States could use force not only to enforce Resolution 687's cease-fire, but also to restore "international peace and security" to the region. In Resolution 1441, the Security Council unanimously found that Iraq was in material breach of these earlier resolutions and that its continuing development of WMD programs, its support for terrorism, and its repression of the civilian population presented an ongoing threat to international peace and security.

These findings triggered Resolution 678's authorization to use force in Iraq. Suspending the cease-fire and resuming hostilities with Iraq was an appropriate response to Iraq's material breaches of Resolution 687. Over the years, Iraq repeatedly refused to respond to diplomatic overtures and other nonmilitary attempts to force compliance with its obligations to disarm and to permit full UN inspections of its WMD program. Military force was necessary to obtain Iraqi compliance with the terms of the cease-fire and to restore international peace and security to the region.

Some have argued, however, that, Resolution 678's authorization had expired. Representatives from France, Germany, and Russia, for example, seemed to take the position that because the current members of the Security Council would not agree to the use of force in the spring of 2003, the 1991 resolution's broad authorization was somehow extinguished. That view is simply wrong as a matter of UN practice and as a matter of law. The UN Security Council has not readily authorized the use of force in the past (indeed, it appears to have done so only in the context of seven conflicts), nor has it rescinded those decisions lightly. When the Security Council has taken the serious step of ending its authorization to use force, it has only done so in one of two ways: either by expressly terminating the prior authorization or by setting an up-front time limit on the authorization.[25] With regard to Bosnia, for example, the Security Council ended the legal authority for the use of force by expressly terminating the previous authorization in a separate resolution, while in Somalia, the Security Council explicitly established a sunset date when it extended the authorization.[26] In fact, when the Security Council has wanted to reserve for itself whether the conditions for termination of its authorization have been met rather than leave the matter to the member states, it has explicitly done so.[27]

[24] *See* SC Res. 678, *supra* note 4, para. 2.

[25] *See, e.g.*, SC Res. 1031 (Dec. 15, 1995), at 3 (Bosnia) (deciding that "the authority to take certain measures conferred upon States by [various UN Security Council resolutions] shall be terminated"); SC Res. 954 (Nov. 4, 1994), at 3 (extending the mandate for the U.N. Mission in Somalia (UNOSOM II) for a "final period" until March 31, 1995); SC Res. 929 (June 22, 1994), at 2 (Rwanda) (specifying that "the mission of Member States cooperating with the Secretary-General will be limited to a period of two months following the adoption of the present resolution," if not earlier).

[26] *See, e.g.*, SC Res. 1031, *supra* note 25; SC Res. 954, *supra* note 25.

[27] *See* SC Res. 940 (July 31, 1994) (Haiti) (deciding "that the multinational force will terminate its mission . . . when a secure and stable environment has been established . . . [as determined] by the Security Council, taking into account recommendations from the Member States of the multinational force").

Security Council practice has been consistent on this point over a substantial period of time. Resolution 678, by contrast, contains no self-imposed time limit, and none of the resolutions relating to Iraq, including Resolution 1441, explicitly terminated the resolution's endorsement of the use of force. Unless the Security Council had clearly stated, using the same language that it has in the past, that it has terminated Resolution 678's authorization for the use of force, any such authorization continued.

In fact, the view of France, Germany, and Russia directly undermines the idea that UN Security Council enactments are law at all. In 1991, the members of the Security Council unanimously agreed to authorize the use of force against Iraq, and reduced that agreement into the written text of Resolution 678. That text has not been subsequently changed, and the conditions for its authorization continued. If the current members of the Security Council disagreed with Resolution 678, they could have repealed it. To argue that Resolution 678's authorization had somehow dissipated, despite its clear text, simply because some of the current members of the Council no longer agreed with it treats Council resolutions as if they merely recorded temporary diplomatic agreement rather than enacted enduring legal texts. The French, German, and Russian view considers Security Council resolutions to be ad hoc executive edicts, rather than legislative acts—a result that would cause considerable uncertainty about their legal force and when they expire.

Some nations attempted to sidestep the inevitable conclusion to this argument by claiming that Resolution 1441 itself somehow eclipsed Resolution 678's authorization to use force. Resolution 1441, however, twice "[r]ecall[ed]" Resolution 678 and explicitly restated that Resolution 678 referred to the ability of member states "to use all necessary means to uphold and implement resolution 660 (1990) and all subsequent relevant resolutions and to restore international peace and security in the area."[28] In fact, rather than negate earlier authorizations, Resolution 1441 instead triggered them. Resolution 1441 "[d]ecides" that Iraq "has been and remains in material breach of its obligations under relevant resolutions," in particular the obligations in Resolution 687 regarding Iraq's WMD program.[29] In addition, the resolution specifies that any false statements or omissions with respect to Iraq's WMD program "shall constitute a further material breach of Iraq's obligations."[30] The resolution also reminds Iraq that the Security Council has repeatedly warned that "serious consequences" will result from the continued violation of its obligations.[31] No member of the Security Council found that Iraq had been in compliance with Resolution 1441, Resolution 678, or Resolution 687.

Resolution 1441's finding that Iraq was in material breach allowed the United States and its allies to terminate the cease-fire created by Resolution 687 and resume the use of force as authorized by Resolution 678. In the multilateral context, it is well-established that a material breach of a treaty by one of the parties entitles a party "specially affected" by the breach to suspend the operation of the treaty in whole or in part vis-à-vis the defaulting state.[32] Even if a state party were not "specially affected," however, a material breach that "radically changes" the position of the parties also permits complete or partial suspension.[33] Resolution 687 explicitly established "a formal cease-fire . . . between Iraq and Kuwait, and the [UN] Member States cooperating with Kuwait in accordance with resolution 678 (1990)."[34] The state parties to the cease-fire agreement were Iraq, Kuwait, the United States, and the other members

[28] *See* SC Res. 1441, *supra* note 2, para. 2.

[29] *Id.*, para. 1.

[30] *Id.*, para. 4.

[31] *Id.*, para. 13.

[32] *See* Vienna Convention on the Law of Treaties, Art. 60(2)(b), *opened for signature* May 23, 1969, 1155 UNTS 331 [hereinafter "Vienna Convention"].

[33] *See id.*, Art. 60(2)(c).

[34] *See* SC Res. 687, *supra* note 6, para. 33.

of the coalition in the Gulf war but not the United Nations itself. During the decade after the end of the first Gulf war, Iraq committed numerous material breaches of the cease-fire, in particular by continuing to develop weapons of mass destruction.[35] Iraq's material breaches of the cease-fire entitled the United States, as a party to the cease-fire, unilaterally to suspend its operation. Under accepted principles of international law, the United States did not need the concurrence of the other parties.[36] Once the cease-fire was suspended, the United States could rely on Resolution 678 to use force against Iraq to implement Resolution 687 and to restore international peace and security to the area.[37]

The March 2003 attack was also justified under the law governing armistices. Resolution 687 was basically an armistice—unlike a peace treaty, it did not terminate the state of war, but merely "suspend[ed] military operations by mutual agreement between the belligerent parties."[38] A cease-fire allows a party to a conflict to resume hostilities under certain conditions. Under the Hague Regulations, "[a]ny serious violation of the armistice by one of the parties gives the other party the right of denouncing it, and even, in cases of urgency, of recommencing hostilities immediately."[39] The missile strikes in 1993 and 1998 serve as clear examples of the suspension of a cease-fire and a resumption of hostilities due to serious violations by Iraq.[40] Because the initial use of force in response to the invasion of Kuwait—Operation Desert Storm—was authorized under Resolution 678, subsequent uses of force against Iraq in response to serious violations of the terms of the cease-fire established by Resolution 687 are authorized as

[35] *See, e.g.*, Bush UN Address, *supra* note 1. Immediately after the cease-fire, Iraq refused to cooperate with UNSCOM and the IAEA. On August 15, 1991, after the adoption of SC Res. 687, the Security Council stated that Iraq's "serious violation" of its disarmament obligations "constitutes a material breach of the relevant provisions of [SC Res. 687] which established a cease-fire and provided the conditions essential to the restoration of peace and security in the region." SC Res. 707, *supra* note 8. Over the next two years, the President of the Security Council issued six different statements reiterating that Iraq's refusal to cooperate with UNSCOM and the IAEA qualified as material breaches of SC Res. 687. *See* the following: Note by the President of the Security Council, UN Doc. S/25970 (June 18, 1993); UN Doc. S/25091 (Jan. 11, 1993); UN Doc. S/25081 (Jan. 8, 1993); UN Doc. S/24240 (July 6, 1992); UN Doc. S/23663 (Feb. 28, 1992); UN Doc. S/23609 (Feb. 19, 1992).

[36] *See* Ruth Wedgwood, *The Enforcement of Security Council Resolution 687: The Threat of Force Against Iraq's Weapons of Mass Destruction*, 92 AJIL 724, 726 (1998) [hereinafter "Wedgwood"]; *cf.* Vienna Convention, Art. 60, *supra* note 32 (one party may suspend a multilateral treaty with respect to itself).

[37] *See* Wedgwood, *supra* note 36, at 726 (Iraq's breach of the terms of the cease-fire in 1997–1998 "allowed the United States to deem the cease-fire in suspension and to resume military operations to enforce its conditions").

[38] Regulations annexed to the Hague Convention on the Law and Customs of War on Land, Oct. 18, 1907, Art. 36, 36 Stat. 2277, 2305 [hereinafter "Hague Regulations"]. *See also* YORAM DINSTEIN, WAR, AGGRESSION AND SELF-DEFENCE 50 (3d ed. 2001) ("A labelling of Resolution 687 as a 'permanent cease-fire' is a contradiction in terms: a cease-fire, by definition, is a transition-period arrangement.") [hereinafter "Dinstein"].

[39] Hague Regulations, Art. 40, *supra* note 38; *see also U.S. Army Field Manual* ¶ 493, *available at* <http://www.adtdl.army.mil/cgi-bin/atdl.dll/fm/27-10/Ch7.htm> (hostilities may be resumed only with "convincing proof of intentional and serious violation of [the armistice's] terms by the other party"). The Hague Regulations do not contain any explanation of what might qualify as "urgency," but the U.S. Army's *Field Manual* sheds some light on the question. According to the Army *Field Manual*, warning must be given to the other side, unless "the delay incident to formal denunciation and warning seems likely to give the violator a substantial advantage of any kind." *Army Field Manual* ¶ 493; *cf. See also* 2 L. OPPENHEIM, INTERNATIONAL LAW, A TREATISE: DISPUTES, WAR, AND NEUTRALITY 556 (H. Lauterpacht ed., 7th ed. 1952) ("since the terms 'serious violation' and 'urgency' lack precise definition, the course to be taken is in practice left to the discretion of the injured party") [hereinafter "Oppenheim, Vol. 2"]. In addition to permitting the resumption of hostilities in response to a serious violation of an armistice, the laws of armed conflict permit the United States to resume hostilities at its discretion—provided that warning is given to Iraq. Hague Regulations, Art. 36. If the parties have not made any stipulation regarding notice, it may be provided at any time, and hostilities may recommence immediately after notification. *See Army Field Manual* ¶ 487; Oppenheim, Vol. 2, *supra* at 556; *see also* Howard S. Levie, *The Nature and Scope of the Armistice Agreement*, 50 AJIL 880, 893 (1956) (although armistices generally do not specify the period of advance notice required, under customary international law, "good faith requires that notice be given of the intention to resume hostilities") (internal quotations and citations omitted).

[40] Dinstein, *supra* note 38, at 50–51. Whether or not required under international law, warnings were given. *See generally* UN Doc. S/25091, *supra* note 35; UN Doc. S/25081, *supra* note 35; Letter to Congressional Leaders Reporting on Iraq's Compliance With United Nations Security Council Resolutions (Jan 19, 1993), 2 PUB. PAPERS OF GEORGE BUSH 2269–70 (1993) [hereinafter "Bush Letter, Jan. 19, 1993"]; President's Radio Address (Dec. 19, 1998), 2 PUB. PAPERS OF WILLIAM J. CLINTON 2197 (1998) [hereinafter "Clinton Papers (1998)"].

well.[41] Thus, because Iraq refused to fully comply with Resolution 687, such as by destroying fully its WMD and their delivery systems, it was in "serious violation" of the cease-fire and the United States was justified in resuming the use of force under Resolution 678.

This understanding of the interaction between Resolution 678 and 687 is supported by state practice in the decade following the end of the first Gulf war. The consistent position of the United States had been that Resolution 678's authorization continued.[42] The United States and Britain, for example, used force against Iraq in 1993 and 1998 in response to Iraq's material breach of Resolution 687. On January 17, 1993, President George H. W. Bush ordered missile strikes against a nuclear facility near Baghdad due to Iraqi infringements of the terms of the cease-fire. Just four days before, President Bush had ordered air attacks on surface-to-missile sites and related facilities in the southern no-fly zone. These attacks, which were joined by Britain and France, appear to have been primarily in response to Iraqi violations of the southern no-fly zone—Iraq had moved surface-to-air missiles into the zone to threaten coalition aircraft—but President Bush also pointed to Iraq's "'failure to live up to the resolutions.'"[43] The president's report to Congress on the attack takes note of a statement by the UN secretary-general explaining that "'the forces that carried out the [January 13th] raid have received a mandate from the Security Council, according to Resolution 687, and the cause of the raid was the violation by Iraq of Resolution 687 concerning the cease-fire .... [T]his action ... conformed to the Charter of the United Nations.'"[44]

In December 1998, the United States launched seventy hours of missile and aircraft strikes against Iraq "in response to Iraqi breaches of its obligations under resolutions of the United Nations Security Council."[45] President Clinton explained that the military action, which targeted facilities actively involved in Iraq's WMD programs or that threatened Iraq's neighbors or U.S. forces, was

> consistent with and has been taken in support of numerous U.N. Security Council resolutions, including Resolutions 678 and 687, which authorize U.N. Member States to use "all necessary means" to implement the Security Council resolutions and to restore peace and security in the region and establish the terms of the cease-fire mandated by the Council, including those related to the destruction of Iraq's WMD programs.[46]

[41] As noted, the Hague Regulations provide either that Iraq must be warned or that such a warning may be avoided because it would be likely to give Saddam Hussein a substantial advantage. There is a good argument that Iraq was warned of a resumption of hostilities by SC Res. 1441. SC Res. 1441 cautions that Iraq's continued violation of its international obligations will result in "serious consequences." *See* SC Res. 1441, *supra* note 2, para. 13. Twice before, military force against Iraq has followed warnings by the Security Council that Iraq's continued intransigence would result in serious consequences. On January 8 and 11, 1993, the president of the Security Council warned Iraq that "serious consequences" would follow if it failed to comply with its international obligations. *See* UN Doc. S/25091, *supra* note 35; UN Doc. S/25081, *supra* note 35.

[42] *See, e.g.*, Letter to Congressional Leaders Reporting on Iraq's Compliance With United Nations Security Council Resolutions (Sept. 16, 1991), 2 PUB. PAPERS OF GEORGE BUSH 1164–65 (1991) (explaining—after the adoption of SC Res. 687—that the United States was willing to take military action to implement SC Res. 678's call for the restoration of international peace and security to the region); *Legal Authority for the Possible Use of Force Against Iraq*, 92 ASIL PROC. 136, 142 (1998) ("In the U.S. Government's view, there is a continuing right to use force [to respond] to such [material] breaches regardless of whether there is further [Security Council] authorization to respond.") (Michael Matheson, Principal Deputy Legal Adviser, Department of State). SC Res. 687 itself explicitly reaffirmed SC Res. 678, as did UN Security Council Resolutions 686 and 949. *See* SC Res. 949 (Oct. 15, 1994); SC Res. 686 (Mar. 2, 1991).

[43] Barton Gellman & Ann Devroy, *Military Action Against Iraq Signaled by Administration*, WASH. POST, Jan. 14, 1993, at A1 (quoting President Bush); *see also* Statement by Marlin Fitzwater, United States Mission to the United Nations Press Release USUN-1(93) (Jan. 13, 1993) ("[T]he Government of Iraq should understand that continued defiance of U.N. Security Council resolutions and related coalition demarches will not be tolerated.").

[44] *See* Bush Letter, Jan. 19, 1993, *supra* note 40.

[45] *See* Letter to Congressional Leaders on the Military Strikes Against Iraq (Dec. 18, 1998), *in* Clinton Papers (1998), *supra* note 40, at 2195.

[46] *Id.* at 2195–96.

As the U.S. ambassador to the United Nations explained to the Security Council, Iraq had acted in "flagrant material breach of resolution 687" by interfering with UNSCOM's inspections, and coalition forces responded under the authority provided by UN resolutions.[47]

In sum, well-established principles of UN Security Council practice, treaty law, and armistice law allowed the United States to suspend the cease-fire in response to Iraq's material breaches of Resolution 687. The United States then could rely on Resolution 678 to use "all necessary means" to bring Iraq into compliance. Nothing in Resolution 1441 suggested that the Security Council needed to adopt any additional resolution to establish the existence of further material breaches to provide the basis for the use of force under Resolution 678. Indeed, Resolution 1441 left intact Resolution 678's reference to the use of force. Resolution 1441 neither revoked Resolution 678's language concerning the use of "all necessary means" against Iraq, nor terminated its effect in any way.

III. Use of Force Against Iraq in Anticipatory Self-Defense

Independent of the support provided by UN Security Council resolutions, authority for the armed intervention in Iraq stemmed from the national right of self-defense. Article 51 of the UN Charter recognizes and affirms, but does not limit, that "inherent" right under international law:

> Nothing in the present Charter shall impair the inherent right of individual or collective self-defense if an armed attack occurs against a Member of the United Nations, until the Security Council has taken measures necessary to maintain international peace and security.[48]

Despite the long-standing recognition of a nation's right to self-defense, some argue that Article 51 has limited the right to permit only a response to an actual "armed attack." Some even argue that an armed attack must occur across national borders to trigger Article 51.[49] Under this interpretation, the UN Charter superseded the existing right under customary international law to take reasonable anticipatory action in self-defense. There is no indication that the drafters of the UN Charter intended to limit the customary law in this way, nor that the United States so understood the Charter when it ratified. Instead, Article 51 merely partially expressed a right that exists independent of the UN Charter.[50]

The customary international law right to use force in anticipatory self-defense is a well-established aspect of the "inherent right" of self-defense. Leading up to the Cuban missile crisis, the Office of Legal Counsel of the Department of Justice explained in an internal memorandum:

[47] *See* UN Security Council Press Release, *Security Council Meets to Discuss Military Strikes Against Iraq; Some Members Challenge Use of Force Without Council Consent* (Dec. 16, 1998), UN Doc. SC/6611, at 1–2, 7. International support for the 1998 airstrikes was reflected by the offers of Argentina, Australia, Canada, the Czech Republic, Denmark, Germany, Hungary, the Netherlands, New Zealand, Portugal, Spain, and the United Kingdom to contribute facilities, equipment or forces to the U.S. military effort, and of Kuwait for the use of its air facilities. Although Britain and Japan spoke in favor of the strikes, the Russian Federation labeled them as "violat[ing] the principles of international law and the principle of the [UN] Charter." *Id.* at 4. Of the Security Council members at the time, China, Costa Rica, Sweden, Brazil, Gambia, Kenya, and Gabon also spoke against the 1998 strikes—some preferring the peaceful settlement of disputes and some criticizing the unilateral use of force. *Id.* at 5–10.

[48] UN Charter Art. 51.

[49] *See, e.g.*, Ian Brownlie, International Law and the Use of Force by States 275–80 (1963); Louis Henkin, How Nations Behave: Law and Foreign Policy 141 (2d ed. 1979).

[50] *See* Myres S. McDougal, *The Soviet-Cuban Quarantine and Self-Defense*, 57 AJIL 597, 599 (1963) ("There is not the slightest evidence that the framers of the United Nations Charter, by inserting one provision which expressly reserves a right of self-defense, had the intent of imposing by this provision new limitations upon the traditional right of states."); Oscar Schachter, *The Right of States to Use Armed Force*, 82 Mich. L. Rev. 1620, 1634–35 (1984); Abraham D. Sofaer, *International Law and Kosovo*, 36 Stan. J. Int'l L. 1, 16 (2000); Thomas M. Franck, Recourse to Force: State Action Against Threats and Armed Attacks 97–99 (2002); *see generally* Christine Gray, International Law and the Use of Force 84–119 (2000).

> The concept of self-defense in international law of course justifies more than activity designed merely to resist an armed attack which is already in progress. Under international law every state has, in the words of [then-Secretary of State] Elihu Root, "the right . . . to protect itself by preventing a condition of affairs in which it will be too late to protect itself."[51]

The classic formulation of the right of anticipatory self-defense arose from the *Caroline* incident. In 1837, the steamer *Caroline* had been supplying armed insurgents against British rule in Canada with reinforcements of men and materials from the United States. In response, a British force from Canada entered U.S. territory at night, seized the *Caroline*, set the ship on fire, and launched it down Niagara Falls, killing two U.S. citizens in the process. The British claimed that they were acting in self-defense, and Secretary of State Daniel Webster called upon the British to show that the

> necessity of self-defence [was] instant, overwhelming, leaving no choice of means, and no moment for deliberation . . . [and that the British force], even supposing the necessity of the moment authorized them to enter the territories of the United States at all, did nothing unreasonable or excessive; since the act, justified by the necessity of self-defence, must be limited by that necessity, and kept clearly within it.[52]

The next year, Lord Ashburton, who had been sent by the British as a special minister to resolve the *Caroline* dispute and other related matters, implicitly accepted this test by justifying Britain's actions in these terms.[53] Webster's formulation was reaffirmed a century later by the International Military Tribunal at Nuremberg, when it ruled that the German invasion of Norway in 1940 was not defensive because it was unnecessary to prevent an "imminent" Allied invasion.[54]

The *Caroline* test has been distilled into two principal requirements. First, the use of force must be necessary because the threat is imminent and, thus, pursuing peaceful alternatives is not an option. Second, the response must be proportionate to the threat.[55] International law does not supply a precise or detailed definition of what it means for a threat to be sufficiently "imminent" to justify the use of force in self-defense as necessary. Although the dictionary definition of "imminent" focuses on the temporal,[56] under international law the concept of imminence must encompass an analysis that goes beyond the temporal proximity of a threat to include the probability that the threat will occur.[57]

In addition to the probability of the threat, the threatened magnitude of harm must be relevant. The advent of nuclear and other sophisticated weapons has dramatically increased the degree of potential harm, and the importance of the temporal factor has diminished. Weapons of mass destruction threaten devastating and indiscriminate long-term damage to large segments of the civilian population and environment.[58] In addition, the danger posed by WMD is exacerbated by the possibility that the means of delivery may be relatively unsophisticated—

[51] Memorandum for the Attorney General, from Norbert A. Schlei, Assistant Attorney General, Office of Legal Counsel, *Re: Legality under International Law of Remedial Action Against Use of Cuba as a Missile Base by the Soviet Union* (Aug. 30, 1962), at 2, *reprinted in* 6 Green Bag 2d 195, 196 (2003).

[52] *See* Letter from Daniel Webster, U.S. Secretary of State, to Henry Fox, British Minister in Washington (Apr. 24, 1841), *in* 29 BRITISH AND FOREIGN STATE PAPERS 1840–1841, *at* 1138 (1857).

[53] *See* Letter from Lord Ashburton to Daniel Webster, U.S. Secretary of State (July 28, 1842), *in* 30 BRITISH AND FOREIGN STATE PAPERS 1841–1842, *at* (1858), also *available at* <http://www.yale.edu/lawweb/avalon/diplomacy/britain/br-1842d.htm>.

[54] *See International Military Tribunal (Nuremberg)—Judgment and Sentences*, 41 AJIL 172, 205 (1947) ("preventive action in foreign territory is justified only in case of 'an instant and overwhelming necessity for self-defense, leaving no choice of means, and no moment for deliberation'") (quoting the *Caroline* case); *see also* D. W. BOWETT, SELF-DEFENCE IN INTERNATIONAL LAW 142–43 (1958).

[55] *See* Dinstein, *supra* note 38, *at* 208–12, 219–20; *see also* McDougal, *supra* note 50, at 597–98.

[56] *See* WEBSTER'S THIRD NEW INTERNATIONAL DICTIONARY (unabridged) 1130 (1993).

[57] *See* Case Concerning the Gabčíkovo-Nagymaros Project (Hungary/Slovakia), 1997 ICJ REP. 7 (Sept. 25).

[58] *See* Legality of the Threat or Use of Nuclear Weapons (Advisory Opinion), 1996 ICJ REP. 95, at ¶ 36 (July 8).

for example, a "dirty bomb" driven into a building by a suicide bomber, or the spread of a biological agent with an ordinary crop duster. At the same time, the development of advanced missile technology has vastly improved the capability for stealth, rendering threats more imminent because there is less time to prevent their launch.

State practice since the development of nuclear weapons and sophisticated delivery systems demonstrates the evolution of the concept of imminence. During the Cuban missile crisis, for example, the United States labeled the secret establishment of long-range nuclear missile bases in Cuba by the Soviet Union as an "immediate" threat to U.S. security and imposed a quarantine on offensive military equipment to Cuba in self-defense.[59] Although the sudden and secret preparation of the missile bases undoubtedly "add[ed] to an already clear and present danger," their positioning in Cuba constituted a less immediate temporal threat of armed attack on the United States than that contemplated by previous applications of the *Caroline* test. There was no indication that the Soviet Union would use them either immediately, or even in the near term.[60]

In the past two decades, the United States has used military force in anticipatory self-defense against Libya, Panama, Iraq, Afghanistan, and the Sudan. The United States justified the 1986 strikes against Libya in large part as necessary to forestall future terrorist attacks.[61] Although several countries criticized the U.S. strikes by supporting a UN resolution condemning the attack as a violation of the UN Charter, Australia, Denmark, France, and the United Kingdom joined the United States in opposing the resolution.[62] The United States again responded in self-defense to an imminent threat to U.S. lives when it took military action in Panama on December 20, 1989.[63] In the midst of the fighting, the Security Council considered a draft resolution that would have labeled the invasion as "a flagrant violation of international law," but Great Britain, France, and Canada joined the United States in opposing the resolution.[64] In 1998 the United States launched cruise missile attacks in self-defense against terrorist training camps and installations in Afghanistan used by Osama bin Laden's organization and against a facility in Sudan being used to produce materials for chemical weapons.[65] The Security Council took no formal action in response.

Self-defense has served specifically as a justification for strikes against Iraq in the decade preceding the March 2003 conflict. In June 1993, for example, the United States justified its strike on Iraqi intelligence headquarters as self-defense because of "compelling evidence" that Iraq had attempted to assassinate President George H. W. Bush. As President Clinton explained:

[59] *See* Proclamation 3504: Interdiction of the Delivery of Offensive Weapons to Cuba (Oct. 23, 1962), PUB. PAPERS OF JOHN F. KENNEDY 809, 810 (1962); White House Statement on Soviet Proposals Relating to International Security (Oct. 27, 1962), *id.*, at 813.

[60] *See* Radio and Television Report to the American People on the Soviet Arms Buildup in Cuba (Oct. 21, 1962), *id.* at 806–07. It should be noted that while the Department of Justice may have believed the quarantine to be a lawful exercise of self-defense, the official justification offered diplomatically relied upon authorization by the Organization of American States.

[61] *See* Letter to the Speaker of the House of Representatives and the President Pro Tempore of the Senate on the United States Air Strike Against Libya (Apr. 16, 1986), 1 PUB. PAPERS OF RONALD REAGAN 478 (1986).

[62] *See* UN Doc. S/PV. 2682 (Apr. 21, 1986).

[63] *See* Letter to the Speaker of the House of Representatives and the President Pro Tempore of the Senate on the United States Military Action in Panama (Dec. 21, 1989), 2 PUB. PAPERS OF GEORGE BUSH 1734 (1989) ("The deployment of U.S. Forces is an exercise of the right of self-defense recognized in Article 51 of the United Nations Charter and was necessary to protect American lives in imminent danger . . . .").

[64] UN Doc. S/21048 (Dec. 22, 1989); UN Doc. S/PV. 2902 (Dec. 23, 1989).

[65] *See* Letter to Congressional Leaders Reporting on Military Action Against Terrorist Sites in Afghanistan and Sudan (Aug. 21, 1998), *in* Clinton Papers, *supra* note 40, at 1464 (1998) ("These strikes were a necessary and proportionate response to the imminent threat of further terrorist attacks against U.S. personnel and facilities. These strikes were intended to prevent and deter additional attacks by a clearly identified terrorist threat.").

> The evidence of the Government of Iraq's violence and terrorism demonstrates that Iraq poses a continuing threat to United States nationals and shows utter disregard for the will of the international community as expressed in Security Council Resolutions and the United Nations Charter. Based on the Government of Iraq's pattern of disregard for international law, I concluded that there was no reasonable prospect that new diplomatic initiatives or economic measures could influence the current Government of Iraq to cease planning future attacks against the United States.[66]

The objective of the strikes was to diminish Iraq's capability to support violence against the United States and others, and "to deter Saddam Hussein from supporting such outlaw behavior in the future."[67] The Security Council rejected the plea of the Iraqi ambassador to condemn the U.S. action as an act of aggression.

The new threat of nuclear weapons apparently is not, however, sufficient to erase completely any requirement of temporality. For example, the international community did not consider the threat posed by an Iraqi nuclear reactor before it had become operational to be sufficient to justify its destruction by Israel in 1981.[68] Nonetheless, the nature of the Iraqi threat has changed significantly in the past twenty years. In 1981, Iraq was permitted to have nuclear materials under the safeguards of the IAEA, and Saddam Hussein had not yet used chemical weapons against Iran and his own people, invaded Kuwait, or spent over a decade flouting his country's international obligations to destroy and cease to develop WMD and their means of delivery. In other words, the imminence of a likely attack by Iraq has increased since 1981 because Iraq has demonstrated a WMD capability and a willingness to use it.

The use of force in anticipatory self-defense must be necessary and proportional to the threat. At least in the realm of WMD, rogue nations, and international terrorism, however, the test for determining whether a threat is sufficiently "imminent" to render the use of force necessary at a particular point has become more nuanced than Secretary Webster's nineteenth-century formulation. Factors to be considered should now include the probability of an attack; the likelihood that this probability will increase, and therefore the need to take advantage of a limited window of opportunity; whether diplomatic alternatives are practical; and the magnitude of the harm that could result from the threat. If a state instead were obligated to wait until the threat were truly imminent in the temporal sense envisioned by Secretary Webster, there is a substantial danger of missing a limited window of opportunity to prevent widespread harm to civilians. Finally, in an age of technologically advanced delivery systems and WMD, international law cannot require that we ignore the potential harm represented by the threat.

Applying the reformulated test for using force in anticipatory self-defense to the potential use of force against Iraq reveals that the threat of a WMD attack by Iraq, either directly or through Iraq's support for terrorism, was sufficiently "imminent" to render the use of force necessary to protect the United States, its citizens, and its allies. The force used was proportionate to the threat posed by Iraq; in other words, it was limited to that which is needed to eliminate the threat, including the destruction of Iraq's WMD capability and removing the source of Iraq's hostile intentions and actions, Saddam Hussein.

[66] *See* Letter to Congressional Leaders on the Strike on Iraqi Headquarters (June 28, 1993), 1 PUB. PAPERS OF WILLIAM J. CLINTON 940 (1993).

[67] *See* Address to the Nation on the Strike on Iraqi Intelligence (June 26, 1993), *id.* at 938 (1993). Similarly, the January 17, 1993, strike on a nuclear facility in Baghdad, while primarily designed to encourage Iraq to comply with its obligations under UN Security Council resolutions, was undertaken in part to prevent the facility from being used again to support Iraq's nuclear weapons program. *See* Bush Letter, Jan. 19, 1993, *supra* note 40, at 2269–70.

[68] *See* SC Res. 487 (June 19, 1981) (unanimously "condemn[ing]" the Israeli strikes as a "clear violation of the Charter of the United Nations and the norms of international conduct.").

IV. CONCLUSION

International law permitted the use of force against Iraq on two independent grounds. First, the Security Council authorized military action against Iraq to implement the terms of the cease-fire that suspended the hostilities of the 1991 Gulf war. Due to Iraq's material breaches of the cease-fire, established principles of international law—both treaty and armistice law—permitted the United States to suspend its terms and to use force to compel Iraqi compliance. Such a use of force was consistent with U.S. practice both with regard to Iraq and with regard to treaties and cease-fires. Second, international law permitted the use of force against Iraq in anticipatory self-defense because of the threat posed by an Iraq armed with WMD and in potential cooperation with international terrorist organizations.

This approach to the use of force against Iraq has significant implications for the future. To be sure, Iraq could be seen as a unique case. Due to the web of Security Council resolutions going back more than a decade, the United States and its allies could draw sufficient legal authority from those enactments alone. No other use of force in the near future may receive such support from the Security Council. On the other hand, Iraq could be said, unfortunately, to represent the coming challenges to international peace and stability as a rogue state that has WMD and supports terrorism. In this type of security environment, the United States and its allies may well have to rely exclusively upon their right to anticipatory self-defense in order to use force against such nations. In order to address the challenge posed by this new threat, the international legal system will have to adapt to take into account the probability of an attack, the magnitude of the possible harm, and the windows of opportunity within which proportionate force may be used.

The use of force in anticipatory self-defense against terrorist groups armed with WMD, or against the rogue nations that support them, will depend on three factors that go beyond mere temporal imminence. First, does a nation have WMD and the inclination to use them? In the case of Iraq, the record made clear that Saddam Hussein both possessed WMD and had used them against external enemies (Iran) and his own citizens. In future cases, the possession of WMD and signs of hostile intent must be taken into account in deciding whether to use force preemptively. That decision will rely, in part, on intelligence about rogue nations' WMD programs, their ability to acquire components and technical knowledge, and their ability to assemble a weapon.

Second, nations will have to use force while taking into account the available window of opportunity. If a state waits until a terrorist attack is on the verge of being launched, it likely will be unable to protect the civilians who are being targeted, especially against suicide bombers who seem immune to traditional methods of deterrence. Terrorists are also difficult to locate and track and seek to escape detection by concealing themselves and their activities among an innocent civilian population. As terrorists burrow more deeply into this population, defensive options may become more limited. Due to these considerations, a state may need to act when it has a window of opportunity to prevent a terrorist attack and simultaneously minimize civilian casualties. Preventing a terrorist attack using WMD may require that the United States and its allies take advantage of a window of opportunity that opens before a rogue nation transfers WMD to a terrorist organization. Were the United States, for example, to wait until a rogue nation has transferred WMD to terrorist groups, it could become extremely difficult to determine where and when WMD would be used given the sporadic nature of terrorist attacks and the terrorist tactic of infiltrating the civilian population.

Third, nations will have to take into account that the degree of harm from a WMD attack would be catastrophic. The combination of the vast potential destructive capacity of WMD and the modest means required for their delivery make them more of a threat than the military forces of many countries. Chemical weapons and biological agents are easy to hide, and small quantities can have a devastating effect on the civilian population. This threat only multiplies if rogue nations were to transfer WMD to terrorists. Thus, even if the probability that a rogue nation would attack the United States directly with WMD were not certain, the exceptionally high degree of harm that would result, combined with a limited window of opportunity and the likelihood that if the United States did not act, the threat would increase, could lead a nation to conclude that military action is necessary in self-defense. Indeed, as President Bush recently cautioned: "If we wait for threats to fully materialize, we will have waited too long."[69]

JOHN YOO[*]

[69] George W. Bush, Commencement Address at the United States Military Academy in West Point (June 1, 2002), 38 WEEKLY COMP. PRES. DOC. 944, 946 (June 10, 2002), *available at* <http://www.whitehouse.gov/news/releases/2002/06/20020601-3.html>.

[*] Visiting Fellow, American Enterprise Institute; Professor of Law, University of California at Berkeley (Boalt Hall) School of Law; Deputy Assistant Attorney General, Office of Legal Counsel, U.S. Department of Justice, 2001–2003. I thank Robert Delahunty and Caroline Krass for their helpful advice, and Gene Hosey for his assistance. The views expressed here are those of the author alone.

[16]

What Was It All About After All? The Causes of the Iraq War

HAKAN TUNÇ

The least understood aspect of the ongoing war in Iraq is what caused the United States to invade the country in the first place. Although various observers and analysts have posited a number of explanations, the George W. Bush administration's reasons for going to war are still subject to intense controversy. This article aims to contribute to this debate by systematically analysing the major underlying factors that influenced the decision to go to war with Iraq. These factors are complex and multifaceted. Yet, I argue that they all point towards an overriding rationale: the war on terrorism.

Following September 11, 2001 (henceforth 9/11), key policy-makers in the Bush administration concluded that toppling Saddam Hussein and his regime was necessary to strike a blow to radical Islamic terrorism on multiple fronts. The administration linked such necessity to its war on terrorism in three main ways: military, psychological and ideological. Eliminating Iraq's weapons of mass destruction (WMD) and thus depriving Al Qaeda of a major source of these weapons was the *military rationale.* The *psychological rationale* involved demonstrating American power and resolve and thereby discrediting the widely held belief in the Middle East about US weakness. With regard to the *ideological rationale*, the Iraq war was seen as a major step towards the political and cultural transformation of the Middle East. Establishing a democratic regime in Baghdad would serve as a model of freedom and liberation for other Arab countries and Iran.

As I argue in detail below, in developing each of these rationales, Bush officials relied on certain assumptions. In the case of Iraq's WMD, the assumption was that Saddam not only intended to harm the US, but that he would also act irrationally to this end. Saddam's alleged irrationality has been largely overlooked by scholars who dismiss the WMD argument as a smokescreen created by the Bush administration in order to hide its other, less politically popular, motives to attack Iraq. The underlying assumption of the psychological rationale was that a US invasion of Iraq would deter individuals, groups and regimes from supporting terrorism, by fostering fear and respect for American authority. Finally, the ideological rationale assumed that the repressive totalitarian/authoritarian political culture in the Middle East is the backbone of Islamic fanaticism and terrorism. Hence, democratization was seen as the best antidote to the dysfunctional political systems and the ills of totalitarianism in the Middle East.

The Al Qaeda attacks on the World Trade Center and New York were crucial in the formulation of this policy. The terrorist attacks transformed the Bush administration's sense of danger and underlined the urgency of offensive strategies. The impact of 9/11 on the decision to go to war in Iraq, however, was neither automatic nor due to the influence of a particular group in the administration. As has been the case with most major US foreign policy decisions, the resolution to go to war in Iraq had no single architect.[1] Rather, it was the end result of distinct arguments put forward by a small group of conservative and neo-conservative 'war hawks', who had been advocating (unsuccessfully) for tough action against Iraq prior to 9/11. These arguments ultimately held sway over top US decision-makers who were in search of a strategy to cope with the new security threat that appeared after 9/11.

The single most important factor contributing to the decision to go to war was the change in President Bush's position towards Iraq after 9/11. Prior to 9/11, Bush had resisted the war hawks' pressure to take action against Iraq. The key question then becomes how Bush, as the ultimate decision-maker, became convinced of the need for regime change in Iraq. In this context, particular attention needs to be paid to the role played by Vice President Richard Cheney and National Security Advisor Condoleezza Rice. This is not only because their opinions are highly valued by Bush, but also because, unlike longtime Iraq hawks such as Secretary of Defense Donald Rumsfeld and the group of neo-conservatives led by Deputy Secretary of State Paul Wolfowitz, both Cheney and Rice became champions of military action against Iraq only after 9/11.[2]

The article begins with a brief discussion of two common alternative explanations for the Bush administration's decision to go to war with Iraq: oil and achieving American dominance. I argue that these explanations are not only logically flawed, they are not substantiated by available evidence. Second, I analyse each of the three military, psychological and ideological rationales for the Iraq war in depth. I do not claim to have direct access to the deliberations of the key players of the administration or to President Bush himself. This analysis is based instead on insiders' accounts, press reports and analysis. Some aspects of the war decision, such as the relative significance of each of the three rationales in Bush's final judgment and the exact reasons why Condoleezza Rice changed her views on Iraq after 9/11, are extremely difficult to know. While this assessment is necessarily incomplete, the three leading rationales it stresses stand out as fundamental. The final section sets out the implications of the Iraq war for US foreign policy, stressing the unique combination of outside forces and decision-making predispositions that led to the invasion of Iraq. This unique combination suggests that it is highly unlikely the US will undertake similar action against other foes such as Iran, North Korea or Syria in the foreseeable future.

Alternative Explanations for the Iraq War

War and Oil

A popular explanation for the Iraq war is that the war was all about oil. In this view, the Bush administration declared war on Saddam in order to enrich American oil companies and/or to seek additional oil from Iraq's vast petroleum reserves. The 'war for oil' argument errs, however, on logical and evidential grounds. As some observers have pointed out, if all the Bush administration cared about was oil, it would have lifted sanctions on Iraq's oil sales and cut deals with Saddam instead of invading the country. Such a policy would have satisfied both the oil companies and America's long-term needs for oil.[3] Moreover, there is no evidence that oil companies lobbied the Bush administration to topple the Saddam regime. In fact, prior to 9/11, the oil lobby in Washington generally favoured the relaxation of US restrictions on Iraq, not Saddam's removal. After 9/11, most oil companies were wary about the Iraq war because they were worried that the war could bring instability to the Gulf.[4]

Last but not least, prior to the Iraq war, it already was widely accepted that any post-Saddam Iraqi government would not be able to stay in power should US oil companies control Iraq's oil sector. Iraqis are extremely sensitive about the issue of oil and treat oil as a national asset.[5] In fact, prior to the Iraq war, Iraqi opposition groups cautioned that US companies would not have any inside track.[6] The 1991 Gulf War had demonstrated precisely this point. Ever since the liberation of Kuwait by US-led coalition forces in 1991, no single US oil company had gained a concession from the Kuwaiti government. The cautious stance that the Bush administration has taken towards Iraqi oil shows that it is well aware of these complexities. In the course of events leading up to the war, the prevailing view in the administration was that keeping oil in Iraqi government hands was absolutely necessary in order that Iraq remain intact under a centralized authority after the war.[7]

Iraq's oil was relevant to the decision to go to war, but not, as many argue, in the Bush administration's desire to control it. First, many in the Bush administration, particularly high-level civilian political appointees in the Pentagon, believed that Iraqi oil would be used to finance the country's reconstruction, thus alleviating the financial burden on the American Treasury. Second, from the Bush administration's perspective, oil in the hands of

Saddam was worrisome, for he could again try to achieve dominance over the Gulf, a region that is vital to world economic stability. As Michael Ignatieff argues, 'Oil was an issue in the [Iraq] war precisely because its revenues distinguished Saddam from the run of other malignant dictators. It was the critical factor that would allow him, sooner or later, to acquire the weapons' that would enable him to pursue aggressive policies within Iraq and in the region.[8]

The Iraq War as an Imperial Project

Many scholars and observers see the Iraq war as part of the Bush administration's (hidden) global agenda to attain American hegemony, and even to create an empire. According to this view, the neo-conservatives and Cheney had long been looking to use America's unequalled military power to make sure the US controlled the world. The proponents of this argument point to the Pentagon's 1992 in-house Defense Planning Guideline (DPG) which was prepared for the then Secretary of Defense Cheney under the guidance of Wolfowitz. At the time, the DPG suggested that the US make sure to block the emergence of any military rival and called for aggressively (and unilaterally if necessary) dealing with certain troubling countries, including Iraq and North Korea among others.[9] Almost a decade later, 9/11 gave Cheney and his fellow neo-conservative a window of opportunity to implement their 'imperialistic' agenda. The Iraq war was conceived as part of that agenda.[10]

There are several problems with this explanation. First, it is not clear how the elimination of the Saddam regime would enhance US military pre-eminence vis-à-vis other great powers, such as China. In other words, from a geo-strategic perspective, it is hard to imagine how the Iraq war factors into big power politics, unless one accepts the unconvincing argument about oil as the primary motive for the Bush administration.[11] The pursuit of regime change in a small power such as Iraq can hardly be a central part of a strategy to achieve US hegemony. China and Russia certainly do not see the Iraq war as a major obstacle in their relations with the Bush administration. Moreover, it is misleading to single out Cheney and his neo-conservative allies as the sole advocates of American pre-eminence. The policy recommendations of the DPG were openly accepted by the Clinton administration after 1992. For instance, President Clinton's Joint Chiefs of Staff declared that the overall goal of the US was to create a 'preeminent' military force capable of 'full-spectrum dominance' in the world.[12]

Second, the argument that the Iraq war was part of a grand American imperial policy does not explain why the Bush administration's foreign policy prior to 9/11 was much less aggressive than the neo-conservatives had wanted it to be.[13] Prior to 9/11, Bush embraced the realism espoused by Rice and Powell who advocated giving priority to relations with China and Russia and staying away from over-commitments to nation-building and mediation, a policy dubbed ABC – Anything But Clinton.[14] In particular, despite great effort, the hawks were unsuccessful in persuading Rice, Cheney and Bush (not to mention Powell) to intervene in Iraq before 9/11. Thus, before 9/11, Bush did not pursue an aggressive policy of regime change in Iraq. Instead, the administration moved to strengthen the containment policy towards the Saddam regime that had been in place since 1991.[15]

9/11 as the Transformative Event for the War

What fundamentally changed the picture and transformed the worldview of the administration was 9/11. The attacks' biggest impact on the Bush administration's policy was to undermine America's sense of security and reinforced the notion that national defence was tied to distant trouble spots abroad, especially Afghanistan and the Middle East. In the words of a prominent historian, 9/11 revealed, 'a homeland security deficit, unlike anything [Americans had] experienced in either of the world wars or the Cold War'.[16]

In this regard, the major consequence of the 9/11 attacks for the Bush administration's foreign policy was the confluence of views between realists and the neo-conservatives. In the 1990s, neo-conservatives argued that the decade of peace and prosperity brought about by the end of the Cold War was an illusion, that the world was a far more dangerous place than most people realized, and that the new challenges to American security would come from dictatorships.[17] Certain policy-makers rebuked this neo-conservative Hobbesian view claiming that it was overly pessimistic and alarmist. While the attacks did not prove the neo-conservatives had been correct in every detail (such as their expectation of an attack on the US from a rogue state), their premise about the dangers of the world came to be accepted. The argument that neo-conservatives were waiting in the wings and used 9/11 to forward their agenda, therefore, does not explain how they were able to influence the views of top officials from their secondary positions. It was 9/11 which caused Bush, Rice and Cheney to move closer to Wolfowitz's view which held that attacking Iraq was an effective strategy in the war against radical Islamic terrorists. As Rice remarked: '9/11 crystallized our vulnerability ... And after 9/11, there is no longer any doubt that today America faces an existential threat to our security'.[18] Below I discuss the motives of the Bush administration in going to war in Iraq in light of its interpretation of 9/11.

Military Rationale for the War: Iraq's WMD as a National Security Threat

The importance of Iraq's WMD capabilities is one of the most controversial issues in the debate over the causes of the war. How did Iraq's alleged

WMD figure in Bush's decision to go to war in Iraq, if at all? Before answering this question several issues need to be clarified.

First, all evidence points to the fact that Bush made his ultimate decision to go to war with Iraq before the end of 2001.[19] This means that in order to understand the degree to which the WMD issue factored in Bush's war decision, one has to look at the few months that followed 9/11. In this regard, the Bush administration's public campaign, which began in earnest in September 2002 and prioritized Iraq's WMD, is largely irrelevant to understanding the weight of the WMD rationale in the war decision.[20] Evidence indicates that the Bush administration chose the WMD issue for its convenience and because it required the least amount of effort to convince the American public of the necessity of the war. For instance, Condoleezza Rice remarked that Iraq's WMD was the only one argument with any 'legs' because at least a dozen resolutions on the issue had already been passed by the UN.[21] Wolfowitz also admitted that among the rationales for the Iraq war, 'For bureaucratic reasons we settled on one issue, weapons of mass destruction, because it was the one reason everyone could agree on'.[22]

Second, by the time Bush made his war decision, it was clear that everybody in the administration believed Iraq had some WMD or some sort of WMD program. This view was shared by members of the former Clinton administration, UN officials and by the international intelligence community. The general consensus regarding Saddam's WMD was reinforced by his systematic resistance to UN inspections, particularly after he expelled the inspectors from Iraq in 1998. Saddam's behaviour appeared to be incomprehensible unless he had WMD to hide.[23]

Third and finally, everyone in the Bush administration believed that Al Qaeda had the resolve to use WMD against the US, if it were capable of obtaining them. Unquestionably, Al Qaeda was aggressively searching for WMD for this purpose and was likely to obtain these weapons from a rogue state.[24] In the months that followed 9/11 and particularly after the Afghan war in the fall and winter of 2001, it became clear that Al Qaeda was interested in WMD and even experimented with some crude forms of chemical weapons. Cheney became consumed with the possibility of a second terrorist attack against the US using WMD, particularly after the anthrax attacks soon after 9/11.[25] Bush also took these reports seriously. According to White House insider sources, the knowledge that Al Qaeda was aggressively searching for WMD and wooing outside support 'transformed the president's thinking about America's enemies, and the horrors that could unfold if any of them made such weapons available to terrorists'.[26]

Thus, the main puzzle concerning the link between the Iraq war and WMD is not about the existence of Iraq's WMD or Al Qaeda's active search of these weapons. Rather, it is about how the Bush administration linked these two issues to each other. As has been previously reported, the administration announced that a major reason for going to war was the possibility that Iraq could pass WMD to Al Qaeda at some future date. It is this factor that is crucial to understanding the role played by Iraq's WMD in the war decision. I argue that many Bush administration officials (particularly Vice-President Cheney) were concerned about such a possibility, even though it is difficult to say how serious that concern was. One can plausibly argue that while Iraq's alleged possession of WMD was not a primary cause for the Iraq war, it was an important factor. Iraq's WMD was important enough to cause top officials to question Saddam's intentions and rationality after 9/11.

Before 9/11, the prevailing view inside the Bush administration was that Saddam was a threat to US allies and troops in the Middle East, but that he was capable to calculate risk when it came to WMD. This assumption was based on the premise that even if Saddam were building WMD, he would not use them against the US or its allies because he knew that this would trigger a massive US military response. In January 2000, Condoleezza Rice wrote an article (widely viewed as the Bush campaign's foreign policy platform) in which she claimed that Iraq's WMD would be 'unusable because any attempt to use them will bring national obliteration'. In the same article, Rice predicted that the Saddam regime would collapse on itself because it was 'living on borrowed time'. Rice concluded 'there need be no sense of panic about [Iraq]'.[27] In other words, Saddam was rational enough to be deterred and contained. Military action against Iraq in order to instigate regime change did not seem attractive. In particular, an invasion of Iraq, in which tens of thousands of American soldiers would participate, was unthinkable given the politically difficult task of convincing the Congress and American public about the necessity of such an action.

During the first eight months of the Bush administration, Rice's view was more or less shared by Secretary of State Colin Powell and Richard Cheney, as well as President Bush himself. The Iraq hawks, including Secretary of Defense Donald Rumsfeld and his neo-conservative allies at the Pentagon (first and foremost Rumsfeld's Deputy Paul Wolfowitz), were on the defensive. These hawks vigorously defended the necessity of military action against Iraq (if not outright invasion), but their argument was ultimately unpersuasive because they defined Saddam's threat in regional terms (i.e., towards Israel and American forces in the Gulf) and had no clear answer about what Saddam would do with his WMD beyond the Middle East.[28]

Thus Powell's argument that the containment of Iraq should become more effective through 'smart sanctions' prevailed before 9/11.[29] For the majority of key foreign policy-makers in the Bush administration, pursuing regime change through overt military action was not seen as a viable alternative to the policy containment, which relied on the assumption of Saddam's rationality.

The 9/11 attacks caused a dramatic shift within the Bush administration regarding prevailing assumptions about Saddam's intentions and risk assessment capability. A new argument that emphasized Saddam's irrationality gained currency among officials who had hitherto been sceptical of taking action against Iraq. Formulated by the neo-conservatives, this argument assumed that Saddam posed an intolerable threat against the US, not only because he displayed extreme hostility towards the US, but also because he was irrational enough to pass WMD to Al Qaeda. Drawing attention to Saddam's hostility was not new, of course, but it was necessary to emphasize it after 9/11 in order to discredit the arguments that Saddam's secularity and Al Qaeda's religious fundamentalism could not co-exist. According to the neo-conservatives, Saddam's hostility against America superseded all his other concerns, so much so that he could easily cooperate with Al Qaeda against the US. For instance, Richard Perle, a prominent neo-conservative then serving as a Pentagon advisor, argued with regard to Saddam: 'The theory [before 9/11] was that secular and religious terrorists were hostile to one another and would not work with each other. That theory, like all such theories, needed to be reexamined'.[30]

More importantly, after 9/11, the neo-conservatives started emphasizing Saddam's irrationality by referring to his tendency to engage in massive miscalculations of power and his reckless propensity to take exceedingly huge risks. Saddam's 'irrational' acts ranged from his invasion of Kuwait in 1990 to the assassination attempt of Bush senior in 1993. This was a dictator, after all, who had invaded Kuwait before completing his development of nuclear weapons. For the neo-conservatives, Saddam had made these miscalculations because he was uninformed about America and frequently misjudged American reactions. Since reason and dialogue did not apply to him, deterrence or containment could not work against him.

In their book, which is considered the neo-conservative manifesto for waging war against Iraq, William Kristol and Lawrence Kaplan stressed Saddam's 'lack of credible information about U.S. intentions, his track record of heedless risk-taking, and most of all, his supreme irrationality'. According to Kristol and Kaplan, 'Sanity, prudence and self-control, needless to say, are not the first qualities that leap to mind when you think of Saddam Hussein'.[31] In short, Saddam was uniquely dangerous. He was not only a vicious despot and a monster, but also a serial miscalculator. For both of these reasons, Saddam's possession of WMD could not be tolerated since he could easily miscalculate again, this time by passing weapons to terrorists. The assumption about Saddam's uniqueness, which combined his enmity toward the US and his irrationality, did not require evidence. Any proven links between Saddam and Al Qaeda were deemed to be unnecessary. For instance, Rumsfeld admitted that there was no new evidence to implicate the Saddam regime after 9/11, but only new assumptions: 'The U.S. did not act in Iraq because we had discovered dramatic new evidence of Iraq's pursuit of weapons of mass murder. We acted because we saw the existing evidence in a new light, through the prism of our experience on September 11'.[32]

After 9/11, it seems that Bush, Rice and Cheney adopted the neo-conservative argument that America could not afford to tolerate Saddam's irrationality in combination with his vicious hostility to the US. Insider accounts indicate that Cheney's views about the need to get rid of Saddam were transformed by 9/11, leading him to emphasize the danger posed by the connection between terrorists and WMD.[33] The most fundamental attitudinal change regarding the Saddam threat, however, seems to have occurred in the thinking of Condoleezza Rice and Bush. Rice began to argue that Saddam could pass WMD to Al Qaeda due to his recklessness. Rice claimed: 'It's not because you have some chain of evidence saying Iraq may have given a weapon to al Qaeda. But it is because Iraq is one of those places that is both *hostile to us*, and, frankly, *irresponsible and cruel* enough to make this available'.[34] A top adviser to the administration at the time summarized the emerging view in the White House by saying that deterring Saddam was no longer enough because he was 'too capable of making a massive miscalculation'.[35] According to insider sources, Bush also became increasingly worried about Saddam's irrationality. In the weeks after 9/11, an adviser to the president revealed that Bush believed Saddam to be 'insane'.[36] Bush himself confessed to journalist Bob Woodward that he changed his views on Saddam after 9/11, when 'all [Saddam's] terrible features became much more threatening'. Bush's view was that 'the containment game' was no longer applicable to Iraq because Saddam was a 'madman'.[37] Echoing the neo-conservative claim, Bush later stated publicly: 'Trusting in the sanity and restraint of Saddam Hussein is not a strategy, and it is not an option'.[38]

Seeing hostile dictatorships as irrational and devoid of objective truth is not new in the history of American foreign policy. Rightly or wrongly, real or imagined, many US administrations before Bush perceived totalitarian adversaries not only as evil, but also incapable of making proper cost-benefit analysis. Such a perception, for instance, emerged within the Truman administration towards the Stalin regime in the Soviet Union after the end of the Second World War. George Kennan, the architect of America's containment policy during the Cold War, argued in his Long Telegram in 1946 that the US could not confront that 'implacable force of evil' on logical terms because Soviet policy 'is seemingly inaccessible to considerations of reality in its basic reactions'. Like Bush officials would argue about Saddam half a century later, Kennan contended that Stalin was 'impervious to logic or reason' because he did not receive 'anything like an objective picture of the outside world'.[39]

Eliminating a major potential source of WMD for Al Qaeda by overthrowing the Saddam regime was only one reason, and likely not the most important one, for going to war with Iraq. The reasons for the Bush administration's war decision went beyond Iraq. As an aid to Cheney admits, 'The imminence of the threat from Iraq's WMD was never the real issue [for us]. WMD were on our minds, but they weren't the key thing. What was really driving us was our overall view of terrorism, and the strategic conditions of the Middle East'.[40]

Psychological Rationale for the War: Demonstration of America's Resolve

A major rationale of the administration for the Iraq war was psychological, in the implied sense that following 9/11, the war on Islamic terror could not be won militarily unless the US changed the 'psyche' of its adversaries. This meant that the US had to reassert itself dramatically in a forceful and audacious manner by demonstrating vigorously American will, strength and resolve. The demonstration of US power in Iraq was seen as necessary in order to change the widespread perception in the Middle East about America's weakness – a perception that had presumably emboldened America's adversaries in the Muslim world.

In a general sense, changing an adversary's psyche is closely associated with credibility. In power politics, a country's credibility is related to its reputation for military capability combined with the political resolve to use that capability in order to promote that country's goals. Enhancing credibility serves two purposes: changing the calculations of adversaries and assuring allies. The former is usually more important than the latter. The concern of decision-makers for the credibility of their respective states is an old one. For instance, credibility was a central concern of the US during the Cold War. The US sought to deter the Soviet Union and, at the same time, reassure America's allies that it would come to their rescue in case of Soviet aggression. A major reason for the escalation of the Vietnam War during the administration of Lyndon Johnson was explained in terms of US credibility. The logic was that if the US were perceived to be lacking military capacity, political resolve or both, the Soviet Union, China or their proxies would act more aggressively.[41]

US credibility was a particularly important factor for decision-makers within the Bush administration, most prominently for Rumsfeld and Cheney. Having developed their careers during the Cold War, both men had long lamented the weakening of American power and resolve in the years following Vietnam. They argued that after the end of the Cold War, the US had put itself at risk because it had failed to demonstrate its will and determination to confront its adversaries. In other words, the US had a credibility problem towards its adversaries. Shortly after he was named Secretary of Defense, Rumsfeld told the then President-elect Bush that the 'word was out' around the world that the US had gone soft, that America had become an easy target.[42]

The 9/11 attacks only strengthened the conviction of Rumsfeld and Cheney that the US urgently needed to demonstrate American strength and resolve in its fight against terrorism. As reported by insiders within the Bush administration, in the view of Rumsfeld and Cheney 'the threat to America posed by terrorism signaled...that this was the time to reassert American will in the world. According to aides to both men, Cheney and Rumsfeld talked often in the days and months after 9/11 about the need to be bold'.[43] For both men, the best solution to reduce the terrorist threat was to go on the offensive. For instance, Cheney argued: 'We need a strategy that puts us on offense ... a strategy that allows us to destroy the terrorists before they launch another attack on the United States'.[44]

Iraq was an ideal place for several reasons: First, the Afghan war against the Taliban and Al Qaeda in the winter of 2001 was not seen as a sufficient enough show of American strength. This view was best exemplified by one of Rumsfeld's confidants just before the Afghan war: 'It will be very tough to get bin Laden in the rocky and mountainous terrain of Afghanistan. How do you send the message of strength as Ronald Reagan sent it, that we don't allow these things – you inflict damage'.[45] Mainly for logistical reasons, the Bush administration could send only a limited number of US servicemen and women (mostly Special Forces and Central Intelligence Agency operatives) to Afghanistan. A top official remarks: 'I remember the day that we put the map on the table, and the color drained from everybody's face. Afghanistan is not the place you would choose to fight'.[46] As a result, the US intervention relied primarily on air power and left the groundwork to the Northern Alliance.

Second, and more importantly, the widespread perception of US weakness was not in Afghanistan, but in the Middle East. In the view of Rumsfeld and Cheney (as well as neo-conservatives), the root cause of such a perception lay in the weak and irresolute American responses to terrorist attacks, primarily in the Middle East, since the 1980s. The examples included:

- Hezbollah's suicide attacks on US Marine headquarters in Lebanon in October 1983;
- The death of US Army rangers in Somalia in October 1993 at the hands of a mob organized by Al Qaeda;
- The attack on the Khobar Towers in Dhahran, Saudi Arabia, in June 1996 that killed 19 American soldiers;

- Al Qaeda's bombing of US embassies in Kenya and Tanzania in August 1998;
- The attack of a group of Al Qaeda suicide bombers on the USS *Cole* in Aden, Yemen, in October 2000.

The responses of past US administrations to these terrorist attacks had been either withdrawal or police action, or firing a few cruise missiles against terrorist targets.[47] For instance, James Woolsey, who was the CIA chief during the Clinton administration, retrospectively describes President Clinton's overall approach to terrorism as follows: 'Do something to show you're concerned. Launch a few missiles in the desert, bop them on the head, and arrest a few people. But just keep kicking the ball down field'.[48]

After 9/11, Cheney and Rumsfeld came to the conclusion that the weak responses of past US administrations had reinforced Al Qaeda's conviction that the US could be defeated by a resurgence of Islamic militancy. Osama bin Laden's various statements in the 1990s, in which he described the US as a 'paper tiger' and a 'weak horse' gave credence to such belief. The perceived US weakness in the Middle East, however, was not restricted to bin Laden. Many radical Islamist groups in the Middle East, such as Hamas and Hezbollah, thought similarly.[49] In the view of Cheney and Rumsfeld, Arabs who hated America were joining terrorist organizations such as Al Qaeda with the conviction that they could defeat the US. In the post-9/11 world, a less than forceful response by the US would reinforce this perception and hasten the unravelling of the region. If the US did not act boldly, Al Qaeda's depiction of it as an essentially weak power would be emphatically confirmed.[50]

Third and finally, Iraq itself had contributed to the perception of the US as a weak power. Simply by staying in power after the Gulf War, Saddam had defied the US and thus contributed to the perception of US weakness in the region. For many individuals in the Arab world, Saddam Hussein was a success story, living proof that one could challenge American power and get away with it. By pursuing a war in Iraq, the hawks aimed to reverse the perception within the Middle East world that the US is a militarily ineffective power. In turn, they believed that such a reversal in perceptions would lessen the appeal of radical Islamic terrorists and rogue regimes by tempering their dreams of destroying America with the fear of US military reprisal.

Bush did not state publicly the psychological rationale before the Iraq war, but there are strong signs that this rationale resonated powerfully in his thinking after 9/11. It seems that he genuinely believed that 9/11 was an ultimate product of past administrations' weak responses to terrorism. Woodward writes: 'President Bush, like many members of his national security team, believed the Clinton administration's response to Osama bin Laden and international terrorism, especially since the embassy bombings in 1998, had been so weak as to be provocative, a virtual invitation to hit the United States again'.[51] As a consequence, Bush claimed that the US had to fight back to change the perception of US weakness. Bush told Woodward that the perception of US weakness was quite problematic for him: 'I do believe that there is the image of America out there that we are so materialistic, that we're almost hedonistic, that we don't have values, and that when struck, we wouldn't fight back. It was clear that bin Laden felt emboldened and didn't feel threatened by the United States'.[52] After the Iraq war, Bush explicitly brought this point home as a major reason for the invasion of Iraq: 'We have learned that terrorist attacks are not caused by the use of strength. They are invited by the perception of weakness. And the surest way to avoid attacks on our own people is to engage the enemy where he lives and plans'.[53]

The psychological rationale explains why the second Bush administration did not attempt to overthrow Saddam Hussein by a method other than outright American military invasion. For instance, no evidence exists that the Bush administration seriously contemplated a proxy war in which armed Iraqi exiles would be supported by a limited involvement of American forces, even though in the immediate aftermath of 9/11, Paul Wolfowitz made such a proposal. What Wolfowitz suggested was a limited war plan against Iraq based on 'an enclave strategy' in which the US military would seize the oil fields in Shiite-dominated southern Iraq and then try to topple Saddam from that enclave.[54] Ultimately, an American invasion of Iraq stood out as the most viable military option, not only because it was most likely to lead to regime change in Iraq, but also because it would create the maximum psychological effect associated with America's demonstration of strength in the Arab/Muslim world.

Ideological Rationale for the War: Democratization of the Middle East

In going to war in Iraq, the Bush administration's decision was not limited to the goals of eliminating Iraq's WMD and to achieving psychological objectives. The most important rationale for the Iraq war was arguably ideological. This rationale, which was formulated by the neo-conservatives and championed by Deputy Secretary of Defense Paul Wolfowitz, entailed spreading democracy throughout the Middle East by establishing a model Arab government in Iraq. Underlying this rationale was the assumption that the war against terror could not be won without defeating the ideas of fanaticism and totalitarianism in the Middle East that fuelled suicidal terrorism in the first place. Freedom and democracy were the best antidotes to totalitarian ideology whether in its religious or secular form. As democratic freedoms flourished in the Arab countries and Iran, suicidal terror would lose its appeal. In other words, this was idealism not for its own sake, but was closely linked to a strategic imperative of reducing terrorist impulses associated with radical ideas.[55] The neo-conservatives believed that replacing the Saddam regime with a democratically elected government was central to a political transformation of the entire Middle East, particularly in Iran, Syria and the Palestinian Authority.

According to this 'reverse domino theory', the democratization of Middle Eastern regimes and the subsequent institutionalization of freedom and human rights will result in two developments that would enhance the security of the US. First, regimes would cease their support for terrorism because of the assumption that democracies are usually US-friendly.[56] Second, in a free and democratic polity, radical political ideas and *jihadist* calls would lose their appeal. Free elections, dialogue and compromise would gain the upper hand. This, in turn, would lead to a decline of terrorism because current frustrations in the Middle East are the outgrowth of the absence of representative institutions within Arab societies. Put differently, by turning the Middle East into a conglomeration of Western-oriented societies in which Al Qaeda-style terror would have no breeding ground, global terror would be neutralized. The goal of spreading democracy and freedom was then a strategic US interest, not only a moralistic policy.

The neo-conservative project of eliminating totalitarian ideologies via an invasion of Iraq appears to have struck a deep cord in Bush's thinking. A senior official who watched Bush and Wolfowitz interact says that they reinforced each other's faith in 'a strategic transformation of the whole region'.[57] Although many critics scoffed at the ideological rationale, believing it to be a smokescreen created by Bush in order to camouflage other secret goals in Iraq, there are strong indications that Bush did, in fact, embrace it.[58] Bush truly believed that a democratic Iraq would open a new vista for other countries in the region, particularly in Iran and Syria, and that winning the war of ideas was the long-term solution to winning the war against terrorism. Bush put this most clearly in an interview when he stated that: 'I think a free Iraq is going to influence Iran . . . I think [war on terrorism] is a long-lasting ideological struggle'.[59] People close to Bush also confirm that he developed an abiding faith after 9/11 that a new Iraq would change the Middle East, even though he rarely stated this view before the Iraq war started.[60]

A crucial remaining question is why Bush adopted the idealistic position of fostering democracy in the Middle East in the aftermath of 9/11. One can list three main reasons as an explanation. The *first* reason is related to Bush's lack of entrenched beliefs in foreign policy when he assumed office in January 2001. Mainly because he was elected with no experience and apparently little interest in international affairs, neither realism nor idealism coloured his foreign policy. During his first eight months in office, Bush's conduct of foreign policy, aside from the rhetoric, was within the confines of traditional conservative policies: a combination of realism and isolationism, but largely pragmatic and without any moralism. In other words, prior to 9/11, Bush did not acquire firm and unshakable beliefs in foreign policy and did not have a strong predisposition towards either realism or idealism.[61] Therefore, his journey to moralist foreign policy after 9/11 was largely unproblematic and painless. This is in contrast to Bush Senior whose long experience in the Foreign Service and the CIA played a role in turning him into a traditional realist. Not surprisingly, former President George Bush Senior showed little penchant to add moralism to his foreign policy during his presidency from 1988 to 1992.[62]

The *second* reason why Bush embraced the idea of democratizing the Middle East is related to the impact that 9/11 had on him. Insider reports indicate that 9/11 was a transformative event for Bush. Before 9/11, Bush seemed to have difficulty in defining the mission of his presidency. As his attention shifted substantially from domestic issues to foreign policy after 9/11, the fight against terrorism became a mission for Bush.[63] Given his zeal to defeat the terrorist threat, Bush became receptive to the 'big idea' of democratizing the Middle East.[64] There is no reason to doubt the conclusion that Bush wanted to establish his legacy 'as the president who brought democracy to the Middle East in his fight against terrorism'.[65] Moreover, no other grand project that addressed the root cause of terrorism was posited by traditional Republican realists after 9/11. Arguably, here the old rule of politics was at play: 'You can't fight something with nothing'. Hence, the neo-conservative idea that effectively linked morality (democratization) to a strategic purpose (defeat of terrorism) looked extremely appealing to Bush.

Third and finally, Bush's religious faith and his tendency to see the world in dichotomous terms may have played a role in his embracing of the neo-conservative idea. Neo-conservative thinking appealed to Bush because its stark distinction between democracy and dictatorship in the world matched Bush's interpretation of the world as a struggle between good and evil. As two observers note:

> For both Bush and the neocons, there emerged recognition of a hitherto unappreciated affinity. In important ways, Bush's religious worldview coincided neatly with the neocons secular worldview ... Thus, in the aftermath of 9/11, President Bush, determined as he was to rid the world of evil, tilted decisively in favor of neocons, determined as they were to make the world free.[66]

Conclusion

This article has argued that the Bush administration went to war with Iraq based on several rationales, all of which were directly related to fighting the war on Islamist terror after 9/11. Taken together, these rationales constitute a comprehensive, even sophisticated, reasoning for going to war in Iraq. That being said, to argue that the Bush administration's reasoning for going to war in Iraq was sophisticated is not to say that its assumptions underlying the different rationales are valid. For instance, it is debatable that Saddam was as irrational as the war hawks claimed. It is also disputable that a demonstration of US power and resolve in Iraq will deter suicidal inclined terrorists and/or the regimes that sponsor them.

Some of the war rationales, the ideological one in particular, make the Iraq war unique in US history. For the first time in its history, the US went to war with the aim of transforming the politics of an entire region and, thus, of reducing the terrorist threat. In this regard, the Iraq war was both an end and a means in the war against terror. It was an end because the US wanted to eliminate a long-time hostile regime and its WMD. It was a means because toppling the Saddam regime by an invasion presented a number of opportunities for the US to win the war on terror against the shadowy enemy in the medium to long run.

Scholars have debated endlessly whether American foreign policy is driven by realism based on power politics or idealism associated with American values and principles. Yet, as one prominent historian notes, the tension we sense in US foreign policy 'is not one between idealism and realism at all, but between competing conceptions of what is both moral *and* realistic'.[67] In its war on terror, the Bush administration combined elements of realism (demonstration of American power and resolve) and idealism (spread of democracy) in the formulation of its foreign policy. In sum, the Bush administration's decision to go to war in Iraq was due to a unique set of concerns and goals.

Many observers have overlooked this uniqueness of circumstances and goals in the war decision against Iraq and instead have interpreted the Iraq war as an attempt by the Bush administration to refashion US foreign policy. According to this view, the Iraq war indicates that a fundamental reassessment of America's role in the world has occurred. This reassessment has precipitated a new strategic emphasis based on unilateralism and preemption. In other words, the Bush administration is 'in search of monsters' to unilaterally and preemptively apply its power against in order to attain global dominance, or even an empire.

This view, however, is inherently misleading as the Bush administration defines pre-emption quite narrowly and reserves unilateralism to special circumstances. For instance, the National Security Strategy (NSS) of 2002, arguably the most important of the Bush administration's foreign policy documents, suggests pre-emption has no role to play with regard to other great powers, such as China, but rather is aimed primarily against terrorist threats and some rogue states. Even for rogue states, pre-emption is seen as a last resort means to thwart these threats. One should also point out that the NSS names only Iraq and North Korea as rogue states that could possibly be the targets of pre-emption. The NSS makes no mention at all of Iran.[68] Since pre-emption against Iraq and North Korea had been a policy option of the Clinton administration (albeit unofficially), it should not be regarded as the invention of the Bush administration. The major difference is that the Bush administration has made the policy overt.

One can go one step further and argue that the Bush administration elevated pre-emption and unilateralism to an official strategy in US foreign policy solely in order to justify the impending war against Iraq. As Iraq became the main (arguably even exclusive) focus of the Bush administration following 9/11 and the war in Afghanistan, administration officials felt it necessary to situate the attack on Iraq in a much larger and better thought-out strategy. In fact, the reading of official statements leads one to conclude that the concepts of pre-emption and unilateralism have been used only within the context of attacking Iraq. For instance, the inclusion of Iran and North Korea in the 'axis of evil' alongside Iraq in early 2002 appears to have been an attempt to avoid unwanted public scrutiny about singling out Iraq at the time. The Pentagon was then at the initial stage of its war plans against Iraq. Moreover, although Bush had probably not made his final decision about whether or not to go to war, he needed an opening salvo to prepare the American public for a potential military confrontation with Iraq. If only Iran and Iraq were mentioned, it would be interpreted as America targeting Islam. North Korea was ideal in this regard because it was non-Muslim and an adversary of the US.

The Bush administration is aware that each crisis, including confrontations with Iran and North Korea, demands a different approach as it is extremely difficult to apply the unique circumstances and rationales that made the Iraq war possible to other cases. In short, one should not overemphasize the importance of the Iraq war as a radical change in US foreign policy. The need to fight terrorism, the primary national security concern, has been at the core of America's foreign policy strategy since 9/11. Imperialistic ambitions and/or a desire to enhance American power vis-à-vis a rising China, Russia or the European Union, have not.

The Iraq war does constitute a significant change in US foreign policy towards the Middle East, a region that has come to be seen as the most problematic since 9/11. Key Bush officials interpreted 9/11 as the result of the failure of past US policies that put modernization ahead of democratization, stressed a peaceful solution to the Israeli–Palestinian problem and supported authoritarian regimes as a means to reduce the threat of radical Islam. After 9/11, top officials of the Bush administration concluded that such status quo-oriented policy and the emphasis on stability in the Middle East for the sake of oil and Israel had not worked. 9/11 was seen as a kind of foam thrown up by a larger wave of problems in the Middle East – suicidal terrorism associated with religious fanaticism, a conviction that America could be defeated, autocratic regimes that either supported or condoned rampant anti-Americanism, lack of freedoms and democracy and, equally important, a Saddam regime that stood alone as a problem case, but meanwhile seemed to characterize most of the ills in the Middle East.

Hence the new policy of the Bush administration towards the Middle East aims to overturn the traditional US approach by giving priority to democratization instead of to modernization and the resolution of the Israeli–Palestinian conflict. The Iraq war should mainly be interpreted in the context of this policy change towards the Middle East. No strategy could guarantee success, but in the view of the Bush administration, inaction following 9/11 would lead to a further deterioration of the security situation in the Middle East.

The Bush administration thus believed that regime change in Iraq offered an exceptional opportunity to correct the problems that were plaguing the region and, simultaneously, threatening US security. This is why Iraq is central to US foreign policy. Failure in Iraq *à la* Vietnam – that is, an embarrassing US withdrawal from Iraq in the face of increasing American casualties and an unbeatable insurgency – would mean not only 'losing Iraq' but also abandoning all possibilities and opportunities that the US has in the region. Given the uniqueness of the Iraq war, it is highly unlikely that the US military would undertake similar action against other foes such as Iran and Syria, in the foreseeable future. Therefore, one should expect the Bush administration to adopt a wait-and-see attitude toward these countries and resort to means other than military

NOTES

1. For a discussion of the role played by different actors and worldviews in the American foreign policy tradition, see Walter Russell Mead, *Special Providence: American Foreign Policy and How It Changed the World* (New York and London: Routledge, 2002). Mead notes: 'American foreign policy rests on a balance of contrasting, competing voices and values – it is a symphony, or tries to be, rather than a solo' (p.54).
2. For Cheney's importance as a policy player in the Bush administration, see Paul Kengor, 'Cheney and Vice Presidential Power', in Gary L. Gregg, II and Mark J. Rozell (eds), *Considering the Bush Presidency* (New York: Oxford University Press, 2004). For an early assessment of Cheney's importance for Bush, see Nicholas Lemann, 'The Quiet Man: Dick Cheney's

Discreet Rise to Unprecedented Power', *New Yorker*, 7 May 2001. Glenn Kessler and Peter Slevin, 'Cheney Is Fulcrum of Foreign Policy', *Washington Post*, 13 Oct. 2002.

3. See, for instance, John B. Judis, 'Why Iraq?', *The American Prospect*, March 2003; David Ignatius, 'War and Oil,' *Washington Post*, 18 Oct. 2002.
4. Peter Grier, 'Is It All About Oil?', *Christian Science Monitor*, 16 Oct. 2002.
5. Trudy Rubin, 'Talk About Oil and Iraq is Just That: A Lot of Talk', *Philadelphia Inquirer*, 12 Jan. 2003.
6. Ignatius, 'War and Oil'.
7. Before the Iraq war, an administration report had warned that 'the U.S. and its partners should make it clear that there are no hidden agendas regarding control of Iraqi oil'. Cited in John B. Judis, 'Over a Barrel', *New Republic*, 20 Jan. 2003.
8. Micheal Ignatieff, 'The Year of Living Dangerously', *New York Times* (Magazine), 14 March 2004.
9. When the original draft of the DPG was leaked to the press, it created a public controversy such that the draft's language and substance was later softened. See 'Pentagon's New World View', *Washington Post*, 24 May 1992.
10. This is the main theme in a recent book edited by Sut Jhally and Jeremy Earp, *Hijacking Catastrophe: 9/11, Fear & the Selling of American Empire* (Northampton, MA: Olive Branch Press, 2004).
11. For a similar point, see Robert S. Snyder, 'The Myth of Preemption: More Than a War Against Iraq', *Orbis* (Fall 2003), p.660.
12. See Melvyn P. Leffler, 'Bush's Foreign Policy', *Foreign Policy* (Sept./Oct. 2004). Also see Andrew J. Bacevich, *American Empire* (Cambridge, MA: Harvard University Press, 2002), Chap. 5.
13. For the frustrations of the neo-conservatives with the Bush administration's foreign policy before 9/11, see Gary Dorrien, *Imperial Designs: Neoconservatism and the New Pax Americana* (New York: Routledge, 2004), pp.141–50.
14. See Michael Hirsh, 'Bush and the World', *Foreign Affairs* (Sept./Oct. 2002).
15. See Ivo H. Daalder and James M. Lindsay, *America Unbound: The Bush Revolution in Foreign Policy* (Washington, DC: Brookings Institution Press, 2003), Chap. 5.
16. John Lewis Gaddis, *Surprise, Security, and the American Experience* (Cambridge, MA: Harvard University Press, 2004), p.73.
17. William Kristol and Robert Kagan (eds), *Present Dangers: Crisis and Opportunity in American Foreign and Defense Policy* (San Francisco, CA: Encounter Books, 2000).
18. Condoleezza Rice, Speech to the Manhattan Institute, 1 Oct. 2002, <www.whitehouse.gov/news/releases/2002/10/20021001-6.html>, accessed 11 Dec. 2004.
19. Daniel Eisenberg, 'We're Taking Him Out', *Time*, 13 May 2002. Also see Glenn Kessler, 'U.S. Decision on Iraq Has Puzzling Past; Opponents of War Wonder When, How Policy Was Set', *Washington Post*, 12 Jan. 2003.
20. A great deal of attention has been paid to how the Bush administration used the available information about Iraq's WMD in order to justify the war to the US public. The list of articles and books written on the Bush administration's public campaign about Iraq's WMD is long. Among others, see Thomas Powers, 'The Vanishing Case for War', *New York Review of Books*, 4 Dec. 2003; Spencer Ackerman and John B. Judis, 'The First Casualty: The Selling of the Iraq War', *New Republic*, 30 June 2003.
21. See Bob Woodward, *Plan of Attack* (New York: Simon & Schuster), p.220.
22. Sam Tannenhaus, 'Bush's Brain Trust', *Vanity Fair*, July 2003.
23. Kenneth Pollack, 'Spies, Lies, and Weapons: What Went Wrong', *Atlantic Monthly*, (Jan./Feb. 2004).
24. See David Albright, Kathryn Buehler and Holly Higgins, 'Bin Laden and the Bomb', *Bulletin of the Atomic Scientists* (Jan–Feb. 2002), p.23.
25. Cited in Glenn Kessler and Peter Slevin, 'Cheney is Fulcrum of Foreign Policy', *Washington Post*, 13 Oct. 2002.
26. Carla Anne Robbins and Jeanne Cummings, 'How Bush Decided That Hussein Must Be Ousted From Atop Iraq', *Wall Street Journal*, 14 June 2002.

27. Condoleezza Rice, 'Promoting the National Interest', *Foreign Affairs* (Jan./Feb. 2000).
28. For instance, see 'Saddam Must Go', *Weekly Standard* (Editorial), 17 Nov. 1997.
29. Kenneth Pollack, *The Threatening Storm, The Case for Invading Iraq* (New York: Random House, 2002), pp.104–8.
30. Interview with Richard Perle, 'Truth, War & Consequences', *Frontline*, PBS, 9 Oct. 2003.
31. William Kristol and Lawrence Kaplan, *The War Over Iraq* (San Francisco, CA: Encounter Books, 2002), p.82.
32. Testimony of US Secretary of Defense Donald H. Rumsfeld before the House Armed Services Committee Regarding Iraq. Washington, DC (18 Sept. 2002), <www.defenselink.mil/speeches/2002/s20020918-secdef2.html>, accessed 20 Dec. 2004.
33. See Michael Elliott and James Carney, 'First Stop, Iraq: How did the U.S. End Up Taking on Saddam?', *Time*, 31 March 2003. Also see Glenn Kessler and Peter Slevin, 'Cheney is Fulcrum of Foreign Policy', *Washington Post*, 13 Oct. 2002.
34. Cited in Carla Anne Robbins and Jeanne Cummings, 'How Bush Decided That Hussein Must Be Ousted From Atop Iraq', *Wall Street Journal*, 14 June 2002. Emphasis added.
35. Evan Thomas, 'Bush Has Saddam in His Sights', *Newsweek*, 4 March 2002. Emphasis added.
36. Michael Elliott and James Carney, 'First Stop, Iraq: How did the U.S. End Up Taking on Saddam?' *Time*, 31 March 2003.
37. Cited in Woodward, *Plan of Attack*, p.27
38. George W. Bush, 'State of the Union Address', 28 Jan. 2003, <www.whitehouse.gov/news/releases/2003/01/20030128-19.html>, accessed 15 Dec. 2004.
39. Cited in Martin Walker, *The Cold War: A History* (New York: Henry Holt and Company, 1993), p.40.
40. Cited in 'The Path To War', *Vanity Fair*, May 2004.
41. See William Curti Wohlforth, *The Elusive Balance: Power and Perception During the Cold War* (Ithaca, NY: Cornell University Press, 1993); John Lewis Gaddis, *Strategies of Containment* (New York: Oxford University Press, 1982).
42. Evan Thomas, 'Rumsfeld's War' *Newsweek*, 16 Sept. 2002.
43. Evan Thomas, 'The 12 Year Itch', *Newsweek*, 31 March 2003. According to an aide's notes, on 9/11 Rumsfeld said, 'hit S.H. @ same time – not only UBL'. Cited in Woodward, *Plan of Attack*, p.25. SH denotes Saddam Hussein and UBL Usama bin Laden. These words of Rumsfeld should be interpreted as a knee-jerk reaction to 9/11, rather than a well thought-out rationale at the time.
44. Cited in 'Dick Cheney: The Man Behind the Curtain', *U.S. News & World Report*, 12 Oct. 2003.
45. Cited in Michael Hirsh, *At War With Ourselves: Why America Is Squandering Its Chances to Build a Better World* (New York: Oxford University Press, 2003), p.51.
46. Cited in Glenn Kessler, 'U.S. Decision On Iraq Has Puzzling Past; Opponents of War Wonder When, How Policy Was Set', *Washington Post*, 12 Jan. 2003.
47. See Jean Bethke Elshtain, *Just War Against Terror* (New York: Basic Books, 2003), pp.155–7.
48. Cited in Norman Podhoretz, 'How to Win World War IV', *Commentary* (Feb. 2002).
49. For examples of perceived US weakness in the Middle East, see Barry Rubin, *The Tragedy of the Middle East* (Oxford University Press, 2002), p.254.
50. This was also the argument of Bernard Lewis and Fouad Ajami, two pre-eminent scholars of the Middle East with close ties to some administration members, particularly Cheney. See David Frum, *The Right Man* (New York: Random House, 2003), p.170.
51. Woodward, *Bush At War*, p.38
52. Cited in Woodward, *Bush At War*, pp.38–9.
53. 'President Bush's Address to the Nation', (Text), *New York Times*, 9 Sept. 2003.
54. For instance, Wolfowitz defended such a limited war plan during the meeting of principals in Camp David on 15 Sept. 2001. See Woodward, *Plan of Attack*, p.22. Also see interview with Deputy Secretary Wolfowitz at the New Yorker Festival, 21 Sept. 2003, New York, at <www.defenselink.mil> (Pentagon's website).
55. For details of the neo-conservatives' thinking about the democratic tsunami created by the Iraqi model for the rest of the Middle East, see Lawrence F. Kaplan and William Kristol,

The War over Iraq: Saddam's Tyranny and America's Mission (San Francisco, CA: Encounter Books, 2003), Chap. 7; Robert Dreyfuss, 'Just the Beginning: Is Iraq the Opening Salvo in a War to Remake the World?', *American Prospect*, April 2003; Nicholas Lemann, 'After Iraq: The Plan to Remake the Middle East', *New Yorker*, 17 Feb. 2003; Joshua Micah Marshall, 'Practice to Deceive: Chaos in the Middle East Isn't the Bush-hawks' Nightmare Scenario – It's Their Plan', *Washington Monthly*, April 2003; 'Bush's Domino Theory', *Christian Science Monitor*, 28 Jan. 2003; David Ignatius, 'The Read on Wolfowitz', *Washington Post*, 17 Jan. 2003; Trudy Rubin, 'Toppling Hussein: A Chance to Rethink Mideast Politics', *Philadelphia Inquirer*, 2 Aug. 2002; John B. Judis, *The Folly of Empire* (New York: Scribner's, 2004).

56. See Charles Krauthammer, 'Democratic Realism: An American Foreign Policy for a Unipolar World', Speech at the American Enterprise Institute, 27 Feb. 2004. Also see, David Wurmser, *Tyranny's Ally: America's Failure to Defeat Saddam Hussein* (Washington, DC: The AEI Press, 1999), p.61.
57. See Bill Keller, 'The Sunshine Warrior', *New York Times* (Magazine), 22 Sept. 2002.
58. For instance, Maureen Dowd, a *New York Times* columnist and a fierce critic of the Bush administration, writes: 'The goal of ending tyranny is available as a retroactive rationale for the war in Iraq, where Americans were originally told that weapons stocks were the primary justification for war'. Dowd, 'A Bunch of Krabby Patties', *New York Times*, 23 Jan. 2005.
59. Interview with *Time*, 6 Sept 2004.
60. For the emerging importance of transforming the Middle East in Bush's thinking after 9/11, see David Frum, *The Right Man.* Before the Iraq war, however, Bush emphasized this rationale on only two occasions, once in a speech at West Point in June 2002 and a second time in a speech at the American Enterprise Institute on 26 Feb. 2003, a few weeks before the beginning of the Iraq war. See, 'President Bush Delivers Graduation Speech at West Point, June 1, 2002' and 'President Discusses the Future of Iraq, American Enterprise Institute, Washington, D.C., February 26, 2003', <www.whitehouse.gov>. Following the toppling of the Saddam regime, Bush frequently stated the link between the Iraq war and democratization of the Middle East. For instance, he told Woodward: 'I truly believe out of this [Iraq war] will come peace'. Cited in Woodward, *Plan of Attack*, p.260.
61. For Bush's idealism, see Michael J. Mazarr, 'George W. Bush, Idealist', *International Affairs*, Vol.79, No.3 (2003).
62. For a glimpse into the minds of Bush Senior and his administration, see George H. W. Bush and Brent Scowcroft, *A World Transformed* (New York: Knopf, 1998).
63. 'In One Month, A Presidency Is Transformed', *The New York Times*, 11 Oct. 2001.
64. Woodward writes: 'Bush was... baffled about the countries that produced ideologies and people with the goal of killing Americans in terrorist attacks. He wondered how the U.S. could reform such societies, and wanted to advocate the promotion of democracy and women's rights in the Muslim world... It was part of change in the president's thinking... since the 9/11 attacks'. Woodward, *Plan of Attack*, p.89.
65. Arthur Schlesinger, Jr., 'Eyeless in Iraq', *New York Review of Books*, 23 Oct. 2003.
66. Andrew J. Bacvich and Elizabeth H. Prodromou, 'God Is Not Neutral: Religion and U.S. Foreign Policy after 9/11', *Orbis*, Vol.48, No.1 (Winter 2004), pp.52–3
67. Walter A. McDougall, *Promised Land, Crusader State: The American Encounter with the World Since 1776* (Boston, MA and New York: Mariner Books, 1997), p.9. Emphasis original.
68. See the National Security Strategy of the United States, Sept. 2002, <www.whitehouse.gov>.

[17]

Does the Iraq War Reflect a Phase Change in Warfare?

Avi Kober

Department of Political Studies and BESA Center for Strategic Studies, Bar-Ilan University, Ramat-Gan, 52900, Israel

Military analysts have offered two competing interpretations of the Iraq War. According to one interpretation, expressed by Revolution in Military Affairs (RMA) adherents, whereas war from 1939 to 1990 was largely a matter of continuity (at least for conventional war-fighting), since 1990 we have been witnessing a discontinuous "phase change" on the way to something radically different or new.[1] Another interpretation has presented the Gulf War, Afghanistan and the Iraq War as rather orthodox theater warfare, based on firepower-maneuver synergy, in which neither element could win alone.[2] The purpose of this article is twofold. At one level it is interested in the pattern of historical change in warfare, trying to find out if the dynamics of the Iraq War reflect a continuous, evolutionary process of change or rather a discontinuous transition in the framework of a process of "punctual equilibrium".[3] At another level it is interested in the practical implications of the war for defense policies and military planning of Western democracies. If one accepts the second interpretation, then it follows that a business-as-usual approach in defense policies and planning is enough. The first interpretation, on the other hand, means that the new form of warfare will only be realized if a radical policy agenda is enacted.

My main argument is that from a historical point of view the Iraq War does reflect a discontinuous change in warfare, as a result of technological and societal changes. Each of these changes has been individually necessary for this discontinuity to happen, but only when taken jointly have they been able to create the sufficient condition for it to materialize. The crux of this discontinuity has been the unprecedented capability to wage surgical wars. It was, however, the 1991 Gulf War rather than the Iraq War that constituted the watershed in this respect. From a policy-oriented point of view this historical process has opened up new opportunities for Western democracies engaged in asymmetrical wars of either low or high intensity nature to conduct their wars more effectively and at a lower cost. The radical technological change has also brought about some military build-up dilemmas.

In this analysis of the Iraq War I have tried to avoid the use of the concepts "revolution" and "evolution". Determining whether a war reflects a revolution or an evolution tends to be semantic, subjective (the case for revolution for one person would be the case for evolution for another), and plagued with vague or tautological definitions of revolution, which are so typical of RMA literature. Even if revolution and evolution were defined abstractly they would be hard to operationalize in a simple, measurable way. The discussion that follows covers the high-intensity conflict (HIC) of 20 March–1 May 2003.[4] It does not deal with the low-intensity, protracted conflict (LIC) in which the Coalition has been engaged since the high-tech war came to an end, a conflict that does not represent any significant change in warfare, and cannot be won by precision strikes and a slim ground component.[5] The article also does not refer to the longer-term domestic, regional and international implications of the war.

Four characteristics of the Iraq War might apparently challenge the external validity of any analysis. First, it was of an extremely asymmetric nature. Although the force ratios in ground forces were in favor of the Iraqis,[6] the Coalition encountered an ill-equipped and low-skilled army[7] that had weakened drastically after the Gulf War and with the loss of Iraq's patron, the former USSR. The Coalition also held a tremendous qualitative edge in training, leadership and equipment, and enjoyed a quantitative advantage in the air and at sea: for example, a ratio of 2.3:1 in fighter aircraft, 5:1 in attack helicopters, 55:1 in bombers, and 5 versus 0 aircraft carriers.[8] Moreover, the war proved that the Coalition had an overwhelming firepower advantage over the Iraqis.[9] The Iraqis did not deploy their forces in a way that would oppose the invasion. The Republican Guard was kept out of the opening battle and was deployed outside Baghdad, where it was vulnerable to Coalition air power. The southern front was not heavily mined. Port facilities were not sabotaged and bridges over the Tigris and Euphrates were not demolished. By focusing on the defense of Baghdad, the Iraqis essentially forfeited about 95 per cent of their land to Coalition forces.[10]

A second aspect that might affect the external validity of the analysis of the war is the fact that the Iraq War was conducted under extremely favorable conditions for the Coalition. The war broke out against the backdrop of an international system in which the US had become the only superpower. Many Arab states supported the Coalition, if only tacitly or passively. Furthermore, Iraq was internationally and even regionally isolated, and Israel kept a low profile. Third, the Iraq War was a high-intensity conflict. As such conflicts constitute only some five per cent of the post-Cold War conflicts, one could question the applicability of this analysis to low-intensity conflicts. Fourth, the Coalition waged a post-heroic war, which is not pervasive in the international system but is rather typical of a small group of Western democracies.[11]

These four features, however, should not undermine the external validity of this study. Whereas in the early 1990s it was unclear whether the 1991 Gulf War was in fact representative of future wars[12] or an exception in a world where the use of force has been undergoing major changes,[13] now the picture is clear. The Iraq War should be considered a link in the chain of wars of its kind: asymmetrical, post-heroically-waged by Western democracies since the early 1990s, and of either type, LIC or HIC.[14]

DISCONTINUITY IN WARFARE

Military–technological changes, radical as they may be, could not be considered a sufficient condition for discontinuity in warfare.[15] For a discontinuity in warfare to happen, a confluence of military–technological changes – technological breakthrough, institutional adaptation and war-fighting innovation[16] – on the one hand, and societal changes on the other, is needed.[17] The combination of the nations-in-arms phenomenon and the early nineteenth-century Industrial Revolution constituted such a discontinuous transition. It soon resulted in the emergence of the so-called modern war, the first expression of which was the American Civil War.[18] Another discontinuous transition occurred as a result of the introduction of the nuclear weapon towards the end of World War II, when for the first time it became technologically feasible to destroy entire societies.[19] A third discontinuous change took place in late twentieth century/ early twenty-first century, when socially constrained, casualty avert Western democracies became able to wage surgical wars thanks to their high-tech capabilities. In this section I will first present the unprecedented technological change reflected by the Iraq War, and then discuss the confluence of the technological and societal transition.

Unprecedented technological capabilities

Many of the technological capabilities demonstrated by the Coalition during the Iraq War were unprecedented.

Precision

One of the goals set out for the US military forces a decade into the twenty-first century by former Secretary of Defense Cohen and former Chairman of the JCS Shalikashvili was that they possess "precision strike" capability from long distances.[20] To realize how far the US has advanced in this respect since World War II, in which no smart weapons were used, one should look at the percentage of smart bombs in the ammunition used in the wars the US has been engaged in since the early 1990s.[21] In the Gulf War, seven to eight per cent of the munitions used were smart bombs. In Kosovo, smart bombs constituted some 35 per cent of the weapons used,[22] and in Afghanistan the rate rose to 56 per cent.[23]

According to US military sources, in the Iraq War 90 per cent of the bombs were smart bombs, although the actual figure is probably closer to 60–68 per cent.[24] In terms of quantity of smart ammunition relative to time, in the Gulf War some 9,250 precision bombs were dropped and 300 cruise missiles were launched during the six-week aerial attack. In contrast, in the Iraq War 18,000 precision bombs were dropped and 800 cruise missiles were launched in less than four weeks. F-117 attack planes launched GBU-27 air-to-land GPS or laser-guided missiles to hit their targets. Of almost 20,000 smart munitions used by the Coalition in the Iraq War, some 33 per cent were GPS-guided and 46 per cent were laser-guided.[25]

Lethality

The most lethal munitions used in the Iraq War were 2,000-pound warhead Joint Direct Attack Munitions (JDAMs). These weapons were introduced for the first time in Kosovo. Five thousand-pound warhead "bunker busters", which were first introduced in Afghanistan and whose destruction force approaches that of tactical nuclear weapons, were also used.

Range

The US's capability of launching ammunitions from long distances has improved radically. In the Iraq War, cruise missiles were launched from long distances – hundreds of miles – and bombers had enough range to carry out long-distance missions hundreds of miles away from bases in Europe.

The sensor-to-shooter loop

In the early nineteenth Century Baron de Jomini presented the obstacles encountered in attempting to communicate intelligence in real time from the enemy camp to the consumer.[26] Advanced command and control systems, as well as improved intelligence, surveillance and reconnaissance (ISR) capabilities have considerably shortened the process of spotting a new target, assigning a weapon to hit it, and finally hitting it – from three days in the Gulf War to less than 40 minutes, both in Afghanistan and Iraq.[27]

High survivability

This is a direct consequence of a combination of stealth and stand-off capabilities (see below). As a result of the dramatic strengthening of firepower, changes have occurred in the importance of key characteristics and variables that have traditionally been typical of the conduct of war. Noteworthy in particular are the rising importance of attrition ratios, concentration of fire, and the "shock and awe" effect of firepower.

The rising importance of attrition ratios

Against the backdrop of an enhanced firepower, not only have attrition ratios become more important than force ratios, but they have also become important force multipliers, particularly for parties enjoying "Dominance in Battlefield Knowledge" (DBK). DBK, part of the backbone of the so-called RMA, can dramatically increase the level of attrition inflicted upon one's enemy,[28] especially in conflicts of asymmetric nature. The loss ratios in the Iraq War were 1 to between 5 and 9 in the Coalition's favor: 471 Coalition fatalities compared to 2,320–5,600 Iraqi soldiers killed.[29] It would be inaccurate, though, to describe these loss ratios as "almost unmatched in modern military history".[30]

In the Gulf War the loss ratios were 60–149:1 (cautious estimates range between 20,000–50,000 Iraqi troops, versus 336 Coalition troops).[31] Attrition rates have also become more meaningful for the achievement of battlefield decision. Unlike the 1973 October War and the Gulf War,[32] in the Iraq War the Republican Guard had lost some 50 per cent of its fighting strength after having been hit by fire before it was ever confronted on the ground.[33] There are still avenues to diminish the attrition inflicted by firepower of a high-tech force like the US military, though. For example, mobile targets and decoys could still frustrate it and SA missiles could be a major problem for air and cruise missiles.

Concentration of fire

In the 1980s the reformist school in the US feared that high-tech capabilities and attrition by fire might overshadow the human factor in military operations, the search for maneuver and military engagement on the ground, and the drive to achieve battlefield decision.[34] One could claim that despite the dramatic strengthening of firepower that was demonstrated in the Gulf War, Afghanistan and the Iraq War, neither firepower nor maneuver has been able to win alone. As much as maneuver continues to be relevant on the battlefield, though, a quick, effective, concentrated fire could turn the armor into a weapon system the main mission of which would be to translate the effect already achieved by firepower into territorial gains.[35]

It is true that in the Iraq War Coalition ground forces were maneuvering and concentrating forces against centers of gravity on the battlefield, but once they got into battle most of the forces confronting them had already been destroyed by fire concentrated against them. Had the precision fire aimed at decapitating key Iraqi leadership figures at the initial stage of the war been successful in eliminating them, the Coalition might have been able to avoid most of what followed.[36]

The "shock and awe" effect of firepower

The image of maneuver as more modern and sophisticated than firepower thanks to the emphasis it places on the adversary's willpower and morale is misleading. Maneuver has traditionally been restricted to the military levels of war, whereas firepower has also been related to grand-strategy (particularly the societal and economical dimensions of war), and as such it could be considered more modern. Furthermore, firepower too can affect the enemy's willpower and morale. The Iraq War was "supposed to" open with a high-tech, 24-hour blitz, based on firepower, to be followed up by a ground assault. The attack was expected to consist of the detonation of non-lethal e-bombs over the Iraqi regime's command and control centers and the launching of over 3,000 laser-guided bombs and missiles against air defense, radar, and missile sites, electronic command and control centers, WMD sites, and leadership targets.[37]

Damaging the country's infrastructure and terrorizing its population, however, was no real option because of the need to win the hearts and minds of the Iraqi people and attain international support. In practice, an air campaign and ground maneuver were launched in tandem aimed at gaining a tactical surprise, seizing the oil fields before they were set on fire, and reaching Baghdad – the grand-strategic center of gravity – as soon as possible. Firepower nevertheless not only accounted for the high attrition inflicted on the Iraqi Army's equipment, but also had a great psychological effect. What seemed at the initial stages of the war as a high-cost tolerance on the part of the Iraqi military – most probably as a result of a false sense of security caused by the absence of a Coalition initial air campaign[38] – ended up with a psychological collapse.

The confluence of technological and societal changes

From a societal point of view, the Iraq War is yet another proof that the process that had been launched in the early nineteenth century, where society became deeply involved in war, has been reversed. Although there is no state–society break as existed before the French Revolution, as troops are still drawn from the society despite their becoming occupational, for Western democracies war has reached a new threshold, whereby high-tech capabilities have enabled them to overcome societal constraints and conduct war both effectively and with relatively low human cost.

This is particularly true for non-existential wars. Western societies are less disposed to pay a high price in war once they perceive of it as being a "war of choice", or a LIC, in which the stakes are not sufficiently high.[39] According to Luttwak their changed demography, by which the emotional capital of most families is invested in just one child, whose death is a tragedy, has made them loss avert.[40] But Western societies also tend to believe that wars result from the evilness of the enemy's political leadership. Given these constraints on the one hand, and their high-tech capabilities on the other, if such countries find themselves in war they would tend to conduct it "post-heroically".

Post-heroic warfare has two rules: first, one is not "allowed" to get killed. Second, one is not "allowed" to kill, at least not civilians.[41] The first rule could be attributed to both demography and low stakes, whereas the second rule is affected by the belief that wars result from the evilness of the enemy's political leadership. If one sees the enemy as leadership, rather than armed forces or society – for example, if the enemy is Saddam, Milosevic, Aideed or Arafat – then the center of gravity should be the enemy's leadership, and killing large numbers of enemy soldiers (as one of the means to achieve battlefield decision), let alone civilians, is not a legitimate act.[42] The Iraq War has demonstrated how post-heroic players could use precision munitions in order to inflict considerable damage on enemy forces or infrastructure while avoiding casualties among enemy civilians and among their own troops.

Avoiding civilian casualties

The Bush administration's policy to avoid civilian casualties[43] dictated hitting only leadership and military targets as well as using a high proportion of smart weapons. As one analyst put it, in post-modern war,

> Americans must be swift, decisive, and victorious in their war-making – but not to the extent that they should kill too many of the enemy. Our GPS bombs must not just be smart, but rather brilliant – and thus distinguish and target (wounding rather than killing) only 80,000 individual Ba'athists and Saddamites within a population of 26 million.[44]

The improved accuracy of weapons enabled the Coalition to launch an air assault while saving civilians' lives, and to achieve a surgical destruction of a totalitarian regime while leaving the country standing.[45] Eighty-two per cent of the targets hit were of counter-force nature.[46]

Links between CIA and Special Forces and Iraqi military and political leaders in many cases avoided direct military confrontation and urban warfare in some of Iraq's cities such as Tikrit and Mosul, which might have inflicted heavy losses on both sides. At other cities such as Baghdad, Nasiriyah, Najaf and Samawah there were urban combat actions. In Baghdad, two extremely violent "Thunder Runs" by the 3rd Infantry Division's 2nd Brigade were required, plus heavy fighting by the division's 3rd Brigade as it subsequently entered the city from the north. US casualties were nonetheless very light by historical standards, but at least hundreds of Iraqis were killed.

Whereas the Vietnam War claimed the lives of at least four per cent of the Vietnamese population, in the 1991 Gulf War only some 0.07 per cent of the Iraqi population were killed. In Kosovo the percentage declined to 0.005 per cent, and in Afghanistan and Iraq it rose a bit to 0.01–0.03 per cent of the population of the respective countries.[47] This happened contrary to the expectation that after 9/11 the US would be less averse to killing enemy civilians. The very low civilian casualty rates mean that it was the 1991 Gulf War rather than the Iraq War that constituted a watershed in this respect. Since the Gulf War the US has managed to leave the lion's share of civilians on the enemy's side intact. In an almost surrealistic report from the Iraqi capital, Charles Krauthammer described "plumes of smoke from precision strikes on Saddam's instruments of power while the city lights remained on and cars casually traversed the streets".[48]

Avoiding casualties among one's own troops

The other side of the coin of post-heroic warfare was that "we should not lose American soldiers abroad; but, to be fair, we should at least suffer a few dead [. . .] to avoid criticism [that] Americans kill without being killed".[49] This new mindset already affected operations in the Gulf War. During the air attack in that war, the Coalition's pilots were directed not to take unnecessary risks, even if doing so might result in poor hits. The phrase "No target is worth an airplane" reflected the attitude that prevailed during that war.[50] And indeed, as a result of this directive, only some 20 per cent of the non-guided bombs hit their target areas, while 90 per cent of the smart bombs hit theirs.[51]

American losses in war since World War II have plummeted. World War II claimed the lives of hundreds of thousands of US troops (more than 290,000). In the Vietnam War, tens of thousands troops were killed (58,000). In the Gulf War, however, only a few hundred troops were killed (300), and in Kosovo and Afghanistan the numbers dropped to single figures – two in Kosovo, and eight in Afghanistan, thanks to the fact that in neither of these two cases did the US commit ground forces. In the Iraq War, up until May 1, 400 American troops lost their lives (the number includes both combat and non-combat deaths),[52] 40 per cent of whom were killed in accidents or so-called friendly fire.

In Vietnam one soldier out of 15 died,[53] whereas in the Iraq War only one soldier out of 625 paid with his/her life. In the Gulf War some 40 troops were dead in each of the four-day ground operations,[54] while in the Iraq War only three soldiers died in each day of the three-week ground operations.[55] These are very low rates given the fact that in the Iraq War the US did commit ground forces, and bearing in mind that in the wars the

US tried to conduct post-heroically, the enemy tried to kill as many American or Coalition troops as possible.

The low casualty rates in the Iraq War were achieved thanks to six main factors, most of them technological. First, the air campaign that was conducted from mid-2002 into the first few months of 2003 under the guise of operations Northern and Southern Watch. American aircraft attacked extensively command and control systems (including the fiber-optic cable system used for military communications), anti-aircraft systems, etc., dropping 606 bombs on 391 carefully selected targets.[56]

Second, the application of Suppression of Air-Defense (SEAD) in the initial stages of the war, in which targets that had been left in central Iraq, were now easily targeted.[57] As a result of the first and second factors, the Coalition troops weakened the Iraqi air defenses to the point of enjoying freedom of action in the skies and clear skies for ground troops.

Third, there was stealth. B-2 bombers and F-117 attack planes made of radar-absorbing material flying virtually invisible to enemy anti-air defenses had already done a good job during the Gulf war. In 1991 the F-117 flew only two per cent of the total attack sorties, but struck nearly 40 per cent of the strategic targets and remained the centerpiece of the strategic air campaign for the entire War.[58] Given the American sensitivity to casualties on both sides, these have been ideal weapon systems for attacking targets in the heart of the heavily defended and populated city of Baghdad.[59] Being aware of the poor performance of the Iraqi air defense in 2003, however, the Air Force planners of the Iraq War sent just 12 F-117s and only four B-2s.[60]

Fourth, American forces had a stand-off launching capability – that is, the ability to attack from a great distance, with low risk to one's troops.

Fifth, by adopting an indirect approach there was available a combination of firepower's physical and psychological effect and a quick push to Baghdad on the ground (some 300 miles in four days) by forces maneuvering in inhabited, open spaces, simultaneously attacking the Iraqi flanks and rear. This created confusion and disorder in the Iraqi ranks, giving them little opportunity to retreat or regroup – all this at the cost of fewer than ten combat deaths. The low Coalition casualty rate could not be explained by the avoidance of urban warfare. Pre-war optimists based their optimism on the belief that urban warfare was unlikely, whereas the pessimists expected city fighting. In fact, the war produced extensive urban warfare, but without the expected Coalition casualties. This is one of the war's most interesting empirical anomalies.

Another aspect of discontinuity in warfare that failed to materialize in the Iraq War pertains to the application of information warfare. A radical view on information warfare has treated it as an alternative to traditional war.[61] Indeed, should developments in the field of information warfare make it possible to bypass any physical, operational confrontation between armed forces, this would constitute a discontinuity in the nature and conduct of war.[62] Causing the collapse of the enemy's civil and military systems by hitting information centers of gravity – something of a "virtual war", a "post-industrial *blitzkrieg*", or an "electronic Pearl Harbor"[63] – could be considered the realization of Sun Tzu's and Liddell Hart's dream "to subdue the enemy without fighting",[64] or "to produce a decision without any serious fighting".[65]

Although both thinkers considered it to be the perfect strategy in general, and the ideal indirect approach in particular, they expressed their skepticism as to the chances of such a dream being realized. Such skepticism is still appropriate. Given the dialectic nature of technological developments throughout the history of war, technological, organizational and doctrinal countermeasures are likely to be taken by states, sub-state players and military forces so as to defend the new points of disadvantage, especially from peer competitors.[66] However, the current situation is one wherein the state leading the RMA – the US – is peerless.

A skeptical view on the impact of information warfare on the nature of war has claimed that although it marks a significant technological advance in warfare, it does not constitute a discontinuity in warfare as it does not affect the fundamental relationship in military affairs, particularly the relationship of politics to war, nor does it supplant the so-called nuclear revolution.[67] Mainstream thinking on the nature and uses of information war in US Army literature, on the other hand, has stressed the need for information superiority in general and obtaining DBK in particular. Information superiority is supposed to help create favorable conditions under which it will be easier to maneuver, attack and achieve battlefield decision.[68] This was clearly the case with the 1991 Gulf War.[69]

The Iraq War was expected to be the first war of the information age.[70] One could therefore expect cyber-war means to be used against centers of gravity such as radar, telephone and computer systems. Had high-power microwave (HPM) or conventional electromagnetic bombs (e-bomb) been used, they could have incapacitated electronic systems without damaging buildings or harming people (unless they were adjacent to or networked with military targets), and isolated the Iraqi leadership from Iraqi troops and people. Cyber attacks, however, did not play a role in the Iraq War, probably because they were still experimental or because their effect was likely to diminish rapidly with distance or just be absorbed harmlessly in the ground. Once absorbed into wiring, one can hardly predict where it would go, and it might have damaged any non-military electronics within range.[71]

A variety of technologies – electromagnetic, acoustic, biotechnical, chemical, mechanical, optical, etc. – that could be considered non-lethal and less-lethal weapons[72] were not used either, probably because while the idea of bloodless war has been the subject of debate among strategists for a long time, it is not yet over. That is, while the advocates of the so-called non-lethal weapons stress the minimal collateral damage they create and present them as politically and morally friendly, or a sword that cuts almost without wounding or killing,[73] the opponents, spearheaded by humanitarian non-governmental organizations such as the International Committee of the Red Cross (ICRC) or Amnesty International, consider non-lethal weapons an oxymoron, with the line between lethal and non-lethal weapons essentially impossible to draw.[74]

IMPLICATIONS FOR DEFENSE POLICIES AND MILITARY PLANNING

In the aftermath of World War II Western countries engaged in non-existential asymmetrical conflicts faced many difficulties in coping with weak adversaries.[75] One of the lessons learnt from post-Cold War conflicts has been the considerably growing prospect of winning such wars. Although post-heroic warfare could easily be presented as a very inefficient way of conducting war, one should not disregard its positive aspects. Contrary to the belief that war must become unpopular with society and therefore be unsustainable,[76] the ability to conduct surgical wars might be conducive to gaining domestic legitimacy for both initiating wars and sustaining them. Except for Somalia, where the intervention got rough after 18 troops were killed, basically, because the perceived stakes were extremely low, post-Cold War post-heroic warfare seems to have been working out quite well. It turned out to be amenable for the Israelis in their LICs, too.

One of the problems in conducting post-heroic LICs, however, stems from the fact that post-heroic strategy, like any other strategy, is interactive in nature. The success for the stronger side in adopting post-heroic strategy will largely depend on the reaction of its enemy. Will the other side "cooperate" with the nature of the warfare being waged against it, or will it rather stick to a heroic pattern and even compel the stronger Western democracies to abandon the post-heroic pattern? As a result of the enemy's refusal to cooperate with post-modern norms, one may find Western democracies abiding only by post-heroic warfare's first rule. In the war in Afghanistan, the level of civilian fatalities was greater than in the 1999 Kosovo War. Through 10 December 2001, the total numbers of attack sorties and weapons expended in Operation Enduring Freedom were far less than those in the 1999 Balkans campaign, Operation Allied Force, yet, the casualty rates were higher than in Kosovo. According to one source, at least 500 civilians were killed in the Kosovo war by the NATO bombardment. In Afghanistan, the bombing campaign appears to have claimed 1,000–3,500 civilian lives.[77] This happened despite the fact that nearly 60 per cent of the 14,000 missiles, bombs and other pieces of ordnance were precision-guided – steered to their targets by laser beams or satellites – compared to less than ten per cent of the bombs expended in the Gulf War.[78]

Military build-up dilemmas

The technological transition reflected by the Iraq War has created some military build-up dilemmas.

The Army's size

On the one hand, the saturation of the battlefield with precision-guided munitions has already forced modern armies to become smaller: the greater the attrition on the battlefield, the more urgent decreasing the size of formations and dispersing them have become,[79] and the greater the need to retain their survivability. The more lethal weapon systems become, the smaller numbers one needs for producing the desired effect.

Furthermore, given the rising cost of military technology, which in the developed industrial states has exceeded the general inflation rate by some three to five per cent,[80] one can save money by reducing one's forces.

In the Iraq War smaller numbers of much more sophisticated weapon systems did produce much more effective fire, did reduce the attrition of the Coalition, and did temper the growing cost of sophistication.[81] The number of US troops deployed on the battlefield was half the number of those deployed in the Gulf War (250,000 and 500,000 respectively).[82]

This reflects the "transformation" of the American military, which places greater reliance on precision strikes and special forces, with a concomitant reduced reliance on large ground forces, as well as the insistence of Secretary of Defense Donald Rumsfeld, his deputy Paul Wolfowitz, Chairman of the JCS General Richard Myers, and Commander of the forces in Iraq General Tommy Franks on a small but smart force. And indeed, the overwhelming heavy-armored offensive across a broad front of the Gulf War (often alluded to as the "Powell Doctrine") was replaced in the Iraq War by a smaller, faster-moving attack force on the ground, capable of launching precision weapons at vast distances and taking advantage of sophisticated reconnaissance systems, combined with overwhelming air power.[83]

The tendency towards smaller but smarter militaries, however, is balanced by a number of factors: first, despite the high cost of precision weapons compared to iron bombs, target destruction cost would be cheaper with precision munitions, as classical bombing missions require far more sorties and might involve heavy support by escort aircraft, electronic warfare aircraft, airborne refueling and intelligence aircraft.[84]

Second, battlefield decision is usually achieved by those who have enough boots on the ground. Third, there is a growing need for specialization, which requires the creation of units of specific destinations. Fourth, it is difficult to shift from a people's army force structure to volunteers' armies – in countries where such process is relevant, e.g. Israel. Fifth, high-tech force is usually incapable of delivering the goods in LICs, where a rapid and relatively bloodless victory is unachievable, as has been exemplified by the post-War LIC in Iraq.

The balance between platforms and systems

It has long been claimed that the systems that platforms carry (e.g. sensors, munitions, electronic devices) have become more important than the platforms themselves.[85] This particularly applies to the heavy platforms (tanks, ships and planes) that consume most of the defense acquisition budget. A counter-claim could be made against over-reliance on systems, based on the argument that even in "systems-centric" warfare platforms still matter, as it is unclear if the military would be able to achieve sufficient "situational awareness" of the battlefield to avoid all threats. Under such circumstances, robust platforms are a hedge against the inevitable friction of battle that drags against any system in wartime.[86]

Although strategists forecasted that the era of the battle tank was coming to an end,[87] one of the main lessons learned from the Iraq War was that the Army still needs a heavy-armored force in order to fight effectively against a well-equipped enemy.[88] Urban warfare in the war, especially in Baghdad, proved that armor is also important for "small" challenges in general and urban warfare in particular. On the other hand, one should not forget that the heavy platform accounted for the fact that although the size of the US force used in Iraq was smaller, it took as much time to bring the equipment to the region by ship and position the forces in preparation for the fight. Moreover, the heavy tank consumes a lot of fuel, it needs a long and huge logistic tail, and its high signature makes it vulnerable (a problem that is hardly resolved by the expensive passive and active protection means). No wonder therefore that the US Army has already cut future funding for heavy armor like the Abrams and that military planners have been looking for a replacement that will be smaller, lighter, more gas-efficient, more easily air-lifted to battlefronts, and easier to deploy.

In the air, the limitations of the tactical aircraft were exposed, particularly their inability to linger long over the battlefield, thereby lowering the risk for their pilots.[89] The poor protection of the AH-64 Apache Longbow attack helicopters from small arms and light anti-aircraft guns that was exposed during their assault on the Republican Guard, re-raised the question as to whether the era of the attack helicopter as a flying tank that enjoys firepower and mobility rather than protection has reached its end. The one time in the war Longbows went out to attack an Iraqi tank regiment, 30 out of the 34 came back all shot up. Two of these helicopters were shot down during the war.[90]

In contrast, quarter-century-old A-10 fixed-wing attack planes did a very good job. Even older, high-flying B-52 bombers, armed with relatively cheap JDAM smart bombs, did it better. A cost-effectiveness problem also exists when it comes to stealth technology. As already pointed out, the F-117 struck nearly 40 per cent of the strategic targets and remained the centerpiece of the strategic air campaign for the entire war. Yet, almost no non-stealth American planes were shot down in the past few wars and no enemy dared to put so much as a single fighter plane up in the sky. If it had, the non-stealthy F-15, F-16 and F-18 fighter planes would probably have downed them. In the future, unmanned aerial vehicles would be able to carry out a variety of missions at lower cost in terms of both money and lives.

Inter-service relationships

The dramatic strengthening of firepower has blurred the traditional division of modern armies to arms and corps. Whereas the Gulf War had been conducted as two sequential, almost separate campaigns in the air and on the ground, under joint forces command, the Iraq War was an integrated war, waged simultaneously and in synchronicity on the ground, at sea and in the air. Jointness makes the whole greater than the sum of its parts. It has penetrated into every level of war, from grand-strategy to tactics, in the spirit of Joint Vision 2020. A research and development process has already begun where the first combat plane is to be built for the Air Force, Navy and Marines, reflecting the fact that the services have been planning and fighting wars more jointly.[91]

"Jointness" has made the "bombing to win" debate, which only a decade ago still seemed timely, much less relevant. In the mid-1990s the air power school[92] offered an up-to-date version of Douhet's notion of decision from the air via grand-strategic and strategic bombing, with only a symbolic number of forces operating on the ground.[93] Today precision munitions and information warfare could serve as substitutes for mass grand-strategic bombardment. Anyway, in the Iraqi case Douhet's vision was irrelevant, as unlike in Kosovo, damaging the societal and economic infrastructure of the country was not a viable option. Rather, minimizing the disruption of civilian life and the task of post-war reconstruction of the country's economy and society were key priorities.

Terrorizing the population was likewise out of the question, since in the spirit of post-heroic approach the Iraqi people were considered victims of the evil regime and had to be spared suffering. Consequently, some 75 per cent of the targets hit by the Coalition belonged to Iraqi ground forces (15,000 strikes out of 20,000). The rest of the targets were connected to regime, communication, air superiority, SSMs, etc.,[94] and it took high tempo maneuver on the ground to seize and hold territory and achieve battlefield decision that could be translated into victory. Even had a Douhet-like strategy been feasible, it would not have produced victory, since given the nature of the Iraqi leadership, air operations, destructive as they might have been for infrastructure, were unlikely to have made it surrender.

The "ground forces school", who believed in a battlefield decision achieved by large formations of ground forces,[95] proved wrong too, as no more than a small number of ground forces was needed to achieve battlefield decision and occupy the country. So was the "intermediate school", which was predicated on the poor record of [grand] strategic bombing, advocating air–land cooperation aimed at targeting theater and tactical targets (by A-10 or F-15 aircraft).[96] Without underestimating the battle-winning effect of close air support of ground operations, one cannot ignore the fact that the fatal blow to the Iraqi war effort was caused by the capture of the Iraqi capital. And indeed, classical military thinkers such as Clausewitz and Jomini had acknowledged the effect of hitting centers of gravity in the rear long ago.[97]

Versatility

The obliteration of the borderlines between arms and corps and the declining relevance of the bombing to win debate notwithstanding, a process has been launched in which the two extremes of military force structure – air power on the one hand and infantry on the other – have gained extra importance. Aircraft and missiles of various kinds and ranges, and infantry, particularly Special Forces, have become more versatile and effective both in HICs and LICs. The combination of air power and Special Forces has become typical of US operations since the 1991 Gulf War.[98] Special Forces units had already conducted covert operations prior to the outbreak of the Iraq War to identify potential military, leadership, and command and control targets.

During the war, Coalition Special Forces units carried out ISR missions; engaged in scouting and designating targets; seized key targets ahead of the main force; hunted missile launchers; prepared airfields; secured oil fields so as to prevent the catastrophe that would result from the Iraqi plan to set them on fire; and lined up contacts and co-ordinated operations with indigenous groups. The CIA and Special Forces units were reported to have opened significant channels to senior Iraqi commanders and political leaders before the outbreak of the war.[99] As a result of the links they had made, some of the Iraqi cities were taken with almost no battle, many Iraqi units remained passive, and many soldiers were ordered to go home and save their lives.[100]

Formations

The increase in the operational effectiveness of the formations, the dramatic improvement of command and control capabilities, the penetration of the integration process into the tactical level, and the high attrition rates on the fire-saturated battlefield tend to limit the size of the basic formation the force structure is based on. At the same time, the process of integration on the battlefield, as well as the variety of challenges the military has to cope with, impose formations without pre-set structure, the components of which are determined on an *ad hoc* basis – by the tasks they are expected to carry out, the tactical context and the demands of combat – as opposed to the traditional homogenized army, in which units largely resemble one another.

The military professional

The capability to conduct war while using stand-off munitions, combined with the very low casualty toll, have accelerated two military–sociological developments on the battlefield of today: the emergence of the technician-soldier and the greater share filled by women in combat or combat-supporting missions. These two developments have been considered among the features of the post-modern army.[101] Most soldiers still conform to the traditional profile, and males still comprise most of the combatants.

The sociological landscape is gradually changing, though. More troops carry out combat or combat-supporting missions while located far from the direct battlefield, launching missiles from submarines or other remote bases. Women, who in World War II or the Vietnam War were hardly active as combatants, with representation not exceeding one per cent, comprised in the Gulf War seven per cent of the personnel involved directly or indirectly in battle. The number grew to 15 per cent in the Iraq War, ranging from a high of 19 per cent in the Air Force to six per cent in the Marine Corps.[102]

The decreasing dependency on bases overseas

Finally, one of the implications of the firepower capabilities, which impinges both on defense policy and military build-up, is a decreasing dependence on host-nation support. Before the Iraq War broke out, US military planners had assumed that bases on Turkish soil would be available for operational purposes, particularly for delivering a direct land assault against northern Iraq. Given the Turkish parliament's refusal to grant US ground forces access to Turkish territory, US thinking had to shift towards mounting a lighter and smaller attack against northern Iraq with airborne and parachuted forces.

This means, on the one hand, that the US global strategic vision can be undermined by the more parochial, regional concerns of its allies (in this case, the implications for Kurdish aspirations).[103] Of the 58 regional airbases within optimal range of the battlefield, only three were available to bomb-dropping aircraft.[104] On the other hand, a strengthened US firepower and airpower reduces its dependence on permission granted by allies located close to the theater of operations to deploy American troops on their territory. The US can now make do with fewer permanent bases abroad, and focus instead on agreements with potential host countries on an *ad hoc* use, which is compatible with its new flexible global basing approach.[105] This also means that the US might be able to project power without having to overextend its forces across the globe, which was damaging to great empires in the past. And indeed, in recent years the number of US military personnel deployed overseas has shrunk, and is now about 60 per cent of the number of ten years ago, stationed in about 1,000 military bases.[106]

The new generation of unmanned weapon systems the US is developing, in the framework of the Falcon (Force Application and Launch from the Continental US) family of weapon systems, would not only enhance its capability of waging post-heroic wars effectively and at a lower moral cost, but would also free the US from dependence on bases and allies across the globe. SLVs (Small Launch Vehicles), later to become HCV (Hypersonic Cruise Vehicles), would be launched via space, and would combine unprecedented range, speed and destructiveness – more than 10,000 miles within two hours (thanks to hyper-sonic speed), and a capability of penetrating bunkers.[107]

CONCLUSION

In this article, I have tried to shed light on the pattern of historical change in warfare. The conclusion is that the Iraq War strongly reflects a process that has emerged the late 1990s–early 2000s. A dramatic strengthening of firepower has accounted for the emergence of the "empty battlefield", which is saturated with fire but "emptier" of forces;[108] the rising importance of attrition ratios, the importance of concentration of fire at the expense of the traditional concentration of forces, and the shock and awe effect of firepower. The discontinuous transition in warfare reflected in that war, however, has been affected by the confluence of technological and societal factors. The unprecedented firepower capabilities have enabled socially constrained, casualty avert Western democracies to both enter wars and sustain them, even ones of lower stakes.

From a military build-up point of view, the following conclusions could be drawn from the Iraq War:

1. The factors pushing towards a smaller but smarter military are balanced by the need to keep the military big enough to carry out non-high-tech missions.
2. Platforms still matter.
3. Inter-service rivalry has become obsolete.
4. Jointness notwithstanding, the aircraft and the infantry enjoy greater versatility and importance.
5. Formations tend to become smaller and their tasks and missions affect their structure.
6. The role played by the technical-soldier and women in combat or combat-supporting duties is growing.
7. One final implication of the firepower capabilities demonstrated in the Iraq War is a decreasing US dependency on bases overseas.

NOTES

1. See, for example, Andrew F. Krepinevich, "Cavalry to Computer: The Pattern of Military Revolutions", *National Interest*, No. 37, Autumn 1994, pp. 30–42; Alvin and Heidi Toffler, *War and Anti-War: Survival at the Dawn of the 21st Century*, Boston: Little, Brown and Co., 1993, p. 32; Michael J. Mazarr, *The Military–Technical Revolution: A Structural Framework*, Washington DC: Center for Strategic and International Studies, 1993, pp. 16–17; Commander James R. Fitzsimonds and Commander Jan M. Van Tol, "Revolutions in Military Affairs", *Joint Force Quarterly*, No. 4, Spring 1994, pp. 24–31.
2. Anthony Cordesman, "The 'Instant Lessons' of the Iraq War", Center for Strategic International Studies (CSIS), 14 May 2003. Biddle raised a similar claim after the Gulf War and Afghanistan. Stephen Biddle, "Victory Misunderstood: What the Gulf War Tells Us About the Future of Conflict", *International Security*, Vol. 21 No. 2, Autumn 1996, pp. 139–179; Stephen Biddle, *Afghanistan and the Future of Warfare: Implications for Army and Defense Policy*, Carlisle: Strategic Studies Institute, 2002, pp. 43–49.
3. "Punctual equilibrium" means long periods of time in which nothing changes (equilibrium), that are punctuated by short, revolutionary transitions, with some forms becoming extinct and replaced by wholly new ones. Eldredge and Gould proposed the theory as a criticism of Darwinian theory of evolution. See Stephen J. Gould and Niles Eldredge, "Punctuated Equilibria: The Tempo and Mode of Evolution Reconsidered", *Paleobiology* 3, 1977, pp. 115–151. See also Richard G. Klein, *The Dawn of Human Culture*, New York: Wiley, 2002; Andre G. Frank and Barry K. Gills, *The World System: Five Hundred Years or Five Thousand?*, New York: Routledge, 1996.
4. On 1 May 2003 President Bush declared that major combat operations had ended.
5. As President Bush put it, "It's a different kind of combat mission, but, nevertheless, it's combat." Dana Milbank and Bradley Graham, "Bush Revises Views On 'Combat' in Iraq", *Washington Post*, 19 August 2003. See also William Safire, "Iraq War III", *New York Times*, 3 November 2003; General Wesley K. Clark, "Iraq: What Went Wrong?", *The New York Review of Books*, Vol. 50 No. 16, 23 October 2003, http://www.nybooks.com/articles/16650.
6. The force ratios between the Iraqi Army and the Coalition were 1.5 to 1 in Iraq's favor in manpower (i.e., 375,000 to 250,000 troops). On the Coalition's side, most of the burden on the ground fell on no more than three Army divisions and one Marine Expeditionary Force (3rd Infantry, 82nd, 101st and 1st Marine). In terms of formations, this meant that the force ratios were actually 5.7 to 1 (23 divisions versus 4) in Iraq's favor, and so were the force ratios in tanks – 4.6 to 1 (2,300 and 500, respectively). Max Boot, "The New American Way of War", *Foreign Affairs*, Vol. 82 No. 4, July/August 2003, p. 44; Isaac Ben-Israel, "The Revolution in Military Affairs and the Operation in Iraq", in Shai Feldman (ed.), *After the War in Iraq*, Tel Aviv: Jaffee Centre for Strategic Studies, 2003, p. 56.
7. According to Biddle, the causal importance of speed, precision and situation awareness has often been overestimated in the public debate on the war, whereas the causal role of the skill differential between the Coalition and the Iraqis as well as the ways in which technology can exploit that differential have been underestimated. Stephen Biddle, "Operation Iraqi Freedom: Outside Perspectives", Statement before the Committee on Armed Services, US House of Representatives, 108th Congress, http://www.house.gov/hasc/openingstatementsandpressreleases/108thcongress/03-10-21biddle.html.
8. Ben-Israel, "The Revolution in Military Affairs and the Operation in Iraq", *op. cit.*, p. 57.
9. Harlan K. Ullman, in Nina Bernstein, "The Strategists Fight a War about the War", *New York Times*, 6 April 2003.
10. John Keegan, "Saddam Was Easily Defeated, Which is Why the War Goes On", *Daily Telegraph*, 4 June 2003; Julian Thompson, "Air Power Was Devastating and Iraqi Forces Lamentable", *The Observer*, 14 April 2003; Michael Eisenstadt, "Analyzing the War's Military Dimensions", in "After the Iraq War: Roundtable Discussion", *MERIA Journal*, Vol. 7 No. 2, June 2003.

11. Edward N. Luttwak, "Where Are the Great Powers?", *Foreign Affairs*, Vol. 73 No. 4, July/August 1994, pp. 23–28; "Toward Post-Heroic Warfare", *Foreign Affairs*, Vol. 74 No. 3, May/June 1995, pp. 109–122; "A Post-Heroic Military Policy", *Foreign Affairs*, Vol. 75 No. 4, July/August 1996, pp. 33–44.
12. Andrew Marshall, "Revolutions in Military Affairs", Statement for the Subcommittee on Acquisition and Technology, Senate Armed Services Committee, 5 May 1995, p. 2.
13. Martin Van Creveld, *The Transformation of War*, New York: The Free Press, 1991.
14. The wars the US has been conducting since the early 1990s (the Gulf War, Somalia, Bosnia, Kosovo, Afghanistan), as well as the conflicts Israel has been engaged in since the late 1970s with Hizbollah in Lebanon and the Palestinians in the Territories, all fit into these categories.
15. "To the extent that attacks become increasingly accurate and lethal [. . .], and forces more distant from or dispersed on the actual battlefield, those trends would represent the continuations of developments that have been under way for centuries [. . .] They may not fundamentally change the nature of battle the way RMA proponents insist." Michael E. O'Hanlon, "Beware the RMA'nia!", paper presented at the National Defense University, 9 September 1998, http://www.brook.edu/printme.wbs?page=/pagedefs/855234f4ccc0ff3a440f9be30a141465.xml.
16. *Ibid.*
17. John Gooch, "Addition, Multiplication and Step Functions: The 'Arithmetic' of Military Revolution in Modern Time", in Stuart A. Cohen (ed.), *Democratic Societies and their Armed Forces*, London: Frank Cass, 2000, pp. 27–46.
18. Michael Howard, "The Forgotten Dimensions of Strategy", in Michael Howard, *The Causes of War and Other Essays*, Cambridge: Harvard University Press, 1984, pp. 101–115.
19. For the nuclear revolution, see for example Bernard Brodie *et al.* (eds), *The Absolute Weapon*, New York: Harcourt Brace, 1946; Robert Jervis, *The Meaning of Nuclear Revolutions*, Ithaca: Cornell University Press, 1989; Michael Mandelbaum, *The Nuclear Revolution*, Cambridge: Cambridge University Press, 1981; Lawrence Freedman, *The Evolution of Nuclear Strategy*, New York: St Martin's Press, 1989.
20. See for example Michael O'Hanlon, *Technological Change and the Future of War*, Washington DC: Brookings Institution, 2000, http://brookings.nap.edu/books/0815764391/html/1.html.
21. Self-guiding, accurate bombs intended to maximize damage to the target while minimizing collateral damage.
22. Ben-Israel, "The Revolution in Military Affairs and the Operation in Iraq", *op. cit.*, p. 64.
23. Ben-Israel, "The Revolution in Military Affairs and the Operation in Iraq", *op. cit.*, p. 64; *Chicago Tribune*, Internet edition, 12 February 2002.
24. 60 per cent according to Borchgrave, "War by Remote Control", *Washington Times*, 28 April 2003; 65 per cent according to Cordesman, "Instant Lessons", p. 18; 68 per cent according to Ben-Israel, "The Revolution in Military Affairs and the Operation in Iraq", *op. cit.*, p. 64.
25. Ben-Israel, "The Revolution in Military Affairs and the Operation in Iraq", *op. cit.*, p. 64.
26. Baron de Jomini, *The Art of War*, Westport: Greenwood, 1971, p. 270.
27. Nineteen minutes in Afghanistan, see Fred Kaplan, "High-Tech US Arsenal Proves Its Worth", *The Boston Globe*, 9 December 2001; 38 minutes in Iraq, see General Richard B. Myers, "The New American Way of War", Remarks before the Navy League's Sea Air Space Exposition, Washington DC, 16 April 2003, http://www.dtic.mil/jcs/chairman/new_american_way_of_war_16apr03.htm.
28. The goals set out for the US military forces a decade into the twenty-first century by former Secretary of Defense Cohen and former Chairman of the JCS Shalikashvili, were that they possess "dominant battlefield knowledge", "full dimensional protection", "dominant maneuver" and "precision strike" ability from long distances. Steven Metz, *Armed Conflict in the 21st Century: The Information Revolution and Post-Modern Warfare*, Carlisle: Strategic Studies Institute, April 2000, pp. 31–33; General John Shalikashvili, *Joint Vision 2010*,

Washington DC: Department of Defense, 1996; O'Hanlon, *Technological Change and the Future of War*; Gordon R. Sullivan and James M. Dubik, "War in the Information Age", *Military Review*, April 1994, pp. 55–56.

29. http://www.cbc.ca/news/iraq/issues_analysis/casualties_postiraqwar.htm; l; Cordesman, "Instant Lessons", p. 140.
30. Victor David Hanson, "Lessons of the War", *Commentary*, Vol. 115 No. 6, June 2003.
31. http://www.cbc.ca/news/iraq/issues_analysis/casualties_postiraqwar.html; http://www.businessweek.com/bwdaily/dnflash/feb2003/nf2003026_0167_db052.htm.
32. In the 1973 October War, by the time the Egyptians admitted defeat, they had lost some 40 per cent of their armor. Avi Kober, *Battlefield Decision: Military Decision in the Arab–Israeli Wars 1948–1982*, Tel Aviv: Ma'arachot, 1995 [Hebrew], pp. 437–440. In the Gulf War, the Iraqi armor's attrition rates in that stage of the war were 50 per cent. International Institute for Strategic Studies, *Strategic Survey 1990–1991*, London: Brassey's, 1991, p. 30; JCSS Study Group, *War in the Gulf: Implications for Israel*, Tel Aviv: Jaffee Center for Strategic Studies, 1992, p. 146. According to DOD's estimates, the Iraqis lost 40 per cent and 90 per cent of their armor in the respective stages. *Conduct of the Persian Gulf War*, Washington DC: Department of Defense, April 1992, pp. 188, 411. Attrition rates of more than 50 per cent and 76 per cent, respectively, were estimated by Keany and Cohen. Thomas A. Keaney and Eliot A. Cohen, *Revolution in Warfare? Air Power in the Persian Gulf*, Annapolis: Naval Institute Press, 1995, chapter 3. In both cases this happened after the enemy had been engaged on the ground.
33. Michael O'Hanlon, "A Remarkable Run to Baghdad", *The Baltimore Sun*, 4 April 2003.
34. See for example, John J. Mearsheimer, "The Military Reform Movement", *Orbis*, Vol. 27 No. 2, Summer 1983, pp. 285–300.
35. Richard M. Ogorkiewicz, "Successors of Today's Tank", *International Defence Review*, Vol. 24 No. 9, 1991, pp. 965–972; Will Fowler, "The Tank is Dead, Long Live the LAV [Light Armored Vehicle]", *Defense and Diplomacy*, Vol. 8 No. 6, 1990, pp. 44–46.
36. Lawrence Freedman, "Lessons from the Bombing of Baghdad, 1991", *The Independent*, 21 March 2003.
37. *Newsweek* Report, usgovinfo.about.com/library/weekly/aaebomb/htm. For the notion of "shock and awe", see Harlan K. Ullman and James P. Wade, *Shock and Awe: Achieving Rapid Dominance*, Washington DC: NDU, December 1996.
38. Christopher Bellamy, "In Scale, Scope and Sweep, This Dwarfs Any Other War Operation", *The Independent*, 22 March 2003.
39. An attempt to develop major tests to be applied when the US is weighing the use of its combat forces abroad, particularly when its vital interests are not involved, was made by Secretary of Defense Weinberger. For a discussion of the doctrine, see Michael I. Handel, *Masters of War*, London: Frank Cass, 2000, pp. 307–325. For a recent study suggesting that the West will not tolerate extensive casualties for relatively unimportant objectives, see Eric V. Larson, *Casualties and Consensus: The Historical Role of Casualties in Domestic Support for US Military Operations*, Santa Monica: Rand, 1996. There are cases, though, wherein it is the politicians, distrustful of their own voters, who would not tolerate casualties, as with US withdrawal from Somalia or Israeli withdrawal from Lebanon. For Somalia, see Steven Kull and I. M. Destler, *Misreading the Public: The Myth of New Isolationalism*, Washington DC: Brookings Institution, 1999. As for Lebanon, in 1998, before Barak was elected Prime Minister, 64 per cent of the Israelis agreed that the security zone in Lebanon made a positive contribution to Israel's security and kept terror out of northern settlements. Asher Arian, *Israeli Public Opinion on National Security 1998*, Tel Aviv: Jaffee Center for Strategic Studies, July 1998, p. 26.
40. For a discussion related to the Iraq War, see Edward Luttwak, "This Is War – It Is No Time for Public Relations", http://www.telegraph.co.uk/opinion/main.jhtml?xml=/opinion/2003/03/30/do3002.xml&sSheet=/opinion/2003/03/30/ixopinion.html.
41. Edward N. Luttwak, "Where Are the Great Powers?", *Foreign Affairs*, Vol. 73 No. 4,

July/August 1994, pp. 23–28; "Toward Post-Heroic Warfare", *Foreign Affairs*, Vol. 74 No. 3, May/June 1995, pp. 109–122; "A Post-Heroic Military Policy", *Foreign Affairs*, Vol. 75 No. 4, July/August 1996, pp. 33–44.

42. Keaney and Cohen, *Revolution in Warfare?*, *op. cit.*, p. 225. And indeed, one cannot ignore the impact of the massive destruction of Iraqi forces retreating from Kuwait to Iraq, along the "Highway of Death", from the air, during the Gulf War on the American decision to stop the war, although it did not deny the Americans battlefield decision. H. Norman Schwarzkopf, *It Doesn't Take a Hero*, New York: Bentam Books, 1992, p. 468.
43. In his St Louis speech of 16 April he said: "In this new era of warfare, we can target a regime, not a nation. Our aim is to track and strike the guilty." Jim Hoagland, "Iraq War Throws Up a New Morality Debate", *The Straits Times Interactive*, 28 April 2003.
44. Victor Davis Hanson, "Postmodern War", *National Review Online*, 15 April 2003.
45. See, for example, Charles Krauthammer, "Gulf War II is First of its Kind", *Washington Post*, 10 April 2003.
46. Fred Kaplan, "Bombing by Numbers", http://slate.msn.com/id/2083605/.
47. The percentage calculus is based on the following numbers: Gulf War: 13,000 out of 19 million Iraqis; Kosovo: 500 out of 10 million Serbs; Afghanistan: 3,500 out of 28 million Afghanis; Iraq War: 3,750 out of 26 million. Karl Schoenberger, "Civilian Casualties will Erode Support, Inflame Hatred Among Iraqis", *KRT International*, 27 March 2003"; Peter Ford, "Surveys Pointing to High Civilian Death Toll in Iraq", *Christian Science Monitor*, 22 May 2003; *Reuters*, 9 July 2003; Carl Conetta, "The Wages of War: Iraqi Combatant and Noncombatant Fatalities in the 2003 Conflict", Project on Defense Alternatives, *Research Monograph* #8, 20 October 2003; http://www.cbc.ca/news/iraq/issues_analysis/casualties_postiraqwar.html, http://www.comw.org/pda/0310rm8.html.
48. Krauthammer, "Gulf War II is First of its Kind", *op. cit.*
49. Hanson, "Postmodern War", *op. cit.*
50. Keaney and Cohen, *Revolution in Warfare?*, *op. cit.*, p. 214.
51. This data is based on US Air Force reports. Shlomo Gazit, "The Gulf War: Main Political and Military Developments", in JCSS Study Group, *War in the Gulf: Implications for Israel*, Tel Aviv: Jaffee Center for Strategic Studies, 1992, p. 33.
52. http://www.cbc.ca/news/iraq/issues_analysis/casualties_postiraqwar.html.
53. Amy Goldstein, Jonathan Weisman and Margot Williams, "Casualties: Low Number, Many Causes", *Washington Post*, 13 April 2003.
54. Cordesman, "Instant Lessons", *op. cit.*, p. 54.
55. *Ibid.*, p. 54.
56. Gordon, "US Attacked Iraqi Defenses Starting in 2002", *op. cit.*
57. Christopher Bellamy, "In Scale, Scope and Sweep, this Dwarfs any Other War Operation", *Independent*, 22 March 2003.
58. Keany and Cohen, *Revolution in Warfare?*, *op. cit.*
59. *Ibid.*
60. Kaplan, "Bombing by Numbers", *op. cit.*
61. R. L. DiNardo and Daniel J. Hughes, "Some Cautionary Thoughts on Information Warfare", *Airpower Journal*, Winter 1995, pp. 69–79.
62. On the so-called Revolution in Military Affairs (RMA), see for example Paul Bracken and Raoul Henri Alcalá, *Whither the RMA: Two Perspectives on Tomorrow's Army*, Carlisle Barracks, PA: Strategic Studies Institute, US Army War College, 1994; Eliot A. Cohen, "A Revolution in Warfare", *Foreign Affairs*, Vol. 75 No. 2, March/April 1996, pp. 37–54; Lawrence Freedman, *The Revolution in Strategic Affairs*, Adelphi Paper 318, London: IISS, 1998; O'Hanlon, *Technological Change and the Future of Warfare*; Steven Metz, "The Next Twist of the RMA", *Parameters*, Autumn 2000, pp. 40–53.
63. Anthony Kimery, "Inviting the Electronical Pearl Harbor", *Military Information Technology*, Vol. 2 No. 4, 1999, pp. 8–12.
64. Sun Tzu, *The Art of War*, *op. cit.*, p. 77.

65. B. H. Liddell Hart, *Strategy*, New York: Praeger, 1967, p. 338.
66. See, for example, Andrew Rathmell, "Strategic Information Warfare: Responding to the Threat", in *Brassey's Defence Yearbook 1998*, London: Brassey's, 1998, pp. 310–324.
67. Bradley A. Thayer, "The Political Effects of Information Warfare: Why New Military Capabilities Cause Old Political Dangers", *Security Studies*, Vol. 10 No. 1, Autumn 2000, pp. 43–85.
68. See, for example, *Information Warfare: A Strategy for Peace . . . The Decisive Edge in War*, Washington DC: The Joint Staff, p. 2.
69. For an analysis of war in the information age as reflected in the Gulf War, see Keaney and Cohen, *Revolution in Warfare?*, *op. cit.*, pp. 216–220.
70. On this, see for example, http://www.usgovinfo.about.com/library/weekly/aaebomb.htm.
71. "E-Bomb Aims to Stun Iraqi Forces", *Associated Press*, 19 March 2003, http://www.wired.com/news/conflict/0,2100,58122,00.html.
72. *Concept for Nonlethal Capabilities in Army Operations*, Fort Monroe, VA: Department of the Army, December 1996.
73. *Ibid.*
74. Steve Wright, "Hypocrisy of 'Non-Lethal' Arms", *Le Monde Diplomatique*, 9 December 1999.
75. No wonder, therefore, that researchers and practitioners have been looking for the recipe for victory in wars of this type. Ivan Arreguin-Taft, "How the Weak Win Wars: A Theory of Asymmetric Conflict", *International Security*, Vol. 26 No. 1, Summer 2001, pp. 93–128.
76. "In planning Operation Desert Storm, minimizing allied and civilian casualties was the highest priority", asserted a publication on the lessons of the Gulf War, issued by the House of Representatives' Committee on Armed Services. *Defense For a New Era: Lessons of the Persian Gulf War*, Washington DC: US Government Printing Office, March 1992, p. 88. The US Army Field Manual of 1993 assumed that, alongside the expectation of seeing a decisive victory achieved by US troops, it is the public's loss-aversion that determines its attitude towards the military operations. *FM 100-5: Operations*, Washington DC: Department of the Army, 1993, 1–3. On the presumption that the US must end future military conflict quickly and at minimum cost, see also Andrew P. N. Erdmann, "The US Presumption of Quick, Costless Wars", *Orbis*, Vol. 43 No. 3, Summer 1999, pp. 363–381.
77. Carl Cornetta, "Operation Enduring Freedom: Why a Higher Rate of Civilian Bombing Casualties", Project on Defense Alternatives (PDA), Briefing Report No. 11, 24 January 2002, http://www.comw.org/pda/0201oef.html. According to another source, they claimed the lives of 3,100–3,600 civilians. Marc W. Herold, "The Bombing of Afghanistan as Reflection of 9/11 and Different Valuations of Life", http://www.cursor.org/stories/heroldon911.htm.
78. See Note 23 above.
79. Henry S. Rowen, *Intelligent Weapons: Implications for Offense and Defense*, Tel-Aviv: Jaffee Center for Strategic Studies, 1988, p. 19; Lance Davidson, "The Impact of Precision Guided Munitions on War", *Brassey's Yearbook 1984*, London: Brassey's, 1984; Eliot A. Cohen, "A Revolution in Warfare", *Foreign Affairs*, Vol. 75 No. 2, March/April 1996, p. 45.
80. Yaacov Lifshitz, "Managing Security Resources in the 2000s", in Haggai Golan, *Israel's Security Web*, Tel-Aviv: Ma'arachot, [Hebrew], 2001, p. 51.
81. James Fergusson, "The Defense Firm in a Changing Politico-Strategic Environment", in Andrew Latham and Nicholas Hooper (eds), *The Future of the Defense Firm: New Challenges, New Directions*, Kluwer: Dordrecht, 1995, pp. 23–35.
82. Boot, "The New American Way of War", *op. cit.*, p. 43.
83. Seymour M. Hersh, "Offense and Defense", *The New Yorker*, 7 April 2003. Critics belonging to the "ground forces school", such as Army Chief-of-Staff General Eric Shinseki or retired General Barry M. McCaffrey, have blamed senior civilian Defense Department officials for holding the view that numbers do not count, although basically some Army senior officers like Shinseki had supported the concept of "a land force more agile and

precise in bringing lethal force to the battlefield". Bernard Weinraub with Thom Shanker, "Rumsfeld's Design for War Criticized on the Battlefield", *New York Times*, 1 April 2003. They have argued that the US entered the war with an unbalanced ground combat force and that the forces it deployed were not sufficient for both decisive operations and occupation of the country. For Shinseki's view, see Weinraub with Thom Shanker, "Rumsfeld's Design for War Criticized on the Battlefield"; for McCaffrey's view, see Vernon Loeb, "Rumsfeld Faulted for Troop Dilution", *Washington Post*, 30 March 2003; Nina Bernstein, "The Strategists Fight a War about the War", *New York Times*, 6 April 2003.

84. Zeev Bonen, "Technology in War: Preliminary Lessons from the Gulf War", in JCSS Study Group, *War in the Gulf*, p. 171.
85. Cohen, "A Revolution in Warfare", *op. cit.*, p. 45.
86. James J. Carafano, "After Iraq: Learning the War's Lessons", The Heritage Foundation Policy Research and Analysis, http://www.heritage.org/research/MiddleEast/bg1664.cfm.
87. See for example, Louis Lavoie, "Is the Tank Dead?", *Journal of Defense and Diplomacy*, Vol. 7 No. 5, May 1989, pp. 16–23. According to Lavoie, the tank is on its way out, along with the combat helicopter, the close air support fighter and armored combat vehicles in general.
88. Richard M. Ogorkiewicz, "Interview with Israel's Major General Tal", *International Defence Review*, Vol. 24 No. 9, 1991, pp. 97–98; Lance Gay, "Battle Tank Still Rolling", http://www.globalsecurity.org/org/news/2003/030417-battletank01.htm.
89. Carafano, "After Iraq: Learning the War's Lessons", *op. cit.*
90. Rowan Scarborough, "Apache Operations: A Lesson in Defeat", *Washington Times*, 22 April 2003.
91. Kaplan, "Bombing by Numbers", *op. cit.*
92. John A. Warden III, "Success in Modern War: A Response to Robert A. Pape's *Bombing to Win*", *Security Studies*, Vol. 7 No. 2, Winter 1997/1998, pp. 172–190.
93. Giulio Douhet, *The Command of the Air*, London: Faber, 1943.
94. Michael R. Gordon, "US Attacked Iraqi Defenses Starting in 2002", *New York Times*, 19 July 2003; Ben-Israel, "The Revolution in Military Affairs and the Operation in Iraq", *op. cit.*, p. 65.
95. Against the backdrop of the role played by the KLA in Kosovo, though without much success, some Army officials preferred to interpret Kosovo as "a confirmation, of sorts, of our joint doctrine, which calls for using air and ground forces together". Bradley Graham, "Air Vs. Ground: The Fight Is On: Air Force, Army Already Battling for Lead Roles in Future Wars", *Washington Post*, 22 June 1999. At the time, even the staunchest air-power advocates in the US were not ready to disband the Army or do away with traditional notions of applying overwhelming force. For the role played by the KLA, see Barry R. Posen, "The War for Kosovo: Serbia's Political-Military Strategy", *International Security*, Vol. 24 No. 4, Spring 2000, pp. 62–66. For a skeptical view of KLA's contribution to NATO's war effort, see Daniel L. Byman and Matthew C. Waxman, "Kosovo and the Great Air Power Debate", *International Security*, Vol. 24 No. 4, Spring 2000, pp. 28–31.
96. Robert A. Pape, *Bombing to Win: Air Power and Coercion in War*, Ithaca: Cornell University Press, 1996; Robert A. Pape, "The Limits of Precision-Guided Air Power", *Security Studies*, Vol. 7 No. 2, Winter 1997/1998, pp. 93–114.
97. Carl von Clausewitz, *On War*, Princeton: Princeton University Press, 1984, p. 596; Henri Jomini, *The Art of War*, Westport: Greenwood, 1971, pp. 87–88.
98. Cordesman, "Instant Lessons", *op. cit.*, p.173.
99. Douglas Jehl with Dexter Filkins, "US Moved to Undermine Iraqi Military Before War", *Washington Post*, 10 August 2003.
100. Richard Bennett, "A War Fought on a Multitude of Levels", www.informationclearinghouse.info/article2292,htm.
101. Charles Moskos, "Toward a Postmodern Military: The United States as a Paradigm", in Charles C. Moskos, John A. Williams and David R. Segal (eds), *The Postmodern Military*, New York: Oxford University Press, 2000, pp. 14–31.

102. Susan Baer, "In Iraq War Women Would Serve Closer to Front Lines than in Past", *Baltimore Sun*, 3 March 2003, www.bayarea.com/m/d/cctimes/news/special_packages/iraq/5352553.htm; Cordesman, "Instant Lessons", *op. cit.*, p. 328; Shauna Curphey, "1 in 7 US Military Personnel in Iraq is Female", *Women's News*, 22 March 2003.
103. Bill Park, "Strategic Location, Political Dislocation: Turkey, the United States, and Northern Iraq", *MERIA Journal*, Vol. 7 No. 2, June 2003.
104. Carafano, "After Iraq: Learning the War's Lessons", Heritage Foundation Backgrounder, 3 July 2003.
105. Kim Berger, "US Moves Towards More Flexible Global Basing", *Jane's Defence Weekly* 7, May 2003, pp. 2–3.
106. Ian Traynor, "How American Power Girds the Globe with a Ring of Steel", *The Guardian*, 21 April 2003.
107. Julian Borger, "US Planning Global-Reach Missiles: Allies Not Required", *Guardian Weekly*, http://www.guardian.co.uk/guardianweekly/story/0,12674,989624,00.html.
108. Chris Bellamy, *The Future of Land Warfare*, New York: St Martin's Press, 1987, pp. 274–277.

ACKNOWLEDGEMENT

I wish to thank Stuart Cohen, Moshe Grundman, Efraim Inbar, Benny Miller and Bradley Thayer for their valuable comments on previous versions of this paper.

Part III
Insurgencies

Yemen

[18]

A Regular Army in Counterinsurgency Operations: Egypt in North Yemen, 1962–1967

☆

David M. Witty*

IN October 1962 the Egyptian army intervened in a civil war in North Yemen to protect a newly established revolutionary government against opponents who were seeking to reinstate the country's traditional, oligarchic government. Egyptian political and military leaders expected to achieve their goal in North Yemen in three months, but they soon found themselves fighting tribal insurgents in a long war that they neither were prepared for nor fully understood. Five years later, when the Egyptians were forced to withdraw from the country as a result of their total defeat by Israel in the 1967 Six-Day War, they appeared no closer to victory in the murky war in Yemen than they had been in 1962.

Egyptian performance in Yemen illustrates the problems that challenge a regular army in a low intensity conflict, a topic that is increasingly relevant for armies today in view of growing commitments in unconventional military situations. Among these problems are the adaptations a regular army must make to conduct counterinsurgency operations, the effect that an intervening force has on the credibility of the government it is supporting, the importance of cultural understanding, the role of superior technology in fighting guerrillas, and the impact of a prolonged counterinsurgency campaign on a regular army's ability to fight a conventional war. The Egyptian experience in Yemen affirms, like the U.S. and Soviet experiences in Vietnam and Afghanistan, that a regular army must make fundamental changes to its conventional doctrine,

* For assistance with this essay, I would like to thank Dr. Thomas Naff, University of Pennsylvania, and Dr. George W. Gawrych, Combat Studies Institute, United States Army Command and General Staff College.

tactics, and procedures to be successful in counterinsurgency. A regular army's prolonged exposure to counterinsurgency operations, whether successful or unsuccessful, can negatively affect its ability to fight a conventional war.

This article is based in large measure on the memoirs and other writings of some of the prominent Egyptian military leaders of the period who have critically analyzed the Egyptian army's performance in Yemen. The most notable of these are General Salah al-Din al-Hadidi, director of Egyptian military intelligence at the beginning of the Yemeni intervention; General Muhammad Fawzi, chief of staff of the Egyptian armed forces from 1964 to 1967; General 'Abd al-Moshin Kamil Murtaji, commander of Egyptian forces in Yemen from 1963 to 1964, and later responsible for Egyptian military and political policy in Yemen; General 'Abd al-Mun'im Khalil, the first chief of operations in Yemen, and a battalion and military district commander there; and Amin Huwaydi, Egyptian minister of war from 1967 to 1968.

Causes of the Egyptian Intervention

On 26 September 1962 North Yemeni army officers mounted a revolutionary coup against the traditional government of Imam Muhammed al-Badr, triggering a civil war between revolutionary forces desiring to safeguard the newly declared republic and tribal forces hoping to restore the old order. Before the coup, North Yemen was a conservative and isolated country ruled by an imam of the Zaydi Islamic sect, a secular and religious leader who strictly enforced Islamic law. Zaydism, the predominant sect of Islam in Yemen, is a form of Shi'ism representing a compromise between the Sunni and Shi'a, the major divisions within Islam. The Zaydis have been loosely called the "Protestants of Islam" and place a special emphasis on their ruler, the imam, a descendant of the Prophet Muhammad.[1]

North Yemen's population was estimated at 4.5 million in the early 1960s, although no formal census had ever been taken. The capital, Sana, located in the center of the country, had a population of about 60,000. Hodeidah, situated on the Red Sea, was the main port, with a

1. W. Montgomery Watt, *Islamic Political Thought: The Basic Concepts* (Edinburgh: Edinburgh University Press, 1964), 115. For general works on the history of North Yemen leading up to the Egyptian intervention, see Harold Ingrams, *The Yemen: Imams, Rulers, and Revolutions* (New York: Praeger, 1964); Robert W. Stookey, *Yemen: The Politics of the Yemen Arab Republic* (Boulder, Colo.: Westview Press, 1978); Manfred W. Wenner, *The Yemen Arab Republic: Development and Change in an Ancient Land* (Boulder, Colo.: Westview Press, 1991); and Mohammed Ahmad Zabarah, *Yemen: Traditionalism vs. Modernity* (New York: Praeger, 1982).

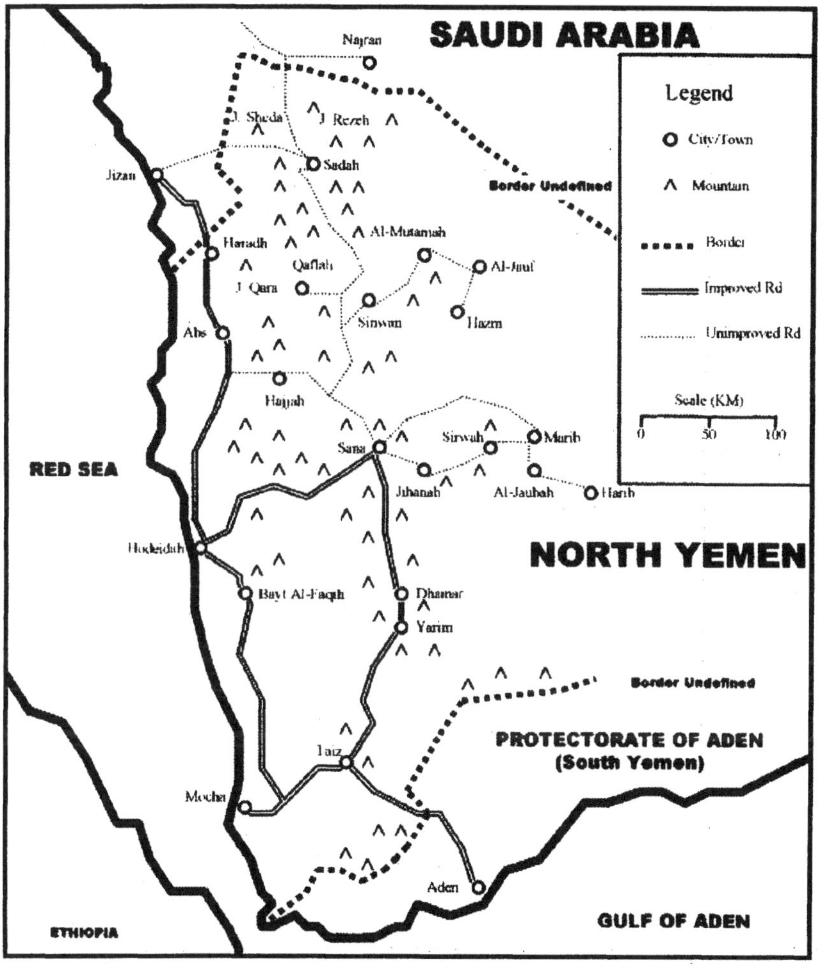

Map 1: North Yemen, 1962.

population of approximately 35,000. Two other important towns were Taiz, the country's main commercial city, in the southwest, with a population of 20,000, and Sadah, a Zaydi stronghold, in the north close to the Saudi border. (See Map 1.)

Urban dwellers, however, represented a small portion of the overall population. Most people belonged to rural or mountain tribes, which were the basic social units in Yemen and tended to be xenophobic and warlike. Of the approximately thirty-five hundred tribes and sub-tribes in Yemen, many had longstanding blood feuds with each other. Further complicating the situation were the country's religious sects. The Zaydi sect of Shiites in Yemen made up four-fifths of the tribal population. Most of them lived in the northern, mountainous half of North Yemen. The Zaydi tribes normally supported the imam but not always out of devotion to Zaydism; a weak imam could expect little backing. Ultimately, tribal support was decided by local issues and ongoing feuds. The Zaydis and their imam were traditionally the strongest force in Yemen and ruled the country except during times of foreign occupation.

Some Yemenis, however, belonged to the Shafi'i school of Sunni Islam. They lived in the south, in urban areas, and along the coast. The Shafi'is were peaceful and open to outside influences. They did not recognize the religious authority of the imam, although they did accept him as a political ruler. The Zaydis limited the activities of the Shafi'is, the merchant class of Yemen, in order to maintain the country's isolation, leading to Shafi'i resentment and tension between the sects.

After World War II, the Imamate, facing internal and external criticism, allowed some foreign influences to penetrate the country. North Yemen became a member of the Arab League and the United Nations, and in 1954 it signed a defense pact with Egypt allowing fourteen Egyptian advisors to train the Yemeni army. In 1958, North Yemen joined the union between Egypt and Syria known as the United Arab Republic (UAR). For Yemen, the union was restricted to defense and foreign policy cooperation because Imam Ahmad ibn Yahya, who ruled North Yemen from 1948 to 1962, wanted no administrative union with Egypt. By joining the UAR, Imam Ahmad hoped to guarantee North Yemen's autonomy from the British Protectorate of Aden in South Yemen, the last British stronghold in the Middle East by that time. But the imam also hoped to preempt Egyptian President Gamal Abdul Nasser from targeting North Yemen with his ideas of Arab revolution and Arab socialism.[2]

Yemen's relationship with Egypt had the opposite effect from what Imam Ahmad had intended. Egypt's revolutionary message calling for the overthrow of conservative Arab monarchies spread to North Yemen through Egyptian soldiers, teachers, doctors, and office workers who came there to work. In 1961, some four hundred to five hundred Yemenis enrolled in Egyptian schools were being exposed to the propaganda

2. Saeed M. Badeeb, *The Saudi-Egyptian Conflict over North Yemen, 1962–1970* (Boulder, Colo.: Westview Press, 1986), 20. Badeeb discusses the events leading to the Egyptian intervention and the war in North Yemen from a Saudi viewpoint. Ali A. Rahmy, *The Egyptian Policy in the Arab World: Intervention in Yemen: 1962–1967, a Case Study* (Washington, D.C.: University Press of America, 1983), 56–57, 60–61. Rahmy was an Egyptian army officer who served in North Yemen during the intervention. For general works concerning Egypt's foreign policy and Nasser, see Arthur Goldschmidt, Jr., *Modern Egypt: The Formation of a Nation State* (Boulder, Colo.: Westview Press, 1988), and Peter Woodward, *Nasser* (New York: Longman, 1992).

of the Arab revolution. Yemeni army officers returning from studies abroad realized the backwardness of their country and became the principal opposition to the imam. Also, many Yemenis had lost respect for the Imamate due to its oppressive rule and the imam's favoritism towards the Zaydi aristocracy.

Imam Ahmad, realizing the negative impact of the union with Egypt on his regime, came to regard the socialist ideology of Egypt as anti-Islamic and openly criticized Nasser. In December 1961 Egypt announced that the union between Egypt and Yemen was over, and Nasser began to attack the imam through Radio Cairo's propaganda programs broadcast throughout the Arab world. Nasser now aimed to convert North Yemen into an Arab revolutionary republic similar to the regime in Egypt. He allowed exiled Yemenis living in Egypt to call openly for the end of the Imamate.

President Nasser saw himself as the leader of the Arab world. Under Nasser, Egypt's foreign policy initially dealt with expelling imperialistic forces and influences from the Middle East, but later it concentrated on creating unity to enable Arab nations to work together to oppose imperialism. The creation of the UAR, the formal union between Egypt and Syria, was the high water mark of Nasser's quest for Arab unity. Although Egyptian arrogance towards the Syrians basically caused the UAR's breakup in 1961, Nasser blamed the demise of the UAR on the forces of Arab reaction, represented by conservative Arab kingdoms such as Saudi Arabia and Jordan. After the UAR dissolved, Nasser shifted his foreign policy to converting these Arab monarchies into revolutionary, republican governments based on the Egyptian model. The rulers of these countries naturally opposed Nasser's Arab revolution, and Saudi Arabia became committed to keeping all Nasserite influences out of the Arabian Peninsula, fearing that they could lead to internal dissension within the kingdom.

Imam Ahmad died of natural causes on 19 September 1962. His son Al-Badr, a progressive, succeeded him and announced a series of reforms, but he was overthrown by a coup on 26 September. The successful coup plotters, led by Colonel Abdullah Sallal, chief of staff of the North Yemeni army, formed a revolutionary Arab republic with Sallal as president. Sallal had intended to kill Al-Badr in the coup, but the imam escaped to the north, rallied the Zaydi tribes to oppose the revolution, and triggered a civil war in North Yemen that would last until 1970. Shortly after the coup, President Sallal asked for Egyptian intervention to protect the new republic from the imam's forces, since Sallal's faction in the Yemini army was weak and no tribes initially appeared to support the republic.[3]

3. There may have been direct Egyptian involvement in the coup. Imam Al-Badr claimed to have seen Egyptian helicopters landing near the Sana airport and the royal

Nasser felt compelled to intervene in Yemen since its revolution was the type he had advocated. A successful Arab republic in Yemen would help Nasser to reassert his influence in the Arab world, which was waning after the breakup of the UAR. Its establishment was also a possible way to spread the revolution to Saudi Arabia or to expel the British from South Yemen. On 29 September Anwar Sadat, speaker of the Egyptian National Assembly, and four operations officers traveled to Yemen to gather information. They returned after twenty-four hours, urging Egyptian intervention to defend the new republic. Nasser agreed, marking the beginning of an Egyptian military involvement in Yemen that would last until 1967. In early October 1962, the first one hundred Egyptian troops arrived in Yemen. At the end of the year, there would be fifteen thousand Egyptian soldiers in Yemen.[4]

Saudi Arabia, Jordan, and Great Britain assisted Imam Al-Badr in his efforts to destroy the new republic because they wished to prevent the spread of Nasser's Arab revolution. Saudi Arabia especially could not tolerate an Egyptian presence in North Yemen. The majority of the imam's external support would come from the Saudis in the form of money, weapons, training, and safe havens inside the Saudi border. (See Map 2)

The Insurgent War in North Yemen

The civil war in Yemen was fought between the supporters and opponents of the revolution. The backers of the revolution were known as the

palace on the morning following the coup. He speculated that they had been launched from the Egyptian embassy or an Egyptian ship. See Badeeb, *Saudi-Egyptian Conflict*, 34–35, 118–19. Before the coup, the opposition in Yemen and Egyptian officials held meetings during which Egypt agreed to protect the new republic after it was established. See Abd al-Ghani Mutahhar, *Yauma Walada al-Yaman Majadahu: Dhikrayat 'an Thaorat Sibtambir 1962* (The Day When Yemen Gave Birth to its Magnificence: Memoirs of the September 1962 Revolution), 2d ed. (Sana: Dar Nobar lil-Tiba'a, 1990), 109–22. Mutahhar was one of the organizers of the September 1962 revolution in North Yemen and a member of the Council of Revolutionary Leadership in the Republican government which was established following the coup.

4. Muhammad Fawzi, *Harb al-Thalath Sanawat, 1967–1970* (The Three Years' War, 1967–1970) (Beirut: Dar al-Wahda, 1983), 22; Salah al-Din al-Hadidi, *Shahid 'ala Harb al-Yaman* (Witness to the Yemeni War) (Al-Qahirah: Maktabat Madbuli, 1984), 21–22; 38; and 'Abd al-Mun'im Khalil, *Hurub Misr al-Mu'asirah fi Awraq Qa'id Maydani: 1939–1945, 1948–1949, 1956, 1962–1967, 1968–1970, 1973* (The Wars of Modern Egypt in the Writings of a Battlefield Commander) (Al-Qahirah: Dar al-Mustaqbal al-'Arabi, 1990), 55. Khalil became the commander of the Egyptian Second Army during the 1973 War with Israel. Mahmud 'Adil Ahmad, *Dhikrayat Harb al-Yaman: 1962–1967* (Memories of the War in Yemen: 1962–1967) (Al-Qahirah: Mataba'at al-Akhwa, 1992), 290. Ahmad was another Egyptian officer who served in North Yemen during the intervention.

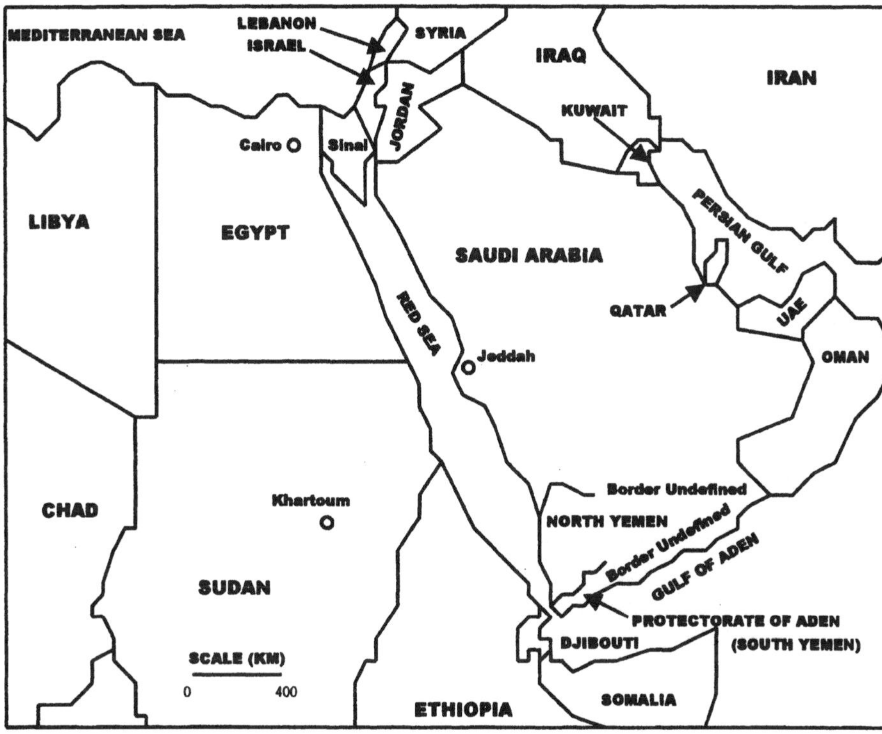

Map 2. Egypt, North Yemen, and the region.

Republicans and included the intellectuals, merchants, urban dwellers, and most of the Shafi'is. Opposing them were the Royalists, who consisted mainly of the Zaydi tribesmen and the imam's extended family. Although most Royalists were Zaydis, and most Republicans were Shafi'is, the war cannot be viewed as being fought along strictly sectarian lines. Sides constantly shifted, especially among the tribesmen, who changed loyalties depending on which faction was being supported by one of their traditional tribal enemies. On the whole, it was often impossible to tell Royalist from Republican.

The war in Yemen was a civil war, but in more defined military terms, it was an insurgency, an armed effort to overthrow an existing government. Insurgent warfare differs from conventional warfare, which implies an open state of hostility between nations and leads to a direct confrontation of their regular armies. In insurgent warfare, the insurgents use guerrilla tactics, which, unlike conventional tactics, are not designed to win quickly but rather to stave off defeat. The guerrillas operate in small bands and avoid confronting the counterinsurgency forces, which usually possess superior firepower. Guerrilla forces fight only when they have an advantage and then withdraw without attempting to hold terrain.[5]

Regular armies are designed, trained, and equipped to fight other regular armies because conventional warfare usually presents the greatest threat to national survival. In counterinsurgency warfare, a regular army must make changes to its organization and tactics since regular forces normally seek to concentrate in time and space to obtain decisive victory quickly, while guerrillas take the opposite approach, dispersing in time and space to avoid defeat. In short, insurgents can win by not losing since they gain credibility and the support of the population by avoiding defeat in the face of a superior force. Regular forces lose when they do not win; they have great superiority but still cannot obtain vic-

5. Dennis M. Drew, "Insurgency and Counterinsurgency: American Military Dilemmas and Doctrinal Proposals," *CADRE Paper, Report No. AU-ARI-CP-88-1* (Maxwell AFB, Ala.: Air University Press, March 1988), 15–16.

tory, losing credibility.[6] Also, the cost to a regular army of maintaining in power a government facing an insurgency can be so great that even though undefeated, the army is exhausted and made ineffective. The Royalists hoped to win in such a way by using a coalition of tribes to defeat the Egyptian army.

The environment of Yemen was well suited for guerrilla operations. The highly compartmentalized and mountainous terrain made movement difficult, and large, regular forces could easily become divided and separated. Due to sheer drop-offs and steep slopes, off-road movement by vehicles often proved impossible. Most of the few roads in Yemen were dominated by high ground, facilitating ambushes. The climate was harsh even by Egyptian standards: highs could reach 130 degrees Fahrenheit in coastal areas, while lows down to 18 degrees Fahrenheit were registered in the mountains. Egyptian troops arriving in North Yemen required a period of acclimation before they could be fully effective in combat operations.

Initial Intervention

In spite of Egypt's prior involvement by way of advisors in Yemen, the Egyptian army was unprepared for the intervention. Egypt had little information about Yemen, and General Hadidi, chief of Egyptian military intelligence at the time, compared the Egyptian leadership to "a person who entered a test unprepared." He joked that at that point, the only thing Egypt knew for certain about North Yemen was that it bordered on Saudi Arabia.[7]

The Egyptians faced several initial difficulties. The army had no maps of Yemen, forcing it to depend on the local population for information about the terrain; the soldiers often could not verify the accuracy of this information since they found it difficult to determine the loyalties of their sources. A lack of understanding about Yemeni society led the army to conclude erroneously that the war was being fought by organized groups and that it could use conventional tactics. The Egyptians also mistakenly believed that some initial Shafi'i backing for the republic meant that there was universal support, while in fact, shifting loyalties made it difficult to determine who supported the republic and who opposed it. Finally, the Egyptian high command did not prepare contingency plans for operations in Yemen or study the situation in detail since

6. David A. Charters and Maurice Tugwell, eds., *Armies in Low-Intensity Conflict: A Comparative Analysis* (London: Brassey's, 1989), 252; and Drew, "Insurgency and Counterinsurgency," 1–2, 20.

7. Hadidi, *Yaman*, 20; and Salah al-Din al-Hadidi, *Shahid 'ala Harb 67* (Witness to the '67 War) (Al-Qahirah: Maktabat Madbuli, 1974), 50.

it believed that the Yemeni problem could be solved quickly.[8] Simply moving and supporting troops would prove a considerable undertaking since North Yemen was almost twenty-five hundred kilometers from Egypt.

Nasser initially appointed Anwar Sadat to coordinate Egyptian political policy in Yemen. Field Marshal Abdul Hakim Amer, first vice president and commander in chief of the Egyptian armed forces, was ultimately responsible for military operations in Yemen. Early in the war, an Egyptian command headed by a lieutenant general directly controlled Egyptian and Republican forces in Yemen and exercised de facto control over the Republican government. Later, another lieutenant general was assigned to coordinate Egyptian military and political policy in Yemen. The Egyptian high command's heavy involvement in Yemen meant that Amer devoted much time to that conflict, either personally directing military operations in Yemen, or studying Yemeni developments from Cairo. His focus on Yemen led him to neglect concerns about a possible future war with Israel.[9] Nasser generally avoided involvement in Egyptian activities in Yemen. He did not visit Yemen until 1964 and mainly left Yemeni policy to Amer and Sadat.

The first one hundred Egyptian troops, a commando company, deployed by ship and arrived at the port of Hodeidah on 5 October 1962, with many others following. The initial purpose of the intervention was to safeguard the republic and consolidate its control of the country, which was expected to be accomplished in three months. Nasser felt that a quick and violent campaign would end threats to the new republic and allow the revolution to spread to Saudi Arabia, a second objective of the Egyptian action. Egypt also wanted to expel the British from Aden, a goal supported by President Sallal after the imam was defeated. The impor-

8. Some Egyptian commanders mistakenly believed that all opposition to the Yemeni republic came from external sources. During the first month of the intervention, an Egyptian brigade commander told a Republican soldier that "We [the Egyptians] arrived here to defend the revolution and the republic from the Saudis and the Jordanians - we were not issued instructions to make war on the Yemenis." Quoted in Ahmad Ahmad Faraj, *Rijal fi Khanadiq al-Difa' 'an al-Thaora* (Men in the Trenches in the Defense of the Revolution) ([Sana?]: n.p., 1995), 146. Faraj was a soldier in the Republican army during the civil war. The Egyptian brigade commander made the remark to Faraj while conducting a reconnaissance prior to his brigade's first action in Yemen. Fawzi, *Harb al-Thalath Sanawat*, 23; Khalil, *Hurub Misr al-Mu'asirah*, 66, 69; Rahmy, *Egyptian Policy*, 109, 157; A. I. Dawisha, "Intervention in the Yemen: An Analysis of Egyptian Perceptions and Policies," *Middle East Journal* 29 (Winter 1975): 49; and Ahmad, *Dhikrayat Harb al-Yaman*, 292.

9. Fawzi, *Harb al-Thalath Sanawat*, 22, 24; Anwar Sadat, *In Search of Identity: An Autobiography* (New York: Harper and Row, 1978), 162; Khalil, *Hurub Misr al-Mu'asirah*, 57; Hadidi, *Yaman*, 37, 63; Ahmad, *Dhikrayat Harb al-Yaman*, 289, 295; and Rahmy, *Egyptian Policy*, 251.

tance of Egypt's three objectives in relation to each other would vary from time to time, leading to changes in Egyptian operational strategy throughout the war. At first, however, all efforts were directed towards simply ensuring the survival of the republic. It soon became apparent, however, that opposition was so stiff that Egypt faced the choice of either withdrawing its troops and allowing the revolution to fail, or dramatically increasing troop levels. The number of troops eventually committed to Yemen would go well beyond what the Egyptian leadership had first imagined.[10]

In October and November 1962, the Egyptians followed a plan known as Operation 9000. Under this plan, the Egyptians and Republicans tried to control all of North Yemen and to keep roads and lines of communications open throughout the country. Egyptian and Republican forces also attempted to secure the Yemeni-Saudi border to prevent the infiltration of Royalist supplies from Saudi Arabia, but with the initiative in the hands of the imam's forces, the Saudis could easily support the Royalists. In order to discourage external support to the Royalists, the Egyptians repeatedly bombed the Saudi border towns of Najran and Jizan, Royalist safe havens, and early in the intervention, the Egyptian air force struck a suspected Royalist base in the British Protectorate of Aden. The air raids had little effect, and the first Egyptian troops arriving in the country were forced to go directly into combat without time to acclimate or gain a proper understanding of the enemy.[11]

On 10 October 1962, the Royalists captured a series of towns in eastern Yemen and advanced west towards Sana. In response, an Egyptian battalion with tanks and armored cars left Sana by road to retake the eastern towns, going by way of Jihanah. On 22 October, an Egyptian paratroop company parachuted to the east of Sirwah to set up a hammer and anvil movement with the battalion moving by road to trap Royalist forces. Due to a lack of maps, the paratroopers became lost in the mountains, and tribesmen killed at least one-third of them. Royalists also ambushed the battalion moving by road, inflicting losses in men and vehicles, and the force mistakenly withdrew to Sana due to a communications mix-up.

10. Hadidi, *Yaman*, 38–39, 43; Fawzi, *Harb al-Thalath Sanawat*, 23; Ahmad, *Dhikrayat Harb al-Yaman*, 291–92; Dawisha, "Intervention in the Yemen," 50–51; Edgar O'Ballance, *The War in Yemen* (Hamden, Conn.: Archon Books, 1971), 81; Thomas J. Abercrombie, "Behind the Veil of Troubled Yemen," *National Geographic* 125 (March 1964): 421; and 'Abd al-Moshin Kamil Murtaji, *Al-Fariq Murtaji Yarwi al-Haqa'iq: Qa'id Jabhat Sina' fi Harb 1967* (Lieutenant General Murtaji Relates the Truth: Commander of the Sinai Front in the 1967 War) (Beirut: Al-Watan al-'Arabi, 1976), 23.

11. Ahmad, *Dhikrayat Harb al-Yaman*, 289, 312, 592; Amin Huwaydi, *Hurub 'Abd al-Nasir* (The Wars of Abdel Nasser) (Beirut: Dar al-Tali'a, 1977), 126; O'Ballance, *War in Yemen*, 87–88, 94; and Hadidi, *Yaman*, 43.

During the early stages of the intervention, communications with field units could be conducted only by using civilian telegraph. Another force of tanks, infantry, and commandos attempted a second drive on Sirwah, but it likewise was unable to get through, leaving the paratroopers isolated. Some of the paratroopers did eventually return to Sana, either walking, riding donkeys, or brought in by tribesmen seeking rewards.[12]

Fighting also began in the north. In the first part of November 1962, Royalist forces moving southeast out of Jizan in Saudi Arabia assaulted the Yemeni town of Haradh and were repulsed by an Egyptian brigade, supported by artillery and aircraft. After the Egyptians withdrew, however, the Royalist continued to make harassing attacks on Haradh. Another Royalist force tried to seize Sadah in the north, but a Republican battalion reinforced the town and prevented its capture, although it remained surrounded. An Egyptian armored column supported by aircraft reached Sadah on 9 November, but Royalist attacks on the town continued throughout the month. In the northeast, the Royalists captured Hazm and Sinwan. Also in November, the Royalists began interdicting supply lines by planting mines and ambushing Egyptian convoys, and for a time they even managed to block the main Egyptian artery, the road between Sana and Hodeidah, where Egyptian ships brought in troops and supplies.[13] (See Map 3.)

With all these setbacks, the Egyptians realized that a quick victory was not at hand and in late November 1962 adopted a new, long-range operational strategy for Yemen. They withdrew into a triangle formed by the cities of Sana, Hodeidah, and Taiz, abandoning areas outside the triangle or turning them over to the Republicans. Within the triangle, which also included the port of Mocha, the Egyptians established bases, stockpiled supplies, and constructed airfields. The Republican army, which had made no major contributions other than defending Sadah, occupied some vital areas outside the triangle. The Egyptians began to train the Republican army so that eventually it could defend the republic aided by only a small Egyptian presence. A strong Egyptian reserve, requiring additional troops from Egypt, was created to strike outside the triangle if key areas were lost. Egypt delayed its plans to spread the revolution to Saudi Arabia.[14]

12. O'Ballance, *War in Yemen*, 92, 94; Ahmad, *Dhikrayat Harb al-Yaman*, 312; Khalil, *Hurub Misr al-Mu'asirah*, 57–58, 66–67; Fawzi, *Harb al-Thalath Sanawat*, 23; and Dana Adams Schmidt, *Yemen: The Unknown War* (London: The Bodley Head, 1968), 65. Schmidt was a Middle Eastern correspondent for the *New York Times* who traveled in North Yemen during the Egyptian intervention.

13. Ahmad, *Dhikrayat Harb al-Yaman*, 312; Khalil, *Hurub Misr al-Mu'asirah*, 56, 58; O'Ballance, *War in Yemen*, 93; Faraj, *Rijal fi Khanadiq*, 142, 146–47; and Mutahhar, *Yauma Walada al-Yaman Majadahu*, 144.

14. Hadidi, *Yaman*, 73–74, 81–83; and Ahmad, *Dhikrayat Harb al-Yaman*, 301.

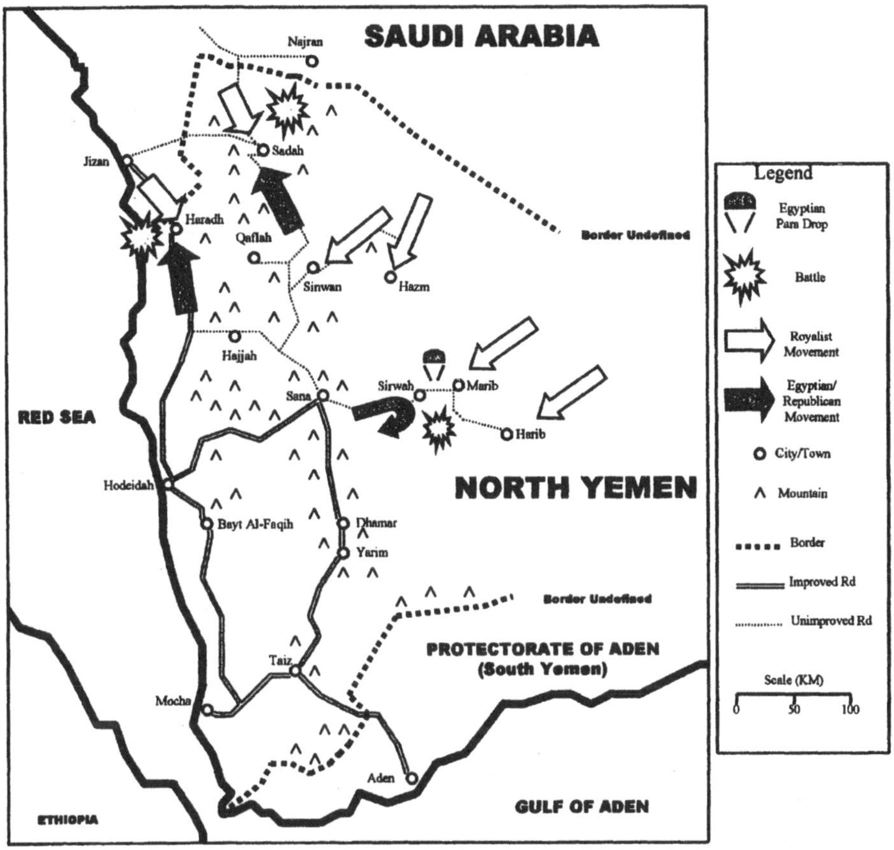

Map 3. Major Operations in 1962.

The triangle strategy was unsuccessful. The Republican army proved ineffective, leaving the Egyptians to fight the war mostly by themselves until 1966. During the period from December 1962 to February 1963, most of the tribes in eastern, northern, and western Yemen joined the imam and captured territory both inside and outside the Egyptian triangle, including Sadah. The Royalists surrounded Sana, containing seven thousand Egyptian troops, except for the vital Hodeidah road, which remained open. Amer and Sadat came to Sana to review operational strategy and to oversee the lifting of the siege. The Egyptian command concluded that its fifteen thousand Egyptian troops in Yemen were insufficient, and sent reinforcements. Abandoning the triangle strategy, the

Egyptians carried out a nationwide Ramadan Offensive from 16 February to 7 March 1963 to regain vital areas of Yemen.[15]

The first phase of the offensive was the seizure of Sadah, almost two hundred kilometers north of Sana. The town's capture would be difficult for the Egyptians due to its remoteness, but by controlling Sadah, and with it the Saudi border area, they could interdict Royalist supply lines from Saudi Arabia and actually invade Saudi territory if necessary. They could not reach Sadah by land since the Sana-Sadah road passed through Royalist-controlled territory, so in one of the most daring operations of the intervention, the Egyptians dropped a paratrooper battalion near Sadah. The paratroopers then prepared a dirt runway where Egyptian aircraft landed an infantry brigade. The move surprised the Royalists, and by 18 February, the Egyptians were in control of Sadah. An Egyptian armored column then began opening the Sana-Sadah road.[16]

The second phase of the offensive was the seizure of eastern Yemen. A force of about 350 infantrymen with tanks and armored cars moved halfway up the Sana-Sadah road and turned due east, moving across the desert to capture Hazm, a center of Royalist activity. It then turned south, and supported by aircraft, took Marib on 25 February, followed by Al-Jaubah, then Harib on 7 March. This was an impressive accomplishment for the Egyptians, who moved without maps through areas devoid of major roads. The Royalists were again surprised, and Egypt gained temporary control of the east. Next, the Egyptians secured the coastal region and the border with Saudi Arabia. In about three weeks, the Egyptians had regained control of most principal towns, all major roads, and more than one-half of the country. The Egyptian command in Yemen believed that it had gained a decisive victory, and President Sallal declared that "We have concluded the military revolution and now we celebrate by laying the corner stone of the peaceful revolution."[17] The Royalists were defeated in the conventional sense, but the war continued. (See Map 4.)

During the rest of 1963, Egypt tried to consolidate its gains. It conducted no other major offensives, although it carried out small air raids on Saudi territory. The Royalists continued to conduct ambushes and harassing attacks, and to infiltrate supplies from Saudi Arabia. Although the Ramadan Offensive had been a success, Egyptian troop levels rose to thirty-six thousand in 1963; the Egyptians could not control the coun-

15. Hadidi, *Yaman*, 81–82, 83; Khalil, *Hurub Misr al-Mu'asirah*, 59–61; O'Ballance, *War in Yemen*, 96–97, 98; and Ahmad, *Dhikrayat Harb al-Yaman*, 289, 313.

16. Hadidi, *Yaman*, 83, 84–85; Ahmad, *Dhikrayat Harb al-Yaman*, 313; O'Ballance, *War in Yemen*, 98; and Schmidt, *Yemen*, 164.

17. Sallal is quoted in Mutahhar, *Yauma Walada al-Yaman Majadahu*, 149. Hadidi, *Yaman*, 84–85, 90; Ahmad, *Dhikrayat Harb al-Yaman*, 313; Schmidt, *Yemen*, 164–65; and O'Ballance, *War in Yemen*, 98.

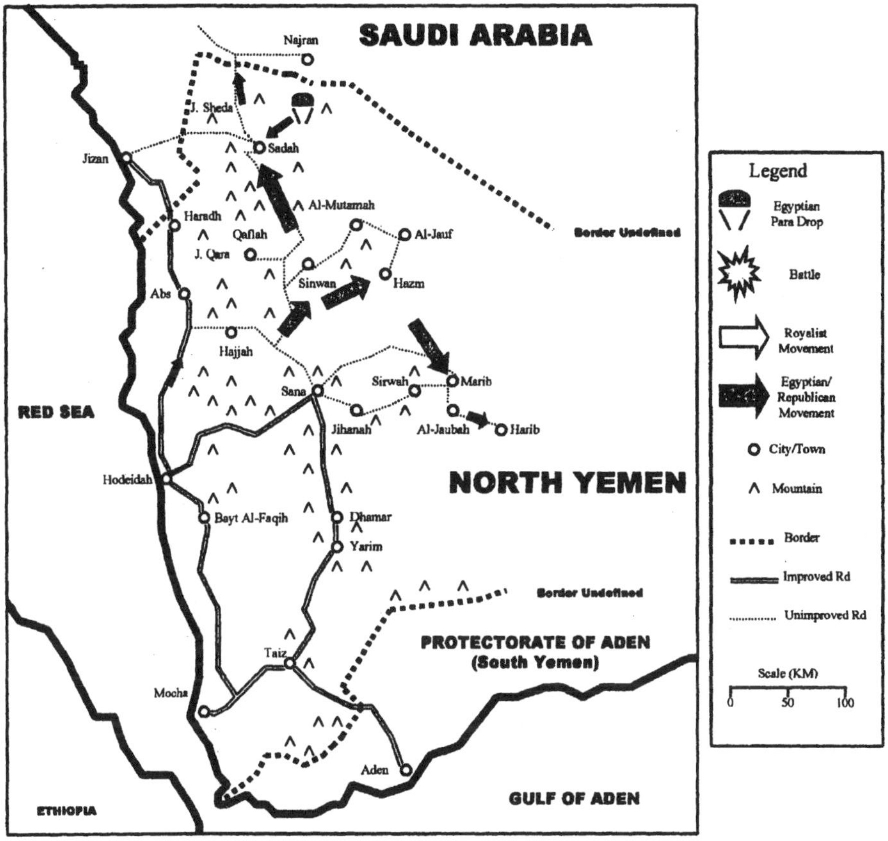

Map 4. Egyptian Ramadan Offensive, 1963.

tryside without physically occupying the land, which in their absence generally reverted to Royalist control. In late 1963, Lieutenant General Anwar al-Qadi, the Egyptian commander in Yemen, was injured by mortar fire when his convoy was ambushed in northeast Yemen, and had to be evacuated to Egypt.[18] Yemen was far from secure for the revolutionary regime in Sana.

18. O'Ballance, *War in Yemen*, 108, 113, 119, 120; Khalil, *Hurub Misr al-Mu'asirah*, 63; Schmidt, *Yemen*, 131; Ahmad, *Dhikrayat Harb al-Yaman*, 290, 293; and Mutahhar, *Yauma Walada al-Yaman Majadahu*, 149.

The Yemeni Quagmire

The Egyptian army got more than it had bargained for in Yemen as the projected three-month intervention turned into a Yemeni quagmire. Faced with its inability to defeat the insurgents, the Egyptian command in Yemen was obliged to call in fresh troops from Egypt. Three years into the intervention, the Egyptian army in Yemen consisted of thirteen infantry brigades, a separate tank brigade, most of Egypt's commando and paratrooper battalions, ten artillery battalions, and various administrative, logistical, and medical units. Due to its rotation policy, almost the entire Egyptian army of the period served in Yemen over the course of the intervention, with some units and individuals rotating to Yemen two or three times. The first infantry brigades that came to Yemen did not have the best equipment, but as numbers increased, some of the best units with the latest Soviet weapons deployed to Yemen, although many of the modern weapons proved unsuitable for fighting insurgents. Later, the Egyptians made modifications to some weapons, but the army still had a tendency to waste large amounts of ammunition fighting guerrillas in Yemen's rugged terrain.[19] A small percentage of the Egyptian air force, which faced no enemy opposition in the air, supported the army in Yemen with fighters, fighter-bombers, transports, and bombers, some of which flew directly from Egyptian bases. The Egyptian navy was used mainly to transport troops and supplies, and to deny the Royalists support from the sea.[20]

During the course of the intervention, morale problems arose within the Egyptian ranks. Initially, the leadership and soldiers of the Egyptian army did not question going to Yemen, but with losses, long service, and the hardships that their families faced in Egypt because of the war effort, some soldiers began to question why they were fighting other Muslims. The Egyptian soldiers were demoralized by the tribal enemy, who seemed barbaric to them, and by their inability to tell friend from foe in the insurgent war. Yemen's terrain made movement slow and difficult, fresh food and water were not readily available, and the Royalists often poisoned wells. Soldiers encountered difficulties in getting mail, newspapers, and magazines from Egypt, and could receive Radio Cairo only rarely in remote Yemeni areas. The climate's variations between hot and

19. Ahmad, *Dhikrayat Harb al-Yaman*, 288, 290–91, 633. The Egyptians modified towed 37-mm antiaircraft guns and 12.7-mm four- barreled machineguns by placing them on trucks so that they could rapidly engage the enemy. They modified some aircraft, such as trainers and transports, for use in a ground attack role in Yemen. Fawzi, *Harb al-Thalath Sanawat*, 23–24, 28; and Hadidi, *Yaman*, 154.

20. Fawzi, *Harb al-Thalath Sanawat*, 24; Huwaydi, *Hurub 'Abd al-Nasir*, 143; and Mohrez Mahmoud Hussini, *Soviet-Egyptian Relations, 1945–85* (New York: St. Martin's Press, 1987), 126. Hussini was a commodore in the Egyptian navy.

cold, dryness and heavy rains, were harsh to the Egyptians, whose uniforms were unsuited to the severe environment.[21]

In response to these morale issues and to avoid opposition, Nasser and Amer established privileges for those serving in Yemen. These included higher pay; assistance in obtaining land, cars, and housing after duty in Yemen; appointment of relatives to government positions; assistance in entering college for the sons of those serving in Yemen; and opportunities to profit from the purchase and resale of black market goods. These privileges angered the common Egyptian citizen who suffered from the cost of the war effort.[22]

One of the high command's stated reasons for the intervention was to train junior officers in combat, but instead, officers, particularly Field Marshal Amer, exploited the situation for personal gain. Some officers used the Protectorate of Aden for the black market purchase of consumer commodities such as refrigerators, stoves, and televisions, and established lucrative businesses marketing these items in Egypt, where they were in short supply. Careerism became so rampant that officers were afraid to question the feasibility of planned operations or issues about morale. Discipline deteriorated, combat was sometimes avoided or falsely reported, and equipment was neglected. Rumors circulated in Egypt as to what the army was actually doing in Yemen.[23]

Field Marshal Amer used the war in Yemen to consolidate his control of the armed forces, increase the influence of the military in Egyptian society, and establish a power base separate from Nasser. Yemen became Amer's fiefdom. Although Amer's motives were aimed strictly at advancing his own position, most of the Egyptian leadership was dedicated to the mission in Yemen, and many prominent Egyptians served there. As noted, Anwar Sadat, president of Egypt from 1970 to 1981, was a central figure in Egyptian involvement in Yemen. Husni Mubarak, who became president of Egypt in 1981, headed an air force squadron which transported troops to Yemen and later commanded long-range bombers used in Yemen. General Anwar al-Qadi, later the Director of Operations during the Six-Day War, commanded Egyptian forces in Yemen from mid-October 1962 through November 1963. He was a capable officer who carefully studied the situation and worked with Yemeni political and

21. Ahmad, *Dhikrayat Harb al-Yaman*, 296–97; and Khalil, *Hurub Misr al-Mu'asirah*, 64–65.

22. Rahmy, *Egyptian Policy*, 246–47; and Hadidi, *Shahid*, 57–58.

23. Huwaydi, *Hurub 'Abd al-Nasir*, 146; Sadat, *In Search of Identity*, 161; and Mohamed Hassanein Heikal, *The Sphinx and the Commissar: The Rise and Fall of Soviet Influence in the Middle East* (New York: Harper and Row, 1978), 148. Heikal was the editor of *Al-Ahram*, Egypt's semiofficial newspaper, and one of Nasser's closest friends and advisors. Khalil, *Hurub Misr al-Mu'asirah*, 64–65, 67–68; Fawzi, *Harb al-Thalath Sanawat*, 27–28; and Hadidi, *Shahid*, 58.

tribal leaders to try to develop a coherent strategy for the war. In May 1963 he met with Nasser and recommended that Egypt withdraw from Yemen, an idea which Nasser quickly rejected.[24] General Murtaji, who commanded the Sinai Front in the Six-Day War, led the Egyptian forces in Yemen for a time and was later responsible for coordinating Egyptian military and political policy in Yemen.

Egyptian Tactics and Strategy in Yemen

The Egyptian army was trained and equipped to fight open desert warfare at the divisional level, which was the kind of war it expected to wage against its primary enemy, Israel. In Yemen, the Egyptians ineffectively used conventional tactics against a dispersed enemy who disappeared at will, and the Egyptian command had difficulty in coordinating its units, which at times were spread throughout the country. In the end, the Egyptians resorted to shows of force and intimidation to pacify the land. Although they did make some improvements to their tactics, General Fawzi, chief of staff of the Egyptian armed forces, believed that the army never adapted to counterinsurgent warfare because the military bureaucracy stifled change.[25] As a result, the Egyptians could say that they won most battles in Yemen, but their army was exhausted in the process.

The Egyptians usually conducted infantry and tank attacks, supported by aircraft and artillery, against Royalist positions. The Egyptians were successful when the enemy remained to fight, but that was seldom, and the Royalists reoccupied territory won by the Egyptians as soon as the army withdrew. Due to Yemen's rough terrain, the Egyptians sometimes conducted hammer and anvil operations by using paratrooper drops to support infantry and tank attacks, but this approach was still a conventional response to counterinsurgency.[26]

24. Sadat, *In Search of Identity*, 162; Hadidi, *Yaman*, 64–65, 155; Rahmy, *Egyptian Policy*, 204; Ahmad, *Dhikrayat Harb al-Yaman*, 295; and Mohamed Abdel Ghani El-Gamasy, *The October War: Memoirs of Field Marshal El-Gamasy of Egypt*, trans. Gillian Potter, Nadra Morcos, and Rosette Frances (Cairo: American University in Cairo Press, 1993), 18. Gamasy was Director of Operations during the 1973 war with Israel.

25. Hadidi, *Yaman*, 47; Khalil, *Hurub Misr al-Mu'asirah*, 56, 59; Rahmy, *Egyptian Policy*, 153–54, 159; and Ahmad, *Dhikrayat Harb al-Yaman*, 293–94. The Egyptians obtained training manuals from India and Pakistan describing mountain warfare tactics, which they applied in Yemen. Fawzi, *Harb al-Thalath Sanawat*, 24. In January 1966, Czech guerrilla warfare experts arrived in Yemen to assist the Egyptians; see Richard E. Bissell, "Soviet Use of Proxies in the Third World: The Case of Yemen," *Soviet Studies: A Quarterly Journal on the USSR and Eastern Europe* 30 (January 1978): 94.

26. Schmidt, *Yemen*, 65; Khalil, *Hurub Misr al-Mu'asirah*, 67; and Ahmad, *Dhikrayat Harb al-Yaman*, 313.

Many Egyptian operations dealt with maintaining lines of communications and supply. The Royalists frequently ambushed Egyptian convoys, and keeping the roads open was a constant challenge. 'Abd al-Mun'im Khalil, a lieutenant colonel and colonel during the intervention, reported that his units spent a great deal of time reopening roads to relieve isolated convoys and garrisons. He often received his orders directly from Amer, indicating the importance of these missions. Eventually the Egyptians learned to secure roads in advance of convoys by occupying overlooking mountain tops, but this tactic was time consuming and exhausting. Later, instead of immediately reopening blocked roads, the Egyptians allowed the enemy time to leave, since the Royalists seldom had the discipline to remain in one place very long. The Egyptians also learned to resupply isolated posts with helicopters and to stockpile supplies in garrisons so that if a road were blocked, it did not have to be reopened at once.[27]

Some of Egypt's best tactics were its efforts to modernize and reform the Yemeni government to enable it to gain popular support. This process was difficult because Yemeni officers and civilians had no political or administrative experience. The level of centralization under the Imamate had been so absolute that most people in Republican-controlled Yemen refused even the smallest order unless President Sallal personally signed it, which was how the imams had operated. Egypt was forced to assume full administration of the country, and Egyptian officers governed most districts, where they became entangled in local issues and disputes. Egyptian advisors established a High Council for the Republican government, which attempted to build a modern state with a modern civil service. They also conducted training courses for Yemeni government employees and helped the Republicans issue a temporary constitution. Egyptians developed Yemen's first national budget, but gathering revenues was a problem since there was no effective tax system, and ultimately Egypt financed the republic. Egyptian reforms were effective to an extent, and by 1965, the Yemeni Republicans were running routine matters under Egyptian supervision, although the most important ministries remained under direct Egyptian control. One negative impact of the close Egyptian supervision was that the Egyptians came to be viewed as occupiers, and many Royalists fought out of hatred of Egypt rather than out of love for the imam. The Egyptian lead in military operations caused some Republican leaders to become apathetic

27. Khalil, *Hurub Misr al-Mu'asirah*, 59–62, 67; Hadidi, *Yaman*, 47, 65; Ahmad, *Dhikrayat Harb al-Yaman*, 294; and O'Ballance, *War in Yemen*, 85.

towards the conduct of the war, which came to be viewed as a conflict between the Egyptians and the Royalists.[28]

Egypt also built schools, hospitals, roads, and airports, and improved the port of Hodeidah. Approximately three hundred Egyptian primary and secondary school teachers came to Yemen, as well as administrative advisors and doctors. Scholarships were offered for Yemeni students to study in Egypt, and cultural delegations distributed books and films. A special Egyptian army section installed water pumps, dug wells, gave agricultural advice, and introduced Yemenis to modern technology. The Egyptians brought in one hundred religious scholars in hopes of convincing the Yemenis that there were no important differences between the Shafi'is and the Zaydis, and Egyptian military chaplains delivered sermons in Yemeni mosques. A delegation from the Al-Azhar religious university and mosque in Egypt supervised Yemen's religious institutions.[29]

The Egyptians trained and modernized the Republican army. Early in the intervention, the Egyptians realized that without a Republican army fighting alongside them, they would appear as occupiers and support for the republic would further diminish. Training the Republican army was also seen as a way to eventually limit the size of the Egyptian presence. Once again, this was a difficult task. The prerevolutionary army of the imams was poorly armed and trained, foraged for its food, and lived in primitive conditions. It was used mainly to collect taxes and perform police duties. After the coup in September 1962, the imam's army disintegrated, with some factions supporting the Republicans, others the Royalists, and some simply deserting. Most officers sided with the Republicans.[30]

The regular Republican army, that is the organized forces other than purely tribal forces, consisted initially of less than seven thousand men and never rose above a strength of ten thousand. Some Egyptians served as staff officers and commanders, and President Sallal had a three-thou-

28. J. C. Hurewitz, *Middle East Politics: The Military Dimensions* (New York: Praeger, 1969), 256–57; Abercrombie, "Behind the Veil," 421; Mutahhar, *Yauma Walada al-Yaman Majadahu*, 150–51; Fawzi, *Harb al-Thalath Sanawat*, 25; O'Ballance, *War in Yemen*, 158; Huwaydi, *Hurub 'Abd al-Nasir*, 132–33; John Law, "Forgotten War in the Desert: A First-Hand Report," *U.S. News and World Report* 58 (24 May 1965): 68; and Faraj, *Rijal fi Khanadiq*, 197.

29. Huwaydi, *Hurub 'Abd al-Nasir*, 133; Fawzi, *Harb al-Thalath Sanawat*, 23; Rahmy, *Egyptian Policy*, 91; Schmidt, *Yemen*, 208; Hadidi, *Yaman*, 67, 134; and Mutahhar, *Yauma Walada al-Yaman Majadahu*, 158.

30. Ahmad, *Dhikrayat Harb al-Yaman*, 289–99, 301; O'Ballance, *War in Yemen*, 42, 85; and Ahmad Saqqaf, *Ana 'Aid min al-Yaman* (I Returned from Yemen) (Kuwait: Mataba'at al-Anba', 1985), 78. Saqqaf was a Kuwaiti official traveling in Yemen during the Egyptian intervention and was responsible for organizing Kuwaiti assistance for the Republican government.

sand-man bodyguard of Egyptian paratroopers. The Republican army was plagued with desertion caused by poor living conditions and friction between religious sects, and at first its combat effectiveness was extremely limited. At the start of the intervention, the Republican army was used only to guard towns, but by the end of summer 1964, it was employed in some combat operations with Egyptian air and artillery support. Only after March 1966 did the Republican army prove effective in conducting independent operations. The Egyptians completely supplied it with clothing, weapons, and money, and not until 1966 did the Republican army begin providing its own food. Republican officers were sent to Egypt to study in military schools, and several of the Yemeni brigades were transported to Egypt for training. The Egyptians opened schools in Sana and Taiz for training Republican soldiers, including a military academy modeled on its Egyptian counterpart, as well as infantry, armor, artillery, and parachute schools.[31]

The Egyptians' biggest headache in Yemen was the tribes, and at the outbreak of the revolution, not a single tribe announced its loyalty to the republic. The Egyptians followed a carrot and stick policy towards the tribes, combining brutality with attempts to bribe them. Early in the war, the Egyptians began paying gold to buy the tribes' loyalty, as well as supplying them with one hundred thousand transistor radios. The Egyptians also provided them with weapons in an attempt to create tribal auxiliaries.[32] Egyptian attempts at bribing the tribes led to a bidding war with the Royalists. Many tribes accepted money from both and fought on whichever side was offering the most at a particular moment.

Four tribal armies were generally loyal to the republic, but their size is impossible to estimate. The most prominent is said to have numbered five thousand men. The tribal armies were usually controlled directly by their sheiks, and a special section of Egyptians served as liaison officers with the tribes to coordinate their activities. Loyal tribes were used to control territory and participate in operations against the Royalists. Sometimes a tribe was hired for a specific operation, but the next week it could just as well be fighting against the republic. One of the more imaginative Egyptian tactics in dealing with the tribes, and which represented a true conversion to the tactics of counterinsurgency, was the

31. O'Ballance, *War in Yemen*, 85, 86; Hurewitz, *Middle East Politics*, 257; Ahmad, *Dhikrayat Harb al-Yaman*, 299–300, 301; Hadidi, *Yaman*, 133; Rahmy, *Egyptian Policy*, 159; Huwaydi, *Hurub 'Abd al-Nasir*, 131; Fawzi, *Harb al-Thalath Sanawat*, 24; and Faraj, *Rijal fi Khanadiq*, 205, 226–27.

32. Khalil, *Hurub Misr al-Mu'asirah*, 66; Huwaydi, *Hurub 'Abd al-Nasir*, 131–32; and Hadidi, *Yaman*, 71–72. Recently, accusations have arisen in Egypt that the Nasser regime might have used part of Egypt's gold reserve to buy the loyalty of tribes in Yemen; see Mona El-Nahhas, "Squabble Over Old Gold," *Al-Ahram Weekly*, 24–30 December 1998.

creation of special teams of officers assigned to work and live with the tribes. These officers, all from Egyptian rural areas and not cities, were incorporated into the tribesmen's daily lives, learned their customs, ate the same food, wore the same clothes, and spoke the local dialect. After winning the confidence of the tribal sheiks, they led the tribes into battle against the Royalists. Where this policy was used, it was successful, but it was too limited in scope to make a major difference.[33] Many tribes who did announce their loyalty to the republic tended to melt away in combat, disappearing with their Egyptian-supplied weapons.

Egypt's brutality in dealing with the tribes, the other side of its tribal policy, was probably the ugliest aspect of the entire war and gave Egypt a negative image in the Western press. Harold Vocke, a West German journalist in Yemen at the time, reported that "the Egyptian Air Force pursues herdsmen and herds as if they were wild game."[34] Another correspondent reported that if a few tribesmen ambushed a single Egyptian vehicle, Egyptian planes would bomb and destroy their villages. A Republican soldier during the war said that if a mine exploded in a road, or if there were an attack on an Egyptian position, Egyptian forces would surround nearby villages and treat them as collectively guilty. Frequent bombing of the countryside inhabited by uncommitted tribes was seen as a method to gain their support, and Egypt bombed Royalist areas daily, often using long-range bombers flying out of Aswan in Egypt. Egyptian aircraft strafed villages, which President Sallal believed was the most effective way to gain their loyalty. Harsh Egyptian treatment of the population cost the republic support and made it difficult for Republican leaders to justify Egyptian actions.[35]

In 1963 the Egyptians began destroying the agricultural lands and wells of the tribes suspected of supporting the imam, and throughout Royalist-controlled areas, most houses were abandoned. By late 1966 and early 1967, many areas were deserted because villages, wells, agricultural lands, and livestock had been destroyed, and people were living in caves. Hundreds of noncombatants had been killed in the Egyptian attacks. The Egyptians also tried to exterminate the imam's family by offering a large reward for any royal family member killed or captured.[36] The brutality of these acts had the opposite effect to what was intended

33. Hadidi, *Yaman*, 82; Ahmad, *Dhikrayat Harb al-Yaman*, 300; Faraj, *Rijal fi Khanadiq*, 159; and Hurewitz, *Middle East Politics*, 257.

34. Harold Vocke, "What Goes on in Yemen," *Atlas: The Magazine of the World Press*, May 1967, 25.

35. Law, "Forgotten War," 68; Faraj, *Rijal fi Khanadiq*, 195–96; and Hadidi, *Yaman*, 48–49.

36. O'Ballance, *War in Yemen*, 117, 133, 157–58; and Badeeb, *Saudi-Egyptian Conflict*, 38.

and caused many tribes to support the Royalists since the imam was seen as trying to rid the country of the Egyptians.

Continued Escalation

In January 1964 the Egyptian command concluded that its forces were too dispersed and adopted a new operational strategy, the "policy of concentration." Egyptian forces would focus on only key regions, but in areas spread throughout the country, a variation of the triangle strategy. They abandoned some zones, but created reserves to strike throughout the country if necessary, and Egyptian troop levels rose to fifty thousand in 1964. As areas were evacuated in accordance with the new strategy, some tribes believed a general Egyptian withdrawal from Yemen was occurring, triggering a spontaneous Royalist offensive. Royalists blocked the Sana-Hodeidah road in January and February, and forces advancing from the north, south, and west, were able to besiege Sana. Other Royalist attacks occurred in the east and west against outlying areas. The Sana-Hodeidah road was reopened after heavy fighting, and in March, Egyptian and loyal tribal forces were able to lift the siege of Sana, but problems with supply continued throughout the country due to ambushes. Sadah could be supplied only by air.[37]

By mid-1964, the Egyptian command believed that if the imam were captured, killed, or forced to flee to Saudi Arabia, tribal loyalties would switch and the war could be ended. Egyptian, loyal tribal, and Republican forces, conducted a major offensive to rid Yemen of the imam. Starting in late June, an armored column drove northwest from Hajjah, while another pushed southwest from Sadah, both converging on the imam's headquarters at Jebel Qara. The imam was forced to flee into the extreme northern region of Yemen, the Jebel Sheda/Jebel Rezeh area. On 15 August, another series of pincer movements pushed northeast from Haradh and northwest from Sadah, finishing at the imam's new location. Although the imam was not captured, he was forced into Saudi Arabia, and thousands of his supporters were killed. The success of the operation in summer 1964 did nothing to settle the war, and minor fighting continued for the rest of the year. In September Egypt withdrew from the northwest regions it had captured in the summer since they were considered pacified. The Royalists reoccupied these areas.[38] (See Map 5)

37. Hadidi, *Yaman*, 105–6; O'Ballance, *War in Yemen*, 122, 126; Mutahhar, *Yauma Walada al-Yaman Majadahu*, 155; and Ahmad, *Dhikrayat Harb al-Yaman*, 209, 314–15.

38. Ahmad, *Dhikrayat Harb al-Yaman*, 315–16; Huwaydi, *Hurub 'Abd al-Nasir*, 133; Mutahhar, *Yauma Walada al-Yaman Majadahu*, 155; O'Ballance, *War in Yemen*, 128, 133; and Hadidi, *Yaman*, 106.

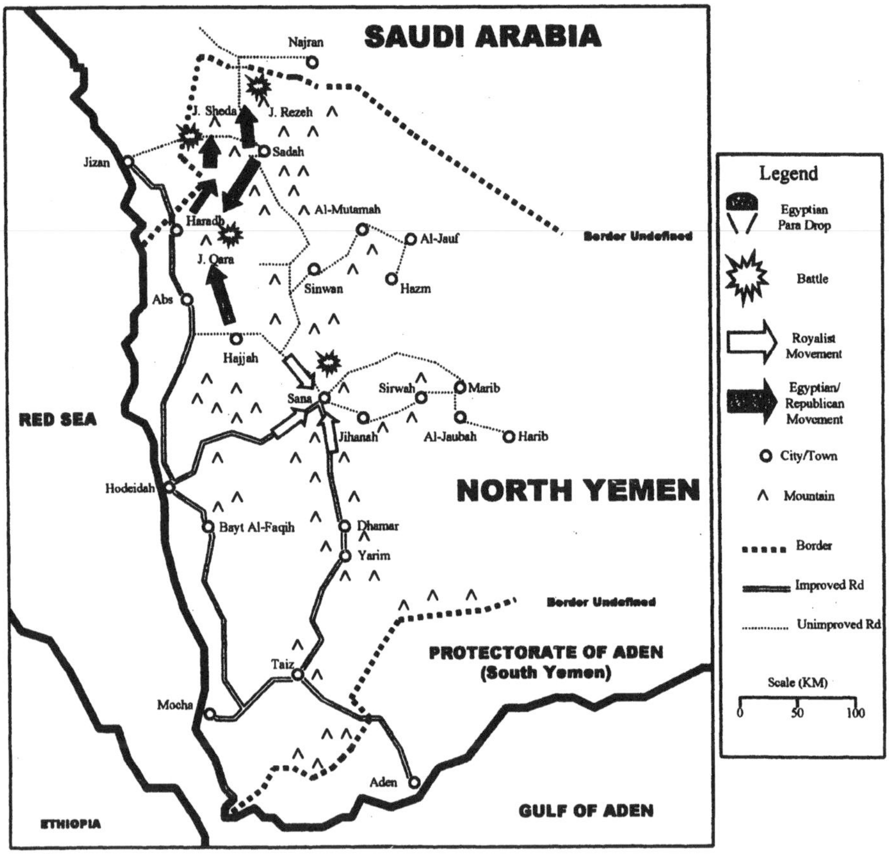

Map 5. Major Operations in 1964.

In 1965 Egypt experienced its worst year in Yemen, as Egyptian troop levels there peaked at seventy thousand, roughly one-third of the total Egyptian armed forces.[39] From January to July 1965, large parts of the republic were lost. At the beginning of the year, the Royalists gained control of the chain of mountains north of Sadah, and other Royalist forces cut the only road into northeast Yemen, entrapping Egyptian forces at Al-Jauf. The Western press labeled this a potential "South Arabian Stalingrad," although Dien Bien Phu or Khe Sahn would have been better comparisons. Three battalions of paratroopers and commandos attempted to relieve the situation, with paratroop drops occurring at Al-

39. Hadidi, *Yaman*, 133; O'Ballance, *War in Yemen*, 155; Rahmy, *Egyptian Policy*, 155; Sadat, *In Search of Identity*, 163; and Gamasy, *The October War*, 17.

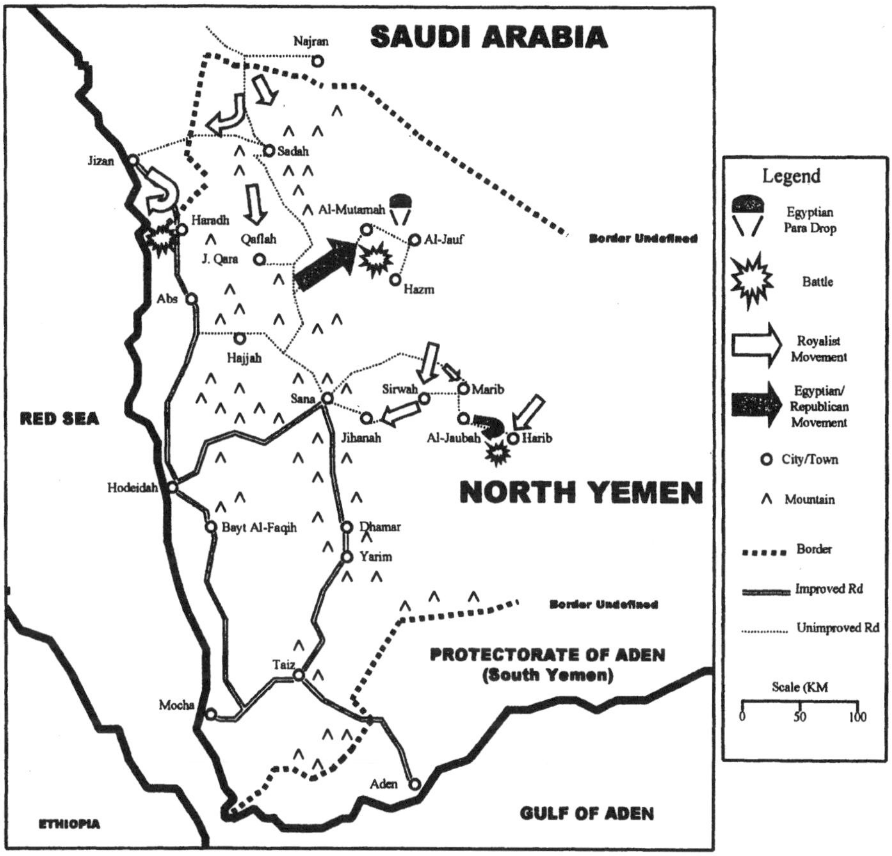

Map 6. Major Operations in 1965.

Mutamah, along the region's main road. The situation was eased in late February when the road was reopened, but frequent ambushes continued. Fighting also occurred throughout the northwest and east, with supply lines, including the Sana-Sadah road, being continuously cut and reopened. In the east, the Royalists captured Harib on 10 March, and an Egyptian-Republican attempt to retake it in April failed. Sirwah fell to the Royalists on 24 May, Jihanah on 24 July, and Marib the next day, where Egyptian and Republican forces had been in control since March 1963. Royalists also attacked Haradh, but Republican forces were able to defend it successfully. In the northwest, Qaflah fell on 14 June.[40] (See Map 6.)

40. Ahmad, *Dhikrayat Harb al-Yaman*, 316; Khalil, *Hurub Misr al-Mu'asirah*, 64, 66; Vocke, "What Goes on in Yemen," 21; Mutahhar, *Yauma Walada al-Yaman Majadahu*, 159; and O'Ballance, *War in Yemen*, 139–40.

By summer 1965 it was clear that if the situation did not dramatically change, the Egyptians would not achieve any of their objectives in Yemen. The Royalists controlled about one-half of the country, and the Egyptians and Republicans were in no position to retake lost territory. As a way to end the war decisively, Egypt began to consider a ground invasion of Saudi Arabia to seize Jizan, the main source of Royalist support. Although Egypt had conducted air raids against Jizan and Najran throughout the war, it had not invaded Saudi territory. The Egyptian command in Yemen now favored this option, and Amer proposed to Nasser that three infantry brigades and one tank brigade invade Saudi Arabia, with the probable objective of seizing only the area around the main Royalist haven. If this option were executed, the Egyptian command believed that it would have to reinforce Yemen with several more divisions. Amer thought that the attack would be successful as long as the United States did not intervene to support Saudi Arabia, an action he felt was unlikely due to the increasing American involvement in Vietnam. Nasser agreed to begin preparations for the invasion but stipulated that this escalation of the war occur gradually, with an initial small incursion into Saudi territory, to be followed by a larger one if the United States did not react. Egyptian forces began to assemble on the Saudi border, and on 23 July 1965, Nasser publicly threatened to invade Saudi Arabia. On 4 August, it was announced that Saudi Arabia had assembled ten thousand troops to repel the invasion.[41]

Nasser also made moves towards the negotiating table to end the conflict peacefully, perhaps realizing the dangers of an invasion of Saudi Arabia. In any case, he was becoming disillusioned with the Yemeni Republicans. After his visit to Yemen in April 1964, Nasser privately expressed his disgust with the Republicans due to their ineffectiveness in operating the government. Sallal, the Republican president, had accomplished little more than promoting himself to field marshal, and even Nasser developed a dislike for him; in 1964, after introducing Sallal to Soviet Premier Nikita Khrushchev during a Red Sea cruise on Nasser's yacht, Nasser remarked to the Soviet leader, "I just wanted you to see what I have to put up with."[42] In hopes of ending the war, Nasser met King Faysal of Saudi Arabia in Jeddah. On 24 August 1965, they signed the Jeddah Agreement, in which both rulers agreed to end the war between the Arabs. Faysal pledged to stop supplying the Royalists, and Nasser promised to withdraw from Yemen. The military situation

41. Ahmad, *Dhikrayat Harb al-Yaman*, 316–18, and O'Ballance, *War in Yemen*, 141, 149.

42. Malcolm H. Kerr, *The Arab Cold War: Gamal Abd al-Nasir and His Rivals, 1958–1970* (London: Oxford University Press, 1971), 112; and Dawisha, "Intervention in the Yemen," 55.

calmed, the planned invasion of Saudi Arabia was cancelled, and fighting died down. In late August, Egyptian troops began withdrawing from the Saudi border, and in September, Egyptian troops began abandoning some areas and concentrating only in the main towns. The Royalists announced a cease-fire, but it was difficult to enforce among the various tribal forces. Egyptian troops began departing from Yemen on 23 November, with plans for the entire evacuation of Yemen to be completed over the next ten months. Troop levels dropped to twenty thousand by mid-1966.[43]

End of the Egyptian Intervention

The final Egyptian withdrawal from Yemen did not take place in 1966 as planned. Full implementation of the Jeddah Agreement depended on the Royalists and Republicans reaching a comprehensive agreement, which did not occur. In December 1965, the Yemeni Republicans and Royalists met at Haradh to form a provisional government until a general plebiscite could be conducted. During the meeting, a dispute broke out as to whether to call the provisional government an Islamic state or a republic; the Royalists also wanted the imam's family included in any government, which the Republicans rejected. These differences ended any chance of peace between the two sides in the immediate future. However, the major factor in Egypt's continued presence in Yemen was the British announcement in February 1966 that they would leave the Protectorate of Aden in South Yemen no later than 1968. Nasser decided against withdrawal from North Yemen and adopted as his primary goal the control of Aden and South Yemen after the British departure. In the same month Nasser said in reference to Yemen, "If anyone believes that, because of the length of time, we have become fed up or tired, then we say we are a struggling and capable people."[44] In March 1966 Nasser announced that the Jeddah Agreement would not be implemented. He added, "We are thus revising our plans so that we may stay in Yemen for five years or longer if necessary. We shall reduce our forces and expenditures in the Yemen, but we shall not leave the posts which are important."[45] Amer declared that Egypt would stay in Yemen for twenty years if necessary.[46]

43. Kerr, *The Arab Cold War*, 108; Ahmad, *Dhikrayat Harb al-Yaman*, 318; Mutahhar, *Yauma Walada al-Yaman Majadahu*, 163; and O'Ballance, *War in Yemen*, 151, 155.

44. Nasser is quoted in Dawisha, "Intervention in the Yemen," 57; Kerr, *The Arab Cold War*, 108–9; and O'Ballance, *War in Yemen*, 165.

45. Nasser is quoted in Rahmy, *Egyptian Policy*, 154.

46. "Captive in Egypt," *The Economist* 221 (8 October 1966): 147.

This was the beginning of another Egyptian operational strategy, "the long breath," intended to allow Egypt to stay in Yemen indefinitely. The Egyptian army reoccupied the triangle area and abandoned areas along the Saudi border and outside the triangle, which either the Republican army or the Royalists occupied. Although the Royalists were still to be fought and the triangle maintained, the new goal of the intervention was to oppose the British in South Yemen. The Egyptians increased their aid for training and arming Adeni dissidents who operated from Taiz and made raids into the Aden Protectorate with Egyptian backing. After the British departed, Nasser wanted the void filled with a Nasserite presence, not a Saudi one. Egypt's army was geared to move rapidly south and occupy Aden when the British left. Since December 1963, Egypt had aided the dissidents of Aden in an operation known as Sallah al-Din, but it now became the main effort and the focus of all propaganda.[47]

The war took on a new character in 1966 and 1967 with "the long breath." The Egyptians did not launch any more major operations, although they did occasionally bomb Saudi border areas as well as the Yemeni countryside, and they conducted punitive attacks against various tribes. Skirmishing and ambushing continued, especially in spring 1966 when the Egyptians abandoned some areas to concentrate in the triangle. The Republican army was becoming more proficient and was able to defend some vital areas outside of the triangle without major Egyptian assistance. When the Sana-Sadah road was cut by the Royalists, five thousand Republican troops, aided only by Egyptian aircraft and artillery, were able to open it. In 1967 the Republican army retook some areas that had been lost to the Royalists since 1965.[48]

According to General Hadidi, the Egyptians might have stayed in Yemen indefinitely if it had not been for the Six-Day War, 5–11 June 1967. Fought by Israel against Egypt, Syria, Iraq, and Jordan, the war caught Egyptian forces divided between two theaters, the Sinai and Yemen. Before the war's outbreak, the Egyptians had been able to transfer only one infantry brigade from Yemen to the Sinai, leaving six infantry brigades, one tank brigade, and one artillery brigade in Yemen. These were some of the army's best units. In the first week of June 1967, the Egyptian army in the Sinai numbered one hundred thousand men. To obtain this level, the Egyptians had mobilized large numbers of reservists, many of whom were untrained. An additional disadvantage was that General Murtaji, appointed the front commander of the Sinai,

47. Hadidi, *Yaman*, 133; Rahmy, *Egyptian Policy*, 155–56; O'Ballance, *War in Yemen*, 127, 156–57; Huwaydi, *Hurub 'Abd al-Nasir*, 134; and Dawisha, "Intervention in the Yemen," 58–59.

48. Ahmad, *Dhikrayat Harb al-Yaman*, 318–19; O'Ballance, *War in Yemen*, 156, 174, 179; and Hadidi, *Yaman*, 133.

had been heavily involved in Yemen but was unfamiliar with the situation in the Sinai. He arrived to take command only one week before the war's start.[49]

The Israelis conducted a surprise air attack against Egypt on the morning of 5 June and destroyed 85 percent of Egypt's combat aircraft on the ground. Next, the Israelis began to penetrate Egyptian lines in the eastern Sinai, and the Egyptian high command's ability to control its divisions and brigades began to deteriorate. By the afternoon of 6 June, Amer, in a state of panic, ordered a withdrawal of the army from the Sinai to the Suez Canal, but specified no times or phases. As a consequence of the unplanned withdrawal, each unit, without air cover, began to race to the canal on its own, and by 8 June, the army had collapsed, with some officers abandoning their men. The Israeli army was able to advance to the Suez Canal almost unopposed. About ten thousand Egyptian soldiers died in the Sinai, some from thirst and exposure during their flight to the Suez Canal. At least another five thousand were captured. The Israelis destroyed or captured 80 percent of the Egyptian army's equipment and obtained a fleet of modern Soviet tanks in the process.[50]

After the defeat in the Sinai, an Egyptian withdrawal from Yemen became a military necessity. Troops began leaving in July 1967, but the Egyptians did not rush their evacuation and continued to pursue their Aden strategy. The last Egyptian troops did not leave Yemen until after the British had left South Yemen. Egyptian General 'Abd al-Kadir Husan was appointed commander in Yemen for the withdrawal, and under his command, the Egyptians continued to bomb the Royalists and the Saudi border, as well as to discipline the tribes. Fighting intensified; the Royalists captured Haradh and began an advance towards Hodeidah, but Egyptian forces retook Haradh on 20 July.[51]

Ultimately, however, the Egyptians had no choice but to withdraw, since the units in Yemen were needed to defend Egypt. From 29 August to 1 September 1967, Nasser, Faysal, and Sallal attended the Khartoum Conference in Sudan. The Arabs showed a new willingness to make peace and a spirit of cooperation in light of the Israeli victory. Nasser

49. Hadidi, *Yaman*, 139, 152; Ahmad, *Dhikrayat Harb al-Yaman*, 633; George W. Gawrych, "The Egyptian Military Defeat of 1967," *Journal of Contemporary History* 26 (1991): 279, 281; and Gamasy, *The October War*, 49.

50. Gawrych, "The Egyptian Military Defeat," 280, 290, 298, 300; Huwaydi, *Hurub 'Abd al-Nasir*, 47; Goldschmidt, *Modern Egypt*, 128; and Anthony McDermott, *Egypt From Nasser to Mubarak: A Flawed Revolution* (New York: Croom Helm, 1988), 160.

51. Hadidi, *Yaman*, 140; Ahmad, *Dhikrayat Harb al-Yaman*, 292, 295, 319; Mutahhar, *Yauma Walada al-Yaman Majadahu*, 167; and O'Ballance, *War in Yemen*, 183.

agreed to withdraw from Yemen, and Faysal pledged to stop supplying the Royalists. Their agreement stipulated that the Yemenis make peace among themselves, but Sallal refused to support it, and the Royalists rejected peace talks until the Egyptians had left Yemen.[52] Saudi Arabia continued to supply the Royalists.

Despite the persisting disagreement among the Yemenis, the Egyptians continued withdrawing, moving their units to Sana and then to Hodeidah for departure to Egypt. The last infantry brigade withdrew in October, although a small number of troops remained until 8 December 1967. The withdrawal was difficult. Some Yemenis felt betrayed by the Egyptian departure. On 1 October, protesters in Sana killed some Egyptian soldiers. Two days later, an angry Yemeni mob in Hodeidah killed one hundred Egyptian troops. The Royalists attacked Egyptian units as they withdrew, and some had to fight their way out. A spirit of doom hung over the Egyptian army during the final phase of the intervention. It had fought for five years in Yemen, and the republic appeared no more secure than in 1962. At home, the Sinai was lost and the army destroyed. General Hadidi described the feelings of the departing soldiers; "The homeland of Egypt was shredded - Israel was perching on our chests."[53]

During the withdrawal, however, one of Nasser's original objectives was partially accomplished, when the British evacuated Aden on 30 November 1967. The group of dissidents that Nasser had supported to fill the vacuum in Aden did not come to power, however; instead, a much more radical element, the Marxist National Liberation Front, did. Although not falling under the influence of Nasser, this group did advocate the overthrow of the Saudi monarchy. But ousting the government of Saudi Arabia was no longer one of Nasser's objectives since now all Arab states needed to stand together to confront Israel.

Many expected the Royalists to quickly destroy the Yemeni Republic after the Egyptians left, but this did not occur. On 5 November 1967, while Sallal was in Iraq seeking foreign aid, the Republican army staged a coup and overthrew his government. Sallal had lost support among the Republicans and had gradually become an unpopular dictator. He was viewed as Nasser's puppet, and with the Egyptian army almost gone, he had nothing to offer. The new Yemeni president was General Hassan al-Amir, who like Sallal, opposed making peace with the Royalists.[54]

52. Ahmad, *Dhikrayat Harb al-Yaman*, 307, 319; and O'Ballance, *War in Yemen*, 185–86.

53. Hadidi, *Yaman*, 146; Ahmad, *Dhikrayat Harb al-Yaman*, 319; Fawzi, *Harb al-Thalath Sanawat*, 25; Faraj, *Rijal fi Khanadiq*, 224–25, 231; and Rahmy, *Egyptian Policy*, 239.

54. Huwaydi, *Hurub 'Abd al-Nasir*, 133; O'Ballance, *War in Yemen*, 188; Bissell, "Soviet Use of Proxies," 98; Faraj, *Rijal fi Khanadiq*, 224–25, 231; and Mutahhar, *Yauma Walada al-Yaman Majadahu*, 153.

In late 1967, the Royalists, confident that they could end the war, launched an offensive that carried them to Sana, and by 29 November, the city was surrounded and under bombardment. The siege of Sana in late 1967 was the biggest Royalist operation of the war, involving over fifty thousand tribesmen, opposed by only three thousand Republican soldiers in Sana. The Sana garrison was able to defend the city, and a Republican force advancing from Hodeidah lifted the siege on 8 February 1968. During the rest of the year, the Republicans were able to gain control of about half of the country. Saudi Arabia stopped supporting the Royalists in March 1968 because the Egyptians had left, and, afterwards, fighting eased since the Royalists had no money to buy tribal loyalties. In addition, the imam had lost support among the Zaydis since he did not always personally lead his forces into battle, was indecisive, and allowed the Saudis to place stipulations on how he used their aid. The Saudis decided to work toward peace in North Yemen and the establishment of a moderate Republican-Royalist government that would not oppose Saudi interests in the area.[55] In spring 1970, the Saudis encouraged negotiations between the Republicans and Royalists, excluding any members of the imam's family. An agreement was reached on forming a joint Royalist-Republican government, and Imam Al-Badr was given political asylum in Great Britain.

One of the more ambiguous aspects of the war is the number of Egyptian deaths. The numbers given in sources consulted for this article range from four thousand to twenty-six thousand.[56] The greatest Egyptian losses probably occurred between October 1962 to March 1963. During this period, Egyptian forces arriving in Yemen went directly into combat without time to acclimate or learn about the enemy. General Fawzi believed that more Egyptians died in Yemen from disease and accidents than from combat. General Hadidi estimated Egyptian losses at over ten thousand. He was probably in the best position to give an accurate assessment because in 1967, he headed the Egyptian military court's investigation of Egypt's defeat in the Six-Day War and its connection to the Yemeni intervention.[57] Imam Al-Badr, in an interview con-

55. Bissell, "Soviet Use of Proxies," 99–100; Ahmad, *Dhikrayat Harb al-Yaman*, 19–22; and Rahmy, *Egyptian Policy*, 240.

56. Badeeb, *Saudi-Egyptian Conflict*, 105; and Ahmad, *Dhikrayat Harb al-Yaman*, 597. Badeeb, writing from the Saudi viewpoint, believed that the Egyptians suffered twenty-six thousand killed in North Yemen, while Ahmad, an Egyptian, estimated Egyptian dead at four thousand. In May 2000, the author visited an Egyptian military cemetery in Sana which the caretaker said contained the graves of 1,737 Egyptians killed during the intervention. According to the caretaker, similar cemeteries are located in Hodeidah, Taiz, and other towns.

57. Fawzi, *Harb al-Thalath Sanawat*, 26; Ahmad, *Dhikrayat Harb al-Yaman*, 592–93; and Hadidi, *Yaman*, 161, 165.

ducted after the war, stated that the number of Yemenis killed was around forty thousand.[58]

Effects on the Egyptian Army

Egyptian performance in the Six-Day War demonstrates the effect of prolonged counterinsurgency warfare on a regular army's ability to fight a conventional war. There is no doubt that the Yemeni intervention had a negative impact on the readiness of the Egyptian army, and it has been suggested that Egypt's total defeat in the Sinai in 1967 resulted from the Egyptian intervention in North Yemen. General Hadidi believed that it was impossible to separate Egyptian performance in the Six-Day War from the effects of the Yemeni intervention on the army.[59]

As indicated, during the Yemeni intervention, the Egyptian high command, particularly Field Marshal Amer, neglected the Sinai Theater, where Egypt expected to fight another war with Israel. And due to the deployment of troops in Yemen, Egypt was forced to rely partially on reservists in the Sinai. Losses suffered by the Egyptian army in Yemen affected its performance in the Six-Day War, and the intervention did not give it the stability it needed to train for a conventional war. Equipment was in bad repair, little training occurred because of budget problems associated with the intervention, and troops were fatigued from repeated rotations to Yemen. In early 1967, fuel and ammunition for training were reduced for economic reasons. After officers returning from Yemen convinced the high command to limit training because the army was exhausted, training was restricted to the small unit level.[60]

The intervention affected the army's attitude towards warfare. Egyptian soldiers came to believe that all wars would be like the one in Yemen, where the Egyptian air force dominated the skies unopposed by enemy aircraft. Soldiers grew accustomed to working under the umbrella of friendly air power and did not employ antiaircraft measures. In the Six-Day War, after the Egyptian air force was destroyed on the ground, the army was unprepared to deal with enemy-controlled skies. In Yemen, the tempo of combat operations was slow. The enemy moved mainly on foot and had few vehicles, and in turn, Egyptian responses could be slow. The enemy in Yemen had little artillery and no armored vehicles, so there was no fighting in the conventional sense as would occur in the

58. Badeeb, *Saudi-Egyptian Conflict*, 120, from an interview conducted with Imam Al-Badr.

59. Hadidi, *67 War*, 47; and Rahmy, *Egyptian Policy*, 250.

60. Huwaydi, *Hurub 'Abd al-Nasir*, 142–43; Gamasy, *The October War*, 37; Fawzi, *Harb al-Thalath Sanawat*, 26, 28; Murtaji, *Al-Fariq Murtaji Yarwi al-Haqa'iq*, 24, 31; Hadidi, *Shahid*, 57; and Hussini, *Soviet-Egyptian Relations*, 176.

Sinai. Small units in Yemen became accustomed to operating independently, while in a conventional war, close coordination between units and the use of combined arms would be vital.[61]

The war in Yemen led to the corruption of the officer corps and affected the army's discipline and attitude. Due to the privileges accorded to those serving in Yemen, many officers used patronage and nepotism to get assigned to Yemen repeatedly. Some officers in Yemen exaggerated their exploits in battle to gain promotions and awards, and developed a personal interest in the continuation of the war. The intervention led to a sense of overconfidence; the Egyptians won most of the encounters in Yemen, and the army came to believe that if it could defeat tribesmen, it could defeat Israel. The Egyptian media exaggerated the victories in Yemen, as did individual officers. Some of the high command became overconfident, especially Amer, who on the eve of the Six-Day War told Nasser that the military was in "tiptop shape." Amer ignored the fact that in 1966 the Council of Operations for the armed forces had studied Egypt's strategic situation and recommended to him that war with Israel be avoided due to the Yemeni intervention. Discipline at the small unit level also deteriorated as a result of the intervention because units in Yemen operated without close supervision.[62]

Many of the Egyptian generals as well as Anwar Sadat believed that the assignment of twenty thousand Egyptian troops to Yemen weakened the army's defense of the Sinai. Yemen absorbed 30 percent of Egypt's infantry, 16 percent of its armor, and 10 percent of its artillery. But given the magnitude of the Israeli victory, these troops would probably not have made a significant difference, and Amin Huwaydi, Egyptian Minister of War in 1967 and 1968, felt that the troops' presence in Yemen saved them from destruction. These units were later vital for defending Egypt since most of the remainder of the army had been destroyed in the Sinai.[63]

Although some of the Egyptian generals tried to use the Yemeni War to exonerate the army's poor performance in 1967, much of what caused the 1967 defeat was poor senior leadership at the highest level. Amer used the armed forces to establish a power base and corrupted the army with his policies, in which loyalty to him became the primary criterion

61. Rahmy, *Egyptian Policy*, 252; Fawzi, *Harb al-Thalath Sanawat*, 27, 28; Hadidi, *67 War*, 58–60; and Murtaji, *Al-Fariq Murtaji Yarwi al-Haqa'iq*, 24.

62. Anwar Sadat, *Those I Have Known* (New York: Continuum Publishing, 1984), 131; Hadidi, *67 War*, 57–58, 60; Sadat, *In Search of Identity*, 172; Gamasy, *The October War*, 37; Murtaji, *Al-Fariq Murtaji Yarwi al-Haqa'iq*, 24; Fawzi, *Harb al-Thalath Sanawat*, 26–28; and Rahmy, *Egyptian Policy*, 252.

63. Gamasy, *The October War*, 49; Murtaji, *Al-Fariq Murtaji Yarwi al-Haqa'iq*, 24; Sadat, *Those I Have Known*, 131; Sadat, *In Search of Identity*, 163; Hadidi, *67 War*, 57; Huwaydi, *Hurub 'Abd al-Nasir*, 147; and Ahmad, *Dhikrayat Harb al-Yaman*, 635.

for promotion and advancement. These policies were in place before the Yemeni intervention, although Amer did use Yemen to strengthen further his position against Nasser.[64]

Conclusion

The Egyptian army was never able to achieve anything more than a stalemate in North Yemen. During the five-year intervention, the Yemeni republic was almost always under attack, and the Egyptians had to expend great resources just to keep the country from being overrun. Egypt's performance in Yemen demonstrates, as did U.S. and Soviet efforts in Vietnam and Afghanistan, respectively, that conventional tactics and a conventional approach to counterinsurgency will fail against a determined opponent with outside support.

Conventional tactics prescribe the concentration of force against the enemy's main strength to gain decisive results. Maximum use is made of a range of ordnance to destroy the enemy and limit friendly casualties. When these tactics are applied to counterinsurgency, they result in the death or alienation of the nearby population. The government is blamed for these actions, and the insurgents gain popular support. Like the Egyptians in Yemen, the United States in Vietnam employed conventional tactics, with large units conducting sweeping operations to clear the enemy from the population. After the completion of these operations, the army would move on and the enemy would return to the area that was just cleared. And in the course of the battle, the countryside was destroyed and the local population embittered towards the United States and the Saigon regime.[65]

In counterinsurgency, most authorities agree that a regular army should avoid large-scale operations and instead focus on small unit patrolling. Operations should concentrate on rapid mobility, going anywhere at anytime, and making all areas unsafe for the insurgents. When small units make contact with guerrillas, they should be assisted by a quick reaction force in pursuing and destroying the guerrillas, taking care to minimize civilian deaths. Once an area is secured, a presence should be maintained there until the support of the local population has been won, or the area will revert to insurgent control, as happened

64. Hadidi, *67 War*, 81; Hadidi, *Yaman*, 155; and George W. Gawrych, "The Egyptian High Command in the 1973 War," *Armed Forces and Society* 13 (Summer 1987): 542.

65. Douglas S. Blaufarb and George K. Tanham, *Who Will Win?* (New York: Taylor and Francis, 1989), 23, 78. During the 1960s, many comparisons were made between Egyptian involvement in Yemen and U.S. involvement in Vietnam; see, for example, Vocke, "What Goes on in Yemen," 25.

repeatedly in Yemen, Afghanistan, and Vietnam. The regular army, using small, elite units, should work to win the confidence of the population and convince them that the government regime offers a better way of life than the insurgents. Eventually, the locals, aided by the army, can form militias to defend themselves from the insurgents.[66]

A regular army has difficulty adopting these tactics for several reasons. First, in the initial stages of intervention, the national and army leadership frequently believe that the insurgent problem can be solved quickly; neither deep study nor major modifications to the army's normal procedures are required. The Soviets expected to be in Afghanistan for no longer than two years, while the Egyptians believed that they could achieve their goals in Yemen in three months. Even the United States recognized the new Yemeni republic in December 1962 since U.S. intelligence predicted that the war would end in a quick Royalist defeat.[67] Second, a regular army often has no preexisting counterinsurgency doctrine to use as a basis for modifying its conventional war fighting organization and methods. The Soviets had no such doctrine prior to their intervention in Afghanistan because they believed that all true insurgent movements were just, progressive, and designed to overthrow capitalists, and thus the Soviet Union would always be on the side of the insurgent.[68] The Egyptians, who adopted Soviet doctrine after 1956, also had no counterinsurgency doctrine and did not believe they would ever fight insurgents. Without a counterinsurgency doctrine as a start point, the Soviets and Egyptians had to experiment through trial and error. Eventually, the Egyptians developed some suitable doctrine, but it was not fully adopted.

Finally, to adapt fully to counterinsurgency, a regular army must make fundamental changes to its traditional methods of operations and organization, which the generals frequently prefer not to do. Even an army that has a preexisting counterinsurgency doctrine and standing counterinsurgency force can have trouble conducting counterinsurgency operations. Before Vietnam, the U.S. army had developed its Special Forces into a counterinsurgency force, and the army had a highly developed counterinsurgency doctrine. Still, most of the U.S. army in

66. Blaufarb and Tanham, *Who Will Win?*, 24; Drew, "Insurgency and Counterinsurgency," 42–43; and Sir Robert G. K. Thompson, "Regular Armies and Insurgency," in *Regular Armies and Insurgency*, ed. Ronald Haycock (London: Croom Helm, 1979), 12–13.

67. Scott R. McMichael, *Stumbling Bear: Soviet Military Performance in Afghanistan* (London: Brassey's, 1993), 10; Fawzi, *Harb al-Thalath Sanawat*, 23; John S. Badeau, *The American Approach to the Arab World* (New York: Harper and Row, 1968), 137–38; and H. W. Brands, *Into the Labyrinth: The United States and the Middle East, 1945–1993* (New York: McGraw-Hill, 1994), 85.

68. McMichael, *Stumbling Bear*, 39–40.

Egyptian military cemetery in Sana containing graves of Egyptian soldiers killed during the intervention.

Vietnam did not follow the doctrine, and there was constant friction there between the U.S. Special Forces and the rest of the U.S. military.[69] It is difficult for a regular army to transform itself into a counterinsurgency force because most regular armies focus on preparing to fight other regular armies in a major conventional war, which is viewed as presenting the greatest threat to national survival. All aspects of most regular armies, their organization, equipment, and training, are geared towards preparing for the next big war.[70]

Armies prefer not to make the required fundamental changes to fight counterinsurgency because this threatens their most basic institutional structures. General Fawzi believed that the Egyptian bureaucracy prevented Egypt from adapting to counterinsurgency, and Amer used the armed forces and the Yemen intervention to consolidate his personal

69. Joseph R. Fischer, "Cut from a Different Cloth: The Origins of U.S. Army Special Forces," *Special Warfare Magazine*, April 1995, 38. For a description of U.S. counterinsurgency doctrine prior to the major involvement in South Vietnam, see Department of the Army, *FM 31-15, Operations Against Irregular Forces* (Washington, D.C.: Government Printing Office, May 1961). Charters and Tugwell, *Armies in Low-Intensity Conflict*, 253.

70. Drew, "Insurgency and Counterinsurgency," 22–23.

Monument located outside of Sana dedicated to Egyptian soldiers killed in North Yemen.

power. Radical changes to normal army functioning would have threatened this goal. In Vietnam, the U.S. military thought that reshaping its tactics and strategy would wreck its hierarchy. An unidentified U.S. general described the reasoning: "I'll be damned," he said, "if I permit the United States Army, its institutions, its doctrine, and its traditions, to be destroyed just to win this lousy war."[71]

Instead of adapting to the methods of counterinsurgent war, regular armies often try to use superior technology to defeat the insurgents, as in Afghanistan, Vietnam, and Yemen. In all three cases, the insurgents were usually armed with only the most basic weapons, but the tanks, airplanes, and artillery of the intervening armies still could not defeat them. When technology fails, regular armies sometimes resort to shows of force, intimidation, and destruction of the countryside in order to subdue the population. These tactics lead to the further loss of popular backing for the government supported by the intervening army.

71. The general is quoted in Blaufarb and Tanham, *Who Will Win?* 90.

DAVID M. WITTY

According to Sir Robert Thompson, the real purpose of a regular army conducting counterinsurgency is not so much to fight insurgents as it is to support the civil government which it has intervened to help. The problem is to make the civil government attractive to the people; the insurgents are only symptoms of larger social and political issues which caused the insurgency.[72] However, attempts by an intervening nation to solve the social and political problems of a nation facing insurgency can lead to some of the most serious mistakes in counterinsurgency warfare. In the United States, it is popularly argued that application of democratic concepts, and social and political reform will end an insurgency.[73] In Vietnam, the United States tried to apply its model of democratic reform, but the Saigon government still lost support. The Egyptians, for their part, tried to modernize Yemen and build a nation patterned on Egypt. Yemen's tribal population, which did not particularly care about Arab revolution and Arab socialism, opposed these changes. In Afghanistan, the Soviet Union tried to introduce socialist concepts to the Muslim population with disastrous results. Counterinsurgency efforts tend to try to reshape the country according to the world view of the intervening force instead of the concerns of the indigenous population of the country facing the insurgency.[74]

Another issue in counterinsurgency is the role of an intervening army in relation to the government it is supporting. In Yemen, Egypt came to be regarded as an occupier. Its reforms of the Yemeni government, based on the Egyptian model, did not win support for the Republicans. Egypt also led and provided most of the forces used in the military operations, further reinforcing the belief that Egypt was an occupying nation. Current U.S. counterinsurgency doctrine correctly states that an intervening force should take a background role and allow the government facing the insurgency and its army, to bear the major responsibilities of reform and fighting. This reinforces the belief that the host government is actually in charge and the intervening force is merely providing assistance.[75] But as with the Egyptians in Yemen, and the Soviets

72. Sir Robert G. K. Thompson, introduction to Blaufarb and Tanham, *Who Will Win?*, v. Thompson was the principal director of the British counterinsurgency effort in Malaya.

73. See, for example, Howard L. Dixon and Charles M. Ayers, "Operational Art in Low Intensity Conflict," *CLIC Papers* (Langley AFB, Va.: Army–Air Force Center for Low Intensity Conflict, 1987), 11.

74. For the tendency of the U.S. military to impose the U.S. model of government on nations facing insurgency, see Lawrence A. Yates, "Military Stability and Support Operations: Analogies, Patterns and Recurring Themes," *Military Review*, July-August 1997, 60.

75. For the role of the U.S. military in supporting counterinsurgency in current U.S. doctrine, see Department of the Army, *FM 100-20, Military Operations In Low Intensity Conflict* (Washington: GPO, December 1990), 2.18–2.20.

in Afghanistan, there will be cases where the host nation's government and army are so weak that the intervening force must bear the full burden, leaving it in a no-win situation. The conflict will come to be regarded as a war between the intervening army and the insurgents, not a war between the country's legal government and the insurgents. Ironically, despite the fact that they had borne the full responsibilities of the wars, when the Egyptians departed from Yemen and the Soviets left Afghanistan, the government forces in both cases were able to fight on unaided for a time, indicating that the very presence of intervening armies might have diminished the effectiveness and credibility of the government forces they were supporting.

In Yemen, the Egyptians did have an advantage over the U.S. and Soviet counterinsurgency efforts in Vietnam and Afghanistan. The Egyptians were Arabs like the population of Yemen. Although the Egyptians spoke a different dialect of Arabic, the Egyptians and the Yemenis could communicate with each other. The Egyptians were Muslims like the insurgents they fought, although there were differences between the Egyptian Sunnis and the Royalist Zaydis. On the whole, however, these differences were far fewer than those between the Americans and Vietnamese, or between the Soviets and the tribal population of Afghanistan. While the cultural differences between the Egyptians and the Yemenis remained great, the Egyptians did have a cultural affinity advantage in their counterinsurgency efforts, and the outcome of the Egyptian intervention was different from the results of the U.S. and Soviet interventions. These met with complete failure and resulted in the establishment of governments which were fundamentally opposed to the philosophies of the intervening nations. In contrast, the Egyptian intervention, which militarily achieved only a stalemate, did prolong the struggle and avert a quick Royalist stifling of the Yemeni revolution. Ultimately, the Egyptian intervention prevented the restoration of the Imamate, and the civil war ended in the establishment of a compromise government composed of both Royalists and Republicans, without the Zaydi imam.

However, the stalemate the Egyptians achieved in Yemen came at a great cost, not just in the blood and resources lost there, but in the Sinai, as well, in June 1967. Yemen is an example of the effects of prolonged counterinsurgency warfare on a regular army's ability to fight a conventional war. The U.S. army left its counterinsurgency war in Vietnam in 1973 in a shambles. Fortunately, it had eighteen years to recover and rebuild before it had to fight a major conventional war. The Egyptian army was not so lucky.

Intifadas in the Occupied Territories

[19]

The Strategic Impasse in Low-Intensity Conflicts: The Gap Between Israeli Counter-Insurgency Strategy and Tactics During the *Al-Aqsa Intifada*

SERGIO CATIGNANI

Department of War Studies, King's College London

ABSTRACT Over the past 15 years Israel has been involved in a bitter counter-insurgency campaign against the Palestinians. Palestinian insurgency, particularly during the current *Al-Aqsa Intifada*, has posed serious challenges to the Israel Defence Forces (IDF). Whilst being able to adapt successfully its tactics to Palestinian terror and urban guerrilla warfare in order to reduce the level of Palestinian violence, the IDF has not been able to achieve a battlefield decision or victory. This has been due to the nature of the Israeli–Palestinian conflict, which ultimately necessitates the provision of a political solution by the Israeli political leadership, which has relied too often on the IDF as a panacea for its own strategic and political indecisiveness vis-à-vis the Palestinian national question.

KEY WORDS: Intifada, Israel, Counter-insurgency

Introduction

During the past 15 years Israel has strategically and tactically responded to local Palestinian insurgency, which has evolved from civil disobedience to outright terrorism and guerrilla warfare, particularly following the now-defunct Oslo peace process. Indeed, over the past 15 years the Israeli Defence Forces (IDF) has tried to transform its conventional army and adopt low-intensity urban warfare strategies and tactics against violent rioters, guerrilla fighters and

Correspondence Address: Sergio Catignani, Department of War Studies, King's College London, 138-142 Strand, London, WC2R 1HH. E-mail: Sergio.catignani@klc.ac.uk

terrorists. This has been particularly the case during the current *Al-Aqsa Intifada*, which began in September 2000.[1]

Notwithstanding the extensive and successful innovations in tactics, weaponry and training adopted in response to Palestinian attacks, the political nature of low-intensity insurgencies and, thus, the inherent strategic restraint imposed by the Israeli political echelon's decision to conduct low-intensity warfare ad nauseam without seriously providing an alternative political resolution to the current Israeli–Palestinian conflict, have prevented the IDF from imposing a clear battlefield decision. Thus, the IDF has been unable to achieve strategic success during the past four years of the Israeli–Palestinian conflict despite the Israeli political leadership's belief that the IDF could achieve such objectives on its own.

Despite the enormous difficulties in fighting Palestinian terror and guerrilla elements within the villages and cities of the West Bank and Gaza Strip, the IDF has managed to develop significant tactical and technological solutions to the Palestinian terrorist and insurgent threat, albeit occasionally at the cost of collateral damage and cases of excessive force vis-à-vis the Palestinian civilian population. Such tactical success, moreover, cannot yield strategic dividends without greater political and strategic direction from the Israeli political echelon. Whether or not such tactical successes will be used by the political echelons to arrive at a strategic solution to the Israeli–Palestinian conflict is to be seen, especially in light of the Israeli perception that there is no serious political negotiator on the Palestinian side.

Strategy

Strategy involves the employment of military forces to achieve a specific political goal. Since war is an instrument of politics, 'limited political aims result in the definition of limited war aims'.[2] However, the use of conventional strategy to low-intensity type conflicts has always been difficult. The 'intermingling with enemy forces, mixing with the civilian population, and extreme dispersion' have significantly challenged conventional armies particularly within the urban arena of warfare.[3]

As shall be seen in the following sections, the IDF has struggled to deal with Palestinian insurgency since the end of the 1980s. Even though the IDF has adapted its tactics, particularly since the late 1990s, to face Palestinian threats head-on within their towns, villages and refugee camps, it has been unable to attain a battlefield decision in the Israeli–Palestinian conflict. In fact, such a goal is unattainable: a strategic doctrine that is geared towards linear conventional threats and that has clear parameters for what constitutes battlefield decision and victory cannot be applied to insurgencies, which for the most part, must be resolved politically. Before elaborating such an argument, though, the article will first look briefly at Israel's strategic doctrine.

Israeli Security Strategy

Since the early 1950s,[4] the IDF has customarily differentiated between two types of military operations: '*bitachon shotef* ("current security") and *bitachon yisodi* ("fundamental security"). The former – which has abbreviated usually to the term *batash* – includes responses to terrorist attacks, retaliatory raids and border skirmishes; the latter to big wars [i.e., conventional], real or potential'.[5] The Israeli strategic doctrine has traditionally focused on fundamental security aspects due to the greater nature of the threat stemming from a conventional war against Israel.

Hence, the basic assumptions underlying the Israeli strategic doctrine, which has focused primarily on the Arab conventional threat, were:

(1) That Israel was and will continue to live in a hostile environment, hence, the belief that it was confronted with wars of 'no choice' (*ein briera*).
(2) That Israel was involved in a conflict in which it finds itself strategically inferior vis-à-vis its Arab enemies both in terms of manpower and resources.
(3) That no matter how decisive results on the battlefield were, Israel would never be able to achieve complete strategic victory.

Hence, Israel's general strategic goal has always been that of maintaining the status quo by deterring major attacks against it. The exception in applying such a strategic goal occurred unsuccessfully in the 1982 Lebanon War when force was not used only to root out the Palestinian Liberation Organization's terror infrastructure in South Lebanon, but also to establish a friendly Lebanese government controlled by the Christian Maronite minority.

In the case of the outbreak of a conflict, returning to the status quo *ante* would be achieved through the attainment of a swift battlefield decision, which as Avi Kober points out, is not synonymous with victory. In fact,

> battlefield decision can be defined in terms of negating the other side's combat capability, victory can be defined in terms of the correlation between what each adversary defines as its political

> and military war objectives ... and what it actually succeeds in achieving during that war.[6]

That is, battlefield decision in the Israeli case can be equated to deterrence by denial: once war is imminent or has broken out, the defending armed forces will try to disrupt as quickly as possible the aggressor's military capability by attacking and disrupting the enemy's centre of gravity.

Indeed, the three pillars of Israeli strategic doctrine have been 'deterrence, early warning, and the winning of a decisive [battlefield] victory'.[7] As stated above, the Israeli strategic doctrine has mostly focused on the conventional threat due to the fact that the Israeli political and military leadership have always seen it as an existential threat. Nonetheless, the three pillars of the Israeli strategic doctrine were called into question in light of the 1973 Yom Kippur War[8] and have proven difficult to put into operation since the Lebanon War when dealing already with low-intensity type operations, which have increased over the past 20 years. In fact, Israel's deterrence policy based on reprisals vis-à-vis sub-conventional threats and guerrilla or terrorist forces has served 'as a means of redress, but has not, generally prevented recurrence of the provocation'.[9]

Despite the IDF's traditional focus on the conventional threat in terms of strategy, order of battle, manpower policy and training, the IDF has been for the most part involved in current security operations particularly since the first *Intifada* (1987–93). Whereas at the beginning of the first *Intifada*, IDF and other security forces such as the *Shabak* (General Security Services, GSS) were involved in anti-riot, policing operations and arrests, towards the end of the first *Intifada* and particularly during the Oslo peace process (1993–2000), the IDF was used increasingly in counter-terror and counter-guerrilla operations.

Moreover, with the outbreak of the *Al-Aqsa Intifada* in September 2000, the IDF has been in constant operational use against terrorist and armed guerrilla fighters – many of which were trained, equipped and mobilized paradoxically under the auspices of the Oslo peace process in order to 'to guarantee public order and internal security'.[10] Over the past 20 years, the IDF has also been involved increasingly in routine policing and anti-riot measures some of which have had a preventative function, such as the widespread creation of checkpoints and the extensive use of military patrols. However, the IDF has also resorted to measures that, although tactically effective, have had more of a punitive purpose and negative strategic results, such as the imposition of wide-ranging closures of the Occupied Territories and the enforcement of protracted curfews within Palestinian cities, towns and villages considered to be either hotbeds of Palestinian militancy or outright terror. The practice of demolishing the houses belonging to terrorists or to their closest relatives has also been considered successful by the IDF as a deterrent for other terrorist attacks in spite of the censure expressed by human rights organizations, by the European Union and other states as well as by Israel's closest ally, the US.

Nonetheless, many others have stated that such collective measures have not had much of a deterrent effect, but have actually galvanized the Palestinian population. For example, Zuhair Kurdi, a journalist with Hebron's Al Amal TV station, has stated somewhat rhetorically that, 'the legal father of the suicide bomber is the Israeli checkpoint, whilst his mother is the house demolition'.[11]

In spite of the pervasive use of IDF personnel and materiel as well as the growing threat and lethality of Palestinian terror and guerrilla operations within Israel and the Occupied Territories, the IDF has not been able to address strategically the predominant and relatively new low-intensity dynamic of the Israeli–Palestinian conflict. Some Israeli military (e.g., the former Chief of Staff [COS] and current Defense Minister Shaul Mofaz and current COS Lieutenant-General Moshe Ya'alon) and political leaders (e.g., Prime Minister Ariel Sharon) have come to define as existential threats that which were regarded traditionally as current security threats (i.e., terror, guerrilla and riot violence).

In June 2000, for example, then Coordinator of Operations in the Territories, Major-General Amos Gilad, declared that 'the purpose of the IDF's campaign was to reduce the level of terror, which in the scope and depth of its damage has become a strategic threat, with the first signs of threatening our existence in terms of our quality of life'.[12] There has not been, however, a major shift in the IDF's strategic approach vis-à-vis low-intensity-type conflict scenarios and the belief that it could impose a battlefield decision or even obtain victory was shared initially by top officers of the IDF General Staff. Thus, COS Ya'alon in the most widely-read daily newspaper, *Yediot Ahronot*, stated in August 2002 for example that, 'the only solution [to the *Al-Aqsa Intifada*] is to achieve an unequivocal victory over the Palestinians' and that such a victory would not come at a low price or immediately.[13]

In fact, Israel's conventional strategic goal of maintaining the status quo through deterrent retaliatory or pre-emptive measures has been erroneously applied to the contemporary Israeli–Palestinian conflict. Whereas in the past maintaining the status quo through deterrence was geared towards the conventional threat and, hence, towards the avoidance of a major conventional attack by Israel's neighboring states, which could have threatened its existence, today Israel's attempts at deterring Palestinian terror and guerrilla attacks have the very limited goal of reducing the level of violence and maintaining a status quo. Such a status quo, though, has become unbearable for the Palestinian population living in the Occupied Territories and has proved increasingly costly to Israeli society.

Preserving the status quo vis-à-vis the conventional threat was not only essential to Israel's existence, but was also a widely-held principle of the Israeli national security consensus. Nonetheless, upholding the status quo today has diminished dramatically the already meager Palestinian standard of living due to extensive curfews and closures – which, together with the rampant corruption of the Palestinian Authority, have wrung out of the Palestinian population any notable source of financial sustenance. It has also eroded the Israeli national security consensus[14] – particularly regarding the continued control of most of the territories seized during the 1967 Six-Day War. The continuation of the conflict also has led to a major economic crisis in Israel that has affected all major government budgets, Israeli living standards, but most crucially, the defense budget. In fact, a recent survey of 7,200 Israelis aged 20 and over from all regions of the country commissioned in late 2003 by the Central Bureau of Statistics found that 35 percent of the survey sample reported that their financial situation had deteriorated in the past five years.[15]

Furthermore, the use of the IDF to lower the level of violence – as well as to convince the Palestinians that Israel would not negotiate under fire – has led to an over-emphasis on the operational and tactical use of the IDF as a panacea for the lack of the military and political leadership's ability to come up with a new strategic paradigm that will bring about a strategic decision vis-à-vis the problem of Palestinian terror and other forms of low-intensity violence. The political echelon's inability to develop a new strategic paradigm, given the experiences of the first *Intifada*, the 1991 Gulf War and the changed strategic environment brought about by the Israeli–Palestinian and Israeli–Jordanian peace accords, was already voiced in 1998 by a Knesset sub-committee chaired by Knesset member Dan Meridor. The enquiry, in fact, stated that:

> The Government of Israel, which is responsible for Israel's national security, has not conducted any substantial and comprehensive discussion regarding the national security policy and its applications... We have not found any integrative and long-term thinking, examination or decisions. It is critical that such examinations be conducted at the national level, and not only by the military or the defense system.[16]

In fact, until very recently Israel has been able to afford the price of not addressing the Palestinian problem strategically, because of the IDF's ability in providing effective short-term tactical solutions to all forms of Palestinian violence.

IDF Tactical and Technical Innovations

The IDF, in fact, has been able to prepare and respond tactically to the various Palestinian threats used against Israeli civilians and soldiers, particularly since the mid- to late 1990s when the Oslo peace process was quickly degenerating into tit-for-tat low-intensity skirmishes and which eventually erupted in September 2000 into a full-scale guerrilla and terror campaign. According to then COS Shaul Mofaz, this was due to the fact that 'we prepared the military for this confrontation. We trained and bought equipment for low-intensity conflict'.[17] Preparations, in fact, had started soon after the September 1996 riots. Colonel Gal Hirsh stated that in early 1997 then COS Amnon Lipkin-Shahak had told him: 'We must prepare for war and continue with the peace process; go there and help General Itzhak Eitan, who was the Chief of the Central Command [i.e., the area responsible for security in the West Bank], help him to prepare units for war'.[18]

Despite the rhetoric of such men, Israel's actual preparedness for the reality of urban warfare involving a very hostile, well-armed group of terrorist-guerrillas entrenched in an extremely complex built-up and densely populated battleground was severely tested. For example, Operation 'Defensive Shield' which was initiated in April 2002 in response to the late-March 'Passover Massacre' suicide bomb attack,[19] was the first major Israeli urban warfare operation to be carried out since the siege of Beirut in 1982. Operation 'Defensive Shield' was initiated in order to seize weapons, arrest terrorists and their support network, destroy weapons factories and suicide bomb workshops and kill suicide bombers. The IDF encountered significant resistance in most Palestinian West Bank cities it entered (e.g., Qalqiliya, Nablus and Tul Karem), but encountered especially stiff opposition in the Jenin refugee camp where 15,000 poverty-stricken civilians packed into 600 square yards.

Around 300 Palestinian guerrilla fighters affiliated to the various Palestinian terrorist organizations (i.e., the PLO-affiliated *Al-Aqsa* Martyrs Brigade, Force-17 and *Fatah Tanzim*, as well as Palestinian Islamic *Jihad* and *Hamas*) were involved in the battle of Jenin. Snipers, mines and booby-traps were literally everywhere: inside cupboards, under sinks, inside sofas, in cars and dumpsters. 'On one street alone, an Israeli [D-9] bulldozer detonated 124 explosive charges, some weighing as much as 250 pounds.'[20] Similar to other armies' urban warfare doctrine, the Israeli army used in Jenin overwhelming numbers: there were approximately 100 soldiers for every Palestinian gunman.

Intelligence

Not only did the Israeli army use overwhelming numbers, it also used most of its highly sophisticated weaponry and intelligence-gathering capabilities in order to maintain the initiative, something armies entering urban theater of operations have always had trouble keeping, because of the high-tempo and level of confusion that urban battlefields usually create. Indeed, reiterating the importance of intelligence in the urban battlefield, IDF COS Ya'alon stated at a recent international conference on low-intensity conflict warfare that, 'creating intelligence dominance is a critical factor for managing and dominating the low-intensity conflict. Qualitative intelligence provides the ability to realize military power properly and precisely'.[21]

Hence, the decision by the IDF General Staff to form in April 2000 the Field Intelligence Corps, which includes combat intelligence-gathering units trained to gather tactical intelligence and to provide it in real-time to combat units during operations. This has enabled information, through the increasing digitization of its armed forces, to go faster to the troops and, in turn, has reduced the element of surprise from looming Palestinian attacks.[22] Thus, despite the greater freedom that IDF commanders and soldiers traditionally have had on the battleground, the improvement in the command, control and communication systems has enhanced the IDF's ability to monitor ground operations and provides real-time operational intelligence to ground troops.

According to an AH-1S Cobra helicopter squadron commander, during the fighting the Israel Air Force 'kept four attack helicopters and two Searcher II reconnaissance Unmanned Aerial Vehicles (UAVs) overhead at all times to ensure commanders knew where their troops on the ground were'.[23] Human intelligence was also provided by *Mistar'aravim* (i.e., '*to become an Arab*') units, who are trained to blend in with the local population, gather operational intelligence and sometimes conduct targeted killings. More importantly and underlining former GSS Chief Ya'acov Peri's conviction that 'there is no substitute for a human source',[24] the GSS had begun by the end of 2000 a large recruitment drive for Palestinian collaborators given the fact that their use had diminished significantly since the start of the Oslo Peace Process.[25] Such collaborators, in fact, have been put to good use, particularly in the targeted assassination of key terrorist leaders, such as *Hamas* 'spiritual' leader, Sheik Yassin and *Hamas* 'political' leader Abdel Aziz al-Rantissi in early 2004.

Furthermore, attack helicopters very often were used to pinpoint and eliminate hostile forces either by using snipers or missile attacks. As one Special Forces captain stated, for example, 'in Jenin I was in a helicopter above everything. I saw it happening. We were snipers; from the helicopter we were supposed to locate a certain area and eliminate hostile elements'.[26]

Circumvention and 'Swarm' Tactics

Such intelligence-gathering efforts, which were initiated weeks prior to the IDF's entry into Jenin, were carried out in order to avoid most armies' traditional urban infantry 'attrition approach' tactics, whereby operations abide by the 'move, make contact, deploy, fire and maneuver model (usually along a linear axis)'.[27] The Israelis, on the other hand, 'were able to deploy out of contact with the enemy by selectively seizing small areas of the camp, drastically reducing exposure to enemy fire and maintaining momentum by only clearing as necessary'.[28] This could be done only through the prior accumulation and analysis of field intelligence.

Various small-unit infantry, armor and air force task forces 'swarmed' around the Palestinian forces from all directions, thus, successfully integrating disparate units and proving their interoperability. Such 'swarm' tactics especially managed to confuse Palestinian guerrilla fighters and terrorists. According to Colonel Gal Hirsh:

> In one battle in the Nablus Kasbah in about 24 hours they lost more than 80 of their gunmen and they never could identify where we were. We used the air force, combined forces and new fighting groups. Even if they were inorganic forces, they became task forces that knew how to fight together.[29]

The extensive use of tanks in major combat operations within urban Palestinian areas since Operation 'Defensive Shield' is due to the fact that the IDF's newest Merkava Mk.4 and upgraded versions of the Merkava Mk.3 have been 'equipped with advanced communication and battle management systems that enable individual tanks or very small groups of tank crews to operate autonomously for extended periods in conjunction with infantry units'.[30]

When faced with dangerous alleyways full of booby-traps and snipers, IDF forces used D-9 bulldozers to create alternative avenues of approach within buildings, albeit at the cost of significant collateral damage. Such circumvention and 'swarm' tactics were particularly used after 9 April when a suicide bomber detonated his explosive belt in a courtyard where 13 IDF infantry soldiers were instantly killed. The need to avoid such targeted and lethal suicide bomb attacks as well as heavy sniping against IDF personnel, led to such an innovation.

Since Operation 'Defensive Shield' the IDF has also learnt that in order to eliminate terror cells embedded in Palestinian towns and villages, it does not have to show overwhelming force, due to the fact that such large-scale 'invasions' into Palestinian areas create considerable international outcry and negative media exposure.[31] The IDF has, in fact, learnt that 'low-signature' operations often are not only more effective, but also domestically and internationally less controversial due to the stealth and rapidity with which they are carried out.

Continuous Improvement of Training Capabilities

Despite the IDF's relative success, the number of both IDF and Palestinian deaths was perceived as being too high.[32] Thus, in order to obviate 'high' casualty rates through better preparation and equipping, the IDF launched a multi-million dollar program in June 2002 to upgrade the IDF's Tze'elim National Training Center in the Negev desert. The center provides Israeli soldiers with significantly better urban warfare training facilities. The digital urban warfare center is, in fact, modeled after Palestinian cities and provides various landscapes in order to train the Israeli combat soldier for all types of contingencies. It includes a downtown area, rural village section, market area with narrow alleys and urban outskirts. The center eventually will help train around 90,000 reservists as well as all conscript ground forces in urban warfare battle skills.[33]

The need to reduce cases of abuse by Israeli soldiers whilst carrying out security duties after Operation 'Defensive Shield,' particularly at checkpoints and roadblocks, has led to the development of an ethical and operational code of behavior. Such a code, which is based on 11 key rules of conduct, has been taught over the past two years to both regular and reservist ground forces units. It provides extensive role-playing exercises that deal with the dilemmas of how to operate security checks on civilian and civilian property and, more importantly, with the dilemmas of what rules of engagement are acceptable within heavily populated civilian areas.[34] This, however, has not eliminated all cases of misconduct given the fact that soldiers at checkpoints have occasionally abused Palestinians, albeit more as a result of operational stress than malice.[35] In any case, a special Checkpoint Unit is being formed presently under the command of a Lieutenant Colonel from the Military Police Corps This unit will be trained in routine checks using advanced technological measures, in Arabic and in civil rights issues in order to improve the conditions of Palestinians trying to enter Israel and to reduce especially the time spent by Palestinians at checkpoints.[36]

Avoiding Non-Combatant Casualties

Efforts at reducing the number of non-combatant deaths during the current *Al-Aqsa Intifada*, while not sacrificing the protection of friendly forces, have led to three major weapons development efforts. The Israeli Ministry of Defence has commissioned, for example, the development of a high-speed automatic anti-sniper gun, called 'Believer' based on thermal imaging techonology[37] as well as the 'Corner Shot' non-line-of-sight weapon system, both of which aim to reduce the soldier's exposure to enemy fire.[38] The IDF has, furthermore, purchased from Israel Military Industries a new infantry weapon system called *Refa'im*, which enables the infantry soldier to launch grenades against targets beyond visual range. More importantly, the rifle 'features a self-destruct mechanism to avoid post-battle casualties'.[39]

In order to avoid killing civilians when targeting Palestinian snipers, who often have used deliberately crowds of women and youth supposedly demonstrating or rioting as their human shields, the IDF has decided to equip not only trained snipers, but also regular infantry with advanced optic sight scopes that have enhanced laser rangefinder capabilities.[40]

The IDF ground forces multi-year plan titled, '*Kela* 2008' [i.e., 'Shot 2008'] is also under way. This plan will significantly expand the IDF's infantry capability by adding five new light-infantry battalions, which will be based permanently in the Territories.[41] Their permanent stationing in the Territories will increase their tactical knowledge of the urban terrain and of the civilian population. Such knowledge and the greater number of infantry brigades will, in turn, reduce the stress soldiers usually suffer in urban warfare operations due to the greater turnover of rest periods that such soldiers will undergo. In fact, not only has the reduction of stress reduced cases of IDF abuse vis-à-vis the Palestinian population, but the 'ongoing connection with a single specific area ... has greatly improved operational successes'.[42]

Bitsuism

The reason for the IDF's successful provision of short-term tactical innovations and success can be ascribed to the organizational culture and military tradition of the IDF, which has demonstrated 'at virtually every level of war (except, perhaps, the highest: that of strategy) ... throughout its history a proclivity for the dashing, the unusual or the creative solution to military problems'.[43] Indeed, within the IDF, preference for the pragmatic *bitsuist* ('doer') over the reflective thinker has led to the IDF's tendency to focus its energies in current security problems often at the expense of long-term vision and innovation. As one IDF organizational psychologist admitted recently in an interview:

> When I got my education in the officers' course, the most central value that you are educated on is 'complete your task'. There's nothing more important than completing your task. Now, this is one of the ten principles of war: stick to your mission. But the full commandment is 'stick to your mission as long as it is directed to the strategic goal'. Remember the goal of the mission, the aim. But Israel tends to forget the second half.[44]

In fact, a typical aspect of the IDF's *bitsuist* organizational culture is 'the concentration of effort on pressing day-to-day problems' whereby commanders are not really 'troubled by the war to come', but find themselves instead in a position where they want 'to be everywhere, to decide everything, to invest the maximum in whatever engages' them.[45] One battalion commander, for example, stated that in 2003 alone, he planned and executed over 240 missions throughout the year.[46]

Such an operational tempo, clearly, reduces the chances of commanders to focus on the long-term strategic consequences that any operation may have on the Israeli–Palestinian conflict. This has been particularly the case during the current *Intifada* where, due to the growing lethality of Palestinian terror and guerrilla groups and the increasing budgetary constraints, the IDF has had to improvise and make due with fewer resources, despite the growing number of military operations.[47]

Another reason for this *bitsuist* tactical and operational emphasis has been the IDF's mantra of avoiding any political considerations and statements when suggesting to the political-security leadership strategic solutions to Palestinian violence. Although the supremacy of the political echelon over the military regarding the decision and nature of military force to be used is a paramount principle of democratic states – including Israel – taken to the extreme, military operations may prove strategically short-sighted if the military practitioners on the ground avoid making 'political-strategic' assessments of their current security operations.

Thus, the IDF has often found itself conducting missions, which although tactically and operationally effective, have had negative political-strategic outcomes due to cases of excessive force and collateral damage that are often repeatedly aired on major news networks around the world. A slight break from such a tradition of overlooking the strategic consequences of day-to-day tactical operations has been initiated by the current COS, Moshe Ya'alon. In fact, he has criticized the Israeli political leadership's decision to use heavy-handed military operations against Palestinian terrorists because of the substantial collateral damage caused when conducting such operations and despite the fact that they may have negative strategic consequences. Ya'alon was quoted in *Yediot Ahronot* as saying that, 'in our tactical decisions, we are operating contrary to our strategic interests'.[48]

Collateral Damage

Despite attempts to use precision-guided missiles and highly-trained snipers to eliminate Palestinian guerrillas and terrorists while avoiding casualties and major collateral damage, the extensive insertion of armored and infantry elements during Operation 'Defensive Shield' into such a compact battleground did bring about 53 Palestinian deaths – most of whom were in any case combatants – and structural damage to over 100 homes. The dilemmas and difficulties IDF soldiers went through when trying to eliminate Palestinian fighters fighting from inhabited civilian homes were poignantly voiced by a reservist from the *Nahal* Infantry Brigade, First Sergeant Sean Sachs:

> One building that I saw in particular was a three-storey building where the bottom was still booby-trapped, the middle floor had a family living in it and the top floor was used as a heavy machinegun post, a 50-calibre machinegun. They were firing at soldiers coming down one of the alleyways. How do you explain to someone that the only way that you can take out a heavy machinegun – which is armor-piercing – is that you have to call in a helicopter and it has to be a pinpoint strike at that building? So you hit that top floor. Just the top floor is damaged and the people on the middle floor are fine. But a cameraman comes and shows that building. And suddenly it's a destroyed building and you are accused of having killed people.[49]

Still, the IDF has on occasion conducted missile strikes with apparent disregard for the possible collateral damage that they would entail. A blatant example is when on 21 July 2004 the Israeli Air Force dropped a one-ton bomb over late *Hamas* leader Salah Shehade's home situated in a heavily-populated residential area of Gaza City. The operation wounded over 150 people, killed 14 civilians, including nine children and led to an official – and embarrassing – apology on the part of the former IDF Chief of Operations, Major-General Dan Harel.[50] The bombing's repercussions are still being felt by former Chief of the Israeli Air Force Major-General Dan Halutz who ordered the strike and who had to justify the morality of his order in late November 2004 before the Israeli High Court of Justice prior to being appointed as IDF Deputy Chief of Staff.

Thus, as a result of such collateral damage, the IDF has over the past two years mapped out most Palestinian cities by developing a system that divides 'the urban battlefield into precise increments and gives each building in a city ... an individual four-digit designation so both land and air forces know exactly which target they are trying to hit'.[51] Such detailed mapping and digital designation of Palestinian urban areas have helped reduce to some extent the cases of operational errors, which have led often in the past to extensive collateral damage.

Yet, despite efforts at reducing collateral damage Palestinian civilians have suffered over the past four years; house demolitions have been a common punitive measure used against families related to terrorists or suspected of supporting terrorism. According to *B'Tselem*, an Israeli human rights organization, almost half of the 628 housing units demolished since the start of the *Al-Aqsa Intifada* belonged to people who had no relation to terrorists. Such deliberate demolitions have led to the unlawful creation of over 1,200 homeless civilians. Many more, furthermore, were destroyed during military operations to unearth arms smuggling tunnels.[52] Operation 'Days of Penitence', which was carried out during the first two weeks of October 2004 in order to root out *Hamas Qassam* rocket attacks against Israeli towns bordering the Gaza Strip, led to the destruction of over 90 homes and to the death of over 100 Palestinians, most of whom were combatants, but which included also 27 youths.[53]

Operation 'Defensive Shield', successive IDF counter-terror operations, but especially the so-far partial construction of the security fence-wall have managed to reduce in any case by 30 percent the number of terrorist attacks committed against Israeli objectives (from 5,301 to 3,338) and by 50 percent the number of terror victims (from 451 to 213) between 2002 and 2003.[54] Moreover, according to a senior IDF officer there was a further '75 percent reduction in attempted suicide attacks so far this year compared to the same period in 2003'.[55]

The Strategic Impasse

And yet such tactical and operational successes have not achieved any major strategic dividends, despite the belief of the higher political and military echelons – for example, Moshe Yaalon, Ariel Sharon and Shaul Mofaz – to the contrary. Indeed, according to Brigadier-General (Ret.) Shlomo Brom:

> On the tactical-operational level, the preparations were excellent, but the problem was, as usual, on the strategic level, because of

> the problem of how to prepare for it. Was it a preparation for a military conflict? Now in a military conflict you have very clear benchmarks for success. You have to – in blunt terms – kill as many of the other side and have as few own casualties as possible... In this type of conflict there is a totally different benchmark, which should be built on the analysis of how it can be stopped, not on how it can be won. Thus, in this type of conflict, my aim should be to minimize as much as possible my casualties and to minimize as much as possible my opponent's casualties, because killing breeds more killing.[56]

Such an observation is particularly poignant given the fact that according to IDF Major-General (Ret.) Yaacov Amidror, '....Palestinians regard victory differently from Israelis or those in the West. They measure success not by achieving positive results for their people, but rather by the amount of suffering inflicted on their enemies'.[57] Thus, no matter how successful the IDF may be at the tactical and operational levels, the Palestinians are willing to conduct terror and guerrilla operations and inflict suffering on the Israeli and Jewish settler population until they achieve a political resolution leading to national independence. This has been a constant problem for conventional armies that have conducted low-intensity campaigns with much larger and better-equipped forces against weaker and poorly-equipped terrorist and guerrilla fighters.

Even when loss ratios have been considerably skewed in favor of conventional forces, irregular forces have gained historically the strategic upper hand on innumerable occasions. For example, in Vietnam the loss ratios (including civilians) were 16 to 1 in the US's favor; in Algeria the loss ratios were 24 to 1 in the French forces' favor; in Afghanistan the loss ratios were at least dozens to 1 in favor of the Soviet forces; and in Mogadishu, Somalia, during an 18-hour battle on 3 October 1993 the loss ratio between US forces and Somali irregulars and civilians was 56 to 1 in the US's favor.[58] Yet, despite their military superiority, all such forces ended the conflict without achieving any battlefield decision or victory as Israel did when it left the South Lebanese Security Zone in April 2000.[59]

IDF approaches to low-intensity operations have continued to maintain by default the traditional conventional goal and ethos of maintaining the status quo by trying to impose a battlefield decision through military means. However, the nature of the current conflict itself makes it very difficult for the IDF to achieve any major strategic dividends despite its extensive tactical and operational successes, because of the political nature of low-intensity conflicts, which are ultimately resolved through political negotiation and diplomacy. The late Prime Minister Yitzhak Rabin was able to reach finally such a conclusion after six long years of Palestinian violence and the IDF's unsuccessful counter-insurgency efforts. Current Prime Minister Ariel Sharon has possibly come to the same conclusion after avoiding any major political options by relying heavily on IDF over the past three years.

Thus, it is crucial that the Israeli political-security leadership provide proactively alternative political solutions to the current conflict, whether through unilateral disengagement in order to kickstart the Road Map, which could provide a two-state solution to the Israeli–Palestinian conflict, the creation of a newly-modified Oslo-type agreement or any other political resolution that will end the current conflict. Now seems a particularly auspicious moment given the fact that the former Palestinian Authority and Palestinian Liberation Organization leader, Yasser Arafat, has died.

The Israeli political leadership will have to understand finally that when it is fixed on maintaining the status quo and the Palestinians are set on modifying it, recurring episodes of fighting between them become inevitable. It is imperative that the Israeli political echelon find a political solution to Palestinian insurgency given the fact that it is able to negotiate from a position of strength – Palestinian terror capabilities, although not motivation, have been dramatically reduced by the IDF and Israeli intelligence organizations – and given the fact that ultimately the occupation over and control of Palestinians will only continue to wear out both societies and, in particular, the IDF.

When commenting on the need to continuously and creatively address threats to Israeli security, David Ben-Gurion, Israel's founding prime minister and 'the father of its strategic doctrine stated that "the most dangerous enemy to Israel's security is the intellectual inertia of those who are responsible for security"'.[60] The IDF has done its part in trying to provide solutions to Palestinian terror and guerrilla tactics, now it is up to the politicians to heed Ben-Gurion's warning against strategic sclerosis.

Acknowledgement

Earlier versions of this paper were presented at the American Political Science Association Annual Meeting (2 September 2004) and at the Inter-University Seminar on Armed Forces and Society Canada Meeting (3 October 2004). I am grateful to Professor John Mearsheimer, Professor Sir Lawrence Freedman and the journal's anonymous referees for their many useful comments. However, any remaining errors are my own.

Notes

1 The IDF and other Israeli security services have had a long tradition in fighting Palestinian terrorism and insurgency both locally and internationally since the early 1950s. However, the scope of this article will focus primarily on the period relating to the *Al-Aqsa Intifada* as, I believe, it has proven to be the greatest challenge to Israeli counter-insurgency efforts since the state's establishment in 1948. For a brief historical overview of Palestinian insurgency and Israeli counter-insurgency see: Sergio Catignani, 'The Security Imperative in Counterterror Operations: The Israeli Fight Against Suicidal Terror', *Terrorism and Political Violence* Vol. 17/1, forthcoming.
2 Beatrice Heuser and Lawrence Freedman, 'Strategy' in Lawrence Freedman (ed.), *War* (Oxford: Oxford UP 1994) p.192.
3 Martin van Creveld, *On Future War* (London: Brassey's 1991) p.208.
4 See Ze'ev Drory, *Israel's Reprisal Policy, 1953–1956* (London: Frank Cass 2004).
5 Eliot A. Cohen *et al.*, *'Knives, Tanks, and Missiles:' Israel's Security Revolution* (Washington, DC: Washington Institute for Near East Policy 1998) p.71.
6 Avi Kober, 'Israeli War Objectives into an Era of Negativism', *Journal of Strategic Studies* 24/2 (June 2001) p.187.
7 Efraim Inbar, 'Israel National Security, 1973–1996', *AAPSS 555* (Jan. 1998) p.71.
8 See Chaim Herzog, *The War of Atonement: The Inside Story of the Yom Kippur War* (London: Greenhill Books 2003) pp.270–91.
9 Uri Bar-Joseph, 'Variations on a Theme: The Conceptualization of Deterrence in Israeli Strategic Thinking', *Security Studies* 7/3 (Spring 1998) p.153.
10 'Public Order and Security', 13 Sept. 1993, *Declaration of Principles on Interim Self-Government Arrangements*, Article VIII, available at <http://www.jmcc.org/research/series/dop.html#declare> (accessed 1 July 2004).
11 Matthew Gutman, 'Destruction, Constructively Speaking', *Jerusalem Post*, 9 Jan. 2003.
12 Col. (Res.) Yehuda Wegman, 'Israel's Security Doctrine and the Trap of "Limited Conflict"', *Jerusalem Viewpoints* 514 (1 March 2004), available at <http://www.jcpa.org/jl/vp514.htm> (accessed 1 May 2004).
13 Rami Hazut, 'The Palestinians Are an Existential Threat: Iraq Is Not', *Yedioth Ahronoth*, 23 Aug. 2002, *Israel Resource Review*, available at <htttp://israelvisit.co.il/cgi-bin/friendly.pl?url=Aug-23-02!IDF> (accessed 14 March 2003).
14 For an account of the demise of national security consensus, see Dan Horowitz, 'The Israeli Concept of National Security' in Avner Yaniv (ed.), *National Security and Democracy in Israel* (London: Lynne Rienner 1993) pp.27–31.
15 Moti Bassok and Eynav Ben Yehuda, 'Survey: 46 percent of households cannot meet monthly outlays', *Ha'aretz*, 10 Aug. 2004.
16 Reuven Gal, 'The Israeli Defense Forces (IDF): A Conservative or an Adaptive Organization?', in Daniel Maman *et al.* (eds.), *Military, State and Society in Israel: Theoretical and Comparative Perspectives* (New Brunswick, NJ: Transaction Publishers 2001) pp.364–5.
17 Steve Rodan, 'Interview: Lt. Gen. Shaul Mofaz', *Jane's Defence Weekly* (17 Oct. 2001), available at <http:jdw.janes.com> (accessed 23 June 2004).
18 Col. Gal Hirsh, Head of the IDF Officers Training School, interview with author, Tel Aviv, Israel, 6 Aug. 2003.
19 The attack killed 27 and wounded over 100 Israeli civilians celebrating the Jewish Passover at a hotel reception in Netanya.

20 Yagil Henkin, 'Urban Warfare and the Lessons of Jenin'. *Azure* 15 (Summer 2003), available at at <http://www.shalem.org.il/azure/15-henkin.htm> (accessed 23 May 2003).

21 'Briefing by the IDF Chief of The General Staff', International Conference on Low-intensity Conflict, 23 March 2004, available at <http://www.idf.il/newsite/English/032304-4.stm> (accessed 23 March 2004).

22 See 'IDF Steps up Intelligence War against Palestinians,' *Jane's Defence Weekly* (5 Jan. 2001), available at <http:jdw.janes.com> (accessed 23 June 2004) and 'Israel's Digital Army', *Foreign Report* (22 May 2003), available at <http://frp.janes.com> (accessed 23 June 2004).

23 David A. Fulghum and Robert Wall, 'Israel's Future Includes Armed, Long-Range UAVs', *Aviation Week and Space Technology* (25 June 2002), available at http://www.aviationnow.com/avnow/search/autosuggest.jsp?docid = 3864&url = http%3A%2F%2Fwww.aviationnow.com%2Favnow%2Fnews%2Fchannel_military.jsp%3Fview%3Dstory%26id%3Dnews%2Fmiuav0625.xml> (accessed 14 July 2002).

24 David Eshel, 'Israel Hones Intelligence Operations to Counter *Intifada*', *Jane's Intelligence Review* (1 Oct. 2002), available at <http://jir.janes.com> (accessed 23 June 2004).

25 According to various estimates the GSS and Military Intelligence 'ran about 7,000 informers throughout the West Bank and Gaza during the first *Intifada*'. See 'Israel Uses *Intifada* Informers to Abet Assassination Campaign', *Jane's Intelligence Review* (1 Dec. 2001), available at <http://jir.janes.com> (accessed 23 June 2004).

26 Capt. E., *Sayeret Matkal* (General Security Services Élite Commando Unit), interview with author, Tel Aviv, Israel, 18 Aug. 2003. See also Barbara Opall-Rome, 'Israeli Gunships, Troops Team for Pinpoint Strikes', *Defence News*, 21 April 2003, available at <http://www.defensenews.com/pgt.php?htd = i_story_1787859.html&tty = topnews> (accessed 20 May 2003).

27 Jeremy Gwinn, 'Jenin and the Fundamentals of Urban Operations', *Infantry Online*, 15 March 2003, available at <http://www.benning.army.mil/OLP/InfantryOnline/issue_21/art_125.htm> (accessed 28 June 2003).

28 Ibid.

29 Col. Gal Hirsh, interview with the author, 6 Aug. 2003.

30 See Barbara Opall-Rome, 'Tanks Fill Wider Role in Israel's Anti-Terror War', *Defense News*, 17 March 2003.

31 On the IDF's attempts at trying to grapple with media exposure during the *Al-Aqsa Intifada*, see Baruch Nevo and Yael Shur, *The IDF and the Press During Hostilities* (Jerusalem: The Israel Democracy Institute 2003).

32 Since the start of the *Al-Aqsa Intifada*, for example, 323 children under the age of 14 have been killed by IDF fire. See Gideon Levy, 'Suffer the Children', *Ha'aretz*, 2 Dec. 2004.

33 See Barbara Opall-Rome, 'Objective: Re-create the Fog of War', *Defence News*, 24 June 2002.

34 For a detailed account of the IDF Code of Conduct, see Amos Guiora, 'Balancing IDF Checkpoints and International Law: Teaching the IDF Code of Conduct', *Jerusalem Issue Brief* 3/8 (19 Nov. 2003), available at <http://www.jcpa.org/brief/brief3-8.htm> (accessed 11 Nov. 2004).

35 Soldiers often find themselves guarding checkpoints from around 8 to 12 hours a day and dealing with up to 5,000 frustrated and hostile Palestinians. See Eitan Rabin, 'Army Cameras Catch Soldiers' Abuse', *Maariv International*, 11 July 2004.

36 See 'IDF Readjusts to the Needs of the Palestinian Population', *IDF Spokesperson's Unit*, 8 July 2004, available at <http://www.1.idf.il/DOVER/site/mainpage.asp?sl = EN&id = 7&docid = 32567.EN> (accessed 11 July 2004).

37 Barbara Opall-Rome, 'Israel Tests Anti-Sniper System in Combat', *Defense News*, 15 July 2002.

38 See Dror Marom, 'Corner Shot Invests $2m in Weapons Systems', *Globes*, 15 Dec. 2003, available at <http://www.globes.co.il/DocsEn/did=751435.htm>(accessed 7 April 2004).

39 See Alon Ben-David, 'Israel Unveils *Refa'im* Rifle Grenade System', *Jane's Defence Weekly* 40/11 (17 Sept. 2003) p.15.

40 See Amir Buhbut, 'On Target', *Maariv International*, 19 March 2004.

41 IDF infantry brigades usually rotate among the three Territorial Commands (North, South and Central).

42 Amos Harel, 'Infantry Boosted in Leaner Army', *Ha'aretz*, 29 July 2003.

43 Cohen (note 5) p.50.

44 Major (Res.) Danny Gal, *MAMDA* (IDF Behavioural Sciences Unit), interview with author in Herzliya, Israel, 12 Aug. 2004.

45 Yaakov Hasdai, '"Doers" and "Thinkers" in the IDF', *The Jerusalem Quarterly* 24 (Summer 1982) pp.16–18.

46 Lt.-Col. A., *Sayeret Egoz* Commander, interview with author, Shrivenham, England, 23 June 2004.

47 Due to the large budget cuts over the past four years imposed by the Ministry of Finance, the IDF's new 'Shot 2008' five-year strategic review will envisage the reduction of its ground forces by more than 25 percent, the withdrawal of older fighting platforms and the drastic cutback in the use of combat reservists for routine security operations amongst other things. See Alon Ben-David, 'Extensive Cuts to Hit Israeli Ground Forces the Most', *Jane's Defence Weekly* 40/2 (16 July 2003) p.17.

48 Quoted in: 'Israel Needs Yaalon', *Foreign Report* (6 Nov. 2003), available at <http://frp.janes.com>(accessed 23 June 2004).

49 First Sergeant (Res.) Sean Sachs, *Nahal Brigade*, interview with author, Tel Aviv, Israel, 11 Aug. 2003.

50 See Pierre Klochendler, 'Israeli General Apologises for Civilian Deaths', CNN, 23 July 2002, available at <http://www.cnn.com/2002/WORLD/meast/07/23/hamas.assassination>(accessed 10 Aug. 2003).

51 David A. Fulghum and Robert Wall, 'Israel Refocuses on Urban Warfare', *Aviation Week & Space Technology* 156/19 (13 May 2002) p.25.

52 See Uri Glickman, 'Report: Over 4,100 Palestinian Homes Razed by IDF Since Start of *Intifada*', *Maariv International*, 15 Nov. 2004.

53 See 'UNRWA Gaza Field Assessment of IDF Operation Days of Penitence', 20 Oct. 2004, available at <http://www.un.org/unrwa/news/incursion_oct04.pdf>(accessed 21 Dec. 2004).

54 See Margot Dudkevitch, '50% Fewer Terror Victims in 2003', *The Jerusalem Post*, 9 Jan. 2004.

55 Arieh O'Sullivan, 'Mofaz Terror in Decline, Pressure to Continue', *The Jerusalem Post*, 22 June 2004.

56 Brig.-Gen. (Ret.) Shlomo Brom, former IDF Head of the Strategic Planning Division, interview with author, Tel Aviv, 13 June 2003.

57 Maj.-Gen. (Ret.) Yaacov Amidror, 'Israel's Strategy for Combating Palestinian Terror', *Joint Forces Quarterly* 32 (Autumn 2002) p.120.

58 Statistics quoted in Avi Kober, 'Has Battlefield Decision Become Obsolete? The Commitment to the Achievement of Battlefield Decision Revisited', *Contemporary Security Policy* 22/2 (Aug. 2001) p.111.

59 See: Gil Merom, *How Democracies Lose Small Wars* (Cambridge: Cambridge UP 2003).

60 Cohen (note 5) p.142.

[20]
The Palestinian H-Bomb: Terror's Winning Strategy

Gal Luft

Never in Israel's history, to paraphrase Churchill, has so much harm been inflicted on so many by so few. Since the onset of the second intifada in late September 2000, dozens of exploding humans -- Palestinian H-bombs -- have rocked the Jewish state and transformed the lives of its people. As little as a year ago, suicide bombings were seen as a gruesome aberration in the Israeli-Palestinian conflict, an expression of religious fanaticism that most Palestinians rejected. But in recent months a new, unsettling reality has emerged: the acceptance and legitimation of the practice among all Palestinian political and military factions.

Increasingly, Palestinians are coming to see suicide attacks as a strategic weapon, a poor man's "smart bomb" that can miraculously balance Israel's technological prowess and conventional military dominance. Palestinians appear to have decided that, used systematically in the context of a political struggle, suicide bombings give them something no other weapon could: the ability to cause Israel devastating and unprecedented pain. The dream of achieving such strategic parity is more powerful than any pressure to cease and desist. It is therefore unlikely that the strategy will be abandoned, even as its continued use pushes the Middle East ever closer to the abyss.

From Mortars to Martyrs

The Palestinian endorsement of suicide bombings as a legitimate tool of war was not hasty. At the start of the second intifada, the Palestinians' preferred method of fighting was based on the strategy that Hezbollah used to drive the Israel Defense Forces (IDF) out of southern Lebanon after 15 years of occupation -- a mix of guerrilla tactics such as ambushes, drive-by shootings, and attacks on IDF outposts. It was thought that the "Lebanonization" of the West Bank and the Gaza Strip would cause the Israeli public to view these territories as security liabilities (as they had with southern Lebanon), and to pressure the government to withdraw once more.

Palestinian leader Yasir Arafat's division of labor was clear. His political wing, Fatah, authorized its paramilitary units, spearheaded by the Tanzim militias along with segments of the security services of the Palestinian Authority (PA), to carry out a guerrilla campaign against Israeli settlements and military targets in the West Bank and Gaza. The militant groups Hamas and Islamic Jihad, meanwhile, were given the liberty to carry out attacks against civilian targets inside Israel.

From the Palestinian perspective, however, the results of the guerrilla campaign in the first year were poor, especially considering the duration of the fighting and the volume of fire. Palestinian forces launched more than 1,500 shooting attacks on Israeli vehicles in the territories but killed 75 people. They attacked IDF outposts more than 6,000 times but killed only 20 soldiers. They fired more than 300 antitank grenades at Israeli targets but failed to kill anyone. To demoralize the settlers, the Palestinians launched more than 500 mortar and rocket attacks at Jewish communities in the territories and, at times, inside Israel, but the artillery proved to be primitive and inaccurate, and only one Israeli was killed.

Israel's response to the guerrilla campaign, moreover, was decisive. Using good intelligence, the Israeli security services targeted individual Palestinian militants and destroyed most of the PA's military infrastructure. Israeli soldiers also moved back into "Area A," the territory that had been turned over as a result of the Oslo peace negotiations to exclusive Palestinian control, to raze suspected mortar activity sites. At first these incursions met with international rebuke, even from the United States. Secretary of State Colin Powell, for example, denounced the first foray into Gaza in April 2001 as "excessive and disproportionate." But over time the temporary incursions became such a common practice that the international community stopped paying attention. Stung by the lack of progress in the struggle, at the end of 2001 Arafat tried a final gambit, attempting to smuggle in a cache of Iranian weapons on board the Karine-A. But Israeli naval commandos seized the ship and turned his ploy into a shameful diplomatic disaster. Thus ended Palestinian emulation of the "Hezbollah model."

Unlike the guerrilla strategy, meanwhile, the terror campaign carried out by Hamas and Islamic Jihad was showing results. The Islamic movements managed to kill or maim more Israelis in 350 stabbings, shootings, and bombings inside Israel than the mainstream Palestinian organizations had in more than 8,000 armed attacks in the West Bank and Gaza. The strongest impact came from 39 suicide attacks that killed 70 Israelis and wounded more than 1,000 others. If one compares this bloodshed with the limited damage caused by the 39 Scud missiles Saddam Hussein launched at Israel in 1991 -- 74 fatalities, most of them caused by heart attacks -- it is not hard to understand why the new methods caused such intoxication.

Palestinians are fully aware of what they have suffered at the hands of the Israeli military in response to the terror campaign, but most view it as a great success nevertheless. They derive comfort and satisfaction from the fact that the Jews are also suffering. The Palestinians view the campaign's greatest achievement as not just the killing of so many Israelis but the decline of Israel's economy, the destruction of its tourism industry, and the demoralization of its people. According to a mid-May poll, two-thirds of Palestinians say that the second intifada's violence has achieved more for them than did the previous years of negotiations.

Legitimizing Terror

Before the outbreak of the second intifada, Palestinians distinguished among attacks on settlers, on Israeli military targets, and on civilians inside Israel. Now, however, those distinctions are disappearing. Although after the Israeli incursions this spring support for attacks against civilians inside Israel dropped 6 points to 52 percent, opposition to arresting those carrying out such attacks rose 10 points to 86 percent -- a figure close to the 89 percent and 92 percent support for attacks on Israeli settlers and soldiers in the territories, respectively.

In the post-9/11 era, however, when deliberate attacks against innocent civilians are anathema to most people, embracing terrorism as a strategy has required the Palestinians to persuade themselves, and others, that what they are doing is legitimate. They have therefore created what they see as a moral equivalence between Israel's harm to the Palestinian civilian population and Palestinian attacks against Israeli civilians, including children.

They have also developed a creative interpretation of what terrorism is, one that stresses ends rather than means. Thus, in December 2001, more than 94 percent of Palestinians told pollsters that they viewed Israeli incursions into Area A as acts of terror, while 82 percent refused to characterize the killing of 21 Israeli youths outside a Tel Aviv disco six months earlier that way. And 94 percent reported that they would characterize a hypothetical Israeli use of chemical or biological weapons against Palestinians as terrorism, whereas only 26 percent would say the same about Palestinian use of those weapons against Israel. Interestingly, the new definition extends beyond the conflict with Israel. Only 41 percent of Palestinians, for example, viewed the September 11 attacks as terrorism, and only 46 percent saw the Lockerbie bombing that way.

The more enchanted Palestinians have become with the achievements of their "martyrs," the more Fatah has found itself under pressure to adopt the suicide weapon. Last year, fearing a loss of popular support if the "street" perceived the Islamists' methods as more effective than Fatah's tack, Fatah leaders decided they had to follow suit. The part of Arafat that wanted to show solidarity with the United States and that was determined to avoid any association with terror against civilians, in other words, succumbed to the anti-Israel rage and political calculations of his lieutenants and the members of what Palestinian pollster Khalil Shikaki has called the "young guard" of Palestinian nationalism.

Fatah's official espousal of "martyrdom" operations took place on November 29, 2001, when two terrorists blew themselves up together on a bus near the Israeli city of Hadera. One, Mustafa Abu Srieh, was from Islamic Jihad; the other, Abdel Karim Abu Nafa, served with the Palestinian police in Jericho. But the bond of blood with the Islamists did not last long, and soon Fatah's al Aqsa Martyrs Brigades and the Islamists found themselves engaged in a diabolical contest over which group could perfect the use of the suicide weapon and be viewed as most valuable to the war effort. Al Aqsa has capitalized on the Islamists' opposition to the participation of women and established squads of willing female suicide bombers named after Wafa Idris, the Palestinian woman who blew up herself and an Israeli man in Jerusalem in January. Islamic Jihad, for its part, has recruited children as young as 13 for suicide missions.

Both Islamists and secular Palestinians have come to see suicide bombing as a weapon against which Israel has no comprehensive defense. To counter the Iraqi Scuds, Israel developed and deployed the Arrow, a $2 billion ballistic missile defense system. Against Palestinian H-bombs, Israel can at best build a fence. The suicide bombers are smarter than Scuds, and Palestinians know that even though in Israel today there are more security guards than teachers or doctors, the bomber will always get through.

Military Blinders

If history is any guide, Israel's military campaign to eradicate the phenomenon of suicide bombing is unlikely to succeed. Other nations that have faced opponents willing to die have learned the hard way that, short of complete annihilation of the enemy, no military solution will solve the problem.

But the Israeli authorities are deeply reluctant to accept this reality. Israeli society seeks absolute security and adheres to the notion that military power can resolve almost any security problem. If the Palestinians put their faith in Allah, Israelis put theirs in a tank. Whether consciously or not, their belief in the utility of force -- evident in the popular "Let the IDF Win" campaign, which advocated a freer hand for the army -- reflects a strategic choice to militarize the conflict rather than politicize it. The IDF's senior leaders repeatedly claim that the smart application of military force can create a new reality on the ground that, in turn, will allow the government to negotiate political agreements under more favorable terms.

It is true that when the IDF was finally allowed to "win," Israel achieved impressive tactical results. Operation Defensive Shield this past April eliminated an entire echelon of terrorist leaders in the West Bank, crippled the PA's financial and operational infrastructure, and reduced PA arsenals. But as at other times in its history, Israel failed to convert its tactical achievements into strategic gains. Its intensive use of military instruments earned it international condemnation, further radicalized Palestinian society, and created an environment of anger conducive to more terrorist activities. By May, unsurprisingly, the suicide bombings had started again.

IDF simulations before the second intifada had predicted that a military reentry into major Palestinian cities would lead to hundreds of Israeli casualties. In fact, however, the incursions into territories under Palestinian control proved to be almost painless. Following the assassination of Israel's tourism minister, Rehavam Ze'evi, in October 2001, the IDF launched a broad assault on the PA, entering all six major West Bank cities. Palestinian resistance was negligible, and only six Israeli soldiers were wounded. Operation Defensive Shield, the second big incursion into Area A, also met relatively weak resistance. Aside from the struggle in the Jenin refugee camp, in which 23 Israeli soldiers were killed, Israeli forces conquered six Palestinian cities and dozens of smaller towns and villages while suffering only three fatalities.

The IDF has interpreted the Palestinian lack of resistance in the cities as a sign of weakness rather than a strategic choice. Israelis view with disdain the Palestinian "victory" celebrations after each incursion comes to an end. They are puzzled by the fact that their enemy fires more bullets into the air than at Israeli troops. What Israel fails to comprehend is the paradigm by which the Palestinians are choosing to conduct their war.

Acknowledging their perpetual conventional inferiority, Arafat's people feel no need to demonstrate strong resistance to Israeli forces. They simply wait for the storm to pass while preparing another batch of "martyrs." Families of suicide bombers now receive more than double the financial compensation than do the families of those killed by other means. Nurturing an ethos of heroism fundamentally opposed to that of the Israelis, the Palestinian war of liberation has elevated the suicide bomber to the highest throne of courage and devotion to the national cause.

Fencing Lessons

Israelis' misunderstanding of the new Palestinian way of war may come back to haunt them. Their perception of their enemy's weakness is likely to embolden them and encourage more broad punitive operations in response to future attacks. But Israel's military responses will eventually exhaust themselves, whereas the Palestinians will still have legions of willing "martyrs."

In fact, despite defiant Israeli rhetoric insisting that there will be no surrender to terrorism, one can already see the opposite happening. Israelis are willing to pay an increasingly high economic and diplomatic price for increasingly short periods of calm. As a result, more and more people support panaceas such as unilateral separation -- the building of walls, fences, and buffer zones to protect Israel's population centers from Palestinian wrath.

Unilateral separation would no doubt make the infiltration of suicide bombers into Israel more difficult, but it would also increase their prestige in the eyes of many in the region. The bombers would be viewed, correctly, as the catalyst that drove the Israelis out of an occupied territory yet again, and the years of agony Palestinians have endured would be sweetened by a genuine sense of victory. Israel's wall policy, perceived as withdrawal, would reassure the Palestinians that war succeeded where diplomacy failed.

As currently conceived, moreover, walling off the territories would not do much to reduce Palestinian grievances. No matter how long the fence, for example, dozens of Jewish settlements scattered on the hills of the West Bank would necessarily remain beyond it. Two-thirds of Israelis, according to recent polls, support the removal of such isolated and indefensible settlements to make the separation more feasible. But despite such views, Israeli Prime Minister Ariel Sharon has reiterated his refusal to dismantle a single settlement. "The fate of Netzarim is the fate of Tel Aviv," he said recently, referring to the tiny, isolated, and fortified Gaza Strip settlement that has been the target of repeated Palestinian attacks.

Defusing the Bomb

Israel finds itself, therefore, at a crucial turning point in its history, but one from which no path seems particularly attractive. It must find some way of defending itself against an enemy so eager to inflict pain that it is willing to bring suffering and death on itself in the process. Retaliation is unlikely to work, but retreat is likely only to bring more of the same.

If there is any way out of this dilemma, it may lie in convincing the Palestinian public that its constructive goals can be achieved only by relinquishing its destructive strategy. Israel should therefore embark on a policy that rewards the Palestinians for genuinely fighting terrorism and avoid any policy that feeds the perception that terrorism works.

The rewards will have to be tangible and meaningful. Israel could, for example, offer the PA the removal of a number of small hilltop settlements in exchange for a period of non-belligerency and unequivocal renunciation of suicide bombing. This cooling-off period could then set the stage for renewed talks on a final-status agreement. Such an approach would indicate to the Palestinian population that Israel is serious about peace and ready to pay the necessary price for it, not only in words but in deeds. Most important, showing that Israel is prepared to confront and rein in its own radical rejectionists would put the onus on the Palestinian leadership to do the same.

Before this intifada, a large majority of Palestinians opposed attacks against civilians inside Israel. They hoped to achieve their aspirations for independence without resort to terror. Figuring out how to make that happen is not only the right thing to do, but it is also the best way to ensure Israel's security. Unless that hope can be revived, the fate of Tel Aviv could indeed become that of Netzarim -- which would be a tragedy for all.

* Details of the poll and survey methodology are available at the Web site of the Palestinian Center for Policy and Survey Research (www.pcpsr.org).

Gal Luft is a former Lieutenant Colonel in the Israel Defense Forces and the author of The Palestinian Security Forces: Between Police and Army.

Hezbollah in Lebanon

[21]

Hizballah and the Israeli Withdrawal from Southern Lebanon

Augustus Richard Norton

Using as its starting point the Israeli withdrawal from southern Lebanon, this article traces Hizballah's development from a radical resistance group known especially for suicide bombings and kidnappings to a highly professional guerrilla force and a political party with a broad constituency and pragmatic leadership. The author examines Hizballah's entry into politics in the early 1990s; its evolving relations with state actors, especially Syria and Iran; its conduct during and following the withdrawal; and its future prospects.

On 24 May 2000, some quarter century after Israel became entangled in Lebanon and fifteen years after it declared a "security zone" covering 10 percent of the territory, the Israeli occupation of southern Lebanon finally came to an end. Although Israeli prime minister Ehud Barak has been at pains to emphasize that the Israel Defense Force (IDF) was not defeated and had left as a result of a calculated government decision, there is no confusing the fact that were it not for the activities of the Lebanese resistance, particularly Hizballah (the Party of God), Israel would have tarried far longer. The scent of defeat hung in the air for months preceding the hurried unilateral withdrawal.

The last months of occupation were also marked by dire predictions from virtually all quarters that the IDF withdrawal would only spark disorder and mayhem. Most of the speculation focused on Hizballah. Arguments were voiced that it was essentially a terrorist group (indeed, it continues to be listed as such by the U.S. State Department); that it was but a pliant instrument in the hands of Syria; and that, lacking real support among the population, it would resist ending its armed militancy so as not to be consigned to marginality on the Lebanese scene. These arguments often led to a single conclusion: that Hizballah would continue attacking Israel.

In fact, of course, this did not happen. Although tension along the Israeli-Lebanese border persists, the summer of 2000 is the first extended period since the days preceding the 1967 war that Lebanese civilians and Palestinian refugees in southern Lebanon have not been living in the cross fire.

Augustus Richard Norton is a professor in the departments of anthropology and international relations at Boston University and the author of a number of books and monographs, including *Hizballah of Lebanon: Extremist Ideals vs Mundane Politics* (New York: Council on Foreign Relations, 1999).

Hizballah's behavior confounded would-be Cassandras precisely because it is not the same organization that it was in the early 1980s. What the doomsayers had missed was that since its creation as a rigidly ideological organization known especially for kidnappings and suicide bombings of Western targets, Hizballah has developed not only into a highly professional guerrilla force, but also into an impressive political organization with a broad and varied constituency, a pragmatic leadership, and a clearheaded strategy. All of these elements, as well as its ability to adapt to a changing political and social context, have been essential to Hizbailah's success.

FORGED IN WAR

Hizballah emerged in the wake of Israel's all-out invasion of Lebanon in 1982. The invasion was a follow-up to the 1978 Litani Operation, which had left Israel in possession of areas adjacent to the border through proxy Lebanese militiamen (eventually named the South Lebanese Army—the SLA). The operation also led to UN Security Council Resolution 425 calling on Israel to "withdraw forthwith its forces" and the deployment of the United Nations Interim Force in Lebanon (UNIFIL). From that time forward, the IDF moved freely back and forth across the border, and at any one time Israel deployed at least 1,000 soldiers in Lebanon, not including intelligence operatives, staff officers, and advisers.

But Israel's 1978 invasion failed in its goal of driving the PLO forces north of the Litani River. The massive 1982 invasion sought to finish the job by destroying the PLO as a coherent political and military force. After a two-month siege of Beirut, the PLO agreed under pressure to withdraw from the city. This placed Israel in control of Lebanon from Beirut southward. The 1982 invasion's second objective of installing a pro-Israeli government in Beirut foundered with the assassination of President Bashir Gemayel in September 1992. Within months, however, the United States applied itself to brokering an agreement between Israel and the Lebanese government of President Amin Gemayel. The lopsided 17 May 1983 agreement virtually ceded southern Lebanon to Israel and all but guaranteed relentless Syrian and local opposition. Meanwhile, Multinational Forces (MNF)—primarily American and French—had been rushed to the country following the September 1982 massacre by Christian militiamen of more than 1,000 Lebanese and Palestinian civilians in the Sabra and Shatila camps surrounded by the IDF. Before long, the MNF lost all semblance of neutrality and came to be seen as accomplices to an unpopular regime, and just another militia—the "international militia," in the vernacular of the period.

It was against this background that Hizballah emerged during the summer of 1982, in the midst of the Israeli invasion, though its existence was not formally declared until 1983. At the time Hizballah came into being, the most dynamic Shi'i force in Lebanon was the reformist Amal movement, founded by the charismatic Iranian-born cleric Musa al-Sadr in the early 1970s and led primarily by Nabih Birri after Sadr's mysterious disappearance in Libya in 1978. But by the summer of 1982, Amal, having crossed swords with the PLO in southern Lebanon, was seeking a modus vivendi with Israel and the United States. The movement's constituency was especially the battered people of the south, many of whom had welcomed the Israeli invaders (expecting, incorrectly, that the IDF would expel the PLO and then leave). Inevitably, Amal's stance came under fire from more radical elements inspired by the Iranian revolution.

There is little doubt that Iran and Syria were deeply involved in the creation of Hizballah. For Iran, the creation of Hizballah was part of its campaign to spread the message of the "Islamic revolution," whereas for Syria the new group was a fortuitous instrument for preserving its interests in Lebanon at a time when, militarily defeated by Israel, it faced the disastrous prospect of sharing its long western border with a state dominated not only by the United States but by Israel as well. Syria's alliance with Iran presented it with the means to strike indirectly at both Israel and the United States as well as at their presumed or real Lebanese allies, including at the time Amal, which, as noted, had been flirting with pax Americana but was eventually put off by Washington's churlish diplomacy in Lebanon. It was thus that in the summer of 1982, Syria acceded to the introduction of a contingent of Pasdaran, Iran's revolutionary guards, in the Biqa' Valley of eastern Lebanon. The contingent, initially 1,000-strong, eventually numbered as many as 1,500 and became the nodal point for the Iranian training, supply, and support of Hizballah under the watchful eye of Iran's ambassador to Syria, Ali Akbar Mohtashemi (today considered a member of the reformist pro-Khatami camp in Iran).

Hizballah began as a coterie of clerics and lay Shi'a inspired by the Iranian revolution of 1978–79. In contrast to Amal, whose new leaders had aspirations to join the Shi'i bourgeoisie, most of the Hizballah leaders were trained in the Shi'i seminaries of Najaf and Karbala in Iraq. If Hizballah's founders were inspired by Iran, their roots in the Hizb al-Da'wa should also be noted. The Shi'i Islamist group al-Da'wa is best known for its opposition to the Ba'thist regime in Iraq and its links to the leading Arab Shi'i jurists of the 1970s, Muhsin al-Hakim and Muhammad Baqr al-Sadr, and the group had branches in Lebanon as well as in other parts of the Shi'i world. Nevertheless, at the beginning, Hizballah was hardly a popular movement, but a conspiracy of a handful of men funded by the nascent Islamic Republic of Iran.

But with Israel showing no intention of withdrawing from Lebanon, and a central government in Beirut whose actions belied its claim to represent all Lebanese, Hizballah soon began attracting large numbers of followers. Secular groups had originally spearheaded the resistance, but by 1984 the "Islamic resistance," led by Hizballah and its rival Amal, clearly dominated the campaign to end the occupation. Hizballah, in particular, shrewdly deployed the perception of its followers as men in search of martyrdom. The use of suicide bombers, especially in the 1980s, unnerved the IDF and provoked reactions that served to further enflame resistance.

Of course, Hizballah has no monopoly on martyrs (many secular resistance fighters died fighting the occupation, some in suicide missions), yet it is true that martyrdom in the service of jihad (resistance struggle) enjoys a particular resonance for Shi'i Muslims. The martyrdom in 680 C.E. of Imam Husayn, the grandson of the Prophet Muhammad, at Karbala in present-day Iraq, has always been the single most important symbolic referent for Shi'i Muslims. Since the Iranian revolution, however, it has acquired new political meaning and been transformed from an exemplary act of suffering and sacrifice into an inspiring model for revolution and action. Husayn's determined struggle against injustice and his sacrifice is ritually and dramatically recreated in annual Muharram ceremonies (Muharram being the Islamic lunar month during which Husayn and his companions were killed by the forces of the Sunni Caliph Yazid). Significantly, the annual rituals commemorating Imam Husayn's martyrdom—revived in Lebanon by leading Shi'i clerics, including Imam Musa al-Sadr—became massive affairs over the course of Israel's occupation of southern Lebanon. Reference to the Israeli occupiers as "Yazidis" equates them to the oppressors of Imam Husayn and thereby invokes the living memory of his death and inspires courage in those who revere him. (It should be noted that the popular religion aspects of Muharram, which include bloody self-flagellation rituals, are proscribed by virtually all Shi'i clerics. Interestingly, while Amal participates enthusiastically in these rituals, Hizballahis tend to adhere to the more doctrinaire interpretations of Shi'ism and limit themselves to orderly marching in the festivals.)

IDEOLOGICAL BORROWINGS FROM IRAN

Throughout the 1980s, Hizballah's actions very much reflected the radicalization of the Shi'i scene under the impact of the Israeli occupation and Iran's implacable opposition to U.S. policies in Israel and Lebanon. During this period, Hizballah moved aggressively to strike at western influence and westerners in Lebanon. Either Hizballah itself or groups linked to it abducted dozens of foreigners and held them hostage for as long as seven years, posing such demands as the freeing of Lebanese prisoners held in Germany, Israel, and Kuwait. During the same period, Hizballah carried out suicide bombings at the U.S. embassy, U.S. embassy annex, Israeli intelligence headquarters in then-occupied Tyre, U.S. marine barracks at the Beirut airport, and French embassy. It is fair to say that the emergence of Hizballah changed the whole tenor of the conflict in Lebanon, and it played a major role in provoking the departure of the American marines from Lebanon and the scuttling of the U.S.-brokered 17 May 1983 agreement between Lebanon and Israel.

While Hizballah pursued its own local agendas in these acts, Iran's early hand in coordinating, if not always controlling, Hizballah was not difficult to discern. Indeed, during the 1980s Hizballah hewed closely to the Iranian line. Iran's influence is particularly clear in Hizballah's remarkable open letter to "The Downtrodden in Lebanon and in the World." The document bears a distinctive made-in-Tehran coloration and, in fact, is reliably reported to have been written by an Iranian who is today very much in the pro-Khatami reform movement. The letter, released in February 1985 to mark the one-year anniversary of the assassination, no doubt by Israel, of a rising young cleric from southern Lebanon, declared the world to be divided between the oppressed and the oppressors, chief among the latter being the United States and its regional ally, Israel. This perspective not only enjoyed resonance among the Shi'a, many of whom had first hand experience with Israeli oppression, but it also legitimized and commended the use of violence against the enemies of Islam, particularly the West.

In 1983, Israel pulled back from the Shouf Mountains overlooking Beirut but continued to occupy all of Lebanon from the 'Awali River southward, including Sidon, Lebanon's fourth largest city. Israeli losses continued to mount, and attempts to create village militias in southern Lebanon foundered (the same scheme had failed in the West Bank). Despite widespread arrests of young and middle-aged men, the momentum of the resistance was increasing. (On a personal note, it was during this period, in December 1984, that Yitzhak Rabin initiated a meeting with me and a few other scholars during which he asked what Israel should do. I told him that Israel must leave Lebanon and that to remain would only further radicalize the population. He replied that, as a politician, he could not risk attacks on Israeli towns.)

At all events, Israel redeployed its forces in Lebanon in January 1985 and officially declared the "security zone." As the IDF redeployed, the "Iron Fist" policy left ruined villages and broken bodies as a calculated lesson to those who would dare resist the occupation. The occupied area, more aptly described as an (in)security zone, predictably became a magnet for yet more resistance attacks.

Hizballah's Social Base

While Hizballah's operations in the south were coordinated with Syria (Hizballah officials freely admitted as much) and substantial Iranian and Syrian support facilitated its development (as did the tacit and increasingly explicit support of the Lebanese government), there is no doubt that Hizballah is very much an indigenous organization. Its fighters—no more than 1,500, of whom two-thirds are part-timers—are local men. They have family ties, jobs if not professions, homes, networks of social support, and hopes and aspirations for Lebanon. In contrast to the Palestinian fedayeen, who were ascendant in parts of the country from 1970 to 1982, Hizballah fighters are totally integrated into Lebanese society. This was brought home to me when I was the guest of a wealthy factory owner in the Biqa' Valley, whose lovely home would have fit nicely in Thousand Oaks, California, or Burke, Virginia. My host and his wife were more interested in discussing issues like education and their children than politics, and the beautifully bound books by prominent Muslim thinkers, such as Muhammad Baqr al-Sadr and Muhammad Husayn Fadlallah, seemed more ornamental than well thumbed. Only later did I learn that one of their sons was an active member of the resistance and that as we dined, he had been out on operations. My point is that this young man, like so many others, did not choose the resistance because he had no alternative, but because it suited his ideals.

Similarly, if Hizballah had acted as a cat's paw of complementary Syrian and Iranian interests in Lebanon, its primary agenda was very much its own: ridding Lebanon of the Israeli occupation. Since the late 1960s, hundreds of thousands of Lebanese have been repeatedly displaced from their homes in the south as a result of Israeli military action. In April 1996 alone, hundreds of thousands of civilians were driven north by IDF bombing, sometimes in response to warnings of only two hours. Israeli strikes on Lebanese power plants darkened much of Lebanon in 1996, 1999, and twice in 2000, causing damages estimated at $300 million. Israel calculated that by punishing the Lebanese in general and destroying national infrastructure, the government (or the Syrian government) would be motivated to trim the sails of the resistance. Israel consistently miscalculated. Israeli attacks had the opposite effect. Every time Beirut was blacked out as a result of Israeli bombing, support for the resistance surged among non-Shi'i Lebanese, especially for its steadfast response to punitive Israeli attacks.

If Hizballah had acted as a cat's paw of foreign interests in Lebanon, its primary agenda was very much its own: ridding Lebanon of the Israeli occupation.

The Grapes of Wrath operation in early 1996 only intensified the perception of Israel as evil incarnate. The 1996 massacre by shelling of more than 100 Muslim and Christian civilians who had taken refuge in the UN base in Qana, an ancient village cited in the Bible as the place where Jesus turned water to wine, became a rallying cry against the occupation and the Zionist state. Close to the UN base, the memorial cemetery where the victims are buried has become a point of pilgrimage for Lebanese of all religious persuasions. The site is festooned with banners accusing Israel of terrorism and genocide and invoking sayings by central figures in Shi'ism (such as Imam Husayn). Many of the banners emphasize the loss of innocent blood and demand vengeance. One sign read, "Qana is the Karbala of the twentieth century; it is a land made holy by the Lord Jesus and contaminated by the Zionist Satan (enemy of God)."

It is in this sense that Hizballah officials agree privately that the movement was in large measure "created" by Israel. Without the raison d'être of opposing the occupation, Hizballah would not have been able to build a broad Shi'i constituency. Hizballah officials frequently observed that if Israel's presence in the south were not resisted, Israel would have little incentive even to consider withdrawing its forces. The untested converse proposition—that by ceasing resistance activities Israel would be induced to withdraw—was widely dismissed as laughably improbable.

LAYING THE GROUND FOR A POLITICAL FUTURE

Throughout the 1980s, nothing in Hizballah's actions or statements hinted at any possibility that the party would adapt to the Lebanese political game, particularly running for election and sitting in the National Assembly. To the contrary, the Lebanese political system was condemned by Hizballah as "rotten to the core" and "unreformable." By the early 1990s, however, things began to change. Several factors explain Hizballah's new approach: the end of the cold war, change in Iran, and the end of the civil war in Lebanon.

Not only did the end of the cold war facilitate the formation in 1990 of a U.S.-dominated war coalition, including Syria, but it also permitted people at least to imagine an end to the Arab-Israeli conflict. If Syria were to sign a peace agreement with Israel, the Hizballahis knew that disarming the resistance would be part of the package. Meanwhile, the death of Ayatollah Ruhollah Khomeini in 1989 signaled the beginning of Iran's Thermidor or postrevolutionary phase. Indeed, even before his death, the internecine fighting between Amal and Hizballah had provoked an attitude shift in Tehran, where the bloodletting was viewed with disgust. President Rafsanjani did not hide his impatience with Hizballah as a militia and publicly declared his intention to shift Iran's orientation in Lebanon to state-to-state relations and the Shi'i community as a whole. Finally, the civil war in Lebanon officially ended with the 1989 Ta'if accord, though the war in effect went on until October 1990, when General Michel Aoun (who objected that Ta'if did not address Syria's preponderant role in Lebanon) was forced to end his so-called uprising by the pressure of Syrian arms.

All of these factors provided the context for a debate that took place in 1992 over whether or not to field candidates in that year's parliamentary elections. A minority led by Hizballah's first secretary-general, Shaykh Subhi Tufayli (expelled from the party in January 1998 for his vehement criticism of his successors), staunchly opposed participation, arguing that Hizballah would be co-opted and sacrifice its ideals. The vast majority supported participation, accurately reflecting the aspirations of Hizballah's constituency, which wanted a more effective voice in the political system. A senior Hizballah official told me in May 2000 that the 1992 decision to participate in the Lebanese elections was one of the two watersheds in Hizballah's history (the other being the April 1996 Understanding discussed below).

The upshot was that Hizballah took part in the 1992 elections and came away with eight seats, the largest single bloc in the 128-member parliament (its allies from other sects won an additional four). From this time forward, Hizballah began to develop a reputation as a serious political party adept at pragmatic parliamentary alliances and tactics, even while intensifying its guerrilla warfare against the occupier.

THE "RULE BOX"

In southern Lebanon, warfare has been defined by tacit rules that, over time, have been formalized. Early in the 1980s, Daoud Suleiman Daoud, then a leading Amal figure, publicly elaborated the view that Israel should be attacked *in Lebanon*, not Israel. Hizballah excoriated Daoud at the time for respecting Israel's "illegal boundaries," but later the leadership came to appreciate the value of such limits in warfare, as did Israel. Thus, despite Syria's support for the resistance, Israel generally took care to avoid targeting Syrian forces and to apologize when it happened unintentionally. Hizballah did the same when unauthorized Katyushas hit Israel in 1998.

One of the most obvious distinctions in war is the difference between soldiers and civilians. When either side fired at civilians, retaliation in kind generally followed. After Israel assassinated Hizballah Secretary-General 'Abbas Musawi in 1992, his successor, Hasan Nasrallah, enunciated a tit-for-tat policy: If Israel hits Lebanese civilian targets, then Hizballah hits Israel. Although the action-reaction cycle was not always transparent, generally the cycle started with an Israeli attack on a civilian target, intentional or otherwise. Hizballah crossed the line as well, for example, in 1995 it shelled Israel in retribution for the killing of a top military official, Rida' Yassin.

After Israel's punitive Operation Accountability and Grapes of Wrath campaigns in 1993 and 1996, the rules of war in southern Lebanon were drawn with increasing specificity. The first formalization of the rules came in an unwritten 1993 understanding. These were further nailed down in a *written* but unsigned understanding between Israel and Lebanon (in fact, Hizballah) in April 1996, following Israel's Grapes of Wrath operation, during which the resistance fired more than 600 Katyushas into Israel, while Israel fired an estimated 25,000 shells at Lebanese targets and flew about 600 combat air sorties. The gist of the April Understanding, reached thanks to intense efforts by France, Iran, Israel, Syria, and the United States, was simple: Israel would not target civilians or civilian targets (e.g., power stations), and Hizballah would not target Israel per se. In principle, each side reserved the right of self-defense within the confines of the rules, though in fact both violated them. The April Understanding also produced an innovation; a monitoring group based in the UNIFIL headquarters in Naqura with American, French, Israeli, Lebanese, and Syrian participation. The monitoring group operated on the basis of unanimity and had no enforcement mechanism, but its very existence helped lend solidity to the rules of the game.

There have been many violations of the understandings, but both sides were highly conscious of their existence, to the extent that Israeli spokespersons sometimes referred to killings of Israeli soldiers in Lebanon as having been "within the rules." With its increasing professionalism and intense motivation, the resistance was more effective operating within the "rule box" than the IDF or the SLA. Indeed, the tempo of "routine" fighting was often broken when Israel, frustrated at its impotence in the face of mounting casualties, intentionally attacked Lebanon's civilian infrastructure. Action and reaction were usually so predictable that Israel was clearly stymied in February 2000, when the resistance wisely chose not to respond to its destruction of three Lebanese power plants and thereby denied the IDF an excuse for continuing to attack Lebanese infrastructure. Whether Hizballah's restraint reflected Syrian or Iranian advice or the party's own calculation is debatable, but the decision lent credibility to the party and furthered the momentum toward withdrawal (withdrawal advocates were able to underline the futility of Israel's occupation and the rationality of the opponent).

Hizballah, even while remaining within the "rule box," had become increasingly effective in keeping its enemies on the defensive. In marked contrast to the late 1980s, when its attacks often involved large losses, the ratio of Hizballah casualties to IDF/SLA casualties dropped from more than 5:1 in 1995 to less than 2:1. Suicide bombers were superseded by coordinated military attacks that benefited from excellent planning and intelligence. Using relatively simple weapons, such as cleverly camouflaged remote control roadside bombs, the resistance impeded the enemy's mobility. Weapons were for the most part standard fare, though in January 2000 Hizballah began using wire-guided antitank missiles (reportedly American TOWs originally supplied in the 1980s to Iran by Israel as part of the Iran-Contra deal), which could literally be guided through the observation ports of heavily fortified and otherwise impregnable positions. Most of the seven Israeli soldiers felled during January and February of this year were killed by TOWs.

As part of Israel's efforts to minimize its losses, the IDF in the last years of occupation largely hunkered down in southern Lebanese strong points, venturing out on occasion on patrol, but mostly sitting captive in sandbag prisons. Even the tour of duty had to be extended to reduce the vulnerabilities inherent in changing over from one unit to another. By 1999, only eight of fifty occupation positions were actually manned by IDF soldiers, the SLA being left to man the remaining forty-two. Hizballah continued to pursue a strategy of relentlessly pummeling the most exposed positions. The extent of their success came home to me in March 2000, when I encountered, while traveling in the south, two "self-service" SLA checkpoints. In both cases, motorists would stop at a barricade, move it, drive a few meters to the next barricade, repeat the process, and then drive on. In transiting the heavily sandbagged position, one could observe the top of the helmeted head of an SLA militiaman who waved warily at the traveler but had no interest in exposing himself to more danger than necessary.

It is noteworthy that up until the eleventh hour, Israel never challenged Hizballah's right to attack its soldiers in Lebanon. Thus, Israel tacitly conceded that the IDF was an occupation force in Lebanon. Only in early 2000, as seven IDF soldiers were killed in Lebanon, were the rules of the game vigorously challenged by Israel under the impact of a public that saw its enemies as terrorists and villains. The shock was as much that the "bad guys" played by rules as that the IDF was being outplayed.

The Occupation Ends Ahead of Schedule

Certainly, the momentum for getting out of Lebanon very much came from Israeli society, which was increasingly intolerant of Israeli losses and rattled by Hizballah's brilliant campaign of psychological warfare begun in the late 1990s. The signal for al-Manar (the Beacon), the Hizballah flag station, was strong enough to reach Israel, offering viewers vivid images of dead and wounded IDF soldiers that were otherwise muted on Israeli television. Videotapes of martyrs in action and successful resistance assaults on SLA positions were conveniently made available to feed public anxiety in Israel.

It was to honor his campaign promises that Israeli prime minister Barak, shortly after his electoral victory, announced, "by July 2000, the army will withdraw to the international border, and it is from the international border that we will defend the north of the country. I don't advise anyone to test us when we draw back and are sitting on the border." Of course, it was clear that unilateral withdrawal was a *default strategy*. Israel clearly preferred to reach an agreement with Syria that would contain a Lebanese component.

But with the resounding failure of the Clinton-Asad summit in Geneva in March 2000, prospects of a Syrian-Israeli agreement in time to forestall a unilateral withdrawal came abruptly to an end. Remarkably, the idea that Israel would withdraw unilaterally was met with skepticism in official Lebanese and Syrian circles. In both countries, the political elite assumed that Israel, given its security preoccupations, would exit the south only in conjunction with a deal between Israel and Syria, and it was only well after Geneva that it seemed to dawn on the leaderships that Israel was serious about leaving. The months that followed were incredible political theater in both Beirut and Damascus, with the players constantly searching for a script and frequently forgetting their parts altogether. If government officials, presidents, and heads of internal security were flustered, the Hizballah leaders kept their cool. They had a clear sense of the problems Israel was facing in the south and were confident about the outcome.

Barak set the calendar by announcing that the IDF would leave by July, but once the withdrawal timetable was underway, there was no reason to linger. Armies are not mechanical toys in which one simply changes the batteries and an army set on coming home is not easily pumped up for battle. As the month of May progressed, the IDF set about consolidating its lines, destroying positions and preparing for the order to leave. Worries persisted in Israel about the proxy force and its viability. Nominally a force of 2,500, the SLA was demoralized by heavy casualties even before Barak's announced intention to leave. Desertions began to mount, and SLA coherence became suspect. The commander, Antoine Lahd, more a figurehead than an operational leader, spent a lot of time in Israel and France. When the SLA's most important operational leader, 'Aql Hashim, was assassinated in January 2000, there were real doubts that the leadership was up to the challenge.

Israel withdrew its forces on 23–24 May 2000, while the SLA simply crumbled. As the IDF hurriedly exited, thousands of Lebanese rushed by foot, donkey, and car to reclaim their villages. By the evening of 24 May, Israel's self-declared "security zone" was no more.

Meanwhile, in the final days of the occupation, anxiety had been mounting in the south. People feared not only a continuing cycle of violence, but that the victorious resistance fighters would wreak vengeance against those who lent support to Israel. I know from conversations in March, April, and May that the Hizballah leadership was very sensitive to these concerns. The last thing that they wanted was to see their anticipated victory stained by vengeance killings and especially intersectarian clashes. Thus, even as the Israeli forces were withdrawing, they sent reassurances to local community leaders, and in the days that followed they held meetings in key locales in the formerly occupied areas. Shaykh Nabil Qaouk (who headed the resistance forces), Nasrallah, and other leading Hizballah leaders held extensive meetings with Christian clerics to reassure them this was a national victory, not a victory by one sect or militia. And, notwithstanding several unfortunate incidents, especially looting in Marja'uyun, the former SLA headquarters, Hizballah forces displayed impressive discipline. Even the ritual pelting of Israel with stones at Fatima Gate, the crossing point just across the border from the Israeli town of Metulla, exemplifies order, not chaos. Hizballah conveniently provides stones in piles, discreetly delineating the throwing areas. By permitting the ritual as an outlet, Hizballah may be reducing the prospects for maverick acts of violence.

By permitting the ritual pelting of Israel with stones at Fatima Gate as an outlet, Hizballah may be reducing the prospects for maverick acts of violence.

Another lingering concern in the south was the fate of the several thousand men and women who had collaborated with Israel, or merely worked in Israel or traveled there for medical care. This issue was the subject of intense discussion in Hizballah's Political Bureau, which argued that the judicial system is the only mechanism for dealing with collaborators. As the trials of collaborators proceeded, Hizballah officials loudly criticized the "light" sentences and successfully pressured for an addendum to some sentences, forbidding convicted collaborators to return to their native villages for a set period. Aside from these criticisms, they basically allowed the system to work. Information compiled by the capable journalist Nicholas Blanford shows a pattern of light and medium sentences, with the typical sentence being one year in jail for militia participation. Rank and file SLA members received sentences ranging from six to eighteen months, and some were forbidden to return to their villages, usually for two years. (After the SLA withdrew from Jazzin the previous year, former militia members received similarly light sentences, and most have now been released from prison.) Individuals who entered Israel were usually sentenced to two months in jail and a fine, but all women in this category were acquitted. In absentia trials imposed harsher sentences of fifteen years and up on SLA leaders, but the big fish had already fled to Israel or wherever a visa would carry them. The only senior official in custody is Major Emile Nasr, who had been an agent for Hizballah for many months preceding Israel's withdrawal. He is now in jail serving concurrent terms that total 120 years imprisonment but is widely expected to be released within five years, if not much sooner.

Throughout the painstaking process of confirming the Israeli withdrawal, Hizballah was at pains to declare its commitment to recovering the last millimeter of Lebanese territory, but it also acknowledged that it would not act hastily to reinitiate violence. In sum, Hizballah's behavior and deference to state authority have worked to its political advantage. It reaped recognition in an unprecedented meeting between Nasrallah and UN Secretary-General Kofi Annan, who praised Hizballah's restraint and its promise of cooperation. The meeting with Annan offers a remarkable contrast with Hizballah's earlier days, when it was hostile to the UN and especially to the UN force in the south.

Without an agreement between Syria and Israel, there will be little pressure on Hizballah to disarm. Syria's calculated strategy is to allow Hizballah to serve as a constant reminder of the consequences of continuing to occupy the Golan Heights. This is a role that Hizballah is happy to play, given its enmity toward Israel. At the same time, it remains profoundly aware of the political costs of bringing destruction down on the heads of its supporters, and this further reduces the prospect that Hizballah will initiate attacks on Israel.

Political Life after Resistance

Hizballah defers to Syrian influence in Lebanon, but there is no neat overlap between their interests. Hizballah will continue to take Syria's interests to heart, but only insofar as they do not jeopardize the party's political support in Lebanon. This implies clear limits on Syria's influence, which both sides no doubt appreciate. At the same time, Hizballah's leaders understand that the party's role in Damascus's eyes is utilitarian and transient, and they are ever aware that alliances of convenience may eventually become inconvenient. As for Iran, President Mohamed Khatami, like his predecessor Rafsanjani, favors broader cultural and social ties to Lebanon rather than privileged support for Hizballah. Such circumstances only reinforce Hizballah's demonstrated commitment to plant its feet firmly in Lebanon.

Though Hizballah has garnered praise throughout Lebanon for its role in ending the occupation and is widely admired for its corruption-free politics, it still inspires foreboding among those Lebanese who see in the party's militancy and armaments a threatening force. Certainly, its conduct following the withdrawal has gone a long way toward allaying concerns. In the months leading up to the Israeli exit, Beirut bon vivants who would no more travel to the south than join Hizballah became rapt television viewers when Secretary-General Nasrallah held forth with rigorously analytical assessments of the situation that were frequently contrasted, both in daily conversation and in the press, with the contorted logic and whistling-in-the-dark bluster of Lebanese government officials. Nasrallah and others have made it a point to emphasize that despite Hizballah's enduring rejection of Israel's legitimacy, the organization will choose its actions pragmatically, not dogmatically. It was in this vein that, despite expressions of solidarity with the Palestinians, Hizballah's leaders have pointedly emphasized that the liberation of Palestine is up to them. As elections approached in August 2000, Hizballah politicians focused less on Israel or regional issues and more on Lebanon's corruption-ridden political system.

Such attitudes are not confined to the leadership, but also reflect the outlook of Hizballah's rank and file and especially the population of the south. It is noteworthy that at this year's Muharram commemoration in April, even as speculation on the possibility of Israeli withdrawal was on every tongue, I did not encounter a single person who expressed any desire to invade Israel or conquer Jerusalem. These are the people on the receiving end of the occupation, and they have had quite enough. A Trotskyite project of permanent revolution does not sit well with this constituency, and the Hizballah leadership is nothing if not sensitive to its support base.

But despite Hizballah's moderated views and whatever aspirations it may have for a broader constituency, it is bound to remain a quintessentially Shi'i Muslim party. Confessional politics continues to rule in Lebanon, despite the 1989 Ta'if accord that was supposed to mark a transition from the old system. The hoped-for economic boom has not materialized, and in the absence of dynamic growth the government remains the major source of patronage, which is allocated along confessional lines. Syria also prefers to play the confessional game well suited to the divide-and-rule strategy it has followed in the country for a quarter century. Given this context, Hizballah will inevitably continue to derive its strength from its support base among the Shi'a, including among the fast growing and largely secular Shi'i middle class. Over the past decade, the Shi'i intelligentsia and middle class have increasingly deserted Amal, often out of contempt for its rampant corruption and the movement's domination by parliamentary Speaker Nabih Birri.

Amal remains Hizballah's main rival. The two groups fought vicious battles in the late 1980s for control of the south and the southern suburbs of the capital. In contrast to Hizballah, Amal has been marked by spotty discipline and a level of enthusiasm often unmatched by training or soldierly expertise, so its role in combating Israel and its allies has been secondary. Amal remains a force to be reckoned with, however, not least because of the patronage resources at the disposal of Birri. Amal also benefits from being Syria's closest ally in Lebanon (having patched up its differences with Damascus in the mid-1980s), and Syria is keen to make sure that Amal is not eclipsed by Hizballah. During the 1996 parliamentary elections, for example, Hizballah, with Syrian encouragement, agreed to run on a joint list with Amal, and hence won fewer seats than it had in the 1992 elections. Similarly, it was undoubtedly under Syrian encouragement that in March 2000 Amal and Hizballah, in preparation for this year's elections, agreed to an electoral alliance that has the effect of guaranteeing parliamentary seats for Amal. Tensions at ground level between the two organizations erupted in violent clashes in July, leaving two Hizballahis dead. It would be surprising if this were the end of it. Of course, Amal's influence largely reflects the power of Birri, whose own fortunes rise and fall with the vicissitudes of Syrian politics and the state of Lebanese-Syrian relations. Though it would be naive to expect dramatic changes in those relations in the short term, the subtle process of redefinition clearly underway is likely to diminish Syrian leverage, which will somewhat erode the power of pro-Syrian politicians like Birri.

In parliament, Hizballahi deputies have earned a reputation for acumen and flexibility. They also have sustained a singular reputation for integrity. And while Hizballah, too, dispenses patronage, what is far more important is the support it has garnered because of its growing network of social institutions in areas where public services, including hospitals and support agencies for widows and the disabled, are notoriously lacking. Hizballah has competently filled this vacuum with high-quality social and health services, which have a reputation for being corruption free and nondiscriminatory. (It should be noted, however, that Hizballah is not the only distinctively Shi'i organization providing such services. Amal does as well, though often on a less impressive scale. The Musa al-Sadr Foundation in Tyre, led by Musa's dynamic sister, al-Sayyida Rabab al-Sadr, runs top-flight vocational programs and day schools. Ayatollah Muhammad Husayn Fadlallah leads an impressive foundation that offers an array of services, including a hospital, and the Supreme Islamic Shi'i Council is similarly engaged.)

Given the afterglow of Israel's withdrawal, it is not surprising that all the candidates fielded by Hizballah won, bringing its representation in the parliament from nine to eleven seats (including one-third of the twenty-seven seats allocated to Shi'i Muslims). Indeed, even in the 1996 elections, Hizballah deftly exploited its resistance role to build political support: one widely distributed Hizballah campaign poster read, "They resist with their blood. Resist with your vote."

No doubt, gratitude for Hizballah's central role in ending the occupation will continue for some time, but this support is neither permanent nor unconditional. Actions by Hizballah that provoke new violence would undermine its base of support, and the party understands this. Many challenges lie ahead for Hizballah, not least sustaining its relations with Syria under circumstances when interests will not coincide nearly as neatly as in the past. The Hizballahis have proven, despite their beginnings, to be Lebanese nationalists par excellence, as illustrated by their enthusiastic support for the deployment of the Lebanese army to the south in August 2000. This implies interesting possible alliances and a future that will likely belie simple assumptions about Islamists and their politics. If nothing else, the party provides an important illustration that pluralist politics are experiential, and neither ideology nor religion is an unerring guide to behavior.

Name Index